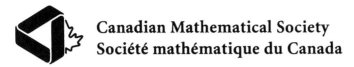

Canadian Mathematical Society
Société mathématique du Canada

Editors-in-Chief
Rédacteurs-en-chef
Jonathan Borwein
Peter Borwein

Springer
New York
Berlin
Heidelberg
Hong Kong
London
Milan
Paris
Tokyo

CMS Books in Mathematics

Ouvrages de mathématiques de la SMC

Jiří Herman   Radan Kučera
Jaromír Šimša

Counting and Configurations

Problems in Combinatorics, Arithmetic, and Geometry

Translated by Karl Dilcher

With 111 Figures

 Springer

Jiří Herman
Brno Grammar School
tř. Kpt. Jaroše 14
658 70 Brno
Czech Republic

Radan Kučera
Department of Mathematics
Faculty of Science
Masaryk University
Janáčkovo nám. 2a
662 95 Brno
Czech Republic

Jaromír Šimša
Mathematical Institute
Academy of Sciences of the
  Czech Republic
Žižkova 22
616 62 Brno
Czech Republic

Translator
Karl Dilcher
Department of Mathematics
  and Statistics
Dalhousie University
Halifax, Nova Scotia B3H 3J5
Canada

Editors-in-Chief
Rédacteurs-en-chef
Jonathan Borwein
Peter Borwein
Centre for Experimental and Constructive Mathematics
Department of Mathematics and Statistics
Simon Fraser University
Burnaby, British Columbia V5A 1S6
Canada
cbs-editors@cms.math.ca

Mathematics Subject Classification (2000): 05-01, 11-01, 51-01, 97D50

Library of Congress Cataloging-in-Publication Data
Herman, Jiří.
  [Metody resení matematickych úloh II. English]
  Counting and configurations : problems in combinatorics, arithmetic, and
geometry /Jiří Herman, Radan Kučera, Jaromír Šimša ; translated by Karl Dilcher.
    p. cm.   (CMS books in mathematics ; 12)
  Includes bibliographical references and index.
  ISBN 0-387-95552-6 (alk. paper)
    1. Combinatorial geometry—Problems, exercises, etc. I. Kučera, Radan.
II. Šimša, Jaromír. III. Title. IV. Series.
  QA167 .H47 2002
  511′.6—dc21   2002026655

ISBN 0-387-95552-6   Printed on acid-free paper.

9 8 7 6 5 4 3 2 1   SPIN 10887276

Typesetting: Pages created by the authors using LATEX 2.09.

www.springer-ny.com

Springer-Verlag New York Berlin Heidelberg
A member of BertelsmannSpringer Science+Business Media GmbH

Preface

This book can be seen as a continuation of *Equations and Inequalities: Elementary Problems and Theorems in Algebra and Number Theory* by the same authors, and published as the first volume in this book series. However, it can be independently read or used as a textbook in its own right. This book is intended as a text for a problem-solving course at the first- or second-year university level, as a text for enrichment classes for talented high-school students, or for mathematics competition training. It can also be used as a source of supplementary material for any course dealing with combinatorics, graph theory, number theory, or geometry, or for any of the discrete mathematics courses that are offered at most American and Canadian universities.

The underlying "philosophy" of this book is the same as that of *Equations and Inequalities*. The following paragraphs are therefore taken from the preface of that book.

There are already many excellent books on the market that can be used for a problem-solving course. However, some are merely collections of problems from a variety of fields and lack cohesion. Others present problems according to topic, but provide little or no theoretical background. Most problem books have a limited number of rather challenging problems. While these problems tend to be quite beautiful, they can appear forbidding and discouraging to a beginning student, even with well-motivated and carefully written solutions. As a consequence, students may decide that problem solving is only for the few high performers in their class, and abandon this important part of their mathematical, and indeed overall, education.

One of the reasons why problem solving is often found to be difficult is the fact that in recent decades there has been less emphasis on technical skills in North American high-school mathematics. Furthermore, such skills are rarely taught at a university, where most courses are quite theoretical or structure-oriented. As a result, a lack of "mathematical fluency" is often evident even in upper years at a university; this reduces the enjoyment of the subject and impairs progress and success.

A second reason is that most students are not used to more complex or multilayered problems. It would certainly be wrong to give the impression that all mathematical problems should succumb to a straightforward approach. Indeed, much of the attractiveness of mathematics lies in the satisfaction derived from solving difficult problems after much effort and several futile attempts. On the other hand, being "stuck" too often and for too long can be very discouraging and counterproductive, and is ultimately a waste of time that could be better spent learning and practicing new techniques.

This book, as well as the earlier *Equations and Inequalities*, attempts to address these issues and offers a partial remedy. This is done by emphasizing basic combinatorial ideas and techniques that are reinforced in numerous examples and exercises. However, even the easiest ones require a small twist. We therefore hope that this process of practicing does not become purely rote, but will retain the reader's interest, raise his or her level of confidence, and encourage attempts to solve some of the more challenging problems.

Another aim of both our books is to familiarize the reader with methods for solving problems in elementary mathematics, accessible to beginning university and advanced high-school students. This can be done in different ways; for instance, the authors of some books introduce several general methods (e.g., induction, analogy, or the pigeonhole principle) and illustrate each one with concrete problems from different areas of mathematics and with varying degrees of difficulty.

Our approach, however, is different. We present a relatively self-contained overview of some parts of elementary mathematics that do not receive much attention in high-school and university education. We give only enough theoretical background to make these topics self-contained and rigorous, and concentrate on solving particular problems.

The chapters of our books are fairly independent of one another, with only a limited number of cross references. Within each chapter, clusters of sections and subsections are tied together either by topic or by the methods needed to solve the examples and exercises. The problem-book character of this text is underlined by the large number of exercises; they can be solved by using a method or methods previously introduced. We suggest that the reader first carefully study all relevant examples before attempting to solve any of the exercises.

The individual problems in the present book (divided into approximately 310 examples and 650 exercises) are of varying degrees of difficulty, from completely straightforward, where the use of a method under consideration will immediately lead to a solution, to much more difficult problems whose solutions will sometimes require considerable effort. The more demanding exercises are marked with an asterisk (*). Answers to all exercises can be found in the final chapter, where additional hints and instructions to the more difficult ones are given as well.

The problems were taken from a variety of mainly Eastern European sources, such as Mathematical Olympiads and other competitions. Many of them are therefore not otherwise easily accessible to the English-speaking reader. An important criterion for the selection of problems was that their solutions should, in principle, be accessible to high-school students. We believe that even with this limitation one can successfully stimulate creative work in mathematics and illustrate its diversity and richness.

One of our objectives has been to stress alternative approaches to solving a given problem. Sometimes we provide several different solutions in one place, while at other times we return to a problem in later parts of the book.

This book is a translation of the second Czech edition of *Metody řešení matematických úloh* II (*Methods for solving mathematical problems II*) published in 1997 at Masaryk University in Brno. Apart from the correction of some minor misprints, the main material has been left unchanged. Three short sections with "competition-type" problems at the ends of the chapters were deleted, and an alphabetical index was added. All the figures from the Czech editions were redrawn for this translation by Karel Horák; we thank him for this important contribution.

The Czech editions of this book have been used by the authors in special enrichment classes for secondary-school students and for Mathematical Olympiad training in the Czech Republic. It has also been used on a regular basis at Masaryk University in courses for future mathematics teachers at secondary schools.

We hope that a course partly or entirely based on this book will work well even in a class with a wide range of mathematical backgrounds and abilities. The book's structure, the worked examples, and the range in level of difficulty of the exercises make it particularly well suited as a source for assigned readings and homework exercises.

Brno, Czech Republic   Jiří Herman
                                                        Radan Kučera
                                                        Jaromír Šimša

Halifax, Nova Scotia, Canada   Karl Dilcher, Translator

Contents

Symbols

In addition to usual mathematical symbols and conventions, the following notation will be used:

$\mathbb{N}_0$	set of all nonnegative integers
$\mathbb{Q}^+$	set of all positive rational numbers
$\mathbb{R}^+$	set of all positive real numbers
$P(n)$	number of arrangements
$P_0(\alpha_1, \ldots, \alpha_n)$	number of arrangements with repetitions
$V(k, n)$	number of variations
$V_0(k, n)$	number of variations with repetitions
$C(k, n)$	number of combinations
$C_0(k, n)$	number of combinations with repetitions
$[x]$	greatest integer not exceeding $x \in \mathbb{R}$
$|A|$	cardinality of a finite set A
$|XY|$	length of a line segment XY
$|\mathcal{K}|$	area of a planar region $\mathcal{K}$
$\triangle ABC$	triangle with vertices A, B, C
$\sphericalangle CAB$	angle with legs AB and AC
$XY \| l$	line segment XY is parallel to line l
$XY \perp l$	line segment XY is perpendicular to line l
$\mathcal{D}(C, r)$	closed disk with center C and radius r
$\overparen{AB}$	circular arc from A to B

1

Combinatorics

In this first chapter we will be concerned with problems that are part of *classical combinatorics* and are based on determining the number of configurations with given properties. The problems in this chapter mainly describe "practical" situations (such as the distribution of cards among different players), that is, the relevant variables are mathematically not sufficiently well formulated. As we begin with the solution of a problem it is therefore important to make sure we know which of the resulting configurations we consider distinct. (In the example of a card game it is normally irrelevant in which order a player is dealt his cards). Our main tool in formalizing these concrete situations will be the notion of an *ordered k-tuple*, with which the reader will already be familiar. At this point we simply note that two k-tuples $(a_1, a_2, \ldots, a_k)$ and $(b_1, b_2, \ldots, b_k)$ are considered equal if and only if the equalities $a_1 = b_1$, $a_2 = b_2$, $\ldots$, $a_k = b_k$ are satisfied.

We begin this chapter with five introductory sections, which will provide deeper insights into the traditional topics of secondary-school combinatorics. The main emphasis will be on consequences of certain combinatorial rules, but not simply on "choosing the right formula," as is unfortunately often the case in school. In the following sections we then deal with more sophisticated combinatorial rules and methods that go beyond the usual secondary-school curriculum.

1   Fundamental Rules

As the three fundamental rules of all combinatorics we consider the *addition rule*, the *multiplication rule*, and the *pigeonhole principle* (or *Dirichlet's principle*). While these rules are very easy to state, they are useful in a number of diverse and often rather complicated situations. These rules are usually not explicitly stated in secondary-school textbooks, but here we will make abundant use of them, especially the addition and multiplication rules.

The *inclusion–exclusion principle* is another fundamental rule of combinatorics. However, it is far more difficult to state and to prove. Therefore, a separate section (Section 6) will be devoted to this principle and its numerous applications.

1.1   The Addition Rule

Let $\{A_1, A_2, \ldots, A_n\}$ be a system of subsets of a finite set A such that these subsets are pairwise disjoint (that is, for each $i, j \in \{1, 2, \ldots, n\}$ with $i \neq j$ we have $A_i \cap A_j = \emptyset$) and their union is all of A; that is, $A = \bigcup_{i=1}^n A_i$. Then we have

$$|A| = |A_1| + |A_2| + \cdots + |A_n| = \sum_{i=1}^n |A_i|. \tag{1}$$

PROOF.   According to the hypothesis, each element $a \in A$ belongs to exactly one of the subsets A_i, $i = 1, 2, \ldots, n$, and therefore it counts exactly once on each side of equation (1). □

1.2   Examples

We illustrate the addition rule with two simple problems.

(i) A square with side length 4 is divided by parallel lines into 16 equal squares. What is the total number of squares in this picture? (See Figure 1).

SOLUTION.   To answer this question, we divide the totality of squares into four sets A_1, A_2, A_3, A_4 such that the set A_i contains all squares of side length i $(i = 1, 2, 3, 4)$. Obviously, $|A_1| = 16$, $|A_2| = 9$, $|A_3| = 4$, $|A_4| = 1$; hence the total number of squares is $16 + 9 + 4 + 1 = 30$. □

(ii) What is the total number of moves a knight can make on an 8×8 chessboard?

SOLUTION.   The number of moves from a given field depends on the location of the field on the chessboard. Therefore, we divide the set of all fields into 5 groups A, B, C, D, E (see Figure 2 for part of the chessboard). From a corner field A a knight can make two moves; from a field of type B it can make three moves, and it can make four from a field C, six from a field D, and eight moves from a field of type E. Hence, according to the addition rule the total number n of moves of a knight on a chessboard is

$$n = 4 \cdot 2 + 8 \cdot 3 + 20 \cdot 4 + 16 \cdot 6 + 16 \cdot 8 = 336. \qquad \square$$

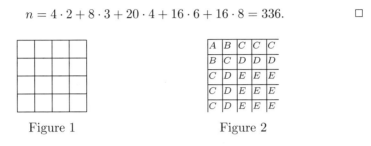

A	B	C	C	C
B	C	D	D	D
C	D	E	E	E
C	D	E	E	E
C	D	E	E	E

Figure 1   Figure 2

1.3 Exercises

   (i) A square with side length n is divided into n^2 equal squares (compare with 1.2.(i)). Determine the total number of squares in this array.

   (ii) Determine the number of all possible moves of a knight on an $n \times n$ chessboard $(n \geq 3)$.

*(iii) Find the number of all possible moves of a queen on an $n \times n$ chessboard $(n \geq 2)$. If you find the general case difficult, then begin with the classical case $n = 8$.

1.4 The Multiplication Rule

We wish to determine the number of all n-tuples $(x_1, x_2, \ldots, x_n)$ constructed according to the following scheme: The element x_1 can be chosen in exactly α_1 ways, and for each $i = 1, 2, \ldots, n-1$ we assume that for a fixed choice $x_1, x_2, \ldots, x_i$ the element x_{i+1} can be chosen in exactly α_{i+1} different ways (where the number α_{i+1} does not depend on the particular choice of the elements $x_1, x_2, \ldots, x_i$). Then the number N of all such ordered n-tuples satisfies

$$N = \alpha_1 \cdot \alpha_2 \cdots \alpha_n. \qquad (2)$$

PROOF.   We use induction on n, but for greater clarity we first deal with the case $n = 2$. In this case all possible ordered pairs (x, y) can be written in an array as follows:

$$
\begin{array}{cccc}
(x_1, y_{11}) & (x_1, y_{12}) & \cdots & (x_1, y_{1\alpha_2}) \\
(x_2, y_{21}) & (x_2, y_{22}) & \cdots & (x_2, y_{2\alpha_2}) \\
\vdots & \vdots & \vdots & \vdots \\
(x_{\alpha_1}, y_{\alpha_1 1}) & (x_{\alpha_1}, y_{\alpha_1 2}) & \cdots & (x_{\alpha_1}, y_{\alpha_1 \alpha_2})
\end{array}
$$

Obviously, the number of pairs is exactly $\alpha_1\alpha_2$.

We now begin with the actual proof. For $n = 1$ the statement is trivial. Let now $n \geq 2$ be an arbitrary integer. We assume that for all $k = 1, 2, \ldots, n - 1$ the assertion is true; we want to show that it also holds for $k = n$. We divide all ordered n-tuples $(x_1, x_2, \ldots, x_n)$ into subsets such that each one contains exactly those n-tuples that agree in their first $n - 1$ positions. These subsets are obviously disjoint, and according to the induction hypothesis there are $\alpha_1 \cdot \alpha_2 \cdots \alpha_{n-1}$ of them. Since the element x_n can be chosen in α_n ways, each subset contains exactly α_n elements, namely ordered n-tuples. According to the addition rule 1.1 (i.e., Section 1.1 of the current section) the number of all ordered n-tuples then satisfies $N = (\alpha_1 \cdot \alpha_2 \cdots \alpha_{n-1}) \cdot \alpha_n = \alpha_1 \cdot \alpha_2 \cdots \alpha_n$. This completes the proof. □

In closing this subsection we consider the case where the choices of the individual elements x_i are independent from each other; that is, each element x_i runs through some finite set A_i. In this case the multiplication rule gives the number of all ordered n-tuples $(x_1, x_2, \ldots, x_n)$, where $x_i \in A_i$ for each $i = 1, 2, \ldots, n$, that is, the number of elements of the *Cartesian product* $A_1 \times A_2 \times \cdots \times A_n$:

$$|A_1 \times A_2 \times \cdots \times A_n| = |A_1| \cdot |A_2| \cdots |A_n|.$$

1.5   Examples

(i) Five routes lead from town A to town B, and three routes lead from B to town C. Find the total number of routes that lead from A to C and pass through B.

SOLUTION.   An arbitrarily chosen route from A to B (5 choices) can be "extended" by three routes from B to C. Hence, according to the multiplication rule, the number of routes in question is $5 \cdot 3 = 15$. □

(ii) In how many ways can one choose one white and one black field on an 8×8 chessboard? In how many ways can this be done if the fields are not to lie in the same row or the same column?

SOLUTION.   A white field can be chosen in 32 ways, and the same is true for a black field. The total number of choices in the first case is therefore $32 \cdot 32 = 1024$.

If in the second case we choose a white field (32 possibilities), then a black field can be chosen only from those rows and columns on which the already selected white field does not lie; that leaves 24 possibilities. The total number of choices is therefore $32 \cdot 24 = 768$. □

(iii) In how many ways can we choose from a complete set of dominoes (28 pieces) two that will fit together (that is, the number of dots on one side of the first piece agrees with the number of dots on a side of the second piece)?

SOLUTION. First we choose one domino piece. That can be done in 28 ways; in 7 of these the chosen piece has the same number of dots on both sides $(00, 11, \ldots, 66)$, while in the remaining 21 cases the numbers of dots are different. Therefore, we divide all admissible pairs of domino pieces into two subsets according to which of the above type the first piece belongs to. In the first case, the second piece can be chosen in 6 different ways (for instance, any of the pieces $01, 02, \ldots, 06$ can be attached to the piece 00); in the second case, the second piece can be chosen in 12 ways (for instance, any of the pieces $00, 02, \ldots, 06, 11, 12, \ldots, 16$ can be attached to the piece 01). According to the multiplication rule, the number of choices is $7 \cdot 6 = 42$ in the first case and $21 \cdot 12 = 252$ in the second case, and by the addition rule the entire number of choices is $42 + 252 = 294$. However, since the order of the chosen pieces is irrelevant, the final answer is half of this, namely 147.   □

1.6 Exercises

(i) From among 12 masculine, 9 feminine, and 10 neuter nouns we want to choose one of each gender. In how many ways can this be done?

(ii) Five paths lead to the top of a mountain. How many choices does a tourist have for a walk up the mountain and back? How many choices when the descent is to be on a different path from that taken on the way up?

(iii) In a fruit basket there are 12 apples and 10 pears. Jirka chooses an apple or a pear, and then Jana chooses one apple and one pear. After which one of Jirka's choices does Jana have a greater number of choices?

In Chapters 2 and 3 we will often use the *pigeonhole principle* (also called *Dirichlet's principle*) which was already discussed in *Equations and Inequalities* ([5], Chapter 3, Section 9). We repeat its statement and proof.

1.7 The Pigeonhole Principle

Theorem. *If at least $nk + 1$ objects (or pigeons) are distributed among k boxes (or pigeonholes), then some box must contain at least $n + 1$ objects.*

PROOF. We prove this by contradiction. We denote by m_i $(i = 1, 2, \ldots, k)$ the number of objects in the ith box and assume that the assertion is false; that is, for each $i = 1, 2, \ldots, k$ we have $m_i \leq n$. Then

$$nk + 1 = m_1 + m_2 + \cdots + m_k \leq \underbrace{n + n + \cdots + n}_{k} = nk,$$

which is impossible.   □

In closing this introductory section we recall the following well-known rule, which is used in set theory to define equality of the number of elements of two sets.

1.8 The Bijection Rule

Theorem. *Let A and B be two finite sets. Then there exists a bijection $\varphi\colon A \to B$ if and only if the sets A and B have the same number of elements.*

2   Standard Concepts

In this section we introduce some concepts that are commonly used in secondary-school mathematics, namely arrangements, variations, and combinations, and we find formulas for determining their numbers. We stress that our goal is not so much to derive these formulas, but rather to illustrate the *methods* used in their derivation. This is all the more important because these methods can be used in a whole range of complicated situations, as we will try to show in the later sections of this chapter.

Since we will work with finite sets, we recall the usual notation $A = \{a_1, a_2, \ldots, a_n\}$ for an n-element set A consisting of the distinct elements $a_1, a_2, \ldots, a_n$.

2.1 Arrangements of n Elements

Let A be a finite n-element set. The ordered n-tuple $(x_1, x_2, \ldots, x_n)$, where each element a of the set A occurs exactly once among the elements $x_1, x_2, \ldots, x_n$ (that is, for each $a \in A$ there is a unique index $i \in \{1, 2, \ldots, n\}$ such that $a = x_i$), will be called an *arrangement of the elements of the set A*, or simply an *arrangement of n elements*.

Two arrangements $(x_1, x_2, \ldots, x_n)$ and $(y_1, y_2, \ldots, y_n)$ of the elements of a set A are considered different if there is a position that contains different elements (that is, there exists an index $i \in \{1, 2, \ldots, n\}$ such that $x_i \neq y_i$). Thus, for example, if $A = \{1, 2, 3, 4\}$, then the quadruples $(1, 3, 4, 2)$ and $(4, 2, 3, 1)$ are two different arrangements of this set.

Let $P(n)$ denote the number of all arrangements of an n-element set A. We will show by induction on n that the value of $P(n)$, which is clearly independent of the nature of the elements of A, is given by the formula

$$P(n) = n \cdot (n-1) \cdots 2 \cdot 1 = n!. \tag{3}$$

Indeed, let $A = \{a_1, a_2, \ldots, a_n\}$ be the given n-element set. If $n = 1$, then $P(1) = 1$, and the assertion is true. Let us assume now that $n \geq 2$ and

$P(n-1) = (n-1)!$. We divide all the $P(n)$ arrangements of the set A into groups $S_1, S_2, \ldots, S_n$ such that S_i contains all those arrangements that begin with a_i ($i = 1, 2, \ldots, n$). Then we have $|S_i| = P(n-1)$, since the subset S_i has as many elements as there are arrangements of the $(n-1)$-element set $A \setminus \{a_i\} = \{a_1, \ldots, a_{i-1}, a_{i+1}, \ldots, a_n\}$. By the addition rule we finally obtain

$$P(n) = \underbrace{P(n-1) + P(n-1) + \cdots + P(n-1)}_{n} = n \cdot (n-1)! = n!,$$

which completes the proof.

2.2 Examples

To illustrate the above, we consider the following four easy examples.

(i) How many four-digit positive integers can be written with the digits 1, 2, 3, 4, such that none of them are repeated?

SOLUTION.   Each such four-digit number can clearly be written as an arrangement of the four-element set $A = \{1, 2, 3, 4\}$; this means that the desired number is $P(4) = 4! = 24$.   □

(ii) How many four-digit positive integers can be written with the digits 0, 1, 2, 3, such that none of them are repeated?

SOLUTION.   Compare this with (i); from the $P(4) = 24$ possibilities we now have to subtract all those that "begin" with the digit 0 (and have therefore only three digits). There are clearly $P(3) = 3! = 6$ of those. The final answer is therefore $4! - 3! = 3 \cdot 3! = 18$.   □

(iii) Five persons, A, B, C, D, E, are to speak at a meeting, each one of them exactly once.

(a) Find the number of all possible orders of their appearance.
(b) As in (a), but with the condition that person B is to speak immediately after A.
(c) As in (a), but with the condition that B does not speak before A.

SOLUTION.   (a) $P(5) = 5! = 120$.
(b) If person B speaks immediately after A, we replace both by a "new speaker" X (who delivers the speeches of both, first that of A, immediately followed by that of B). We therefore have to determine the number of orders in which the four speakers X, C, D, E get to appear, and this number is $P(4) = 4! = 24$.
(c) Exactly one of the following two cases can occur in each fixed arrangement of speakers: Either A speaks before B, or B speaks before A.

Thus the set of all arrangements is divided into two subsets. Since this subdivision does not depend on which one of the two speakers is denoted by A and which one by B, each of the two subsets contains the same number of arrangements, namely $\frac{1}{2}P(5) = 60$.   □

(iv) Find the number of all bijections from a finite n-element set A to itself.

SOLUTION.   Let $A = \{a_1, a_2, \ldots, a_n\}$. If $\varphi \colon A \to A$ is an arbitrary bijection, it can be written in the following usual way:

$$\begin{pmatrix} a_1 & a_2 & \ldots & a_n \\ b_1 & b_2 & \ldots & b_n \end{pmatrix},$$

where $b_i = \varphi(a_i)$ for $i = 1, 2, \ldots, n$, and the second row contains an arrangement of the elements of A. Conversely, if $(b_1, b_2, \ldots, b_n)$ is an arbitrary arrangement of the elements of A, then the above scheme determines a bijection $\varphi \colon A \to A$. Thus the number of all bijections is equal to the number of all arrangements of the set A, namely $P(n) = n!$.   □

Let us note that for a bijection of a finite set to itself the term *permutation* is often used in algebra. From the last example it is now clear why, especially in the older literature in combinatorics, the term *permutation* is also often used instead of *arrangement* of n elements.

2.3 Exercises

(i) Find the number of all even 5-digit positive integers written with 1, 2, 3, 4, 5, such that none of the digits is repeated.

(ii) Find the number of all 6-digit positive integers written with 0, 1, 2, 3, 4, 5, such that none of the digits is repeated.

(iii) In how many different ways can one put 32 playing cards on top of each other?

(iv) In how many ways can 5 men and 5 women sit at a round table such that no two persons of the same gender sit next to each other?

(v) In how many ways can 4 Canadians, 5 Czechs, and 3 Germans be lined up such that the citizens of each country stand beside each other?

(vi) In how many ways can n rooks be placed on an $n \times n$ chessboard such that no two can threaten each other?

2.4 Variations Without Repetitions

Let $k \leq n$ and let A be a finite n-element set. An arrangement of some k-element subset of the set A is called a k-element variation from the elements of the n-element set A (for short, a k-element variation from n elements). We have therefore an ordered k-tuple $(x_1, x_2, \ldots, x_k)$ of elements from A, where each element of A occurs at most once (in fact, exactly k elements occur once, the remaining $n - k$ not at all).

If, for instance, $A = \{a, b, c, d, e\}$, then the ordered triple (c, b, e) is one of the 3-element variations of elements from A.

Theorem. *Let $V(k, n)$ denote the number of all k-element variations from n elements. Then*

$$V(k, n) = \frac{n!}{(n - k)!} = \underbrace{n \cdot (n - 1) \cdot (n - 2) \cdots (n - k + 1)}_{k \text{ factors}}. \qquad (4)$$

PROOF.   The identity (4) can be proved in various ways. Here we present two such proofs.

(a) We use the multiplication rule 1.4. The element x_1 from the first position of the ordered n-tuple $(x_1, x_2, \ldots, x_n)$ can obviously be chosen in n ways. Once it is chosen and fixed, we have $n - 1$ choices for the element x_2. In general, if the elements $x_1, x_2, \ldots, x_i$ $(i = 1, 2, \ldots, k - 1)$ from the first i places of the ordered k-tuple $(x_1, x_2, \ldots, x_k)$ are fixed, then the element x_{i+1} can be chosen in exactly $n - i$ ways. Therefore, the number $V(k, n)$ of all such ordered k-tuples is, according to 1.4,

$$V(k, n) = n \cdot (n - 1) \cdots (n - k + 1).$$

(b) We consider all $P(n) = n!$ arrangements of the elements of A and divide them into subsets in such a way that each one contains exactly those arrangements that agree in their first k places. These subsets are obviously disjoint. Furthermore, there are exactly $V(k, n)$ of them, since the initial k places of each arrangement of the set A is some k-element variation from n elements, and each k-element variation from n elements forms the beginning of some arrangement. It therefore remains to determine how many arrangements each subset contains: This number is $P(n - k) = (n - k)!$, namely, the number of ways of ordering the remaining $n - k$ elements in the final $n - k$ places. By the addition rule 1.1 we then obtain

$$P(n) = V(k, n) \cdot P(n - k),$$

and thus

$$V(k, n) = \frac{P(n)}{P(n - k)} = \frac{n!}{(n - k)!}. \qquad \square$$

To conclude this subsection, we note that in the case $k = n$ each n-element variation from n elements is some arrangement of these n elements, and therefore we have $V(n, n) = P(n)$. This, of course, is in agreement with identity (4), since by convention we have $0! = 1$.

2.5 Examples

(i) Seven members were elected as the executive committee of a student society. In how many ways can a president, vice president, secretary, and treasurer be chosen from among them?

SOLUTION.   In choosing any quadruple from among the 7 members of the committee, it matters which members will occupy what position. Therefore, the total number n of all possibilities is equal to the number of all 4-element variations from 7 elements, that is,

$$n = V(4,7) = \frac{7!}{3!} = 7 \cdot 6 \cdot 5 \cdot 4 = 840. \qquad \square$$

(ii) How many 4-digit positive integers with distinct digits can be constructed with 0, 1, 2, 3, 4, 5? How many of them are even?

SOLUTION.   Let n_1 be the number of all 4-digit numbers with distinct digits and written with 0, 1, ..., 5, and let n_2 be the number of all those that are even. Then

$$n_1 = 5 \cdot V(3,5) = 5 \cdot 5 \cdot 4 \cdot 3 = 300,$$

since the leading digit of the given number can be chosen in 5 ways from among 1, 2, ..., 5, and the remaining three digits can then be chosen in $V(3,5)$ ways.

To determine the number n_2, we divide all even numbers into two subsets S_1, S_2, such that S_1 contains those numbers that end in a 0, of which there are obviously $V(3,5) = 5 \cdot 4 \cdot 3 = 60$, while S_2 contains those that end in a 2 or a 4. By the multiplication rule there are $4 \cdot 4 \cdot 3 \cdot 2 = 96$ such numbers. Altogether,

$$n_2 = |S_1| + |S_2| = 60 + 96 = 156. \qquad \square$$

(iii) Sixteen soccer teams compete for the Czech national title. In how many ways can the first three places be occupied?

SOLUTION.   It is clear that we have to determine the number of all ordered triples taken from among 16 elements, and this is $V(3,16) = 16 \cdot 15 \cdot 14 = 3360.$ $\qquad \square$

2.6 Exercises

(i) Find the number of all even positive integers written with the digits 2, 3, 4, 5, 6, such that no digit may occur more than once.

(ii) How many bilingual but one-directional dictionaries are required so that each one of a group of five languages can be directly translated into all the other languages in that group?

(iii) Show that for $k < n$ we have $V(k+1, n) = (n-k) \cdot V(k, n).$

(iv) Show directly (that is, without using formula (4)) that for each $k \leq n$ we have $V(k+1, n+1) = (n+1) \cdot V(k, n).$

2.7   Combinations Without Repetitions

Let $k \leq n$ and let A be a finite n-element set. In many situations it is convenient to consider k-element collections consisting of elements from A such that each element $a \in A$ occurs at most once and the order of the elements doesn't matter, in other words, to consider k-element *subsets* of the set A. A k-element subset of the set A will be called a k-*combination from the elements of A* (or, in short, a k-*combination from n elements*).

Thus, for example, $\{a, b, c\}$, $\{a, b, d\}$, $\{a, c, d\}$, and $\{b, c, d\}$ are (all the) 3-combinations from $A = \{a, b, c, d\}$.

Theorem. *Let $C(k, n)$ denote the number of all k-combinations from n elements. Then*

$$C(k, n) = \frac{n!}{(n - k)!k!}. \tag{5}$$

Before proving formula (5) we remark that the term on the right-hand side of (5) is usually (and not only in combinatorics) called a *binomial coefficient* or *combinatorial number* and denoted by $\binom{n}{k}$; this is usually read as "n choose k" (see also [5], Chapter 1, Section 1).

PROOF.   We obtain the theorem by using the identity (4), with the help of which we rewrite (5) as

$$C(k, n) = \frac{V(k, n)}{k!}. \tag{6}$$

It is now easy to verify (6). We subdivide all $V(k, n)$ k-element variations from n elements into groups in such a way that each group contains exactly those variations that consist of the same k elements (and differ only in the ordering of the elements). Then each group has exactly $k!$ variations; the number of these groups is clearly equal to the number of all k-element subsets chosen from the given n elements, and this is equal to $C(k, n)$. By the addition rule we then have

$$V(k, n) = k! \cdot C(k, n),$$

and this implies the identity (6).   □

2.8   Examples

(i) In the Czech lottery game "Lucky Ten" one has to choose 10 numbers out of 80. The draw then consists of choosing 20 numbers of these 80, and you win if your 10 numbers are among the 20 drawn.

Find the number of possible draws, and how many different sets of 10 numbers one can choose. Furthermore, determine the number of draws such that a given set of 10 numbers is a winning selection, and finally, find the number of winning sets of 10 numbers for a given draw of 20 numbers.

SOLUTION.   In the draw a 20-element subset of the set $\{1, 2, \ldots, 80\}$ is chosen, so there are exactly $C(20, 80) = \binom{80}{20}$ choices. Similarly, the number of sets of 10 chosen by the player is $C(10, 80) = \binom{80}{10}$. If 10 numbers are chosen, one can add another 10 to them, from out of the remaining 70 numbers; this can be done in $C(10, 70) = \binom{70}{10}$ ways. Finally, for a given draw of 20 numbers, there are exactly $C(10, 20) = \binom{20}{10}$ winning sets of 10 numbers.   □

(ii) Six points lie in a plane in such a way that no three of them lie on the same straight line. How many straight lines do these point determine?

SOLUTION.   Since each straight line is uniquely determined by a pair of points through which it runs, the number of all lines is equal to the number of all pairs of points that can be chosen from the given set of six. Obviously, this is independent of the order of the points chosen. Therefore, the total number of straight lines is equal to

$$C(2, 6) = \binom{6}{2} = 15. \qquad \qquad □$$

(iii) Five identical balls are distributed among seven children. In how many ways can this be done if each child gets at most one ball?

SOLUTION.   A given distribution of balls is equivalent to choosing five children from among the seven. The number of choices of these five "lucky" children is then

$$C(5, 7) = \binom{7}{5} = 21. \qquad \qquad □$$

(iv) From among seven boys and four girls we want to choose a six-member volleyball team that has at least two girls. In how many ways can this be done?

SOLUTION.   We divide all possible compositions of the team into three groups S_2, S_3, S_4 such that on each team in group S_i, $i = 2, 3, 4$, there are exactly i girls. These i girls can be chosen in $\binom{4}{i}$ ways, and the remaining $(6 - i)$ team members are chosen from among the boys in $\binom{7}{6-i}$ ways. By the multiplication rule we then have

$$|S_i| = \binom{4}{i}\binom{7}{6-i},$$

and the number of all different allowable choices is

$$n = |S_2| + |S_3| + |S_4| = \binom{4}{2}\binom{7}{4} + \binom{4}{3}\binom{7}{3} + \binom{4}{4}\binom{7}{2}$$
$$= 6 \cdot 35 + 4 \cdot 35 + 1 \cdot 21 = 371. \qquad \qquad □$$

(**v**) At a dance there are 12 boys and 15 girls. In how many ways can they form four couples consisting of different sexes?

SOLUTION.   A quadruple of four males can be formed in $C(4, 12) = \binom{12}{4}$ ways, and similarly, four females can be chosen in $C(4, 15) = \binom{15}{4}$ ways. From among these boys and girls one can clearly form $P(4) = 4!$ couples. By the multiplication rule the number of ways of choosing four couples is therefore

$$\binom{12}{4} \cdot \binom{15}{4} \cdot 4! = 16\,216\,200. \qquad \square$$

2.9   Exercises

(**i**) Five friends meet and shake hands. Find the total number of handshakes.

(**ii**) From among 50 participants of a meeting a five-member delegation is to be chosen. In how many ways can this be done?

(**iii**) A military company consists of 3 officers, 6 noncommissioned officers, and 60 soldiers. In how many ways can they form a unit consisting of one officer, two noncommissioned officers, and 20 soldiers? What is the answer when the chosen officer is to be the commanding officer and one of the noncommissioned officers his deputy?

(**iv**) Five girls and three boys want to play volleyball. In how many ways can they be divided into two teams of four such that each team contains at least one boy?

(**v**) Six one-crown coins are placed into ten numbered envelopes such that each envelope contains at most one coin. In how many ways can this be done?

(**vi**) A mother has five identical apples and two identical pears. Every day she gives one of these fruits to her child. In how many ways can this be done?

(**vii**) In how many ways can one choose three different positive integers less than or equal to 30 such that their sum is an even number?

*(**viii**) In how many ways can one choose three different numbers from 1, 2, ..., 100 such that one of them is the arithmetic mean of the other two?

(**ix**) In how many ways can one place k rooks on an $n \times n$ chessboard ($k \leq n$) such that no two rooks threaten each other?

2.10   Variations with Repetitions

So far in this section we have formed certain collections (ordered or unordered) of elements of a given set A in which each element occurred *at most once.* In this way we defined arrangements, variations, and combinations *without repetitions.* In what follows we remove this condition; we will consider configurations in which elements may be repeated.

Definition. Let A be a finite N-element set. Then an ordered k-tuple $(x_1, x_2, \ldots, x_k)$, where $x_i \in A$ for all $i = 1, 2, \ldots, k$, is called a *k-element variation with repetition from the n elements of the set A* (or shorter, a *k-element variation with repetition from n elements*).

Note that we do not require that the elements $x_1, x_2, \ldots, x_k$ be distinct. For instance (a, a, b, a), (b, a, a, b) are two 4-element variations with repetition from the elements of the set $A = \{a, b\}$.

If we denote by $V_0(k, n)$ the number of all k-element variations with repetition from n elements, then

$$V_0(k, n) = n^k. \tag{7}$$

The proof of this identity is easy: It is enough to convince ourselves that each of the k positions of a k-element variation with repetition can be occupied by any of the n elements, and that the choices in each of the positions are independent from each other. Therefore, by use of the multiplication rule we obtain

$$V_0(k, n) = \underbrace{n \cdot n \cdot n \cdots n}_{k \text{ factors}}.$$

Thus, for example, the number of all 4-element variations with repetition from the elements of the set $A = \{a, b\}$ is equal to $2^4 = 16$.

2.11 Examples

(i) The Czech soccer pool is based on the outcome of 8 games each week. Find the number of possible bets.

SOLUTION.   The bets are on the outcomes of 8 games which can be either a win for the home team, a win for the visiting team, or a tie. Therefore, the number of all possibilities is equal to $V_0(8, 3) = 3^8 = 6561$.   □

(ii) Find the number of all subsets of a finite n-element set A.

FIRST SOLUTION.   If $A = \{x_1, x_2, \ldots, x_n\}$, then each subset $B \subseteq A$ can be written uniquely as an n-element variation with repetition $(\varepsilon_1, \varepsilon_2, \ldots, \varepsilon_n)$ from the elements $\{0, 1\}$, where $\varepsilon_i = 1$ if and only if $x_i \in B$. Hence the number N of all subsets of A satisfies

$$N = V_0(n, 2) = 2^n.   □$$

SECOND SOLUTION.   By identity (5) the number of k-element subsets of an n-element set ($k = 0, 1, \ldots, n$) is given by $\binom{n}{k}$, and therefore

$$N = \binom{n}{0} + \binom{n}{1} + \cdots + \binom{n}{n}.   □$$

By comparing both solutions we obtain the identity

$$\binom{n}{0} + \binom{n}{1} + \cdots + \binom{n}{n} = 2^n,$$

which was also derived in [5], Chapter 1, Section 1.3, by means of the binomial theorem.

(iii) Find the number of all five-digit positive integers written with the digits 0, 1, 2, 3, 4, 5, 6, if we allow the repetition of digits.

SOLUTION. Let us first consider all nonnegative integers of at most five digits, formed with the given digits. Their number n_1 is

$$n_1 = V_0(5,7) = 7^5,$$

since the digits of each such number forms a 5-element variation with repetition from the set $\{0, 1, \ldots, 6\}$ (for numbers with fewer than five digits we add an appropriate number of zeros to the beginning). Now we determine the number n_2 of integers with fewer than five digits, namely, those whose 5-digit representation begins with a zero. Clearly, we have

$$n_2 = V_0(4,7) = 7^4,$$

and for the number n of all 5-digit integers we then obtain

$$n = n_1 - n_2 = 7^5 - 7^4 = 6 \cdot 7^4 = 14\,406. \qquad \square$$

We remark that this problem can also be solved by a direct application of the multiplication rule: The first digit can be chosen in 6 ways (any of the digits 1, 2, ..., 6), and each of the further digits can be chosen in 7 ways (now we can take any of the digits 0, 1, ..., 6). Since the choices of these digits are independent, we have

$$n = 6 \cdot 7 \cdot 7 \cdot 7 \cdot 7 = 6 \cdot 7^4.$$

(iv) In a barnyard there are three chickens, four ducks, and two geese. In how many ways can one choose groups from among them that contain at least one chicken, at least one duck, and at least one goose?

SOLUTION. Each such group consists of a nonempty set of chickens (which, by (ii), can be chosen in $2^3 - 1 = 7$ ways), a nonempty set of ducks (with exactly $2^4 - 1 = 15$ choices), and a nonempty set of geese ($2^2 - 1 = 3$ choices). Since these three sets can be chosen independently, by the multiplication rule the desired number of groups is equal to $7 \cdot 15 \cdot 3 = 315$. ⊔

(v) Consider all nonnegative integers less than 10^6. Which number is greater, the number of those integers that do not contain a 9 as a digit or the number of those that contain at least one 9 as a digit?

SOLUTION.   Let n_1 denote the number of those nonnegative integers less than 10^6 that can be written without a 9 as a digit, and n_2 the number of those that contain at least one 9 as a digit. Then obviously, $n_1 + n_2 = 10^6$. It will be easier to first determine n_1: If we add an appropriate number of zeros to the beginning of integers with fewer digits, then n_1 is equal to the number of 6-digit decimal representations with the nine digits 0, 1, ..., 8, where the order is essential, but digits may be repeated. Therefore, we have by the multiplication rule $n_1 = 9^6 = 531\,441 > 10^6 - 9^6 = n_2$.   □

2.12 Exercises

(i) King Arthur wants to send six urgent messages to his knights. Each one of his three messengers can deliver any of the messages. In how many ways can King Arthur divide the messages among the messengers?

(ii) Find the number of different positive four-digit integers that can be written with the digits 0, 1, 2, 3, 4, 5, 6, 7, if the digits can be repeated.

(iii) How many symbols can one form in Morse code when dots and dashes are combined in groups from one to four elements?

(iv) How many positive integers less than 10^5 can be written with only the digits 7 and 9?

(v) Find the number of positive three-digit integers in base-12 representation.

(vi) Find the number of all four-digit positive integers that are divisible by four and are formed with the digits 1, 2, 3, 4, 5, 6.

(vii) Find the number of all four-digit positive integers that are divisible by four and are formed with the digits 0, 1, 2, 3, 4, 5.

(vii) Find the number of all four-digit positive integers that are less than 3000 and are formed with the digits 0, 1, 2, 3, 4, 5.

(ix) A club with n members holds a vote for president, where each member is a candidate, and each member has exactly one vote (possibly for him- or herself). How many ways of voting are there?

*(x) Let M be a finite n-element set. Find the number of all pairs of sets A, B such that $A \subseteq B \subseteq M$.

2.13 Multisets

In a number of situations it is convenient to consider unordered collections of elements where some elements can occur repeatedly. Since these are not *sets* in the usual sense, we introduce the different term *multiset*, which we can formally define as follows: A multiset $\mathcal{A}$ is a pair (A, α), where A is a nonempty finite set and $\alpha : A \to \mathbb{N}_0$ is a mapping. For an $a \in A$ the number $\alpha(a)$ is called the *multiplicity* of the element a in the multiset $\mathcal{A}$. If furthermore $\alpha(a) \neq 0$, then we say that a *is an element* of the multiset $\mathcal{A}$. The sum

$$N = \sum_{a \in A} \alpha(a)$$

is called the *number of elements* of the multiset $\mathcal{A} = (A, \alpha)$; we also say that $\mathcal{A}$ is an N-element multiset (but note that N is not normally the number of *distinct* elements of $\mathcal{A}$, that is, the number of elements $a \in A$ with $\alpha(a) \neq 0$).

Let us now agree on writing down each multiset $\mathcal{A} = (A, \alpha)$ *by enumeration*. This is done by writing $\mathcal{A} = [a_1, a_2, \ldots, a_N]$, where $a_i \in A$ ($1 \leq i \leq N$) and for each $a \in A$ we have $a_i = a$ for exactly $\alpha(a)$ different values of the index i. Thus, for example, if $A = \{a, b, c\}$, $\alpha(a) = 2$, $\alpha(b) = 1$, and $\alpha(c) = 0$, then we can write $\mathcal{A} = [a, a, b]$, $\mathcal{A} = [a, b, a]$, or $\mathcal{A} = [b, a, a]$. Therefore, an N-element multiset is often also called an unordered N-tuple. Note that if we give the multiset (A, α) by enumeration alone, we cannot recognize the set A (we don't know whether or not A contains elements a with multiplicity $\alpha(a) = 0$). However, the particular set A will always be clear from the context.

2.14 Arrangements with Repetitions

Let $\mathcal{A}$ be an N-element multiset. An ordered N-tuple $(x_1, x_2, \ldots, x_N)$ will be called an N-*element arrangement with repetitions of the elements of the multiset* $\mathcal{A}$ if $\mathcal{A} = [x_1, x_2, \ldots, x_N]$ in the sense of Section 2.13.

Thus, for example, (a, b, a, b, b) and (b, b, a, b, a) are two different 5-element arrangements with repetitions of elements of the multiset (A, α), where $A = \{a, b\}$, $\alpha(a) = 2$, and $\alpha(b) = 3$. It is easy to see that in the case $\mathcal{A} = (A, \alpha)$ with $\alpha(a) = 1$ for all $a \in A$, the above definition reduces to the definition of arrangements of elements of a set A, as given in Section 2.1.

A different terminology is also often used in the literature: Let $(x_1, x_2, \ldots, x_N)$ be an N-element arrangement with repetition of the elements of the multiset $\mathcal{A} = (A, \alpha)$, where $A = \{a_1, a_2, \ldots, a_n\}$. We say that the elements x_i, x_j are of the *same* (namely the kth) *type* when $x_i = a_k = x_j$; otherwise (when $x_i \neq x_j$), we say that the elements are of *different types*. The ordered N-tuple $(x_1, x_2, \ldots, x_N)$ is then called an *arrangement with repetitions of α_1 elements of the first type, α_2 elements of the second type, $\ldots$, α_n elements of the nth type* (for clarity we write α_i instead of $\alpha(a_i)$, for $i = 1, 2, \ldots, n$); the number of all these arrangements will be denoted by $P_0(\alpha_1, \alpha_2, \ldots, \alpha_n)$.

Theorem.

$$P_0(\alpha_1, \alpha_2, \ldots, \alpha_n) = \frac{(\alpha_1 + \alpha_2 + \cdots + \alpha_n)!}{\alpha_1! \alpha_2! \cdots \alpha_n!}. \qquad (8)$$

PROOF.   We begin with the assumption that elements of the same type
are also distinct, and consider all $N! = (\alpha_1 + \alpha_2 + \cdots + \alpha_n)!$ arrangements
(without repetition) of these N elements. Then we subdivide these arrangements into groups in such a way that two members of the same group differ
only in that elements of the same type are interchanged. Each group can
then be represented by one fixed arrangement with repetitions. Since the
elements of the first type can be interchanged in $\alpha_1!$ ways from a fixed position, the elements of the second type in $\alpha_2!$ ways, etc., and the elements
of the nth type in $\alpha_n!$ ways, each group contains exactly $\alpha_1! \cdot \alpha_2! \cdots \alpha_n!$
arrangements, and by the addition rule we have

$$N! = (\alpha_1 + \alpha_2 + \cdots + \alpha_n)! = (\alpha_1! \cdot \alpha_2! \cdots \alpha_n!)P_0(\alpha_1, \alpha_2, \ldots, \alpha_n),$$

which implies the identity (8).   □

2.15   Examples of Arrangements with Repetitions

(i) In how many ways can the following white chess figures be distributed
on the first row of a chessboard: two knights, two bishops, two rooks, the
king, and the queen?

SOLUTION.   It is clear that we have to find the number of all arrangements
of two elements of the first type (knights), two elements of the second type
(bishops), two of the third type (rooks), one of the fourth type (king), and
one of the fifth type (queen). By identity (8) we have

$$P_0(2, 2, 2, 1, 1) = \frac{8!}{2!2!2!1!1!} = 5040. \qquad \square$$

(ii) An *anagram* of a word is an ordered collection of letters of this
word, where each letter occurs as often as in the given word. How many
anagrams are there of the word MATEMATIKA (the Czech word for
"mathematics")?

SOLUTION.   Each anagram is clearly given by an arrangement of three
letters A, two letters M, two letters T, one letter E, one letter I, and one
letter K. Therefore, the desired number of all anagrams is

$$P_0(3, 2, 2, 1, 1, 1) = \frac{10!}{3!(2!)^2(1!)^3} = 151\,200. \qquad \square$$

(iii) Find the number of all anagrams of the word PARABOLA under the
condition that consonants and vowels alternate.

SOLUTION.   Each admissible anagram starts either with a vowel or with a
consonant; the positions for all further vowels, respectively consonants, are
then determined. The consonants can be put in their allowable positions

in $P(4) = 4!$ ways, and for placing the vowels there are $P_0(3,1) = \frac{4!}{3!}$ possibilities. The entire number of admissible anagrams is therefore, by the multiplication rule, equal to $2 \cdot 4! \cdot \frac{4!}{3!} = 192$. □

(iv) Find the number of all anagrams of the word ROKOKO, if we do not allow all three letters O to be adjacent to each other.

SOLUTION.   We denote the number of all anagrams of ROKOKO by n_1, and by n_2 the number of all those anagrams that have all three letters O adjacent to each other. Then the desired number n of anagrams is clearly $n = n_1 - n_2$. The numbers n_1, n_2 are now easy to determine: $n_1 = P_0(3,2,1) = \frac{6!}{3!2!1!} = 60$; $n_2 = P_0(2,1,1) = \frac{4!}{2!1!1!} = 12$ (here we have considered the three connected letters O as one new "letter"). So finally, $n = 60 - 12 = 48$. □

2.16   Exercises

(i) Eight students are to be put up in a student residence in three rooms, two of which have three beds, and one has two beds. In how many ways can the students by distributed over the three rooms?

(ii) An express train is to consist of two baggage cars, one dining car, four sleepers, and three coach cars. In how many different ways can the train be put together?

(iii) Combining the identities (5) and (8) gives $P_0(m,n) = C(m, m+n)$. Can you prove this last identity with an easy direct argument?

(iv) A mother has two apples of the same kind, three of the same pears, and four of the same oranges. Every day she gives her child one of these fruits. In how many ways can this be done?

(v) Find the number of all integers greater than $7 \cdot 10^5$ and consisting of the digits 2, 4, 7, and such that the digit 2 occurs twice, the digit 4 occurs once, and the digit 7 occurs three times.

(vi) Do Exercise (v), with $7 \cdot 10^5$ replaced by $4 \cdot 10^5$.

(vii) Find the number of all anagrams of the word INGREDIENT. How many of them begin and end in a vowel?

(viii) In how many ways can the letters of the word MISSISSIPPI be written such that in the resulting anagrams not all four letters I stand together?

*(ix) A university applicant has to pass four entrance exams, which means getting 2, 3, or 4 points for each exam. Furthermore, a total of at least 13 points must be achieved. How many possible exam results are there that result in the student being accepted?

2.17   Combinations with Repetitions

Let A be a finite n-element set. A k-element multiset (A, α) in the sense of Section 2.13 is called a *k-element combination with repetitions of the elements of the set A*, or also (in the older terminology introduced in Section 2.14) a *k-element combination with repetition of elements of n types*. Thus, for example, $[a, a, b, b, b]$ is a 5-element combination with repetitions of the elements of the set $\{a, b, c\}$. Two multisets (A, α) and (A, β) are considered to be *different* combinations with repetitions if $\alpha(a) \neq \beta(a)$ for some $a \in A$.

Theorem. *The number of all k-element combinations with repetitions of elements of n types, denoted by $C_0(k, n)$, is equal to*

$$C_0(k, n) = P_0(k, n - 1) = \frac{(k + n - 1)!}{k!(n - 1)!}. \tag{9}$$

PROOF.   To each k-element combination with repetitions (A, α) of elements of the set $A = \{a_1, a_2, \ldots, a_n\}$ we associate a sequence of zeros and ones as follows: First we write $\alpha(a_1)$ ones, then one zero, then $\alpha(a_2)$ ones, then one zero again, etc., up to $\alpha(a_n)$ ones. (If $\alpha(a_i) = 0$ for some i, then there will be no "1" in the corresponding position.) Thus we obtain an ordered p-tuple of 0's and 1's, where

$$p = \alpha(a_1) + 1 + \alpha(a_2) + 1 + \cdots + \alpha(a_{n-1}) + 1 + \alpha(a_n)$$
$$= (\alpha(a_1) + \alpha(a_2) + \cdots + \alpha(a_n)) + n - 1 = k + n - 1.$$

This "encodes" the multiset (A, α) as an ordered $(k+n-1)$-tuple consisting of k ones and $(n - 1)$ zeros. Conversely, we can "decode" this particular $(k + n - 1)$-tuple and uniquely get the multiset (A, α) back. It is furthermore clear that each ordered $(k + n - 1)$-tuple of k ones and $(n - 1)$ zeros is the "cipher" of some k-element combination with repetitions of the elements of the set A. (For instance, the combination $[a, a, b, b, b]$ of elements of the set $\{a, b, c\}$ corresponds to the septuple $(1, 1, 0, 1, 1, 1, 0)$; the septuple $(1, 0, 0, 1, 1, 1, 1)$ corresponds to the combination $[a, c, c, c, c]$.) Hence by Section 1.7 the number $C_0(k, n)$ is equal to the number of all the $(k+n-1)$-tuples mentioned above, and therefore to the number of all arrangements with repetition of the collection of k ones and $(n - 1)$ zeros, of which there are $P_0(k, n - 1)$. The identity (8) now gives (9), and the proof is complete.   □

In closing we note that this result is often written in terms of binomial coefficients in the form

$$C_0(k, n) = \binom{k + n - 1}{k} = \binom{k + n - 1}{n - 1}. \tag{10}$$

2.18 Examples

With the help of several problems we now illustrate a way of determining the number of combinations with repetitions. It should be pointed out that the "direct" use of identity (10) can lead to errors, since it is easy to make incorrect substitutions into the binomial coefficients. It is therefore useful to proceed in the solutions of these problems in a way similar to the proof of (9).

(i) A stationery store in Brno sells 12 different postcards of this city. In how many ways can a tourist purchase 8 postcards he wants to send to different addresses, where multiple copies of a postcard can be purchased?

SOLUTION.   Although this problem can be solved by a direct use of (10) (since each purchase corresponds to an 8-element combination with repetitions of postcards of 12 types), for the sake of reinforcing the method of proof in Section 2.17, we prefer that approach. We encode each possible purchase with sequences of ones and zeros as follows: We write as many ones as postcards of the first type purchased; then, separated by a zero, we write down as many ones as postcards of the second type purchased, etc. A given purchase is therefore "encoded" by an ordered 19-tuple consisting of 8 ones and 11 zeros. Clearly, the number of these sequences is

$$P_0(8, 11) = \frac{19!}{8!11!} = \binom{19}{8},$$

and this is also the number of all possible purchases.   □

(ii) We distribute 15 (identical) tennis balls among 6 children.
- **(a)** Find the number n_1 of all possible distributions.
- **(b)** Find the number n_2 of all distributions in which each child receives at least one ball.

SOLUTION.   If we consider all balls as indistinguishable, then two distributions are distinct if some child receives a different number of balls. Once again we encode each distribution with a sequence of zeros and ones: We write down as many ones as the first child receives balls; then, separated by a single zero, we write down as many ones as the second child receives balls, etc. In this way we have a 20-tuple consisting of 15 ones and 5 zeros; there are exactly $P_0(15, 5)$ such sequences, which means that $n_1 = P_0(15, 5) = \binom{20}{5}$. To determine the number n_2 of "more just" distributions, where none of the children remains empty-handed, we use the following trick: We first give all the children one ball each, and then we distribute the remaining 9 balls without any restrictions. Using the easy argument of part (a), we then find that this can be done in $P_0(9, 5)$ ways. Therefore, $n_2 = P_0(9, 5) = \binom{14}{5}$.   □

(iii) In how many ways can three persons divide 7 identical pears and 5 identical apples among themselves without cutting them up?

SOLUTION. We first find the number of ways of distributing the pears among the three persons; as in (ii) we see that this number is $P_0(7,2) = \binom{9}{2}$. Similarly, the number of ways of distributing the apples is $P_0(5,2) = \binom{7}{2}$. By the multiplication rule the two types of fruits can be distributed in $\binom{9}{2} \cdot \binom{7}{2} = 756$ ways.   □

(iv) For any fixed $n \in \mathbb{N}$ find the number of all solutions of the equation

$$x_1 + x_2 + \cdots + x_k = n$$

(a) in the set of all nonnegative integers.
(b) in the set of all positive integers.

SOLUTION. (a) We assume that the ordered k-tuple $(c_1, c_2, \ldots, c_k)$ of nonnegative integers is a solution of the given equation; that is, $\sum_{i=1}^{k} c_i = n$. Once again we encode this solution with a sequence of zeros and ones: First we write c_1 ones, then, separated by a zero, we write c_2 ones, etc. This sequence then contains n ones and $(k-1)$ zeros. Since each solution can be expressed as such a sequence and, conversely, each sequence of n ones and $(k-1)$ zeros corresponds to a solution of the given equation, it is enough to find the number of all such sequences. But this number is $P_0(n, k-1) = \binom{n+k-1}{n}$.

(b) The positive integers $x_1, x_2, \ldots, x_k$ are solutions of the given equation if and only if the nonnegative numbers $y_i = x_i - 1$ $(i = 1, 2, \ldots, k)$ are solutions of the equation

$$y_1 + y_2 + \cdots + y_k = n - k;$$

note that this is the same trick as the one used in finding the number of "just" distribution of balls in Example (ii). We have therefore reduced the problem to the situation of part (a): There are solutions only when $n - k \geq 0$, that is, $k \leq n$, and their number is $P_0(n-k, k-1) = \binom{n-1}{k-1}$.   □

(v) Given the positive integers n and k, find the number of k-element sequences of integers $(a_1, a_2, \ldots, a_k)$ that satisfy the condition $1 \leq a_1 \leq a_2 \leq \cdots \leq a_k \leq n$. (The $k-1$ inequalities in the middle mean that we are dealing with *nondecreasing* sequences.)

FIRST SOLUTION. If $(a_1, a_2, \ldots, a_k)$ is an admissible sequence, then the numbers $x_1 = a_1 - 1$, $x_2 = a_2 - a_1$, $x_3 = a_3 - a_2$, $\ldots$, $x_k = a_k - a_{k-1}$, $x_{k+1} = n - a_k$ are nonnegative integers, and we have

$$x_1 + x_2 + \cdots + x_k + x_{k+1} = (a_1 - 1) + (a_2 - a_1) +$$
$$\cdots + (a_k - a_{k-1}) + (n - a_k) = n - 1.$$

Conversely, if the nonnegative integers $x_1, x_2, \ldots, x_{k+1}$ are a solution of the equation

$$x_1 + x_2 + \cdots + x_{k+1} = n - 1,$$

then the numbers $b_1 = 1+x_1$, $b_2 = 1+x_1+x_2, \ldots, b_k = 1+x_1+x_2+\cdots+x_k$ form a k-element nondecreasing integer sequence with $b_1 \geq 1$ and $b_k \leq n$. This means that the number of all such sequences is equal to the number of all solutions of the above equation in nonnegative integers $x_1, x_2, \ldots, x_{k+1}$, and by Example (iv)(a) this number is exactly $P_0(n-1, k) = \binom{n+k-1}{k}$. $\quad\square$

SECOND SOLUTION.   We begin with the realization that each nondecreasing sequence is uniquely determined by a multiset of its terms (in the sense of Section 2.13). Therefore, the number of sequences under consideration is equal to the number of k-element combinations with repetitions of the elements of the set $\{1, 2, \ldots, n\}$; this number is, by (9), equal to $C_0(k, n) = \binom{n+k-1}{k}$. $\quad\square$

2.19   Exercises

(i) A coffee shop sells five different kinds of coffee. With how many choices can one buy 12 bags of coffee? In how many ways can this be done if we want at least two bags of each kind?

(ii) In how many ways can one distribute 12 one-crown coins over 7 numbered envelopes such that none of them remains empty?

(iii) 30 people vote on a choice of 5 recommendations, with each person voting for exactly one recommendation. In how many possible ways can these secret votes be distributed over the recommendations?

(iv) In how many ways can 2 white and 7 red balls be distributed over 9 different drawers
(a) if none of the drawers is to remain empty?
(b) if some drawers may remain empty?

(v) If you have an unlimited supply of nickels, dimes, and quarters, in how many ways can you choose 20 coins?

For each of the equations and inequalities (vi)–(x) find the number of solutions (a) in nonnegative integers, (b) in positive integers:

(vi) $x_1 + x_2 + x_3 + x_4 = 9$.   (vii) $x_1 + x_2 + x_3 + x_4 \leq 9$.
(viii) $x_1 + x_2 + x_3 + x_4 < 9$.   *(ix) $x_1 + x_2 + \cdots + x_k \leq n$.
*(x) $x_1 + x_2 + \cdots + x_k < n$.

(xi) Given the integers n and k, find the number of k-term sequences of integers $(c_1, c_2, \ldots, c_k)$ such that $0 \leq c_1 \leq c_2 \leq \cdots \leq c_k \leq n$.

3   Problems with Boundary Conditions

In the previous section we defined some standard combinatorial concepts and gave examples of problems in which their numbers were to be determined in relatively easy situations. We will now proceed to solving somewhat harder problems, where at first sight it is not clear how those standard concepts could be used, and problems where some *boundary conditions* come into play. While the examples in 3.1 and 3.2 are thematically varied, the later parts of this section will predominantly be dealing with problems in the same area: Sections 3.3 and 3.4 will be devoted to *digit representations* in the decimal system, and in 3.5 we will solve some combinatorial problems in *number theory*. An interesting and important problem is the so-called *queuing problem*; the concluding sections, 3.7–3.10, are devoted to this and related problems.

3.1   Examples

(i) A train compartment has two facing rows of 5 seats, with one seat each by the window. There are 9 passengers, three of whom want to sit facing the engine, while the direction doesn't matter to the remaining six passengers, who include a little boy and his mother; however, the boy wants to sit by the window and next to his mother. In how many ways can the passengers sit so that all are content?

SOLUTION.   We divide all possible seatings into two groups S_1, S_2 according to whether the boy and his mother face the engine, or have their backs to the engine. Let us first determine the number $|S_1|$: The three passengers who want to face the engine can sit in 3! different ways, and the remaining four passengers can sit on the five backwards-facing seats in $\frac{5!}{1!}$ ways; therefore, $|S_1| = 3! \cdot 5!$. To find the number $|S_2|$, we use the following argument: We first place the three passengers in the seats facing the engine, which can be done in $\frac{5!}{2!}$ ways; opposite them are the boy and his mother, sitting in uniquely determined seats. With each of these distributions there remain five seats that can be taken by the remaining four passengers; this can be done in $\frac{5!}{1!} = 5!$ ways. So together this gives $|S_2| = \frac{5!}{2!} \cdot 5!$. The total number of possible seating arrangements is therefore $|S_1| + |S_2| = 3! \cdot 5! + \frac{5!}{2!} \cdot 5! = 7920$.   □

(ii) Find the number of all arrangements of m ones and n zeros such that no two ones are adjacent to each other.

SOLUTION.   It is clear that when $m > n + 1$, no such arrangement exists. Let now $m \leq n + 1$; then the number of desired arrangements is equal to the number of ways the given m ones can be placed in $n + 1$ positions, namely the $n - 1$ spaces between the zeros, and in front of the first and behind the last zero. (One could say that the n zeros form dividing walls between the $n+1$ compartments from which we have to choose m that are to contain one "1" each.) Therefore, the desired number of arrangements is $\binom{n+1}{m}$. □

(iii) In how many anagrams of the word LOCOMOTIVE are no two letters O next to each other?

SOLUTION.   We first arrange the remaining letters L, C, M, T, I, V, E in $P(7) = 7!$ ways. In each of these arrangements the letters form barriers between 8 compartments from which we can choose three to contain the letter O (see (ii)); this can be done in $\binom{8}{3}$ ways. By the multiplication rule the total number of allowable anagrams is therefore $\binom{8}{3} \cdot 7! = 282\,240$. □

(iv) We want to place m men and n women at a round table with $(m+n)$ chairs in such a way that no two men sit next to each other. In how many ways can this be done?

SOLUTION.   If we select places for the men and for the women in the required fashion, then the men can take these places in $P(m) = m!$ ways, and the women in $P(n) = n!$ ways. It therefore remains to find the number of ways in which we can select m chairs (for the men) of which no two are adjacent, from the given $(m+n)$ chairs. If we associate with each "male" chair the digit 1, and with each "female" chair the digit 0, then there is an obvious connection with Example (ii); however, the digits are not written in a row, but in a circle. We now reduce the problem to a "linear" one by using the following trick: We choose an arbitrary fixed chair, denote it by X, and divide all choices of chairs into two disjoint groups according to whether X is for a man or for a woman. In the first case (that is, if X corresponds to a 1) both neighboring chairs correspond to a 0, and by (ii) there are exactly $\binom{n-1}{m-1}$ choices of chairs for the men. In the second case the group has exactly $\binom{n}{m}$ members. All this is under the assumption that $m \leq n$; otherwise, the number of possible seatings is zero. So, if we denote the number of all possible seatings by $P(m,n)$, we can conclude that

$$P(m,n) = m!n! \left[\binom{n-1}{m-1} + \binom{n}{m} \right] = m!n! \cdot \frac{n+m}{n} \binom{n}{m}$$

for $m \leq n$, while $P(m,n) = 0$ for $m > n$. □

(v) A deck of cards has eight cards each in four suits. In how many ways can one choose six cards such that cards of all four suits are among them?

SOLUTION.   We divide all allowable sextuples of cards into two groups S_1, S_2 as follows: S_1 contains all those sextuples that contain two cards each of two suits and one card each of the remaining two suits; S_2 contains those sextuples in which one suit is represented by three cards, with one card each in the remaining suits. To determine $|S_1|$, we note that the suits with two cards can be chosen in $\binom{4}{2}$ ways, and the two cards of each of these suit can be chosen in $\binom{8}{2}$ ways. If we choose one card of a given suit, this can be done in $\binom{8}{1}$ ways, and so by the multiplication rule we have

$$|S_1| = \binom{4}{2} \cdot \binom{8}{2} \cdot \binom{8}{2} \cdot \binom{8}{1} \cdot \binom{8}{1} = \binom{4}{2} \cdot \binom{8}{2}^2 \cdot 8^2.$$

In a similar way we determine the number of sextuples in S_2: The suit for the card triple can be chosen in $\binom{4}{1}$ ways, and the three cards of this suit in $\binom{8}{3}$ ways; the one card in each of the remaining three suits can be chosen in $\binom{8}{1}$ ways, and therefore

$$|S_2| = \binom{4}{3} \cdot \binom{8}{3} \cdot \binom{8}{1} \cdot \binom{8}{1} \cdot \binom{8}{1} = \binom{4}{3} \cdot \binom{8}{3} \cdot 8^3.$$

Altogether, the number of allowable sextuples of cards is then

$$|S_1| + |S_2| = 8^2 \left[\binom{4}{2} \cdot \binom{8}{2}^2 + \binom{4}{3} \cdot \binom{8}{3} \cdot 8 \right] = 415\,744. \qquad \square$$

(vi) We are given the quotation OKO ZA OKO, ZUB ZA ZUB (which in this Czech version is more suitable as an example than the English, "An eye for an eye, a tooth for a tooth").

(a) In how many ways can some letters be chosen from the string if their order does not matter?

(b) In how many ways can three letters be chosen if their order does not matter?

(c) In how many ways can three letters be chosen if their order does matter?

SOLUTION.   (a) In the given quotation the letters O and Z occur four times, and the remaining letters K, A, U, B occur twice. Among any set of letters chosen, the letter O may not occur at all, or it may occur once, twice, three times, or four times, and so there are five possibilities for its choice. The same is also true for the letter Z, and similarly, for the choices of the remaining four letters there are three possibilities each. By the multiplication rule we therefore have a total of $5^2 \cdot 3^4 = 2025$ choices, where in one case no letters were taken at all, so there are 2024 possibilities for at least one letter to be chosen.

(b) A triple of letters with all three letters distinct can be chosen in $\binom{6}{3} = 20$ ways; a triple in which exactly two are distinct can be chosen in $6 \cdot 5 = 30$ ways, and three of the same letters can be chosen in only two ways. The total number of triples is therefore 52.

(c) If we take the order of letters in these three-letter "words" into account, then their number will be $6 \cdot 20 + 3 \cdot 30 + 2 = 212$.   □

(vii) In how many ways can the letters of DOREMI be rearranged such that in each anagram the vowels occur in alphabetical order?

SOLUTION.   We subdivide all 6! anagrams into groups in such a way that all those anagrams that differ only in the order of their vowels are in the same group. Each group contains 3! anagrams, exactly one of which has the vowels in alphabetical order. Therefore, the number of desired anagrams is $n = \frac{6!}{3!} = 120$.   □

(viii) We want to line up 6 Canadians, 7 Finns, and 10 Turks in such a way that every Canadian stands between a Turk and a Finn, and Turks and Finns never stand next to each other. In how many ways can this be done?

SOLUTION.   In the first part of the solution we will not distinguish between the members of a given nationality. By F_i we denote a group of some Finns standing together, by T_i a group of some Turks standing together, and the letter c denotes a Canadian. It follows from the conditions of the problem that we can have either one of the following two distributions: $F_1 c T_1 c F_2 c T_2 c F_3 c T_3 c F_4$, $T_1 c F_1 c T_2 c F_2 c T_3 c F_3 c T_4$. In the first case we have to distribute the 7 Finns over 4 nonempty groups F_i (this can be done in $\binom{6}{3} = 20$ ways), and the 10 Turks over 3 nonempty groups T_j (which can be done in $\binom{9}{2} = 36$ ways). In the second case we distribute 7 Finns over 3 nonempty groups in $\binom{6}{2} = 15$ ways and 10 Turks over 4 nonempty groups in $\binom{9}{3} = 84$ ways. Hence there are $\binom{6}{3} \cdot \binom{9}{2} + \binom{6}{2} \cdot \binom{9}{3} = 1980$ ways in total of placing all persons, not taking into account which member of a given nationality stands on which place reserved for that nationality. Now, for a given "national" arrangement the individual persons can be lined up in 6!7!10! ways, so that the total number of all possible line-ups is $1980 \cdot 6! \cdot 7! \cdot 10!$.   □

3.2   Exercises

(i) How does the result of Example 3.1.(i) change if we assume that the compartment
(a) has one more person without a preference (so that there are 10 passengers)?
(b) contains two fewer persons without a preference (so that there are 7 passengers)?

(ii) In how many ways can one rearrange the letters in the magic word ABRACADABRA such that no two letters A stand next to each other?

(iii) How many anagrams does the word TERAKOTA have, when
(a) vowels and consonants alternate?
(b) no two vowels are adjacent?

(iv) Find the number of anagrams in Example 3.1.(iii) with the additional condition that all consonants of an anagram are in alphabetical order.

(v) A lion tamer wants to walk a line-up of 5 lions and 4 tigers into the arena such that no tiger immediately follows another tiger. In how many ways can this be done?

(vi) A bookshelf contains 14 books. In how many ways can we select six books, no two of which are next to each other?

*(vii) Twelve knights sit at the round table in King Arthur's court. The relationships between them are complicated: Each one has two enemies, and these are exactly his immediate neighbors at the table. We have to choose five knights for a mission to free a princess from a dungeon. In how many ways can these five be chosen such that no two are hostile to each other?

*(viii) $2n$ chairs are placed around a round table. In how many ways can one select r chairs $(1 \leq r \leq n)$ such that no two of them are next to each other?

(ix) In how many ways can one choose eight cards from a deck of 32 (eight cards in each of the four suits) such that
(a) no ace is among the chosen cards?
(b) the ace of spades is among the chosen cards?
(c) there is at least one ace among the chosen cards?
(d) there are at least two aces among the chosen cards?
(e) two cards from each suit were chosen?
(f) there are cards of all four suits among the chosen cards?
(g) cards from exactly three different suits were chosen?
(h) cards from at least three different suits were chosen?

(x) In how many ways can eight boys and four girls be distributed over two volleyball teams of 6 players each such that each team contains at least one girl?

(xi) In how many ways can one choose four letters from the barbaric Czech word BARBAR such that
(a) the order of the letters chosen is irrelevant?
(b) the order of the letters chosen is essential?

(xii) How many anagrams can one form from ABECEDA (Czech for "alphabet") such that the consonants are in alphabetical order?

(xiii) How many anagrams can one form from the word BATOLE (baby) such that both the vowels and, independently, the consonants are in alphabetical order?

(xiv) How many anagrams can one form from the word STUDNA (well) such that between the two vowels there are two consonants?

(xv) Work out Example 3.1.(viii) in the case that there are only five Canadians among the seven Finns and ten Turks.

(xvi) In how many ways can one choose k integers from the set $M = \{1, 2, \ldots, n\}$ such that the difference between any two is not 1?

3.3   Problems on Digit Representations

One of the most popular topics in elementary combinatorics, especially at the advanced high-school level, involves finding the number of integers with a given number of digits and under certain extra conditions. We recall that in the previous section we already dealt with a few common examples of this type (in particular, see the relatively difficult Examples 2.11.(v) and 2.12.(iv)). A number of problems concerning digit representations will also be solved in Section 6.

We note that in all problems we will be dealing with the *decimal (positional) system*. We begin with the following eight problems.

(i) Find the number of 6-digit positive integers consisting of three even and three odd digits.

SOLUTION.   First we choose places for the even digits; this can be done in $\binom{6}{3} = 20$ ways. Once such a choice has been made, we can place one of five even (respectively odd) digits in each of the six places. Therefore, by the multiplication rule the number of these possibilities is $\binom{6}{3} \cdot 5^6 = 312\,500$. However, this contains the cases where the "beginning" digit is 0, so that the corresponding integers have fewer than 6 digits. With a similar argument as above we find that the number of these integers is $\binom{5}{2} \cdot 5^5 = 31\,250$, so that the total number of the desired integers is $312\,500 - 31\,250 = 281\,250$.   □

(ii) How many different 10-digit positive integers containing the digits 1, 2, 3 are there if we assume that the digit 3 is used exactly twice in each integer? How many of these integers are divisible by 9?

SOLUTION.   The position for the two 3's can be chosen in $\binom{10}{2}$ ways. For each of the remaining eight places, one of two possibilities can occur: It can contain the digit 1, or the digit 2. Hence the total number of desired integers is $\binom{10}{2} \cdot 2^8 = 11\,520$. Next, if n is one of these $11\,520$ integers, we denote its digit sum by $S(n)$. Then by the given conditions we have the following inequalities: $2 \cdot 3 + 8 \cdot 1 \leq S(n) \leq 2 \cdot 3 + 8 \cdot 2$, or $14 \leq S(n) \leq 22$. Since n is divisible by 9 if and only if $S(n)$ is divisible by 9, we must have $S(n) = 18$. This means that, apart from the two occurrences of the digit 3, n must have the digit 2 four times and the digit 1 four times. Therefore, the number of all such integers divisible by 9 is $P_0(2, 4, 4) = \frac{10!}{2!(4!)^2} = 3150$.   □

(iii) How many 4-digit integers can be formed from the digits of the number 123124?

SOLUTION.   Each desired integer has either four different digits (there are $P(4) = 4! = 24$ of them), or two pairs of identical digits (they are $P_0(2, 2) = \frac{4!}{(2!)^2} = 6$ in number), or a pair of identical digits, with the remaining two digits distinct. A pair of identical digits can be chosen in 2 ways, the remaining two distinct digits then in $\binom{3}{2} = 3$ ways, and therefore the number of digits formed in this way is $2 \cdot 3 \cdot P_0(2, 1, 1) = 6 \cdot \frac{4!}{2!} = 72$. Hence the total number of desired integers is $24 + 6 + 72 = 102$.   □

(iv) Find the number of 4-digit integers with digit sum 4.

SOLUTION.   If the digit sum of the desired integer is 4, then it can be formed only from the following five multisets of digits: $[4, 0, 0, 0]$, $[3, 1, 0, 0]$, $[2, 2, 0, 0]$, $[2, 1, 1, 0]$, and $[1, 1, 1, 1]$. The quadruples $[4, 0, 0, 0]$ and $[1, 1, 1, 1]$ give only one 4-digit number each, while from $[3, 1, 0, 0]$ we obtain $2 \cdot P_0(2, 1) = 6$ integers, and $[2, 2, 0, 0]$ gives $P_0(2, 1) = 3$ integers; the multiset $[2, 1, 1, 0]$, finally, gives rise to $P_0(2, 1) + P(3) = 9$ integers. The total number of 4-digit integers with digit sum 4 is therefore 20.   □

(v) Find the sum of all 4-digit positive integers formed with the digits 1, 2, 3, 4. Do this in the following two cases:
(a) The digits of each integer are distinct.
(b) Some digits may occur more than once.

SOLUTION.   (a) Each of the digits 1, 2, 3, 4 is in the units position for $\frac{P(4)}{4} = 3! = 6$ integers, and for the same number of integers each digit is in the tens, hundreds, and thousands position. The "contribution" of each of the digits $k \in \{1, 2, 3, 4\}$ to the desired sum is therefore $k \cdot 6 \cdot (1 + 10 + 100 + 1000)$. Hence the desired sum S_1 is equal to

$$S_1 = (1 + 2 + 3 + 4) \cdot 6 \cdot (1 + 10 + 100 + 1000) = 10 \cdot 6 \cdot 1111 = 66\,660.$$

(b) As in part (a) we determine the "contribution" of each digit $k \in \{1, 2, 3, 4\}$ to the entire sum S_2: Each digit occurs in the units (respectively tens, hundreds, thousands) position in $\frac{4^4}{4} = 4^3$ cases. Therefore, we obtain

$$S_2 = (1+2+3+4) \cdot 4^3 \cdot (1+10+100+1000) = 10 \cdot 64 \cdot 1111 = 711\,040.\ \square$$

(vi) Find the sum of all 5-digit integers consisting of the digits 0, 1, 2, 3, 4, such that none of the digits is repeated.

SOLUTION.   We determine this sum as in Example (v), including the "contributions" of the integers beginning in 0. Thus we begin with $4! \cdot (0 + 1 + \cdots + 4) \cdot 11111 = 2\,666\,640$, and then we subtract the following contributions: Each of the digits $k \in \{1, 2, 3, 4\}$ occurs in the units (respectively tens, hundreds, thousands) position in exactly $3! = 6$ integers that begin with the digit 0. We must therefore subtract a total of $6 \cdot (1 + 2 + 3 + 4) \cdot (1 + 10 + 100 + 1000) = 66\,660$, and so the desired sum is $2\,599\,980$.   □

(vii) Find the number of all 4-digit integers formed with exactly two distinct digits.

SOLUTION.   Two distinct digits can be chosen in $\binom{10}{2} = 45$ ways, and with the two selected digits we can form 2^4 different 4-digit integers, from which, however, we must exclude the two cases that contain only one of the digits. This gives $45 \cdot (2^4 - 2) = 630$ integers. Among those, however, there are integers starting with the digit 0; their number is $\binom{9}{1} \cdot (2^3 - 1) = 63$. The desired number is therefore $630 - 63 = 567$.   □

(viii) A k-digit integer $a_{k-1} \cdot 10^{k-1} + a_{k-2} \cdot 10^{k-2} + \cdots + a_1 \cdot 10^1 + a_0$, where $k \geq 1$, $a_j \in \{0, 1, \ldots, 9\}$ for $j = 0, 1, \ldots, k-1$, and $a_{k-1} \neq 0$, will be called *ascending* if it satisfies $a_i \leq a_{i-1}$ for all $i = 1, 2, \ldots, k-1$. (For example, for $k = 5$ the integer $24\,667$ is ascending, but $12\,134$ is not). Find the number of all 5-digit ascending integers.

SOLUTION.   The sequence of digits $\overline{a_4 a_3 a_2 a_1 a_0}$ of a 5-digit ascending integer is a 5-term nondecreasing sequence of positive integers, where $a_4 \leq a_3 \leq \cdots \leq a_0 \leq 9$. Therefore, according to 2.18.(v) applied with $k = 5$ and $n = 9$, the number of desired integers is $\binom{13}{5} = 1287$.   □

3.4   *Exercises*

(i) Find the number of all 5-digit positive integers that contain exactly three even digits.

(ii) Find the number of all $2n$-digit positive integers composed of n even and n odd digits.

(iii) How many odd positive integers can be formed from the digits of the number 3694 when each digit is used at most once?

(iv) How many 6-digit positive integers are there whose digit sums are even?

(v) How many 6-digit positive integers are there whose digit sums are divisible by 10?

(vi) Consider all 8-digit positive integers composed of the digits 1, 2, 3, 4 such that each of the digits 3 and 4 occurs exactly twice.
(a) What is the total number of these integers?
(b) How many of them are divisible by 4?
(c) How many of them are divisible by 3?

(**vii**) How many 4-digit positive integers can be formed from the digits of the integer 123 123?

(**viii**) How many 5-digit positive integers can be formed from the digits of the integer 12 334 444?

(**ix**) How many 4-digit positive integers can be formed from the digits of the integer 3 332 210?

*(**x**) How many 5-digit positive integers can be formed from the digits of the integer 11 223 334 if we require in addition that no three digits 3 stand together.

(**xi**) Find the number of 4-digit positive integers with digit sum 5.

(**xii**) Find the number of 4-digit positive integers composed of the digits 0, 1, 2, 3, 4 and divisible by 9, if the digits may be repeated.

(**xiii**) Find the sum of all 3-digit positive integers composed of the digits 1, 2, 3, 4, 5, 6 in the following two cases:
(a) All digits of the integers are distinct.
(b) The digits in any of the integers may be repeated.

(**xiv**) Find the sum of all 4-digit positive integers that are obtained by interchanging the order of the digits of the following integers:
(a) 1225.   (b) 1333.   (c) 1144.

(**xv**) Find the sum of all 4-digit positive integers composed of the digits 0, 1, 2, 3, 4, 5 in the following two cases:
(a) All digits of the integers are distinct.
(b) The digits in any of the integers may be repeated.

(**xvi**) Find the sum of all even 4-digit positive integers composed of the digits 0, 1, 2, 3, 4, 5 if the digits in any of the integers may be repeated.

(**xvii**) Find the number of all 5-digit positive integers composed of exactly two distinct digits.

*(**xviii**) Find the number of all even 4-digit positive integers composed of exactly two distinct digits.

*(**xix**) Find the number of all 6-digit positive integers composed of exactly three distinct digits.

(**xx**) Find the number of all k-digit ascending integers (see 3.3.(viii)).

(**xxi**) A k-digit positive integer is called *strictly ascending* if the inequality $a_i \le a_{i-1}$ in 3.3.(viii) is strict, that is, if $a_i < a_{i-1}$ for all $i = 1, 2, \ldots, k-1$. Find the number of all k-digit strictly ascending integers.

3.5   Problems on Integers

Many problems with combinatorial character belong thematically to the *theory of numbers*, to which Chapter 3 of [5] was devoted. One whole section in that chapter (Section 9, *Dirichlet's principle*) even dealt with combinatorial methods. Also, we have already solved some number-theoretic problems earlier in this chapter; after all, we can consider the problems in Sections 3.3 and 3.4 to be of this type. And finally, let us point to the problems on the number of solutions of the simple Diophantine equation

$$x_1 + x_2 + \cdots + x_k = n$$

in 2.18.(iv) and the related Exercises 2.19.(vi)–(x). Belonging to this general area are also problems of finding the number $\tau(n)$ of all positive divisors of a positive integer n given in its canonical form as a product of powers of primes,

$$n = p_1^{\alpha_1} \cdot p_2^{\alpha_2} \cdots p_k^{\alpha_k},$$

where $p_1, p_2, \ldots, p_k$ are distinct primes, and $\alpha_i \in \mathbb{N}$ for all $i = 1, 2, \ldots, k$. In Chapter 3, Section 2.6.(i), of [5] it was shown that

$$\tau(n) = (\alpha_1 + 1) \cdot (\alpha_2 + 1) \cdots (\alpha_k + 1). \tag{11}$$

We now turn to several problems.

(i) In how many ways can one select three integers from $1, 2, \ldots, 100$ such that their sum is divisible by 3?

SOLUTION.   The sum of three integers is divisible by 3 only in the following cases: Either all three integers leave the same remainder upon division by three, or each of them leaves a different remainder. Among the integers $1, 2, \ldots, 100$ there are 33 that are divisible by 3; furthermore, 34 leave remainder 1 upon division by 3, and 33 leave remainder 2. Therefore, the total number of choices is $\binom{33}{3} + \binom{34}{3} + \binom{33}{3} + 33 \cdot 34 \cdot 33 = 53\,922.$ □

(ii) Find the number of integer solutions of the inequality $|x| + |y| \leq 1000$.

SOLUTION.   Since $|x| + |y| \leq 1000$, we certainly have $|x| \leq 1000$. If $k \in \{0, 1, \ldots, 1000\}$ is fixed and $|x| = k$, then the inequality $|y| \leq 1000 - |x| = 1000 - k$ has exactly the $2 \cdot (1000 - k) + 1$ solutions $y = -1000 + k, -999 + k, \ldots, -1, 0, 1, \ldots, 1000 - k$. Since $|x| = k$ holds exactly when $x = -k$ or $x = k$, the number of solutions of the given inequality is

$$(2(1000 - 0) + 1) + 2 \cdot \sum_{k=1}^{1000} (2(1000 - k) + 1)$$

$$= 2001 + 4 \cdot \sum_{k=1}^{1000} (1000 - k) + 2000 = 2\,002\,001.$$

We note that this result can also be obtained by a different approach: At least one of the integers x, y is zero in $2 \cdot 2 \cdot 1000 + 1 = 4001$ solutions; if both integers x, y are nonzero, then we can add another $4 \cdot \binom{1000}{2}$ solutions by Exercise 2.19.(ix)(b), for $k = 2$ and $n = 1000$. In addition, we have to take into account the four different choices of signs for the integers x and y. □

(iii) Find the number of all solutions in $\mathbb{N}_0$ of the equation $x + y = n$, under the condition that $x \leq y$.

SOLUTION.   Let r denote the desired number of solutions. The equation $x + y = n$ without the boundary condition $x \leq y$ has $2r$ solutions when n is odd, and has $2r - 1$ solutions when n is even, and obviously in both cases this number of solutions is equal to $(n+1)$. Hence for odd n we have $r = \frac{n+1}{2}$, while for even n it is $r = \frac{n+2}{2}$, and so in any case $r = \left[\frac{n}{2}\right] + 1$.

This problem can also be solved differently: If $x + y = n$ holds for some $x, y \in \mathbb{N}_0$, then $x \leq y$ if and only if $x \leq \frac{n}{2}$. There are exactly $\left[\frac{n}{2}\right] + 1$ such x, and for each of them the number $y = n - x$ is uniquely determined. □

(iv) Find the number of all solutions in $\mathbb{N}_0$ of the equation $x + y + z = n$ under the condition that $x \leq y \leq z$.

SOLUTION.   We denote by r_1, r_2, r_3, and r_4 the number of all solutions of the given equation under the condition that, respectively, $x < y < z$, $x = y < z$, $x < y = z$, and $x = y = z$. It is clear that the number of all solutions without boundary conditions is $\binom{n+2}{2} = \frac{(n+2)(n+1)}{2}$, and with the numbers r_i this can be expressed as

$$6r_1 + 3r_2 + 3r_3 + r_4 = \frac{(n+2)(n+1)}{2}.$$

From this we determine r_1 after finding r_4 and $r_2 + r_3$. It is easy to find r_4: If $3 \mid n$, then $r_4 = 1$; otherwise, $r_4 = 0$; so in any case $r_4 = \left[\frac{n}{3}\right] - \left[\frac{n-1}{3}\right]$. Next, to determine $r_2 + r_3 + r_4$ we note that this number is equal to the number of solutions in $\mathbb{N}_0$ of the equation $2u + v = n$. With the substitution $s = u$, $t = u + v$, this reduces to the equation $s + t = n$, whose solutions in $\mathbb{N}_0$ have to satisfy the condition $s \leq t$. By Example (iii), its number of solutions is exactly $\left[\frac{n}{2}\right] + 1$, and so we have

$$r_2 + r_3 + r_4 = \left[\frac{n}{2}\right] + 1,$$

which implies $r_2 + r_3 = \left[\frac{n}{2}\right] + 1 - r_4 = \left[\frac{n}{2}\right] - \left[\frac{n}{3}\right] + \left[\frac{n-1}{3}\right] + 1$, and thus

$$r_1 + r_2 + r_3 + r_4 = \frac{1}{6}\left(\frac{(n+2)(n+1)}{2} - 3r_2 - 3r_3 - r_4\right) + r_2 + r_3 + r_4$$

$$= \frac{1}{6}\left(\frac{(n+2)(n+1)}{2} + 3(r_2 + r_3) + 5r_4\right)$$

$$= \frac{(n+2)(n+1)}{12} + \frac{1}{2} + \frac{1}{2}\left[\frac{n}{2}\right] + \frac{1}{3}\left[\frac{n}{3}\right] - \frac{1}{3}\left[\frac{n-1}{3}\right],$$

and this is the desired number of solutions.   □

(v) Find the number of ways in which a given integer $n > 1$ can be written as a product $n = x \cdot y$, where $x, y \in \mathbb{N}$ and $x \mid y$.

SOLUTION.   We first write the integers n, x, y in their prime decompositions

$$n = p_1^{\omega_1} \cdot p_2^{\omega_2} \cdots p_k^{\omega_k}, \quad x = p_1^{\alpha_1} \cdot p_2^{\alpha_2} \cdots p_k^{\alpha_k}, \quad y = p_1^{\beta_1} \cdot p_2^{\beta_2} \cdots p_k^{\beta_k},$$

where $p_1, p_2, \ldots, p_k$ are distinct primes, and $\omega_i \in \mathbb{N}$, $\alpha_i \in \mathbb{N}_0$, $\beta_i \in \mathbb{N}_0$ for $i = 1, 2, \ldots, k$. For the conditions of the problem to hold, we require

$$\omega_i = \alpha_i + \beta_i \quad \text{and} \quad \alpha_i \le \beta_i$$

for all $i = 1, 2, \ldots, k$. By the multiplication rule and the result of Example (iii), the desired number of representations of n is therefore

$$\left(\left[\frac{\omega_1}{2} \right] + 1 \right) \left(\left[\frac{\omega_2}{2} \right] + 1 \right) \cdots \left(\left[\frac{\omega_k}{2} \right] + 1 \right) = \prod_{i=1}^{k} \left(\left[\frac{\omega_i}{2} \right] + 1 \right). \qquad □$$

(vi) Find the number of ways in which a given integer $n > 1$ can be written as a product $n = x \cdot y \cdot z$, where $x, y, z \in \mathbb{N}$ and $x \mid y$ as well as $y \mid z$.

SOLUTION.   We use the notation in (v) and in addition we set

$$z = p_1^{\gamma_1} \cdot p_2^{\gamma_2} \cdots p_k^{\gamma_k},$$

where $\gamma_i \in \mathbb{N}_0$ for all $i = 1, 2, \ldots, k$. For the conditions of the problem to hold, we require

$$\omega_i = \alpha_i + \beta_i + \gamma_i \quad \text{and} \quad \alpha_i \le \beta_i \le \gamma_i$$

for all $i = 1, 2, \ldots, k$. From the result of Example (iv), and using the multiplication rule, we find that the number N of desired representations of n satisfies

$$N = \prod_{i=1}^{k} \left(\frac{(\omega_i + 2)(\omega_i + 1)}{12} + \frac{1}{2} + \frac{1}{2} \left[\frac{\omega_i}{2} \right] + \frac{1}{3} \left[\frac{\omega_i}{3} \right] - \frac{1}{3} \left[\frac{\omega_i - 1}{3} \right] \right). \qquad □$$

(vii) Find the number of ways in which the integer 1728 can be decomposed into a product of three factors $x_1, x_2, x_3 \in \mathbb{N}$ in the following cases:
(a) Decompositions that differ in the order of factors are counted as distinct.
(b) The order of factors is irrelevant.

SOLUTION.   Since $1728 = 2^6 \cdot 3^3$, we set $x_1 = 2^{\alpha_1} \cdot 3^{\alpha_2}$, $x_2 = 2^{\beta_1} \cdot 3^{\beta_2}$, $x_3 = 2^{\gamma_1} \cdot 3^{\gamma_2}$, where $\alpha_i, \beta_i, \gamma_i \in \mathbb{N}_0$ for $i = 1, 2$. We denote by N_1 the number of "ordered" decompositions from part (a), and by N_2 the number of "unordered" decompositions from (b). We clearly require

$$\alpha_1 + \beta_1 + \gamma_1 = 6, \quad \alpha_2 + \beta_2 + \gamma_2 = 3.$$

The first equation has $\binom{8}{2} = 28$ solutions, and the second one has $\binom{5}{2} = 10$ solutions; hence $N_1 = \binom{8}{2} \cdot \binom{5}{2} = 280$. Each "ordered" decomposition can be placed in one of three classes R_1, R_2, R_3, where R_1 contains all decompositions into three identical factors, R_2 contains all decompositions with two identical factors (and the third one different), and R_3, finally, has all the decompositions into three distinct factors. If we set $r_i = |R_i|$ for $i = 1, 2, 3$, then

$$N_1 = r_1 + 3r_2 + 6r_3, \quad N_2 = r_1 + r_2 + r_3.$$

Clearly, $r_1 = 1$; furthermore, $r_1 + r_2$ gives the number of "unordered" decompositions in which at least two factors are identical, and there are as many of them as there are solutions in $\mathbb{N}_0$ of the system

$$2\alpha' + \beta' = 6, \quad 2\alpha'' + \beta'' = 3;$$

this number is $4 \cdot 2 = 8$. Hence $r_1 + r_2 = 8$, that is, $r_2 = 7$ and for r_3 we obtain $r_3 = \frac{1}{6}(280 - 1 - 3 \cdot 7) = 43$. Therefore, $N_2 = 1 + 7 + 43 = 51$.   □

In closing this subsection we remark that other important number-theoretic problems will occur in Sections 6 and 7, and that in Section 8 we will be dealing with problems of *decompositions of positive integers into several summands* in connection with recurrence relations.

3.6 Exercises

In Exercises (i)–(viii) find the number of ways in which one can choose k integers from the set $\{1, 2, \ldots, n\}$ such that their sum is divisible by d:

(i) $n = 100$, $k = 2$, $d = 3$.   (ii) $k = 2$, $d = 3$.

(iii) $n = 3m$, $k = 3$, $d = 3$.   (iv) $k = 3$, $d = 3$.

(v) $n = 100$, $k = 4$, $d = 3$.   *(vi) $k = 4$, $d = 3$.

(vii) $n = 100$, $k = 4$, $d = 4$.   *(viii) $k = 4$, $d = 4$.

For each of the inequalities (ix)–(xiii) find the number of integer solutions:

(ix) $|x| + |y| < 100$.   (x) $|x| + |y| \leq n$.

(xi) $|x| + |y| + |z| \leq 100$.   (xii) $|x| + |y| + |z| \leq n$.

*(xiii) $|x_1| + |x_2| + \cdots + |x_k| \leq n$.

For each n find the number of solutions in $\mathbb{N}_0$ of the equations or inequalities (xiv)–(xix):

(**xiv**) $3x + 2y + z = n$.   (**xv**) $3x + 2y \leq n$.

(**xvi**) $2x + y + z = n$.   (**xvii**) $2x + y + z + v = n$.

(**xvii**) $3x + y + z = n$.   *(**xix**) $3x + 2y = n$.

In the remaining exercises, the integer $n > 1$ is given by

$$n = p_1^{\omega_1} \cdot p_2^{\omega_2} \cdots p_k^{\omega_k},$$

where $p_1 < p_2 < \cdots < p_k$ are primes and $\omega_1, \omega_2, \ldots, \omega_k \in \mathbb{N}$. In Exercises (xx)–(xxiv) find the number of all decompositions of the integer n into a product of m positive integers $x_1, x_2, \ldots, x_m$ such that the stated conditions are satisfied. In all cases give two answers corresponding to whether decompositions that differ only in the order of their factors are considered
(a) different,   (b) identical.
(The parentheses in (xxi)–(xxiii) denote the greatest common divisor.)

(**xx**) $m = 2$.

(**xxi**) $m = 2$, $(x_1, x_2) = 1$.

(**xxii**) $m = 3$, $(x_1, x_2) = (x_2, x_3) = (x_3, x_1) = 1$.

(**xxiii**) $m = 3$, $(x_1, x_2, x_3) = 1$.

*(**xxiv**) $m = 3$.

Find the number of solutions in $\mathbb{N}$ for each of the following equations:

(**xxv**) $n = x^2 y$.   (**xxvi**) $n = x^3 y$.   (**xxvii**) $n = x^3 y^2 z$.

The remainder of this section is devoted to problems that are thematically connected to what in the combinatorial literature is known as *queuing problems*. We will return to this again later, in Section 5.8.

3.7 Queuing at a Box Office

Suppose that at a movie theater box office there is a line of $(m+k)$ people, where m people have a ten-dollar bill and k people have a five-dollar bill. Further suppose that a ticket costs 5 dollars and that at the beginning of the ticket sales there is no money in the cash register. In how many ways can the moviegoers line up at the box office such that they can proceed without delays; that is, nobody has to wait until the cashier has enough change? (We will be interested only in the distribution of "five-dollar bills" and "ten-dollar bills," and not in the individual persons.)

SOLUTION.   We "encode" each queue with the help of ones and fives in such a way that we associate a one with each person who has a ten-dollar bill, and a five with each person who has a five-dollar bill. Then the given problem can be reformulated to an equivalent one as follows: Find the number of $(m + k)$-term sequences consisting of m ones and k fives with the property that for each $i = 1, 2, \ldots, m + k$, among the first i terms of this sequence the number of fives is greater than or equal to the number of ones. The number of such sequences will be denoted by $f(m, k)$.

It is clear that the problem has at least one solution if and only if $m \leq k$; otherwise, the queue will certainly come to a halt, which means that $f(m, k) = 0$ for $m > k$. For $m \leq k$ we first determine the number of all "bad" compositions of the queue. Each such queue is encoded as a sequence of m ones and k fives with the following property: For some $i = 1, 2, \ldots, m + k$ among the first i terms of this sequence the number of ones is greater than the number of fives. We denote the set of these sequences by X. Since the number of all $(m+k)$-term sequences of m ones and k fives is exactly $P_0(m, k) = \binom{m+k}{m}$, we have

$$f(m, k) = \binom{m+k}{m} - |X|.$$

Next we show that the set X has the same number of elements as the set Y consisting of all sequences of $(m - 1)$ ones and $(k + 1)$ fives; we do this by defining a bijection $X \to Y$. (If you find this construction not very comprehensible, you may look at its geometrical interpretation in Section 5.8.) To this end we find for an arbitrary sequence $a = (a_1, a_2, \ldots, a_{m+k}) \in X$ the smallest index d such that among the elements $a_1, a_2, \ldots, a_d$ there are more ones than fives. Then clearly $a_d = 1$, and among the elements $a_1, a_2, \ldots, a_{d-1}$ there are as many ones as fives, so the sequence

$$b = (6 - a_1, 6 - a_2, \ldots, 6 - a_{d-1}, 5, a_{d+1}, \ldots, a_{m+k})$$

consists of $(m - 1)$ ones and $(k + 1)$ fives, which means that $b \in Y$. Furthermore, in a sequence $b = (b_1, b_2, \ldots, b_{m+k})$ constructed in this way, we can "recover" the index d: It is the smallest index i such that among the elements $b_1, b_2, \ldots, b_i$ there are more fives than ones. Since $m \leq k$, such an index d exists for *every* sequence $b = (b_1, b_2, \ldots, b_{m+k}) \in Y$, which we can therefore transform back into the sequence

$$a = (6 - b_1, 6 - b_2, \ldots, 6 - b_{d-1}, 1, b_{d+1}, \ldots, b_{m+k}),$$

a sequence that belongs to X. Thus the rule $a \mapsto b$, which we have just described, is a bijection $X \to Y$.

Finally, since Y has $P_0(m - 1, k + 1) = \binom{m+k}{m-1}$ elements, we obtain

$$\begin{aligned}
f(m, k) &= \binom{m + k}{m} - |X| = \binom{m + k}{m} - |Y| \\
&= \binom{m + k}{m} - \binom{m + k}{m - 1} = \frac{k-m+1}{k+1}\binom{m + k}{m},
\end{aligned} \tag{12}$$

which completes the solution. $\square$

It is worth mentioning that in the case $m = k$, that is, when the number of persons in the queue with five-dollar is the same as the number with ten-dollar bills, the number of ways the queue can be arranged without problems is equal to

$$f(k, k) = \frac{1}{k+1} \binom{2k}{k}. \tag{13}$$

3.8 Example

Let us assume that the box office in Problem 3.7 was prepared and had q five-dollar bills for change. In how many cases does the queue proceed without problems when it contains m persons with ten-dollar bills and k persons with five-dollar bills?

SOLUTION.   We denote the number of "good" cases by $g(m, k, q)$. If $m \le q$, then the queue never stops, since the number of five-dollar bills available for change will always be enough for all the holders of ten-dollar bills. Hence in this case we obtain $g(m, k, q) = \binom{m+k}{m}$. It is also clear that when $m > q + k$, we have $g(m, k, q) = 0$; therefore, it remains to consider the case $q < m \le k + q$.

We now may as well assume that the reason for q five-dollar bills being in the cash register was that the queue started with q people who had five-dollar bills. We can therefore reformulate the problem as follows: Find the number of all "good" arrangements of a queue with $(m + k + q)$ persons, of whom $k + q$ have five-dollar bills and m have ten-dollar bills, and such that the first q persons all have five-dollar bills. We use again the device from the solution of the previous problem: Find the number of all $(m + k + q)$-term sequences with m ones and $k + q$ fives such that there are fives in the first q positions and that furthermore satisfy the condition that for each $i = 1, 2, \ldots, m + k + q$ there are at least as many fives as ones among the first i terms of the sequence.

As before, we first evaluate the number $|X|$ of elements of the set X of all "bad" sequences. As in the previous proof, for each such sequence $a = (a_1, a_2, \ldots, a_{m+k+q}) \in X$ there exists a smallest index d such that after d persons the queue comes to a halt. Therefore, $a_d = 1$, and among the terms $a_1, a_2, \ldots, a_{d-1}$ of this sequence there are as many ones as fives, and so the sequence

$$c = (6 - a_1, 6 - a_2, \ldots, 6 - a_{d-1}, 5, a_{d+1}, \ldots, a_{m+k+q})$$
$$= (c_1, c_2, \qquad , c_{m+k+q})$$

consists of $(m - 1)$ ones and $(k + q + 1)$ fives, and furthermore we have $c_1 = c_2 = \cdots = c_q = 1$. If we denote by Z the set of all such sequences c, we can verify as in 3.7 that the mapping $a \mapsto c$ constructed in this way is in fact a bijection. Therefore,

$$|X| = |Z| = P_0(m - 1 - q, k + q + 1) = \binom{m+k}{m-q-1},$$

and it follows that

$$g(m, k, q) = \binom{m+k}{m} - \binom{m+k}{m-q-1}. \tag{14}$$

Not surprisingly, for $q = 0$ the identity (14) agrees with (12).   □

3.9 Exercises

(i) Write down all "good" sequences from 3.7, for $k = 3$, $m = 2$.

(ii) Write down all "good" sequences from 3.8, for $k = m = 2$, $q = 1$.

(iii) Karel sells a newspaper (for one crown) on the street, and has no money at all for change. Soon there is a queue of 30 people, of whom 20 have a one-crown coin, but 10 have only a two-crown coin. In how many ways can the people line up so that Karel always has change? (Distinguish only between the distribution of "one-crown" and "two-crown" types, and not between the individual persons.)

(iv) Solve Exercise (iii) under the assumption that Karel was far-sighted and had five one-crown coins for change with him before he started selling the papers.

(v) Alenka receives one crown each day from her mother. Sometimes she buys herself some ice cream for a crown, and sometimes she saves the money. Her dad wants her to save half of the money so that eventually she can buy something nice. From time to time he checks her piggy bank and gets angry with her if half or more of the expected money is not there. In how many ways can Alenka buy ice cream during the first 20 days so that she can eat as much as possible without getting into trouble? She is not allowed more than one ice cream per day.

(vi) At noon all the luggage lockers at a train station are taken. During the afternoon people arrive one by one, r of them remove their luggage from a locker, and d of them deposit luggage. In how many different ways can these people arrive and always find a free locker without having to wait (independent of when their trains go)?

(vii) Find the number of all $(m+k)$-term sequences consisting of m ones and k fives, such that before each term (except the first one) there are more fives than ones in the following cases:
(a) $m < k$.   (b) $m = k$.

It is a fairly typical situation in combinatorics that two problems that at first sight appear to be entirely different can in fact be reduced one to the other. We will illustrate this with the following problem.

3.10 Example

In how many ways can $2k$ people of different heights be placed into a double row such that they are ordered according to height in each row, and such that each person in the first row is taller than the one in the second row standing immediately behind this person? (The double row consists of two rows of equal length.)

SOLUTION.   We will show that this problem can be reduced to the queuing problem in Section 3.7. To do this, we first line up the $2k$ people in the two rows in the required fashion, and then give a five-dollar bill to every person in the first row, and a ten-dollar bill to everybody in the second row. Then we let them all line up in one row according to their height, with the tallest person first; thus we get a queue with k five-dollar bills and k ten-dollar bills. It is easy to show that this queue proceeds "without stopping": Consider the person who was in the ith position in the second row ($1 \leq i \leq k$). Among the holders of the ten-dollar bills there are only $(i-1)$ who are taller than the chosen person, but there are at least i holders of five-dollar bills who are taller than this person, namely, the person who stood immediately in front, and all $i-1$ persons who were taller than this last one. Therefore, if our queue goes past the box office, there will always be a five-dollar bill in the cash register, and change is guaranteed.

Conversely, suppose we are given a "good" queue of k persons with five-dollar bills and k persons with ten-dollar bills. Without loss of generality we may assume that these persons are ordered according to height, from the tallest to the smallest. We now put all holders of five-dollar bills in the first row, still according to height, and similarly all the holders of ten-dollar bills in the second row. It is clear that the double row created in this way satisfies the conditions of the problem.

This all shows that the number of all the allowable arrangements is given by equation (13), namely, $\frac{1}{k+1}\binom{2k}{k}$.   □

3.11 Exercise

For a given k find the number of all k-term nondecreasing sequences $(a_1, a_2, \ldots, a_k)$ of positive integers such that for each $i = 1, 2, \ldots, k$ we have $a_i \leq i$ (see also 2.18.(v)).

4 Distributions into Bins

In a number of situations in the previous sections, when we constructed various collections (both ordered and unordered) of elements, we determined the number of such collections that can be obtained under certain conditions. In these cases we never cared about the fate of those elements that were not included in our choices. The problems we are going to consider now will be somewhat different: *All* the elements of a given multiset will now be divided into several classes; our task will then be to find all possible ways of subdividing the elements, and especially to determine the *number* of these subdivisions.

Obviously, one can consider several variants of such subdivisions. Sometimes, the number of elements in the different classes is prescribed. Sometimes, the order of elements in the classes is essential (such as in creating integers from a set of digits or building anagrams of a set of letters), while at other times the order is irrelevant, such as in a game of cards, where the order in which a player receives the dealt-out cards doesn't matter; what counts in the end is only his hand, that is, the collection of cards as the game begins.

It is also important to know whether we distinguish between individual elements, and whether we distinguish between classes into which these elements were placed. Finally, in some problems it is essential to know whether some bins (or classes) remain empty. If, as in Example 2.18.(ii), we divide identical balls among a number of children without any restrictions, then among all the possibilities there will be some that are "extremely unfair," namely, cases where one or more of the children get nothing. It is therefore reasonable also to determine the number of fairer divisions, where each child receives at least a predetermined number of balls.

We have therefore a large number of interesting problems to which this section will be devoted. In 4.1–4.3 we deal with situations where distinguishable objects are subdivided and placed into bins, which may be identical or distinct, and where in addition, the number of objects in the individual bins is predetermined. Then, in 4.4–4.6, we concentrate on the distribution of indistinguishable objects over different bins, and the distribution of various objects over different bins. However, there are a number of problems for which the methods of solution would lie outside the framework of this section. For this reason, problems on the distribution of distinguishable objects over indistinguishable bins will be relegated to Section 6; then in Section 8 we will deal with the distribution of indistinguishable objects over indistinguishable bins. (Example 8.1.(iii) is an equivalent reformulation of this problem).

4.1 Distributions of Different Objects over Bins with Predetermined Numbers of Objects

As an introduction into this topic, we begin with three easy problems.

(i) In a game of dominoes, four players equally divide 28 pieces among themselves. In how many ways can this be done?

FIRST SOLUTION.   An arbitrary possible distribution of the pieces among the four players (which we will call A, B, C, D) can be described by an ordering of the 28 pieces achieved as follows: First, player A puts down his 7 pieces (this can be done in 7! ways); to these, player B adds his pieces (again 7! ways), then so does player C, and finally player D. In this way we must obviously obtain all of the $P(28) = 28!$ orderings of all the pieces. Now, since each distribution leads to $(7!)^4$ different orderings, the desired number of ways of distributing the pieces is $\frac{28!}{(7!)^4}$.   □

SECOND SOLUTION.   The first player chooses 7 pieces from among the 28, disregarding the order. This can be done in $\binom{28}{7}$ ways. The second player also chooses 7 pieces, but only from the remaining 21 pieces, for which there are $\binom{21}{7}$ possibilities; similarly, the third player has $\binom{14}{7}$ choices. Finally, for the fourth player there remains only $\binom{7}{7} = 1$ choice; he has to take whatever remains. By the multiplication rule we therefore obtain for the number of all distributions,

$$\binom{28}{7} \cdot \binom{21}{7} \cdot \binom{14}{7} \cdot \binom{7}{7} = \frac{28!}{21!7!} \cdot \frac{21!}{14!7!} \cdot \frac{14!}{7!7!} = \frac{28!}{(7!)^4}. \qquad \square$$

(ii) The three top contestants in the Mathematical Olympiad are to receive as book prizes 6 different books. In how many ways can they be distributed if the winner should receive three books, the second two books, and the third prize should be one book?

SOLUTION.   Just as in (i) we order the 6 books in one of the 6! ways: The winner received the first three books, the second-place winner received the next two, and the third-place winner got the remaining one. Therefore, the number of all possible ways of distributing the books is $\frac{6!}{3!2!} = 60$ (again, the order in which the winner and the second-place student received the books was irrelevant).

Once again, we obtain the same result with a different argument: The winner can choose his books in any one of $\binom{6}{3}$ ways, and the second-place winner can choose his from the remaining books in $\binom{3}{2}$ ways. By the multiplication rule the total number of possibilities is then $\binom{6}{3} \cdot \binom{3}{2} = 20 \cdot 3 = 60$.   □

(iii) Sixteen contestants take part in a judo tournament. In how many ways can pairs of opponents for the first round be drawn?

SOLUTION.   As in the previous examples we can first order all 16 contestants in one of 16! ways and then subdivide them into eight 2-element sets. We thus obtain $\frac{16!}{(2!)^8}$ choices of (unordered) pairs, where the order of the pairs is still taken into account. However, since the order of the 8 pairs is irrelevant, we can again subdivide all the choices into classes such that each class contains all those choices which have the same pairs of contestants, only in a different order. Clearly, each class has exactly 8! choices, and so the first round of the tournament can be drawn in $\frac{16!}{2^8 \cdot 8!} = 2\,027\,025$ different ways.   □

If we generalize the arguments used in the solutions of the previous three problems, we can derive a formula for the number of all distributions of objects over bins in the following situations.

Suppose that we are given N different objects and k different bins $T_1, T_2, \ldots, T_k$, and we have to place n_1 objects into the bin T_1, n_2 objects into T_2, etc., and n_k objects into T_k, with $n_1 + n_2 + \cdots + n_k = N$. Furthermore, suppose that the order of elements in the individual bins is irrelevant. Then the number $p_1(n_1, n_2, \ldots, n_k)$ of all such distributions satisfies

$$p_1(n_1, n_2, \ldots, n_k) = \frac{N!}{n_1! \cdot n_2! \cdots n_k!}. \tag{15}$$

Let us now consider those distributions where the initial setup of the bins T_* does not matter, that is, where we assign the indices $1, 2, \ldots, k$ after we distribute the objects according to the above conditions, namely, so that the bin T_i is to contain n_i objects for each $i = 1, 2, \ldots, k$. We further assume that among the numbers $n_1, n_2, \ldots, n_k$ there are exactly α different numbers $m_1, m_2, \ldots, m_\alpha$ so that m_1 occurs j_1 times (that is, into each of j_1 bins we have to put m_1 objects), m_2 occurs j_2 times, $\ldots$, m_α occurs j_α times; so obviously, $j_1 + j_2 + \cdots + j_\alpha = k$. Then j_i bins can be interchanged among each other in $j_i!$ ways (for $i = 1, 2, \ldots, \alpha$). In view of equation (15), and by use of the multiplication rule, the number $p_2(n_1, n_2, \ldots, n_k)$ of distributions of N objects over k bins, where the order of the bins is irrelevant, is

$$p_2(n_1, n_2, \ldots, n_k) = \frac{N!}{n_1! n_2! \cdots n_k! j_1! j_2! \cdots j_\alpha!} = \frac{N!}{(\prod_{i=1}^{k} n_i!)(\prod_{i=1}^{\alpha} j_i!)}. \tag{16}$$

Thus, for example, if we want to put five photographs into three envelopes, two of them with two pictures each and one envelope with one pictures then the number of ways in which this can be done is

$$p_2(2, 2, 1) = \frac{5!}{(2!2!1!)(2!1!)} = 15.$$

Let us now consider the special case where all the bins are to contain the same number of objects, that is, $n_1 = n_2 = \cdots = n_k = n$; then $N = k \cdot n$. For the number $P_1(n_1, k)$ of distributions in the case where the order of the bins matters, the identity (15) then reduces to

$$P_1(n, k) = \frac{(kn)!}{(n!)^k}. \tag{17}$$

Similarly, for the number $P_2(n, k)$ of such distributions when the order of the bins is irrelevant, the identity (16) reduces to

$$P_2(n, k) = \frac{(kn)!}{k!(n!)^k}, \tag{18}$$

since here we have $\alpha = 1$, $m_1 = n$, and $j_1 = k$.

4.2 Examples

(i) A group of 10 married couples is getting ready for an excursion in small boats; in particular, they have to split into five groups of four persons each.

   (a) In how many ways can they split in such a way that on each boat there will be two men and two women?

   (b) In how many of these cases will there be a certain Mr. X on the same boat with his wife?

   (c) In how many of these cases will there be two men, X and Y, each on the same boat with his wife?

SOLUTION. (a) The men can clearly be divided into pairs in $P_2(2,5) = \frac{10!}{5!(2!)^5}$ ways, while the number of possibilities for the distribution of the women is $P_1(2,5) = \frac{10!}{(2!)^5}$, since once the pairs of men have occupied their places on the boats, the order of the boats matters for the pairs of women. By the multiplication rule the number of all possibilities is therefore $\frac{10!}{5!(2!)^5} \cdot \frac{10!}{(2!)^5} = \frac{(10!)^2}{5! \cdot 2^{10}}$.

   (b) To the boat occupied by Mr. and Mrs. X we add another man and another woman; that can be done in any one of $9 \cdot 9$ ways. The remaining 8 couples can be distributed over the remaining 4 boats in $\frac{(8!)^2}{4!2^8}$ different ways, similarly to (a). Therefore, the number of all possibilities is $9^2 \cdot \frac{(8!)^2}{4!2^8} = \frac{(9!)^2}{4! \cdot 2^8}$.

   (c) In this case we subdivide all admissible seatings on the boats into two groups S_1, S_2. The group S_1 will contain all those seatings where both couples X and Y are in the same boat; in this case it remains to distribute the remaining 8 couples over the remaining 4 boats, which can be done in $\frac{(8!)^2}{4!2^8}$ ways. The group S_2 will contain those seatings where the couples X, Y occupy different boats. These two boats can take on two more men in $8 \cdot 7$ ways, and also two more women in the same number of ways. The remaining 6 couples are distributed over the remaining 3 boats in $\frac{(6!)^2}{3!2^6}$ ways, again as before. Therefore, the total number of admissible seatings is $\frac{(8!)^2}{4!2^8} + (8 \cdot 7)^2 \cdot \frac{(6!)^2}{3!2^6} = \frac{17(8!)^2}{4!2^8}$.   □

(ii) In how many ways can 52 cards (13 cards each in 4 suits) be dealt to four players such that each one obtains three cards each in three suits and four cards in the fourth suit?

SOLUTION.   The cards of each suit can be dealt in the pattern $4+3+3+3$ in $p_2(4,3,3,3) = \frac{13!}{4!(3!)^4}$ ways. The groups of 4 cards then can be dealt to the four players in $4!$ ways, and the triples of cards in $(3!)^4$ ways. So altogether, the number of ways is

$$\left[\frac{13!}{4!(3!)^4}\right]^4 4!(3!)^4 = \frac{(13!)^4}{(4!)^3(3!)^{12}}.$$ □

(iii) In how many ways can one form 6 "words" of 4 letters each from 24 different letters, when each letter can be used exactly once?

SOLUTION. In comparison with the previous problems we now have four-element groups in each of which the *order of the elements* is essential. If we line up (in any one of 6! ways) the six "words" already formed, we obtain each one of the 24! orderings of all the letters. Since the order of the "words" does not matter, we obtain $\frac{24!}{6!}$ as final answer.   □

4.3 Exercises

(i) The right-hand side of equation (15) is similar to the right-hand side of (8) in Section 2 for the number of arrangements with repetitions. Find the connection between these two formulas.

(ii) In the card game of "skat" each of three players receives 10 cards from a deck of 32, with two cards remaining on the table. Find the number of all possible distributions of the cards.

(iii) Three players play the Czech card game "mariáš," where 32 cards are dealt in such a way that the first player receives 12, and the other two receive 10 cards each. At the beginning of the game the first player puts two cards aside, so that the three players proceed with 10 cards each.

(a) In how many ways can the cards be dealt?
(b) What is the number of distributions of cards at the beginning of the game?

(iv) In how many ways can 30 workers be subdivided into three teams of equal numbers? In how many ways can they be subdivided into ten teams of three?

(v) In how many ways can $3n$ different books be divided among three persons such that each one obtains the same number of books?

(vi) In the situation of Example 4.1.(iii), find the number of all possible outcomes of the first round of the tournament.

(vii) We are given $\frac{n(n+1)}{2}$ different objects, which we divide into n groups such that for each $i = 1, 2, \ldots, n$ there is a group containing exactly i objects, where the order of the objects within the groups is irrelevant. Find the number of all possible subdivisions

(a) if the order of the groups is essential.
(b) if the order of the groups is irrelevant.

In Exercises (viii)–(x) find the number of all divisions of n different books into k packages if the order of the packages is irrelevant, in the following situations:

(viii) $n = 10$, $k = 5$, and each package contains 2 books.

(xi) $n = 9$, $k = 3$, and each package contains 3 books.

(x) $n = 9$, $k = 5$, and four packages contain 2 books each.

(xi) In how many ways can a deck of 32 cards be divided into two parts of 16 each such that each part contains two aces?

(xii) In how many ways can the first round of a table tennis tournament with 9 players be drawn such that each of 4 pairs of competitors play a game, and the remaining one automatically advances to the second round?

(xiii) In the situation of Exercise (xii) find the number of all possible outcomes of the first round.

(xiv) Suppose that 16 contestants meet in a judo tournament. The competition is by elimination: For the first round 8 pairs of adversaries are drawn; the winners proceed to the second round (while the losers are eliminated from the competition), etc., up to the fourth round, where just one final fight takes place that will determine the winner of the tournament. Find the number of all possible outcomes of all the rounds in the tournament, where the order of the fights in each round is irrelevant.

(xv) Find the number of all possible outcomes of all the rounds in the tournament of (xiv) if there are 17 contestants. (If the number of contestants in a round is odd, then one is drawn to proceed automatically to the next round.)

(xvi) Consider the tournament in (xv) under different rules: In the first round two contestants are drawn who fight against each other. The winner then joins the remaining 15 for the second round, where again a pair of contestants is chosen at random from among the 16, etc., until the 16th round, when in the final fight the winner of the tournament is determined. Find the number of all possible outcomes of all the rounds under these rules.

(xvii) There are 12 different signal flags, which are flown in triples on four flagpoles. The meaning of a signal depends on the order in which the flags are raised on an individual pole, and on which pole a given triple of flags appears. How many signals can be expressed in this way?

(xviii) A deck of 52 cards in four suits is dealt to 13 players such that each one gets 4 cards.

(a) In how many ways can this be done?

(b) In how many ways can this be done if we require that each player receive one card each from each suit?

(c) In how many ways can this be done if one player is to receive cards of all suits, while all the other players receive cards of only one suit?

4.4 Distributions of Identical Objects over Distinct Bins

Let us now assume that we are given n indistinguishable objects (say, identical balls), which we have to distribute over k different bins; for clarity we label these bins with the positive integers $1, \ldots, k$. We will be interested in the number of ways this distribution can be done in different situations, such as the possibility of several bins to remain empty, or a prescribed minimum number of objects each bin has to contain, or in the other direction, an upper bound for the number of objects in the individual bins.

We will first determine the number $q_1(n, k)$ of distributions of n identical balls over k bins if one or more bins may remain empty. Between the k bins we can imagine $(k - 1)$ dividing walls; therefore, it suffices to line up the n balls in a row and place between them the $(k - 1)$ dividing walls in an arbitrary fashion. If we encode each such line-up in a sequence of n ones (the balls) and $(k - 1)$ zeros (the dividing walls), then we obtain

$$q_1(n, k) = P_0(n, k - 1) = \binom{n + k - 1}{n} = C_0(n, k). \tag{19}$$

Next we consider the situation where none of the bins is allowed to remain empty, or, more generally, where each one has to contain at least r balls. We denote the number of all such distributions by $q_2(n, k, r)$. Clearly, if $kr > n$, then we have $q_2(n, k, r) = 0$, so we assume that $kr \leq n$. At the beginning of such a distribution we place r balls into each bin; the remaining $(n - kr)$ balls can then be distributed over the bins without any restrictions. Hence we have

$$q_2(n, k, r) = P_0(n - kr, k - 1) = \binom{n - kr + k - 1}{k - 1}. \tag{20}$$

If, in particular, $r = 1$, that is, if each bin contains at least one ball, then for the number $Q_2(n, k) = q_2(n, k, 1)$ of all distributions we get

$$Q_2(n, k) = \binom{n - 1}{k - 1}. \tag{21}$$

Equation (21) can be derived in a different way as well: If n balls are lined up, then there are $(n - 1)$ gaps between them. We choose $(k - 1)$ of those and place dividing walls there. Then we put all the balls that are before the first dividing wall into the first bin, those between the first and the second dividing walls into the second bin, etc., and finally we put the balls that are behind the last dividing wall into the kth bin. Therefore, the number $Q_2(n, k)$ of distributions of n identical balls over k bins such that no bin remains empty is equal to the number of ways one can choose $(k - 1)$ gaps from among $(n - 1)$ gaps, which is $\binom{n-1}{k-1}$.

We now let $q_3(n, k)$ denote the number of all distributions of n balls over k bins such that no bin may contain more than one ball. It is clear that if $n > k$, then $q_3(n, k) = 0$. In the case $n \leq k$ it suffices to select from the k bins those n that contain one ball each; therefore, $q_3(n, k)$ is equal to the number of n-element subsets of the set $\{1, 2, \ldots, k\}$ of the labels of the bins. We have thus proved the identity

$$q_3(n, k) = \binom{k}{n}. \tag{22}$$

The problem of distributing identical balls over bins, each of which may contain at most s balls ($s \geq 2$), is more complicated. We therefore limit our treatment of this to determining the numbers of such distributions in several concrete situations, namely, 4.5.(iv) and 4.6.(viii)–(ix). A general solution will then be given in Section 6.

4.5 Examples

(i) We hand out 7 identical apples and 5 identical pears to 12 people. In how many ways can this be done if
(a) each person gets one fruit?
(b) one or more persons may get nothing?

SOLUTION.   If each person gets one fruit, then it is clearly sufficient to determine the number of ways in which 7 persons out of 12 obtain an apple; by (22) with $k = 12$ and $n = 7$, this number is $\binom{12}{7} = 792$. In the second case we first consider the distribution of the apples, and then separately that of the pears. By (19) with $k = 12$ and $n = 7$, the apples can be distributed in $\binom{18}{7} = 31\,824$ ways, and the distribution of pears (now we set $k = 12$ and $n = 5$ in (19)) can be achieved in $\binom{16}{5} = 4368$ ways. By the multiplication rule, the number of ways in which both kinds of fruit can be distributed among these 12 persons is therefore $31\,824 \cdot 4368 = 139\,007\,232$.   □

(ii) In how many ways can three people divide among themselves six identical apples, one orange, one lemon, one banana, one pear, one plum, and one apricot?

SOLUTION.   The apples can be distributed in $\binom{8}{2}$ ways, and each of the remaining fruits clearly in 3 ways. By the multiplication rule, the total number of distributions is therefore $\binom{8}{2} \cdot 3^6 = 20\,412$.   □

(iii) In the previous example, find the number of all possible distributions of fruits if we require that each person obtain exactly four fruits.

SOLUTION.   First we divide the fruits into three sets with four elements each; their orders will not matter. The apples can be divided in five different ways: $[2, 2, 2]$; $[3, 2, 1]$; $[3, 3, 0]$; $[4, 1, 1]$; $[4, 2, 0]$. In the first case we can distribute the other six different fruits according to the identity (16) in exactly $p_2(2, 2, 2) = \frac{6!}{(2!)^3 \cdot 3!} = 15$ ways. Similarly, in the second case the number of distributions is $p_2(1, 2, 3) = \frac{6!}{2! \cdot 3!} = 60$, in the third case $p_2(1, 1, 4) = \frac{6!}{4! \cdot 2!} = 15$, in the fourth case $p_2(3, 3) = \frac{6!}{(3!)^2 \cdot 2!} = 10$, and finally, in the fifth case it is $p_2(2, 4) = \frac{6!}{2! \cdot 4!} = 15$. These are, altogether, 115 possibilities; for each of them there are 3! ways of distributing the three sets among the three persons (you should convince yourself that the sets are always distinct). The desired number of distributions is therefore $6 \cdot 115 = 690$.   □

(iv) In how many ways can 6 identical balls be placed in 4 different bins such that no bin contains balls of both colors?

SOLUTION.   Each allowable distribution belongs to one of the groups S_1, S_2, where all those for which three bins contain two balls each are in S_1, while in S_2 there are those for which two bins contain two balls each, and the remaining two contain one ball each. It is now easy to determine that $|S_1| = \binom{4}{3} = 4$ and $|S_2| = \binom{4}{2} = 6$, so there exist 10 allowable distributions.   □

(v) In how many ways can three white and three red balls be distributed over 5 different bins such that each bin contains balls of only one color?

SOLUTION.   Suppose that p bins contain white balls, and q contain red balls. Then we have $1 \le p \le 3$, $1 \le q \le 3$, and $p + q \le 5$. The three white balls can be placed into the p bins, each of which is to contain at least one ball, in $\binom{2}{p-1}$ ways; this follows from (21) with $n = 3$ and $k = p$. Similarly, the black balls can be placed into their q bins in $\binom{2}{q-1}$ ways. Since $(p + q)$ bins are nonempty, $5 - (p + q)$ of them are empty; therefore, the p bins for the white balls and the q bins for the red balls can be chosen from the given 5 bins in $P_0(p, q, 5 - (p + q))$ ways, and so for the number N of all allowable distributions we obtain

$$N = \sum_{p, q} P_0(p, q, 5 - (p + q)) \binom{2}{p - 1} \binom{2}{q - 1},$$

where the sum ranges over all pairs of integers (p, q) satisfying the conditions $1 \le p \le 3$, $1 \le q \le 3$, and $p + q \le 5$. This means that

$(p, q) \in \{(1, 1); (1, 2); (2, 1); (1, 3); (3, 1); (2, 2); (2, 3); (3, 2)\}$, and thus

$$N = P_0(1, 1, 3)\binom{2}{0}^2 + 2P_0(1, 2, 2)\binom{2}{0}\binom{2}{1} + 2P_0(1, 3, 1)\binom{2}{0}\binom{2}{2}$$

$$+ P_0(2, 2, 1)\binom{2}{1}^2 + 2P_0(2, 3)\binom{2}{1}\binom{2}{2} = 340. \qquad \square$$

(vi) Suppose we have 6 blue and 4 white flags for forming signals. All the flags are to be flown on 5 flagpoles, some of which may remain without a flag. How many signals can we create if the order of the flags on each pole is important?

SOLUTION.   We encode each signal with a finite sequence of zeros, ones, and twos as follows: Each blue flag corresponds to the digit 1, each white flag to a 2, and a "dividing wall" (gap between two flagpoles) will be encoded with a 0. Then we encode a given signal in the following way: We form an ordered set of digits 1 and 2 that gives the order of the flags on the first pole; then, separated by a zero, we form the code for the second flagpole, etc. It is easy to see that this "code" describes a bijection between the set of all signals and the set of all ordered 14-tuples of six digits 1, four digits 2, and four digits 0. Hence the number of all possible signals is $P_0(6, 4, 4) = \frac{14!}{6!4!4!} = 210\,210$. $\qquad \square$

4.6   Exercises

(i) In how many ways can 3 one-crown coins and 10 two-crown coins be distributed over 4 different change purses?

(ii) In how many ways can one distribute n_1 objects of the first type, n_2 objects of the second type, $\ldots, n_k$ objects of the kth type, over two different bins?

(iii) Use the result of Exercise (ii) to prove identity (11), Section 3.5.

(iv) In how many ways can 4 identical blue and 6 identical white balls be distributed over 10 numbered bins if
    (a) some of the bins may remain empty?
    (b) each bin must receive at least one ball?
    ***(c)** exactly two bins have to remain empty?

(v) In how many ways can 6 identical white, 2 identical blue, and 2 identical red balls be distributed over 10 numbered bins if
    (a) some of the bins may remain empty?
    (b) each bin must receive at least one ball?

(vi) In how many ways can 3 identical red, 3 identical white, and 3 identical blue balls be distributed over 8 numbered bins if
    (a) some of the bins may remain empty?

*(b) each bin must receive at least one ball?

(vii) On a table there are 9 identical apples and one each of 9 other different kinds of fruit. In how many ways can 3 persons divide the fruits among themselves such that each person gets 6 pieces of fruit?

(viii) In how many ways can 10 tennis balls be distributed among three children such that each of them gets at most 4 balls?

*(ix) In how many ways can 5 identical apples and 5 identical pears be handed out to 4 children such that
(a) each child gets at most three pieces of fruit?
(b) no child gets both an apple and a pear?

(x) In the situation of 4.5.(vi), find the number of all signals that can be formed if in addition we have 5 red flags.

(xi) From the letters of the magic formula ABRACADABRA we form "sentences of three words." In how many ways can this be done if each "sentence" contains all 11 letters?

(xii) We are given n white balls and one black ball. In how many ways can p of these balls be distributed over $(n+1)$ numbered bins such that each one receives at most one ball?

*(xiii) We are given $2n$ objects of the first kind, $2n$ objects of the second kind, and $2n$ objects of the third kind. In how many ways can they be distributed over two bins such that each bin contains $3n$ objects?

*(xiv) In the situation of (xiii) we are given another $2n$ objects of the fourth kind. In how many ways can the $8n$ objects be distributed such that each of the two bins receives the same number of objects?

*(xv) We are given 5 identical white, 5 identical black, and 5 identical red balls, which we distribute over 3 different bins such that each bin contains exactly 5 balls. In how many ways can this be done?

4.7 Distributions of Different Objects over Distinct Bins

It is easy to determine the number $r_1(n, k)$ of ways in which n *distinct objects* can be distributed over k *distinct bins* if there are no further conditions:

$$r_1(n, k) = k^n. \tag{23}$$

Indeed, each object can be placed in any of the k bins, and the result is then obtained by use of the multiplication rule. For instance, 6 coins of different denominations can be placed into two pockets in $2^6 = 64$ ways. In two of these possibilities one of the pockets remains empty; hence the number of ways to distribute these coins such that none of the pockets remains empty is $64 - 2 = 62$. In Section 6.7 we will give the general solution of the problem of finding the number $r_2(n, k)$ of all distributions of n distinct objects over k distinct bins if *none of the bins remains empty*.

However, it is not difficult to find the number $r_3(n, k)$ of all distributions of n distinct objects over k distinct bins in the case where each bin can contain *at most one* object. If $n > k$, then clearly $r_3(n, k) = 0$; if $n \leq k$, then

$$r_3(n, k) = k \cdot (k - 1) \cdots (k - n + 1) = \frac{k!}{(k - n)!}, \tag{24}$$

which can easily be derived by use of the multiplication rule: For the first object we can choose any of the k bins, for the second one any of the $(k-1)$ remaining free bins, ..., and finally, for placing the nth object there are $(k - n + 1)$ possibilities, since this is the number of bins that remain empty after all the other objects have already been distributed. The reader will be able to find a connection between (24) and the expression (4) in Section 2.4 for the number of variations without repetitions, and similarly a connection between the identities (23) and (7) in Section 2.10.

The identity (23) is used in the case where the order of the objects in a particular bin is irrelevant. However, in some situations this order is important. We will therefore determine the number $r_4(n, k)$ of all possibilities of distributing n distinct objects over k distinct bins if the *order of the objects in the individual bins is important*. Let us assume for a moment that the objects are indistinguishable; then by (19) the n objects can be distributed over the k bins in $\binom{n+k-1}{n}$ ways. Now, in each of these distributions the objects can be rearranged in $n!$ ways. Thus we obtain

$$r_4(n, k) = n! \cdot \binom{n + k - 1}{n} = \frac{(n + k - 1)!}{(k - 1)!}. \tag{25}$$

This result, however, can also be obtained by a different argument: We add to our n distinct objects $(k - 1)$ "dividing walls" and consider all arrangements of these $(n + k - 1)$ elements. Each such arrangement determines a unique way of distributing n objects over k bins: We place all objects located before the first dividing wall into the first bin, ..., all objects located behind the $(k-1)$th dividing wall will be placed into the last bin (if two dividing walls are located next to each other, then the corresponding bin will be empty). Therefore $r_4(n, k)$ is equal to the number of all arrangements of n different objects and $(k - 1)$ "identical" walls, and thus

$$r_4(n, k) = P_n(\underbrace{1, 1, \ldots, 1}_{n \text{ factors}}, k - 1) = \frac{(n + k - 1)!}{1! \cdots 1!(k - 1)!} = \frac{(n + k - 1)!}{(k - 1)!}.$$

Finally, we consider yet another question, namely, that of finding the number of distributions of n distinct objects over k distinct bins if the order of elements in the bins is essential and if *none of the bins is allowed to remain empty*. We denote the number of such distributions by $r_5(n, k)$ and use the same argument that led to (25). If in a first phase we assume that the objects are indistinguishable and distribute them over k bins such that each bin contains at least one object, then by (21) this can be done in exactly $\binom{n-1}{k-1}$ ways. If we now take into account the fact that the objects are actually distinguishable, then for $n \geq k$ we obtain

$$r_5(n, k) = n! \cdot \binom{n-1}{k-1}. \tag{26}$$

4.8   Examples

(i) Five persons step into an elevator. In how many ways can they get off the elevators on eight different floors? How many possibilities are there if at each floor at most one person gets off?

SOLUTION.   In the first case each of the five persons can get off at any of the eight floors, so the number of all possibilities is $r_1(5, 8) = 8^5 = 32\,768$. The second question is equally easy to answer: The number of all possibilities in this case is $r_3(5, 8) = \frac{8!}{3!} = 6720$.   □

(ii) In how many ways can one place 20 different books on a bookcase with 5 shelves, assuming that all the books fit on all the shelves?

SOLUTION.   If we add to the 20 books 4 identical "dividing walls," then these objects can be ordered in a total of $\frac{24!}{4!}$ ways. The books before the first dividing wall are now placed on the first shelf, those between the first and the second dividing wall go on the second shelf, ..., and the books behind the last dividing wall are placed on the fifth shelf; in this way we have obtained a distribution of the 20 books on 5 shelves. The total number of such distributions is therefore $r_4(20, 5) = \frac{24!}{4!}$.   □

(iii) In how many ways can 33 different books be handed to 3 persons A, B, C such that A and B together get twice as many as C?

SOLUTION.   We first choose 11 books for person C; this can be done in $\binom{33}{11}$ ways. The remaining 22 books can be divided among persons A and B in 2^{22} ways, and so by the multiplication rule the total number of possibilities is $\binom{33}{11} \cdot 2^{22}$.   □

4.9   Exercises

(i) In how many ways can 8 different photographs be distributed among 4 children? In how many ways can this be done if each child is to get 2 photographs?

(ii) In how many ways can 6 "words" be formed from 24 different letters such that each letter is used exactly once? (Assume that the order of the "words" is irrelevant.)

(iii) In how many ways can 33 different books be handed to 3 persons such that some two of them together get twice as many as the third person?

(iv) Suppose we have 7 different signal flags, which we place on 3 flagpoles. Now the meaning of the signal depends on the order of the flags on the individual poles. Find the number of all possible signals in the following situations:

    (a) All flags are used, but some poles may remain empty.
    (b) All flags are used, and no pole remains empty.
    (c) Some flags are used, and some poles may remain empty.
    (d) Some flags are used, and no pole remains empty.

*(v) Solve Exercise (iv) in the general case where n different signal flags and k flagpoles are used.

5   Proving Identities

In the first two sections of our previous book [5] we introduced a large number of identities involving factorials, and especially *combinatorial numbers* (or *binomial coefficients*). There we announced that we would at some point give different methods of proof (that is, different from those in [5]) for these identities, namely, proofs based on *combinatorial interpretations*. A few such proofs were given in the previous sections of the present book, the most notable one being a proof of the identity

$$\binom{n}{0} + \binom{n}{1} + \binom{n}{2} + \cdots + \binom{n}{n} = 2^n, \tag{27}$$

given in 2.11.(ii). Also, the result of Exercise 2.12.(x) can be used as a combinatorial interpretation of the identity

$$\binom{n}{0} \cdot 2^0 + \binom{n}{1} \cdot 2^1 + \binom{n}{2} \cdot 2^2 + \cdots + \binom{n}{n} \cdot 2^n = 3^n \tag{28}$$

if we determine the number of all pairs (A, B) of subsets of an n-element set M that satisfy $A \subseteq B$. Indeed, for each $k = 0, 1, 2, \ldots, n$ there exist 2^k subsets A of the k-element set B that can be chosen from the n-element set M in $\binom{n}{k}$ ways. Therefore, the desired number N of all pairs (A, B) is equal to $\sum_{k=0}^{n} \binom{n}{k} \cdot 2^k$. But the number N can also be determined differently. Indeed, for each element $m \in M$ and each pair (A, B) of sets that satisfy $A \subseteq B \subseteq M$, we have exactly one of the following possibilities: $m \in A$

and $m \in B$, $m \notin A$ and $m \in B$, or $m \notin A$ and $m \notin B$. Hence, by the multiplication rule we have $N = 3^n$, which proves (28).

In the remainder of this section we will systematically study "combinatorial" proofs of such identities. In Section 5.1 we look for *bijections* between two sets, the numbers of elements of which can be expressed with binomial coefficients. In 5.2 we will count items in appropriately ordered *lists* of subsets or other collections of elements. Many relationships among binomial coefficients can be obtained by way of *decompositions* of collections *into classes* (Section 5.4). A very strong and universal method for deriving appropriate identities is the *method of words*, to which 5.6 and 5.7 will be devoted. In closing this section we will treat the method of *walks in a rectangular grid*, which is particularly appropriate for deducing combinatorial identities because of its visual geometrical character. For the proofs of several identities the *inclusion–exclusion principle* is required, and therefore a number of identities will also be obtained in the following sections.

5.1   Bijections

We have already used the *bijection rule* 1.8 in this chapter in a number of situations; in order to determine the number of elements of a set A we constructed a bijection from A to B in the case where finding the number $|B|$ was considerably easier. (In this connection, see especially the proof of Identity (9) in Section 2.17, or the solution to the problem of 3.7.) A similar method can be used for proving a number of identities; we will begin here with two of them.

(i) For $0 \le k \le n$ show that

$$\binom{n}{k} = \binom{n}{n-k}. \tag{29}$$

SOLUTION.   The number $\binom{n}{k}$ counts all k-element subsets of an n-element set A. To each such subset B we associate its complement $B' = A \setminus B$. Thus we have constructed a mapping from the collection of all k-element subsets of the set A to the collection of all $(n-k)$-element subsets of A, and it is easy to see that this mapping is a bijection. This proves (29).   □

(ii) Show that for $n \ge 1$,

$$\binom{n}{0} - \binom{n}{1} + \binom{n}{2} - \binom{n}{3} + \cdots + (-1)^n \binom{n}{n} = 0. \tag{30}$$

SOLUTION.   We rewrite (30) as

$$\binom{n}{0} + \binom{n}{2} + \binom{n}{4} + \cdots = \binom{n}{1} + \binom{n}{3} + \binom{n}{5} + \cdots. \tag{31}$$

(Note that the sums on both sides of (31) are finite; the dots stand for the terms $\binom{n}{k}$, which are included in the sums as long as $k \leq n$.) Then the sum on the left (respectively the right) side of (31) gives the number of all subsets of the n-element set $A = \{a_1, a_2, \ldots, a_n\}$ with *even* (respectively *odd*) numbers of elements. To each subset $B \subseteq A$ that has an even number of elements we associate a set $f(B) \subseteq A$ with an odd number of elements, according to the rule

$$f(B) = \begin{cases} B \cup \{a_1\} & \text{if } a_1 \notin B, \\ B \setminus \{a_1\} & \text{if } a_1 \in B. \end{cases}$$

It is clear that f is a bijection between the collection of all subsets of A with an even number of elements and the collection of all subsets of A that have an odd number of elements (the inverse mapping is defined by the same rule). This proves (31), and thus also (30). □

5.2 Lists

Several identities can be derived by determining in two different ways the number of occurrences of certain *items* (elements, sets, sequences, etc.) in an appropriately established *list*, as in the enumeration of a multiset in the sense of Section 2.13. We illustrate this method with the following three examples.

(i) For $k \leq n$ show that

$$(k+1) \cdot \binom{n+1}{k+1} = (n+1) \cdot \binom{n}{k}. \tag{32}$$

SOLUTION. The expression $(k+1) \cdot \binom{n+1}{k+1}$ gives the sum of the number of elements of all $(k+1)$-element subsets of an $(n+1)$-element set $A = \{a_1, a_2, \ldots, a_{n+1}\}$. We write down all these subsets and try to determine the number of times each element $a_i \in A$ occurs in this list: It is in each set of the form $\{a_i\} \cup X$, where X is an arbitrary k-element subset of the n-element set $A \setminus \{a_i\}$, which means that it occurs $\binom{n}{k}$ times. By summing over all $i = 1, 2, \ldots, n+1$, we obtain (32). □

(ii) Show that for $0 \leq k \leq m \leq n$,

$$\binom{n}{k} \cdot \binom{n-k}{m-k} = \binom{m}{k} \cdot \binom{n}{m}. \tag{33}$$

SOLUTION. To an arbitrarily chosen k-element subset B of an n-element set A we write the list of all m-element sets of the form $B \cup X$, where X is any $(m-k)$-element subset of the $(n-k)$-element set $A \setminus B$; this list has $\binom{n-k}{m-k}$ entries. If we unite all these lists obtained for all $\binom{n}{k}$ sets B, we obtain a list of $\binom{n}{k} \cdot \binom{n-k}{m-k}$ m-element subsets A in which all m-element subsets $C \subseteq A$ occur exactly $\binom{m}{k}$ times (namely, exactly as many times as there are k-element sets B with the property $B \subseteq C$). The list has therefore $\binom{m}{k} \cdot \binom{n}{m}$ items, which proves (33).   □

(iii) Show that

$$\binom{n}{1} + 2\binom{n}{2} + 3\binom{n}{3} + \cdots + n \cdot \binom{n}{n} = n \cdot 2^{n-1}. \tag{34}$$

SOLUTION.   As in (i), the expression $k \cdot \binom{n}{k}$ gives for each $k = 1, 2, \ldots, n$ the number of elements of all k-element subsets of the n-element set $A = \{a_1, a_2, \ldots, a_n\}$. Therefore the left-hand side of (34) gives the sum of the numbers of elements in all nonempty subsets of the set A. If we write down a list of all subsets of A, it suffices to determine in how many subsets each of the elements of the set A occurs: It occurs exactly once in the collection of one-element subsets, it occurs $\binom{n-1}{1}$ times in the collection of two-element subsets, $\ldots$, $\binom{n-1}{n-1}$ times in the collection of n-element subsets; and so in total the number of occurrences of each element is equal to

$$\binom{n-1}{0} + \binom{n-1}{1} + \cdots + \binom{n-1}{n-1} = 2^{n-1},$$

by (27). This complete the proof of (34).

It is worth mentioning that the right-hand side of this identity can also be found in the following way: Any subset of A that contains, for instance, the element a_1 can be obtained from a subset (including the empty one) of the set $A \setminus \{a_1\}$, united with $\{a_1\}$. Therefore, the element a_1 is in exactly as many subsets of A as the set $A \setminus \{a_1\}$ has subsets, and this number is 2^{n-1}.   □

5.3 Exercises

(i) Use the method from 5.2 to show that for $0 \le k < n$ we have

$$(n - k) \cdot \binom{n}{k} = (k + 1) \cdot \binom{n}{k+1}. \tag{35}$$

*(ii)** Suppose that for each ordered pair (A, B) of subsets of an n-element set X we determine the number of elements in the intersection $A \cap B$. Show that the sum of these numbers is $n \cdot 4^{n-1}$.

5.4 Decompositions into Classes

In many cases we can derive relationships between binomial coefficients if we subdivide certain collections into classes and then use the addition rule. Thus, for example, one of the basic identities of binomial coefficients, namely

$$\binom{n}{k} + \binom{n}{k+1} = \binom{n+1}{k+1},\tag{36}$$

which is valid for $k < n$, can be derived as follows: We subdivide all $(k+1)$-element subsets of the set $A = \{a_1, a_2, \ldots, a_{n+1}\}$ into two collections S_1 and S_2 such that S_1 contains all those subsets of which a_1 is an element, and S_2 contains those $(k+1)$-element subsets of which a_1 is not an element. Clearly, $|S_1| + |S_2| = \binom{n+1}{k+1}$, since each $(k+1)$-element subset $B \subseteq A$ belongs to exactly one of the two collections. Now, S_1 contains exactly $\binom{n}{k}$ subsets, since upon adjoining the element a_1 to an arbitrary k-element subset of the set $A \setminus \{a_1\}$ we get a subset belonging to S_1. Furthermore, $|S_2| = \binom{n}{k+1}$, since here we count the $(k+1)$-element subsets of $A \setminus \{a_1\}$. This proves (36).

We will use similar arguments in verifying the following three identities.

(i) Show that

$$\binom{n}{0}^2 + \binom{n}{1}^2 + \binom{n}{2}^2 + \cdots + \binom{n}{n}^2 = \binom{2n}{n}.\tag{37}$$

SOLUTION. Let $B = \{b_1, b_2, \ldots, b_n\}$ and $C = \{c_1, c_2, \ldots, c_n\}$ be two disjoint n-element sets, and put $A = B \cup C$. We now place each of the $\binom{2n}{n}$ n-element subsets of the set A into one of $(n+1)$ collections $T_0, T_1, \ldots, T_n$, such that T_i contains all those n-element subsets of A that have exactly i elements of the set B. The i elements of B can be chosen in $\binom{n}{i}$ ways, and then the remaining $(n-i)$ elements can be chosen from C in $\binom{n}{n-i}$ ways. Therefore, by the multiplication rule, we have $|T_i| = \binom{n}{i}\binom{n}{n-i} = \binom{n}{i}^2$ for all $i = 0, 1, \ldots, n$. Taken together, this means that $\binom{2n}{n} = \sum_{i=0}^{n} |T_i| = \sum_{i=0}^{n} \binom{n}{i}^2$. □

(ii) Show that

$$\binom{m}{m} + \binom{m+1}{m} + \binom{m+2}{m} + \cdots + \binom{m+n-1}{m} = \binom{m+n}{m+1}.\tag{38}$$

SOLUTION. We subdivide all $(m+1)$-element subsets of the set $\{1, 2, \ldots, m+n\}$, of which there are $\binom{m+n}{m+1}$, into classes according to their largest elements: When M is the largest element, where $m+1 \leq M \leq m+n$, then there are exactly $\binom{M-1}{m}$ such subsets, since each of them is of the form $\{M\} \cup X$, where X runs through all m-element subsets of the set $\{1, 2, \ldots, M-1\}$. □

(iii) If we divide combinations with repetitions into classes according to the number of types that occur in a given combination, we can prove the identity

$$\binom{n}{1}\binom{m-1}{0} + \binom{n}{2}\binom{m-1}{1} + \cdots + \binom{n}{n}\binom{m-1}{n-1} = \binom{m+n-1}{m},$$

$$(39)$$

valid for $n \le m$. To do this, we consider all m-element combinations of elements of n types, $C_0(m,n) = \binom{m+n-1}{m}$ in number, and divide them into n groups $T_1, T_2, \ldots, T_n$ such that the group T_i contains exactly those combinations that are formed from elements of exactly i types. These types can be chosen in $\binom{n}{i}$ ways, and for the chosen i types there are then $\binom{m-1}{i-1}$ m-element combinations in which each of the i types occurs with at least one element. Hence $|T_i| = \binom{n}{i} \cdot \binom{m-1}{i-1}$, which means that (39) holds.

5.5   Exercises

Use the method of 5.4 to prove the following identities.

(i) If $m \le p$, then

$$\binom{p}{0}\cdot\binom{m}{0} + \binom{p}{1}\cdot\binom{m}{1} + \cdots + \binom{p}{m}\cdot\binom{m}{m} = \binom{m+p}{m}. \quad (40)$$

(ii) If $m \le n$, then

$$\binom{n}{1}\cdot\binom{m-1}{0} + \binom{n}{2}\cdot\binom{m-1}{1} + \cdots$$

$$+ \binom{n}{m}\cdot\binom{m-1}{m-1} = \binom{m+n-1}{m}. \quad (41)$$

5.6   Words

A widely used method for deriving identities with binomial coefficients consists in forming *words* of several letters, where the length is usually predetermined. We will explain the essence of this method by solving the following six problems.

(i) To prove the identity (27), we consider all words of length n formed with two different letters, say A and B. There are clearly 2^n such words, which we divide into $(n+1)$ groups $S_0, S_1, \ldots, S_n$ such that S_k contains all those words in which there are exactly k copies of the letter A. Since k places for the letter A can be chosen from n places in $\binom{n}{k}$ ways, we have $|S_k| = \binom{n}{k}$, and therefore $2^n = \sum_{k=0}^{n} |S_k| = \sum_{k=0}^{n} \binom{n}{k}$.

(ii) For a proof of (32) we consider all words of length $n+1$ formed from k copies of the letter A, $(n-k)$ copies of the letter B, and one copy of C. We begin by ordering the letters A and B; this can be done in $\binom{n}{k}$ ways.

Then we place the letter C in one of the $(n+1)$ possible positions (namely, $(n-1)$ positions between the other letters, and one each on either end). Hence there are $(n+1) \cdot \binom{n}{k}$ possibilities in total. But the number of these words can also be determined differently: From the $(n+1)$ positions we choose $(k+1)$ for placing the letters A and the single letter C, and the remaining positions will then be occupied by the copies of the letter B. In the $(k+1)$ chosen positions, the letter C can be placed in $(k+1)$ ways. Therefore, the number of all such words is $(k+1) \cdot \binom{n+1}{k+1}$, which proves (32).

(iii) The identities (30) and (31) can be verified in the following way: Just as in (i) we consider all words of length n consisting of the letters A and B. We have to show that the number of words formed with an even number of copies of the letter A is the same as the number of words in which the number of A's is odd. But this is an easy induction: An arbitrary word of length $(n-1)$ can be made into a word of length n in two ways by adding a letter to the right-hand end, namely, adjoining either the letter A or the letter B. In the one case we are dealing with the odd variant, and in the other case, with the even one.

(iv) Let us now prove the identity (34). We form all words of length n that have an arbitrary number of the letter A, an arbitrary number of the letter B, but exactly one C. All such words can be obtained in such a way that to an arbitrary word of length $n-1$ formed with the letters A and B (there are 2^{n-1} of those) we add the letter C in one of the n possible positions. Therefore, the number of such words is $n \cdot 2^{n-1}$. Now we determine this number in a different way: We divide all words into n groups $T_1, T_2, \ldots, T_n$ such that the group T_k contains all those words in which $(k-1)$ copies of the letter A and the one copy of C are used. Then we have $|T_k| = k \cdot \binom{n}{k}$, since the positions for these k letters can be chosen in $\binom{n}{k}$ ways, with the letters B occupying the remaining $(n-k)$ positions, and the letter C can be placed in k ways in the k chosen positions. Therefore, we have $\sum_{k=1}^{n} |T_k| = \sum_{k=1}^{n} k \cdot \binom{n}{k}$.

(v) To prove the identity (39) we form all words of length $(m+n-1)$ composed of m letters A and $(n-1)$ letters B, $\binom{m+n-1}{m}$ in number, and divide them into n groups $S_1, S_2, \ldots, S_n$ such that the group S_i contains exactly those words that have in the first n places exactly i copies of the letter A. If we choose the positions for the letters A in one of the $\binom{n}{i}$ ways, the remaining $(m-i)$ copies of A can be distributed over the remaining $(m-1)$ places of the word in $\binom{m-1}{m-i} = \binom{m-1}{i-1}$ ways. Hence $|S_i| = \binom{n}{i} \cdot \binom{m-1}{i-1}$, and by summing over all $i = 1, 2, \ldots, n$ we obtain (39).

(vi) We will now show that for all $n \geq 2$ we have

$$\sum_{k=2}^{n} k \cdot (k-1) \cdot \binom{n}{k} = n(n-1) \cdot 2^{n-2}. \tag{42}$$

To do this, we consider all words of length n formed with arbitrary numbers of copies of the letters A and B, and one copy each of the letters C and D. It is not difficult to find the number of such words to be $n \cdot (n-1) \cdot 2^{n-2}$. We now divide all these words into $(n-1)$ groups $T_2, T_3, \ldots, T_n$ such that for each $k = 2, 3, \ldots, n$ the group T_k contains all those words in which $(k-2)$ copies of the letter A are used. If we select one of the $\binom{n}{k}$ ways in which from the n places k positions for the letters A, C, and D can be chosen, then we can place the letters C and D in exactly $k \cdot (k-1)$ ways in these positions. Therefore, we have $|T_k| = k \cdot (k-1) \cdot \binom{n}{k}$, which proves the identity.

It is worth noting that (42) can also be proved by combining the number of ways in which a committee, a chairman, and a treasurer can be elected from a group of m people, under the condition that the chairman and the treasurer are different members of the committee.

5.7 Exercises

Use the method of 5.6 to prove the following identities:

(i) (29)　　　　　　**(ii)** (35)　　　　　　**(iii)** (33)

(iv) (28)　　　　　　**(v)** (37)　　　　　　**(vi)** (41)

*(vii) $\binom{n}{1} - 2\binom{n}{2} + 3\binom{n}{3} - 4\binom{n}{4} + \cdots + (-1)^{n-1}n\binom{n}{n} = 0$　$(n \geq 2)$.

(viii) $\binom{n}{0}\binom{n}{m} + \binom{n}{1}\binom{n-1}{m-1} + \cdots + \binom{n}{m}\binom{n-m}{0} = \binom{n}{m}2^m$　$(m \leq n)$.

*(ix) $\binom{n}{0}\binom{n}{m} - \binom{n}{1}\binom{n-1}{m-1} + \cdots + (-1)^m\binom{n}{m}\binom{n-m}{0} = 0$　$(0 < m \leq n)$.

*(x) $1^2 \cdot \binom{n}{1} + 2^2 \cdot \binom{n}{2} + \cdots + n^2 \cdot \binom{n}{n} = n(n+1) \cdot 2^{n-2}$.

*(xi) $\binom{n}{0}\binom{m+n-1}{m} - \binom{n}{1}\binom{n+m-2}{m-1} + \cdots + (-1)^\alpha\binom{n}{\alpha}\binom{n+m-1-\alpha}{m-\alpha} = 0$,
where $\alpha = \min(m, n) \geq 1$.

*(xii) $\binom{n}{0}\binom{0}{0}2^n + \binom{n}{2}\binom{2}{1}2^{n-2} + \binom{n}{4}\binom{4}{2}2^{n-4} + \cdots + \binom{n}{2\alpha}\binom{2\alpha}{\alpha}2^{n-2\alpha} = \binom{2n}{n}$,
where $\alpha = [\frac{n}{2}]$.

5.8 Walks in Rectangular Grids

In this subsection we will give a *geometrical* interpretation of some of the combinatorial identities studied earlier.

Let us assume that in the plane with Cartesian coordinates we are given straight lines with the equations $x = r$, $y = s$, where $r, s \in \mathbb{Z}$. We will now study "walks" along the given lines in such a rectangular grid between two points A, B with integer coordinates. We will consider only "shortest" walks, that is, each such walk from the point $A[a_1, a_2]$ to $B[b_1, b_2]$ (where we assume that for $i = 1, 2$ we have $a_i, b_i \in \mathbb{Z}$, $a_i \leq b_i$) will consist of a finite number of steps of the types "to the right" and "up." The step "to the right" is the transition from the point $X[r, s]$ to the point $Y[r+1, s]$, and similarly, "up" stands for the transition from $X[r, s]$ to the point $Z[r, s+1]$. Figure 3 shows a shortest walk from the point $A[1, 2]$ to $B[5, 4]$; this walk consists of 6 steps, namely, 4 "to the right" and 2 "up."

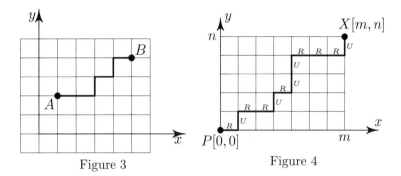

Figure 3   Figure 4

It is not difficult to determine the number $c(P, X)$ of all the shortest walks from the point $P[0, 0]$ to the point $X[m, n]$. To do this, we encode each walk with the help of a word of m letters R and n letters U (each step "to the right" will correspond to the letter R, and each step "up" will correspond to U; see Figure 4). Then each shortest walk corresponds to a unique such word; conversely, each word of m letters R and n letters U encodes a unique shortest walk from P to X. But it is now easy to determine the number of such words, and we have

$$c(P, X) = P_0(m, n) = \binom{m+n}{m} = \binom{m+n}{n}. \tag{43}$$

It is also easy to generalize the identity (43) for the number $c(A, B)$ of shortest walks from the point $A[a_1, a_2]$ to the point $B[b_1, b_2]$, where again we assume that for $i = 1, 2$ we have $a_i, b_i \in \mathbb{Z}$, $a_i \le b_i$:

$$c(A, B) = P_0(b_1 - a_1, b_2 - a_2) = \binom{b_1 + b_2 - a_1 - a_2}{b_1 - a_1}. \tag{44}$$

We now use walks in rectangular grids to prove several identities. In doing this, we should be conscious of the fact that the usefulness of this method lies only in that it helps us visualize proofs (we will especially appreciate this in Example (v) below and in Exercises 5.9.(v) and (vi)). In fact, instead of walks in rectangular grids we could study words with the letters V and N, and we would obtain the same results. Furthermore, if these words had three or more letters, then one would have to study walks in "higher-dimensional grids," which are not so easy to visualize.

(i) Each of the walks from the point $P[0, 0]$ to the point $Z[k + 1, n - k]$ goes through exactly one of the points $X[k, n - k]$, $Y[k + 1, n - k - 1]$ (see Figure 5). Since the continuation of the walk to Z is unique in each case (from X with a step "to the right," and from Y with a step "up"), we have $c(P, Z) = c(P, X) + c(P, Y)$. Using (43) now, we get the identity (36).

(ii) We obtain the identity (27) if we consider all the walks of length n that originate at the point $P[0, 0]$. There are 2^n such walks, and each one ends at some point $B_k[k, n - k]$, $k = 0, 1, \ldots, n$ (this is illustrated in Figure 6 for $n = 5$). Since $c(P, B_k) = \binom{n}{k}$, it follows that

$$2^n = \sum_{k=0}^n c(P, B_k) = \sum_{k=0}^n \binom{n}{k}.$$

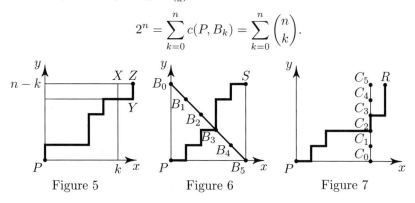

Figure 5   Figure 6   Figure 7

(iii) We now divide the $\binom{2n}{n}$ walks from the point $P[0, 0]$ to $S[n, n]$ into $(n + 1)$ classes $T_0, T_1, \ldots, T_n$ such that the class T_k contains the walks that pass through the point $B_k[k, n - k]$. Since according to (44) we have $c(B_k, S) = P_0(n - k, k) = \binom{n}{n-k}$ for each $k = 0, 1, \ldots, n$ and since each walk from P to S can be obtained by "extending" one of the $\binom{n}{k}$ walks from P to B_k through one of the $\binom{n}{n-k}$ walks from B_k to S, we obtain, with the multiplication rule and the addition rule,

$$c(P, S) = \sum_{k=0}^n c(P, B_k) \cdot c(B_k, S) = \sum_{k=0}^n \binom{n}{k}\binom{n}{n-k} = \sum_{k=0}^n \binom{n}{k}^2;$$

with (43) we then get (37).

(iv) Let us now also prove the identity (38). To do this, we consider all walks from $P[0, 0]$ to $R[m + 1, n - 1]$, which by (43) are exactly $\binom{m+n}{m+1}$ in number. We place a "barrier" of points $C_i[m, i]$, $i = 0, 1, \ldots, n - 1$ (see Figure 7). Each of the walks under consideration leaves the "barrier" at exactly one of the points C_i; then it continues in a unique way toward the point R (first with one step "to the right," possibly followed by further steps "up"). Hence we have

$$\binom{m + n}{m + 1} = c(P, R) = \sum_{i=0}^{n-1} c(P, C_i) = \sum_{i=0}^{n-1} \binom{m + i}{m}.$$

(v) The method of walks in a rectangular grid can also be used to give a very visual solution to the problem of queuing at a box office from Section 3.7. We can visualize each person with a five-dollar bill as a step "to the right," and each person with a ten-dollar bill as a step "up." Then a queue with m ten-dollar bills and k five-dollar bills will be represented by some walk from the point $P[0,0]$ to $T[k,m]$. Every "good" arrangement of the queue can then be visualized by a walk, namely, a broken line that does not touch the straight line $r = RS$, where $R[-1,0]$ and $S[0,1]$ (see Figure 8); that is, it does not "go above" the line p. We now use the notation of Section 3.7, and just as we did there we first determine the number $|X|$ of all "bad" arrangements of the queue. For $m \leq k$ we then have

$$f(m,k) = c(P,T) - |X| = \binom{m+k}{m} - |X|.$$

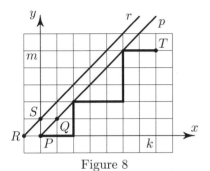   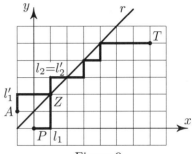

Figure 8   Figure 9

Each "bad" arrangement of the queue can be visualized in the rectangular grid by a broken line l from the point P to the point T that has at least one point in common with the straight line r (see Figure 9). We denote by Z the first such common point (this Z is therefore the one among the common points of l and r that has the smallest first coordinate) and divide the line l into two parts $l_1 = PZ$ and $l_2 = ZT$. We set $A[-1,1]$ and associate to the line $l = PT$ a line $l' = AT$ such that the segment l_1 is replaced by the segment $l'_1 = AZ$ that is symmetric to l_1 about the straight line r; the segment l_2 will be left unchanged, so $l'_2 = l_2$. Since the points A and T lie in opposite half-planes with respect to the boundary line r, each broken line AT crosses r in at least one point. If we choose from among these intersection points the point Z with the smallest first coordinate, replace the segment AZ by its reflection PZ about the axis r, and adjoin the unchanged "remainder" ZT, then we obtain a walk that represents a "bad" arrangement of the queue.

It is therefore clear that the mapping just described is a bijection between the set of all walks from P to T representing all "bad" arrangements of the queue and the set of all walks from A to T; hence with (44) we have

$$|X| = c(A,T) = P_0(k+1, m-1) = \binom{m+k}{m-1},$$

which proves the identity (12) of Section 3.7.

5.9 *Exercises*

Use the method of 5.8 to prove the following two identities:

(i) (40).

(ii) $\binom{m+n}{m} = \sum_{i=0}^{n} \binom{k+i}{k}\binom{m+n-k-i-1}{m-k-1}$ for $k < m$.

Use the method of 5.8 for a different proof of each of the following two problems, which were solved earlier:

(iii) Example 3.8.   (iv) Exercise 3.9.(vii).

*(v) Show that for each $k \geq m$ we have

$$\binom{m+k}{m-1} = \sum_{s=0}^{m-1} \frac{1}{s+1}\binom{2s}{s}\binom{m+k-2s-1}{m-s-1}.$$

*(vi) Show that for each $k \geq 1$ we have

$$\binom{2k}{k} = \sum_{s=1}^{k} \frac{k+1}{s(k-s+1)}\binom{2s-2}{s-1}\binom{2k-2s}{k-s}.$$

6   The Inclusion–Exclusion Principle

This section is devoted to the *inclusion–exclusion principle*, one of the most fundamental rules of combinatorics, which is useful in a large number of situations. In the first few subsections we state and prove this principle, and then we give a fairly large number of applications; we also return to problems that were already formulated in previous sections, but which we were unable to solve at that time. At the end of this section the inclusion–exclusion principle will be generalized.

Before we turn to formulating the principle itself, we explain its essence by way of the two easiest situations. If M_1, M_2 are two finite sets, then it is immediately clear that for the number of elements in the union $M_1 \cup M_2$ we have

$$|M_1 \cup M_2| = |M_1| + |M_2| - |M_1 \cap M_2|. \tag{45}$$

We can write down a similar formula for the number of elements in the union of three finite sets M_1, M_2, M_3. Indeed, we have

$$\begin{aligned}|M_1 \cup M_2 \cup M_3| = {} &|M_1| + |M_2| + |M_3| \\ &- |M_1 \cap M_2| - |M_1 \cap M_3| - |M_2 \cap M_3| \\ &+ |M_1 \cap M_2 \cap M_3|.\end{aligned} \tag{46}$$

It is easy to justify the validity of the formula (46): We choose any element $m \in M_1 \cup M_2 \cup M_3$ and determine the number of times it occurs on the right-hand side of (46). To do this, we distinguish three cases: If the element m belongs to exactly one of the sets M_1, M_2, M_3, say $m \in M_1$, then it is counted only in the summand $|M_1|$. If m lies in exactly two of the sets (say $m \in M_1$ and $m \in M_2$), then it is counted in the summands $|M_1|$, $|M_2|$, and $|M_1 \cap M_2|$, in the first two with a plus sign, and in the third one with a minus sign, and so it is counted altogether only once. Finally, if $m \in M_1$, $m \in M_2$, and $m \in M_3$, then we also have $m \in M_1 \cap M_2$, $m \in M_1 \cap M_3$, $m \in M_2 \cap M_3$, and $m \in M_1 \cap M_2 \cap M_3$. So the element m is counted in each of the summands on the right-hand side of (46), in total four times with the plus sign, and three times with the minus sign, so once again it is counted exactly once. This proves (46).

These two situations that we have just analyzed in detail are special cases of the *inclusion–exclusion principle*, which we now state for n finite sets $M_1, M_2, \ldots, M_n$; it will be an expression of the number of elements of the union of these sets in terms of the number of elements of the sets and their intersections. For the proof we simply generalize the argument used in the special case $n = 3$.

6.1   The Inclusion–Exclusion Principle

Theorem. *Let $M_1, M_2, \ldots, M_n$ be finite sets. Then*

$$
\begin{aligned}
|M_1 \cup M_2 \cup \cdots \cup M_n| = {}& |M_1| + |M_2| + \cdots + |M_n| \\
& - |M_1 \cap M_2| - |M_1 \cap M_3| - \cdots - |M_{n-1} \cap M_n| \\
& + |M_1 \cap M_2 \cap M_3| + \cdots + |M_{n-2} \cap M_{n-1} \cap M_n| \\
& \qquad \cdots \\
& + (-1)^{n+1}|M_1 \cap M_2 \cap \cdots \cap M_n| \\
= {}& \sum (-1)^{r+1}|M_{j_1} \cap M_{j_2} \cap \cdots \cap M_{j_r}|,
\end{aligned}
\tag{47}
$$

where the sum is taken over all nonempty subsets $\{j_1, j_2, \ldots, j_r\}$ of the index set $\{1, 2, \ldots, n\}$.

We have ordered the sum in such a way that the summands corresponding to the i-element subsets are placed in the ith row, which therefore contains exactly $\binom{n}{i}$ summands; the total number of summands on the right-hand side of (47) is then clearly $2^n - 1$.

PROOF.   We choose an arbitrary element $m \in M_1 \cup M_2 \cup \cdots \cup M_n$ and denote by $M_{k_1}, M_{k_2}, \ldots, M_{k_s}$ those of the n given sets that contain the element m, where $s \in \{1, 2, \ldots, n\}$. Now we consider the question as to which of the sets of the type $M_{j_1} \cap M_{j_2} \cap \cdots \cap M_{j_r}$ contain the element m: This is clearly the case exactly for those that satisfy

$$
\emptyset \neq \{j_1, j_2, \ldots, j_r\} \subseteq \{k_1, k_2, \ldots, k_s\}.
$$

For each $r \in \{1, 2, \ldots, s\}$, therefore, the element m is counted in exactly $\binom{s}{r}$ summands of the type $|M_{j_1} \cap M_{j_2} \cap \cdots \cap M_{j_r}|$, and for $r > s$ in none. Therefore, the whole contribution of m to the right-hand side of (47) is

$$\binom{s}{1} - \binom{s}{2} + \cdots + (-1)^{s+1}\binom{s}{s} = 1,$$

by identity (30) of Section 5.1. On the left-hand side of (47) the element M is also counted exactly once, and this concludes the proof.   □

This proof is a more precise formulation of the following underlying idea: In the sum $|M_1| + |M_2| + \cdots + |M_n|$ all those elements are counted exactly once that lie in exactly one of the sets M_i, while those elements that belong to several sets at once are counted several times. If we subtract $\sum |M_i \cap M_j|$, then we correctly count all those elements that belong to exactly two sets; but those belonging to three or more sets will be reduced too much. Now, this will be corrected (for those elements that belong to exactly three sets) if we add the expression $\sum |M_i \cap M_j \cap M_k|$, etc.

From these remarks it is clear why the identity (47) is called the *inclusion–exclusion principle*.

6.2 Remark

In a completely analogous way we can prove the following generalization of the inclusion–exclusion principle. Let $M_1, M_2, \ldots, M_n$ be finite sets and let $f : M_1 \cup M_2 \cup \cdots \cup M_n \to \mathbb{R}$ be a mapping. For an arbitrary subset $L \subseteq M_1 \cup M_2 \cup \cdots \cup M_n$ we set

$$F(L) = \sum_{x \in L} f(x),$$

where we set $F(\emptyset) = 0$. Then

$$F(M_1 \cup M_2 \cup \cdots \cup M_n) = \sum (-1)^{r+1} F(M_{j_1} \cap M_{j_2} \cap \cdots \cap M_{j_r}), \quad (48)$$

where the sum is taken over all nonempty subsets $\{j_1, j_2, \ldots, j_r\}$ of the index set $\{1, 2, \ldots, n\}$. We will use this identity later, in Section 6.13.

The **proof** can be done as in identity (47), that is, by considering the contribution of an arbitrary element $m \in M_1 \cup M_2 \cup \cdots \cup M_n$ to both sides of (48) (see 6.4.(vi)). We can also convince ourselves that if we set $f(x) = 1$ for each $x \in M_1 \cup M_2 \cup \cdots \cup M_n$, then for each $L \subseteq M_1 \cup M_2 \cup \cdots \cup M_n$ we have $f(L) = |L|$, and (48) reduces to (47).

6.3 Examples

The inclusion–exclusion principle is most useful in situations where we have to determine the number of objects that have *at least one* of a given list of properties. In other words, if for each $i = 1, 2, \ldots, n$ we denote by M_i the set of all objects that have the ith property, we want to determine the number $|M_1 \cup M_2 \cup \cdots \cup M_n|$. A direct way of finding this number is often not possible, while it is relatively easy to determine the summands of the form $|M_{j_1} \cap M_{j_2} \cap \cdots \cap M_{j_r}|$ on the right-hand side of (47); these terms express the number of objects that satisfy at the same time the j_1th, j_2th, $\ldots$, j_rth properties (and possibly some of the other properties).

In the following Examples (i)–(iii) we illustrate the inclusion–exclusion principle in some simple situations.

(i) A scientific research institute in the Czech Republic has 67 members. Forty-seven of them speak English, 35 speak German, and 20 speak French. Furthermore, 23 of them know both English and German, 12 know English and French, and 11 speak German and French; finally, 5 members speak all three languages. Find the number of people who speak none of these three foreign languages.

SOLUTION.   We denote by E (respectively G, respectively F) the set of persons who speak English (respectively German, respectively French), and by E' the set of those who do not speak English (similarly G', F'). We have to determine the number $|E' \cap G' \cap F'|$. Clearly,

$$|E' \cap G' \cap F'| = 67 - |E \cup G \cup F|,$$

where the number $|E \cup G \cup F|$ will be determined by use of the inclusion–exclusion principle for three sets, that is, by using the identity (46). By the given data we have $|E| = 47$, $|G| = 35$, $|F| = 20$, $|E \cap G| = 23$, $|E \cap F| = 12$, $|G \cap F| = 11$, and $|E \cap G \cap F| = 5$. Therefore,

$$\begin{aligned} |E \cup G \cup F| &= |E| + |G| + |F| - |E \cap G| - |E \cap F| - |G \cap F| \\ &\quad + |E \cap G \cap F| = 47 + 35 + 20 - 23 - 12 - 11 + 5 = 61, \end{aligned}$$

which means that $|E' \cap G' \cap F'| = 67 - 61 = 6$.   □

(ii) Show that the following report about a grade level in a certain elementary school is faulty: There are 45 children, 30 of whom are boys. Thirty children have good grades, and among these there are 16 boys. Twenty-eight children do sports; 18 of them are boys, and 17 have good grades. Fifteen boys have good grades and also do sports.

SOLUTION.   Let R denote the set of all children in the grade level, $A \subseteq R$ the set of all boys, $B \subseteq R$ the set of children with good grades, and $C \subseteq R$ the set of children who do sports. Let us now determine the number n of all nonsportive girls who do not have good grades:

$$\begin{aligned} n &= |R| - |A \cup B \cup C| \\ &= |R| - |A| - |B| - |C| + |A \cap B| + |A \cap C| + |B \cap C| - |A \cap B \cap C| \\ &= 45 - 30 - 30 - 28 + 16 + 18 + 17 - 15 = -7. \end{aligned}$$

This is impossible, and therefore the report must be faulty.   □

(iii) How many anagrams can be formed from the word TIKTAK (that's a clock "speaking" Czech) if no two of the same letters are to be next to each other?

SOLUTION.   Let n be the desired number of anagrams, and n_0 the number of all anagrams of the word TIKTAK. Let M_1 (respectively M_2) be the set of all anagrams in which the two letters T (respectively K) stand together. Then by (45) we have

$$n = n_0 - |M_1 \cup M_2| = n_0 - |M_1| - |M_2| + |M_1 \cap M_2|.$$

It is easy to determine that $n_0 = P_0(2,2,1,1) = \frac{6!}{2^2}$ and $|M_1| = |M_2| = P_0(2,1,1,1) = \frac{5!}{2}$, $|M_1 \cap M_2| = P(4) = 4!$. Hence

$$n = \frac{6!}{2^2} - 2 \cdot \frac{5!}{2} + 4! = 84.$$ □

6.4   Exercises

(i) All the members of a department in a Czech research institute speak at least one of the foreign languages English, German, and French. Six members speak English, six speak German, and seven speak French; four speak English and German, three speak German and French, two speak French and English, and one department member speaks all three languages.
(a) How many people work in the department?
(b) How many of them speak only English?
(c) How many speak only French?

(ii) At a meeting with all the parents, the teacher of a class makes the following announcement: "In our class, which has 30 students, four after-school clubs are active; each one meets once a week. On Mondays, 19 students meet; on Tuesdays 13 students; on Wednesdays 18; and on Thursdays 11. No student takes part in more than two clubs, and no club has more than five students in common with another club." Is the teacher telling the truth?

(iii) How many anagrams can be formed from BARBAR such that none of them has two of the same letters in a row?

(iv) In how many ways can n rooks be placed on an $n \times n$ chessboard such that each free field is threatened by at least one of the rooks?

*(v) Prove Identity (47) by induction.

*(vi) Prove Identity (48).

(vii) Use the inclusion–exclusion principle to solve part (g) of Exercise 3.2.(ix).

6.5 A Special Case of the Inclusion-Exclusion Principle

In many cases the use of identity (47) is simplified by the fact that all the $\binom{n}{r}$ summands $|M_{j_1} \cap M_{j_2} \cap \cdots \cap M_{j_r}|$ are identical; let us denote their value by $m(r)$. Then it is clear that (47) reduces to

$$|M_1 \cup M_2 \cup \cdots \cup M_n| = \sum_{r=1}^{n} (-1)^{r+1} \binom{n}{r} m(r). \tag{49}$$

We illustrate this case with Examples (i)–(iii) below. However, in Example (iv) we will see that (47) cannot always be reduced to (49). At this point we also mention Example 6.13.(ii), which can also not be solved with identity (49).

(i) Find the number of all arrangements with repetitions of two elements of the first type, two elements of the second type, ..., two elements of the nth type, such that no two elements of the same type are adjacent to each other.

SOLUTION. Let a_n denote the desired number of arrangements, and for $i = 1, 2, \ldots, n$ let M_i be the set of all arrangements in which the elements of the ith type are adjacent. Then we have

$$a_n = \frac{(2n)!}{(2!)^n} - |M_1 \cup M_2 \cup \cdots \cup M_n|,$$

and for determining the number $|M_1 \cup M_2 \cup \cdots \cup M_n|$ we use the inclusion–exclusion principle in the form (49). So, let $\emptyset \neq \{j_1, j_2, \ldots, j_r\} \subseteq \{1, 2, \ldots, n\}$. Then the set $M_{j_1} \cap M_{j_2} \cap \cdots \cap M_{j_r}$ contains those arrangements in which the elements of a fixed collection of r types (namely, the j_1th, the j_2th, ..., j_rth type) are adjacent to each other; each such pair can therefore be united in one new object, and the remaining $(2n - 2r)$ elements can be arranged arbitrarily. The particular choice of the r types is clearly irrelevant, and so we have

$$m(r) = |M_{j_1} \cap M_{j_2} \cap \cdots \cap M_{j_r}| = \frac{(2n - r)!}{2^{n-r}},$$

since we are, in fact, dealing with all arrangements with repetitions of $(n - r)$ pairs of elements of the same type and r different "new objects." Using (49) we now get

$$|M_1 \cup M_2 \cup \cdots \cup M_n| = \sum_{r=1}^{n} (-1)^{r+1} \binom{n}{r} m(r) = \sum_{r=1}^{n} (-1)^{r+1} \binom{n}{r} \frac{(2n - r)!}{2^{n-r}},$$

and therefore

$$a_n = \frac{(2n)!}{2^n} - \sum_{r=1}^{n} (-1)^{r+1} \binom{n}{r} \frac{(2n - r)!}{2^{n-r}} = \sum_{r=0}^{n} (-1)^r \binom{n}{r} \frac{(2n - r)!}{2^{n-r}}, \tag{50}$$

which solves the problem.   □

(ii) Find the number of all k-element variations with repetitions from elements of n different types such that every variation contains at least one element of each type.

SOLUTION.   We denote the desired number by $v_0(k, n)$. If $k < n$, then $v_0(k, n) = 0$; so we assume that $k \geq n$ and let M_i denote the set of all k-element variations with repetitions that contain no elements of the ith type $(i = 1, 2, \ldots, n)$. Hence

$$v_0(k, n) = V_0(k, n) - |M_1 \cup M_2 \cup \cdots \cup M_n| = n^k - |M_1 \cup M_2 \cup \cdots \cup M_n|.$$

If $\emptyset \neq \{j_1, j_2, \ldots, j_r\} \subseteq \{1, 2, \ldots, n\}$, then the set $M_{j_1} \cap M_{j_2} \cap \cdots \cap M_{j_r}$ contains those variations that have no elements from a fixed collection of r types. Therefore, we have

$$|M_{j_1} \cap M_{j_2} \cap \cdots \cap M_{j_r}| = m(r) = (n - r)^k,$$

and thus

$$|M_1 \cup M_2 \cup \cdots \cup M_n| = \sum_{r=1}^{n} (-1)^{r+1} \binom{n}{r} (n - r)^k,$$

so finally,

$$v_0(k, n) = n^k - \sum_{r=1}^{n} (-1)^{r+1} \binom{n}{r} (n - r)^k = \sum_{r=0}^{n} (-1)^r \binom{n}{r} (n - r)^k, \quad (51)$$

and we are done.   □

We can proceed in the same fashion also in the case where $k < n$; in this case we have $v_0(k, n) = 0$. Hence we obtain for $k < n$ the identity

$$\sum_{r=0}^{n} (-1)^r \binom{n}{r} (n - r)^k = 0. \quad (52)$$

We obtain another interesting identity if we set $k = n$ in (51). Then $v_0(n, n)$ gives the number of all n-element variations with repetitions from elements of n types such that an element of each kind occurs at least once, which means *exactly* once. So this is the number of all *arrangements* of the elements of an n-element set, of which there are $P(n) = n!$. Hence

$$\sum_{r=0}^{n} (-1)^r \binom{n}{r} (n - r)^n = n!. \quad (53)$$

(iii) Let $A = \{a_1, a_2, \ldots, a_n\}$ be an n-element set $(n \geq 2)$. Find the number c_n of all arrangements of the elements of A in which the element a_{i+1} does not immediately follow the element a_i, $i \in \{1, 2, \ldots, n-1\}$.

SOLUTION.   For $i = 1, 2, \ldots, n-1$ let M_i denote the set of all those arrangements of the elements of A in which a_{i+1} immediately follows a_i. We clearly have

$$c_n = n! - |M_1 \cup M_2 \cup \cdots \cup M_{n-1}|,$$

and the number $|M_1 \cup M_2 \cup \cdots \cup M_{n-1}|$ can now be determined with the help of (49). For $\emptyset \neq \{j_1, j_2, \ldots, j_r\} \subseteq \{1, 2, \ldots, n-1\}$ the set $M_{j_1} \cap M_{j_2} \cap \cdots \cap M_{j_r}$ contains all those arrangements of elements of the set A in which for $i = 1, 2, \ldots, r$ the element a_{j_i+1} immediately follows a_{j_i}. These pairs of neighboring elements can be put into longer "blocks" of consecutive elements: Thus, for example, when $j_1 = 2$, $j_2 = 3$, $j_3 = 5$, $j_4 = 6$, $j_5 = 7$, and $j_6 = 10$, we obtain the three blocks (a_2, a_3, a_4), (a_5, a_6, a_7, a_8), and (a_{10}, a_{11}). We clearly obtain all desired arrangements from $M_{j_1} \cap M_{j_2} \cap \cdots \cap M_{j_r}$ by forming all arrangements of these blocks and of the remaining elements not belonging to the blocks. If in a given arrangement there are b blocks, then they contain together $(b + r)$ elements from A, and therefore $(n - b - r)$ elements are not contained in them. Thus the number

$$|M_{j_1} \cap M_{j_2} \cap \cdots \cap M_{j_r}| = (b + n - b - r)! = (n - r)!$$

does not depend on the number of blocks and is the same for each r-element subset $\{j_1, j_2, \ldots, j_r\}$ of the set $\{1, 2, \ldots, n-1\}$. Hence we have

$$m(r) = (n - r)!,$$

and the identity (49) becomes

$$|M_1 \cup M_2 \cup \cdots \cup M_{n-1}| = \sum_{r=1}^{n-1} (-1)^{r+1} \binom{n-1}{r} (n-r)!.$$

Altogether, we have therefore

$$c_n = n! - \sum_{r=1}^{n-1} (-1)^{r+1} \binom{n-1}{r} (n-r)!$$

$$= \sum_{r=0}^{n-1} (-1)^r \binom{n-1}{r} (n-r)! = (n-1)! \sum_{r=0}^{n-1} (-1)^r \frac{n-r}{r!}, \qquad (54)$$

and this completes the solution.   □

(iv) We are given the nonnegative integers $\alpha_1, \alpha_2, \ldots, \alpha_n$. Find the number b_n of all those k-element combinations with repetitions from elements of n different types such that for each $i = 1, 2, \ldots, n$ the elements of the

ith type are repeated no more than α_i times in the combinations. (For $\alpha_1 = \alpha_2 = \cdots = \alpha_n = 1$ we obtain combinations without repetitions, and for $\alpha_1 = \alpha_2 = \cdots = \alpha_n = k$ we get all combinations with repetitions.)

SOLUTION.   Let M denote the set of all k-element combinations with repetitions from the elements of the given n types, and for each $i = 1, 2, \ldots, n$ let M_i be the set of all combinations in M in which the elements of the ith type occur with at least $(\alpha_i + 1)$ copies. Then obviously,

$$b_n = |M| - |M_1 \cup M_2 \cup \cdots \cup M_n|.$$

By identity (10) in Section 2.17 we have $|M| = \binom{k+n-1}{n-1}$, and we will now determine the number $|M_1 \cup M_2 \cup \cdots \cup M_n|$ with the help of (47). Let $\{j_1, j_2, \ldots, j_r\}$ be a subset of $\{1, 2, \ldots, n\}$. Then $M_{j_1} \cap M_{j_2} \cap \cdots \cap M_{j_r}$ is the set of all those combinations in which the elements of the j_1th, j_2th, $\ldots$, j_rth type occur, respectively, at least $(\alpha_{j_1} + 1)$ times, $(\alpha_{j_2} + 1)$ times, $\ldots$, $(\alpha_{j_r} + 1)$ times, and there are clearly as many of those as there are $(k - \sum_{i=1}^{r}(\alpha_{j_i} + 1))$-element combinations with repetitions from elements of n types. This means that

$$|M_{j_1} \cap M_{j_2} \cap \cdots \cap M_{j_r}| = \binom{k - \sum_{i=1}^{r} \alpha_{j_i} - r + n - 1}{n - 1},$$

provided that the inequality

$$k - \sum_{i=1}^{r} \alpha_{j_i} - r \geq 0 \tag{55}$$

holds; otherwise, this summand is zero. Hence by (47) we have

$$|M_1 \cup M_2 \cup \cdots \cup M_n| = \sum (-1)^{r+1} \binom{k - \sum_{i=1}^{r} \alpha_{j_i} - r + n - 1}{n - 1},$$

where the sum is taken over all nonempty subsets $\{j_1, j_2, \ldots, j_r\} \subseteq \{1, \ldots, n\}$ such that (55) holds. So altogether,

$$b_n = \sum (-1)^r \binom{k - \sum_{i=1}^{r} \alpha_{j_i} - r + n - 1}{n - 1}, \tag{56}$$

where the sum is taken over all subsets $\{j_1, j_2, \ldots, j_r\} \subseteq \{1, \ldots, n\}$ such that (55) holds; this includes the empty set, which corresponds to the summand $|M| = \binom{k+n-1}{n-1}$.   □

6.6   Exercises

(i) Fifty-four students of a school take part in 11 different clubs, each of which has at least 15 members. No student is a member of more than three clubs, but any three clubs have at least one student in common. Show that there are two clubs that have at least 6 students in common.

(ii) Find the number of all anagrams of the word TERAKOTA such that no two identical letters are adjacent to each other.

(iii) Solve Exercise (ii) for anagrams formed from the letters of the sentence ALSO SEND LEONA.

(iv) In how many ways can we place 3 Canadians, 3 Finns, and 3 Turks in a row such that no three persons of the same nationality are next to each other?

(v) In how many ways can n couples sit in $2n$ chairs in the last row of a movie theater such that all the couples are separated?

(vi) Six persons are in an elevator. In how many ways can they get off at four floors if at each floor at least one person gets off?

(vii) A bookbinder has to bind 12 different books in red, brown, and blue covers. In how many ways can this be done if each color is used for at least one book?

(viii) Five companies are to be simultaneously audited. In how many ways can eight auditors be divided up for this job?

(ix) Find the number of all mappings from a finite set P *onto* a finite set Q.

*(x) Find the number of all mappings from all subsets of a finite set P *onto* a finite set Q.

(xi) A caravan consisting of 9 camels travels through a desert. In how many ways can the caravan be assembled after a rest stop such that in front of each of the last 8 camels there is a different one than before?

(xii) A bookshelf contains 100 books arranged in a row. In how many ways can they be rearranged such that no two covers that were previously touching will be next to each other? (The books can be shifted, but may not be turned.)

(xiii) Given $1 \le r < n$, how many arrangements of the elements of the set $A = \{a_1, a_2, \ldots, a_n\}$ are there such that for each $i = 1, 2, \ldots, r$ the element a_{i+1} does not immediately follow a_i?

*(xiv) How many arrangements of the elements of the set $A = \{a_1, a_2, \ldots, a_n\}$ are there such that for each $i = 1, 2, \ldots, n-1$ the element a_{i+1} does not immediately follow a_i, and a_1 does not immediately follow a_n?

*(xv) There are n children riding on a carousel. They agree that for the next ride they will change their places in such a way that in front of each child there is a different one than before. In how many ways can this be done?

(**xvi**) From a bag containing 3 yellow, 2 blue, 10 black, and 15 green marbles we take a handful of ten marbles. How many different handfuls are there?

(**xvii**) Find the number of all solutions in nonnegative single-digit integers of the equations
(a) $x_1 + x_2 + \cdots + x_5 = 12$.   (b) $x_1 + x_2 + \cdots + x_k = n$.

*(**xviii**) Find the number of placements of n rooks on an $n \times n$ chessboard such that at least one of them is not threatened.

*(**xix**) Find the number of placements of k rooks on an $n \times n$ chessboard such that at least one of them is not threatened and no two are in the same column.

*(**xx**) In how many ways can n rooks be placed on an $n \times n$ chessboard such that each field on the board is threatened by at least one of the rooks? (We make the assumption that a rook does not threaten the field on which it stands.)

6.7 Distributions into Bins

In Section 4 we systematically studied problems concerning the distribution of objects into bins. However, several questions remained open, as we pointed out in that section. In this subsection we will now settle these debts in the following topics: In (i) we determine the number $q_4(n, k, s)$ of distributions of n identical balls over k numbered bins if each bin cannot contain more than s balls (see 4.4); in (ii) we answer the question raised in 4.7, namely, to find the number $r_2(n, k)$ of ways to distribute n distinct objects over k distinct bins if none of the bins is to remain empty; finally, in (iii) we will deal with the distribution of n distinct objects over k indistinguishable bins.

(**i**) Find the number $q_4(n, k, s)$ of all distributions of n identical balls over k numbered bins if each bin cannot contain more than s balls ($s \geq 2$).

SOLUTION.   It is clear that when $n > ks$, we have $q_4(n, k, s) = 0$; we therefore assume that $n \leq ks$. For each $i = 1, 2, \ldots, k$ let M_i denote the set of all distributions of n balls over k bins, where the ith bin contains at least $(s + 1)$ balls. Since by identity (19) in Section 4.4 n balls can be distributed over k bins without restrictions in exactly $q_1(n, k) = \binom{n+k-1}{k-1}$ ways, we have

$$q_4(n, k, s) = \binom{n+k-1}{k-1} - |M_1 \cup M_2 \cup \cdots \cup M_k|.$$

We will now determine the number $|M_1 \cup M_2 \cup \cdots \cup M_k|$ by use of the inclusion–exclusion principle. The set $M_{j_1} \cap M_{j_2} \cap \cdots \cap M_{j_r}$ contains exactly those distributions for which each of the r fixed bins numbered $j_1, j_2, \ldots, j_r$

contains at least $(s+1)$ balls. Therefore, this set is empty when $r(s+1) > n$, while in the case $r(s+1) \leq n$ the number of its elements is

$$|M_{j_1} \cap M_{j_2} \cap \cdots \cap M_{j_r}| = C_0(n - r(s+1), k) = P_0(n - r(s+1), k-1)$$
$$= \binom{n + k - rs - r - 1}{k - 1}.$$

In view of the inequality above, we see that upon substitution into (49) we need only consider those summands for which $r \leq \frac{n}{s+1}$. Hence, if we set $\alpha = \left[\frac{n}{s+1}\right]$, then we obtain

$$|M_1 \cup M_2 \cup \cdots \cup M_k| = \sum_{r=1}^{\alpha} (-1)^{r+1} \binom{k}{r} \binom{n + k - rs - r - 1}{k - 1}.$$

Under the condition that $n \leq ks$, we therefore have

$$q_4(n, k, s) = \binom{n + k - 1}{k - 1} - \sum_{r=1}^{\alpha} (-1)^{r+1} \binom{k}{r} \binom{n + k - rs - r - 1}{k - 1}$$
$$= \sum_{r=0}^{\alpha} (-1)^r \binom{k}{r} \binom{n + k - rs - r - 1}{k - 1}, \tag{57}$$

where $\alpha = \left[\frac{n}{s+1}\right]$.   □

(ii) Find the number $r_2(n, k)$ of all distributions of n distinct objects over k distinct bins if none of the bins is to remain empty.

SOLUTION.   For $i = 1, 2, \ldots, k$ we denote by M_i the set of all distributions for which the ith bin remains empty. Since by (23) in Section 4.7 we can distribute n distinct objects over k distinct bins without restrictions in $r_1(n, k) = k^n$ ways, we have

$$r_2(n, k) = k^n - |M_1 \cup M_2 \cup \cdots \cup M_k|,$$

where $|M_{j_1} \cap M_{j_2} \cap \cdots \cap M_{j_r}| = (k - r)^n$. So (49) implies

$$|M_1 \cup M_2 \cup \cdots \cup M_k| = \sum_{r=1}^{k} (-1)^{r+1} \binom{k}{r} (k - r)^n;$$

hence

$$r_2(n, k) = k^n - \sum_{r=1}^{k} (-1)^{r+1} \binom{k}{r} (k - r)^n = \sum_{r=0}^{k} (-1)^r \binom{k}{r} (k - r)^n, \tag{58}$$

and we are done.   □

(iii) Find the number of all distributions of n distinct objects over k indistinguishable bins.

SOLUTION. Let us first determine the number $S(n,k)$ of all such distributions where none of the bins remains empty. We assume for the moment that the bins are numbered $1,\ldots,k$. All these distributions, the number $r_2(n,k)$ of which we determined in 6.7.(ii), we then subdivide into different classes as follows. Two distributions will be put into the same class if one can be obtained from the other by shuffling the bins (and thus, by not distinguishing among the different bins). We thus obtain $S(n,k)$ classes. Each class contains $k!$ distributions, since the k indistinguishable bins can be labeled in $k!$ ways, and the bins are nonempty and contain distinguishable objects. Therefore, we have

$$r_2(n,k) = k! \cdot S(n,k),$$

and thus, by (58),

$$S(n,k) = \frac{r_2(n,k)}{k!} = \frac{1}{k!}\sum_{r=0}^{k}(-1)^r\binom{k}{r}(k-r)^n. \tag{59}$$

Let now $S_0(n,k)$ denote the number of all those distributions of n distinct objects over k indistinguishable bins, where one or more of the bins may remain empty. But for any $i = 1,2,\ldots,k$ the number of distributions for which exactly i bins are nonempty is equal to $S(n,i)$, and therefore

$$S_0(n,k) = S(n,1) + S(n,2) + \cdots + S(n,k) = \sum_{i=1}^{k} S(n,i), \tag{60}$$

which concludes this problem.   □

It is worth mentioning that the numbers $S(n,k)$ defined by (58) are often called *Stirling numbers of the second kind* in the combinatorial literature.

6.8   Exercises

(i) Use the method of 6.7.(i) to solve the problems 4.5.(iv) and 4.6.(viii).

(ii) Find a connection between 6.7.(i) and the following problem: Determine the number of all solutions in $\mathbb{N}_0$ of the equation

$$x_1 + x_2 + \cdots + x_k = n$$

under the assumption that for each $i = 1,2,\ldots,k$ we have $x_i \le s$, where $s \ge 2$ is a given integer. Then use (57) to give another solution to Exercise 6.6.(xvii).

(iii) In how many ways can six different photographs be distributed over four indistinguishable envelopes if
(a) no envelope remains empty?
(b) one or more of the envelopes may remain empty?

(iv) Find a connection between the identities (58) and (51).

(v) We want to distribute k different objects among n women and m men. In how many ways can this be done under the condition that each woman obtains at least one object?

6.9 Problems on Digit Representations

In this subsection we will apply the inclusion–exclusion principle to the solutions of four problems concerning *digit representations* of positive integers in the decimal (base-10) positional system.

(i) Find the number of nonnegative integers less than 10^6 that contain each of the digits 1, 2, 3, 4.

SOLUTION.   Let N denote the desired number, and n the number of nonnegative integers less than 10^6 that do not contain one or more of the digits $1, 2, 3, 4$. Since the digit representation of an integer with fewer digits can be filled up to become a "six-digit" number by adding an appropriate number of zeros to the beginning, we have

$$N = 10^6 - n.$$

For $i = 1, 2, 3, 4$ we now let M_i be the set of all at most six-digit integers that do not have the digit i in their representation; then $n = |M_1 \cup M_2 \cup M_3 \cup M_4|$. For each nonempty subset $\{j_1, j_2, \ldots, j_r\}$ of $\{1, 2, 3, 4\}$ we clearly have $|M_{j_1} \cap M_{j_2} \cap \cdots \cap M_{j_r}| = (10 - r)^6$, and so (49) implies

$$n = \binom{4}{1} \cdot 9^6 - \binom{4}{2} \cdot 8^6 + \binom{4}{3} \cdot 7^6 - \binom{4}{4} \cdot 6^6 = 976\,840,$$

which means that $N = 1\,000\,000 - 976\,840 = 23\,160$.   □

(ii) In how many ways can the digits of the number $1\,112\,223\,456$ be rearranged such that no three of the same digits are adjacent to each other?

SOLUTION.   Let n denote the desired number of integers, and for $i = 1, 2$ let M_i be the set of all ten-digit integers formed from the given digits such that the three digits i are adjacent to each other. Since from the ten given digits we can form $P_0(3, 3, 1, 1, 1, 1) = \frac{10!}{(3!)^2} = 100\,800$ ten-digit integers, we have

$$n = 100\,800 - |M_1 \cup M_2|.$$

It is not difficult to determine that $|M_1| = |M_2| = P_0(3, 1, 1, 1, 1, 1) = \frac{8!}{3!} = 6720$ and $|M_1 \cap M_2| = P(6) = 6! = 720$, which implies

$$n = 100\,800 - 2 \cdot 6720 + 720 = 88\,080.$$ □

(iii) How many six-digit integers can be formed from the digits of the integer $1\,122\,334\,455$ if we furthermore require that no two of the same digits can be adjacent to each other?

SOLUTION.   We let n denote the desired number of six-digit integers and subdivide the set of all allowable integers into three groups T_1, T_2, T_3 such that the group T_i contains those integers whose digit representations have exactly i pairs of identical digits (why is $1 \leq i \leq 3$?).

To determine the number $|T_1|$ we first choose, in one of $\binom{5}{1} = 5$ ways, the digit $c \in \{1, 2, 3, 4, 5\}$ that occurs twice in the representation of an integer $a \in T_1$, so that all remaining digits are used once. We now fix this digit c and denote by $C \subseteq T_1$ the set of all integers with the required property that contain two copies of the digit c. Then

$$|C| = P_0(2, 1, 1, 1, 1) - P(5) = \frac{6!}{2!} - 5! = 240,$$

since in $P(5)$ of the cases both copies of c stand together. Therefore,

$$|T_1| = 5 \cdot |C| = 5 \cdot 240 = 1200.$$

To determine $|T_2|$, we begin by choosing, in one of $\binom{5}{2} = 10$ ways, the digits c_1, c_2 that occur twice in the representation of an integer $b \in T_2$; the remaining two digits c_3, c_4 can then be chosen in $\binom{3}{2} = 3$ ways. Let $A \subseteq T_2$ denote the set of all six-digit integers whose representations have two copies each of the digits c_1, c_2, and one copy each of c_3, c_4. If we furthermore denote by M_i, for $i = 1, 2$, the set of all six-digit integers built from $c_1, c_1, c_2, c_2, c_3, c_4$ such that the two digits c_i are adjacent, then we have

$$|A| = P_0(2, 2, 1, 1) - |M_1 \cup M_2| = P_0(2, 2, 1, 1) - |M_1| - |M_2| + |M_1 \cap M_2|$$

by the inclusion–exclusion principle. It is now easy to find the numbers $|M_1|$, $|M_2|$, and $|M_1 \cap M_2|$: We have $|M_1| = |M_2| = P_0(2, 1, 1, 1) = \frac{5!}{2!}$ and $|M_1 \cap M_2| = P(4) = 4!$, and so

$$|A| = \frac{6!}{(2!)^2} - 2 \cdot \frac{5!}{2!} + 4! = 84,$$

which means that $|T_2| = 10 \cdot 3 \cdot |A| = 10 \cdot 3 \cdot 84 = 2520$.

In a similar fashion we also determine $|T_3|$: The digits c_1, c_2, c_3, from which all the integers in T_3 are formed, can be chosen in $\binom{5}{3} = 10$ ways. For $i = 1, 2, 3$ we let Q_i denote the set of all six-digit integers consisting of pairs of the digits c_1, c_2, c_3 such that the two copies of c_i are adjacent. We then get

$$|T_3| = 10 \cdot (P_0(2, 2, 2) - |Q_1 \cup Q_2 \cup Q_3|),$$

and with (46) we obtain

$$|Q_1 \cup Q_2 \cup Q_3| = 3 \cdot P_0(2,2,1) - 3 \cdot P_0(2,1,1) + P(3)$$
$$= 3 \cdot \frac{5!}{2!2!} - 3 \cdot \frac{4!}{2!} + 3! = 60,$$

so that $|T_3| = 10 \cdot \left(\frac{6!}{2^3} - 60\right) = 300$. Altogether, we have then

$$n = |T_1| + |T_2| + |T_3| = 1200 + 2520 + 300 = 4020. \qquad \square$$

(iv) How many six-digit positive integers have exactly three different digits?

SOLUTION.   We divide the admissible integers into two sets A, B: We put into set A (respectively B) all those integers whose digit representation does not (respectively does) contain the digit 0. Since three digits can be chosen from among nine in $\binom{9}{3} = 84$ ways, the set A can be decomposed into 84 disjoint subsets S_i ($i = 1, 2, \ldots, 84$) according to which triple of digits is used in the representations; clearly, all these subsets have the same number of elements. Now let S_i contain the integers written with the digits c_1, c_2, c_3; that is, any integer $s \in S_i$ is written only with these digits, and each of the digits occurs at least once. Furthermore, let M_1 denote the set of all six-digit integers written only with the digits c_2 and c_3 (now we also allow integers written with only one of these digits), and let M_2 and M_3 be defined analogously. Then we have

$$|S_i| = 3^6 - |M_1 \cup M_2 \cup M_3|,$$

and since $|M_1| = |M_2| = |M_3| = 2^6$, $|M_1 \cap M_2| = |M_1 \cap M_3| = |M_2 \cap M_3| = 1^6 = 1$, and $|M_1 \cap M_2 \cap M_3| = 0$, the identity (46) gives

$$|S_i| = 3^6 - 3 \cdot 2^6 + 3 \cdot 1 = 540, \quad |A| = 84 \cdot |S_i| = 84 \cdot 540 = 45\,360.$$

We now determine the number $|B|$: The digit representation of an integer $b \in B$ contains two more digits apart from zero; they can be chosen in $\binom{9}{2} = 36$ ways. Let us assume for clarity that we have chosen the digits 1 and 2, and let C_1 (respectively C_2) denote the set of all integers $c \in B$ that begin with the digit 1 (respectively 2). Clearly, we have $|C_1| = |C_2|$, so it suffices to determine $|C_1|$, that is, to find the number of ways in which the digits for the remaining five places can be chosen. The digits that can be used are 0, 1, 2, where 0 and 2 have to occur at least once. In a similar way as for $|A|$ we now obtain

$$|C_1| = 3^5 - \binom{2}{1} \cdot 2^5 + \binom{2}{2} \cdot 1^5 = 180.$$

Hence we have

$$|B| = 36 \cdot 2 \cdot |C_1| = 36 \cdot 2 \cdot 180 = 12\,960,$$

and altogether the number of all six-digit integers consisting of three different digits is equal to $45\,360 + 12\,960 = 58\,320$. \qquad $\square$

6.10 Exercises

(i) How many positive integers less than 10^5 contain each of the digits 1, 3, 5, 7?

(ii) How many eight-digit integers can be formed from the digits of the number m such that none of them has two identical digits adjacent to each other, where
(a) $m = 12\,341\,234$?   (b) $m = 11\,223\,345$?   (c) $m = 11\,223\,450$?

(iii) How many five-digit integers can be formed from the digits of $1\,122\,334\,455$ if we require that none of them have two identical digits adjacent to each other?

*(iv) Find the number of m-digit integers that can be written with exactly k different digits. (After solving this in general, take $m = 4$ and $k = 2$, and compare with Example 3.3.(vii), where this particular case was solved by a different method.)

6.11 Problems on Fixed Points of Permutations

Let A be a finite set and $\varphi\colon A \to A$ an arbitrary permutation (bijection) on A (see 2.2.(iv)). We say that the element $a \in A$ is a *fixed point* of the permutation φ if $\varphi(a) = a$. (For example, for the identity permutation id_A every element $a \in A$ is a fixed point.)

We will now classify all permutations on the set A according to the number of their fixed points. For the remainder of this section we assume that A is an n-element set with elements $a_1, a_2, \ldots, a_n$.

(i) A permutation with the property that each element of A is a fixed point is unique, namely, the identity permutation. Therefore, only the identity has n fixed points.

(ii) Let us now determine the number d_n of permutations of A that have *no fixed points*. For $i = 1, 2, \ldots, n$ we let M_i denote the set of all those permutations of the set A for which the element a_i is a fixed point (a permutation $\varphi \in M_i$ may or may not have other fixed points). We clearly have

$$d_n = n! - |M_1 \cup M_2 \cup \cdots \cup M_n|,$$

and we use the identity (49) to determine the number $|M_1 \cup M_2 \cup \cdots \cup M_n|$. Since the set $M_{j_1} \cap M_{j_2} \cap \cdots \cap M_{j_r}$ (where $\emptyset \neq \{j_1, j_2, \ldots, j_r\} \subseteq \{1, 2, \ldots, n\}$) contains all those permutations φ for which $\varphi(a_{j_1}) = a_{j_1}$, $\varphi(a_{j_2}) = a_{j_2}$, $\ldots$, $\varphi(a_{j_r}) = a_{j_r}$, while the images of the remaining $(n - r)$ elements can be arbitrarily chosen, we have

$$|M_{j_1} \cap M_{j_2} \cap \cdots \cap M_{j_r}| = (n - r)!.$$

This implies

$$|M_1 \cup M_2 \cup \cdots \cup M_n| = \sum_{r=1}^{n}(-1)^{r+1}\binom{n}{r}(n-r)!,$$

and thus

$$d_n = n! - \sum_{r=1}^{n}(-1)^{r+1}\binom{n}{r}(n-r)!$$

$$= \sum_{r=0}^{n}(-1)^r\binom{n}{r}(n-r)! = n!\sum_{r=0}^{n}\frac{(-1)^r}{r!}. \tag{61}$$

We remark that in view of 2.2.(iv) the identity (61) also gives the number of all arrangements of the set $A = \{a_1, a_2, \ldots, a_n\}$ such that for no $i = 1, 2, \ldots, n$ is the element a_i in the ith place. This formulation of the problem is often found in the literature.

(iii) We will now determine the number of elements of the set T_s of all those permutations of A that have exactly s fixed points, $0 \leq s \leq n$. By (i) and (ii) we know that $|T_n| = 1$ and $|T_0| = d_n$, where the number d_n is given by (61); you should convince yourself why $|T_{n-1}| = 0$. In order to determine the number $|T_s|$ for the remaining s, we place (for a fixed s) each permutation $f \in T_s$ into one of the $\binom{n}{s}$ classes according to which s elements of A are the fixed points. If $T_{s,X}$ is the class of permutations whose set of fixed points is given by the s-element set $X \subseteq A$, then each permutation $f \in T_{s,X}$ is uniquely determined by its values on $A \setminus X$, where $|A \setminus X| = n-s$ and $f(a_i) \neq a_i$ for any $a_i \in A \setminus X$. By (ii) we have therefore $|T_{s,X}| = d_{n-s}$, and thus $|T_s| = \binom{n}{s}d_{n-s}$. If we also set $d_0 = 1$, then this formula holds for all $s \in \{0, 1, \ldots, n\}$.

Let us add another interesting result: Since the sets $T_0, T_1, \ldots, T_n$ form a subdivision of the $n!$-element set of all permutations of A, the numbers d_s given by (61) satisfy the identity

$$n! = \sum_{s=0}^{n}|T_s| = \sum_{s=0}^{n}\binom{n}{s}d_s$$

for all $n \geq 1$, which makes it possible to determine the numbers $d_1, d_2, \ldots$ recursively without use of the identity (61).

6.12 Exercises

(i) Use the recurrence relation from the end of the previous paragraph to determine d_i, $i \leq 5$.

(ii) A secretary wrote 5 different letters and addressed 5 envelopes for the recipients of the letters. The letters got mixed up, and the secretary decides to place them at random into the envelopes.

      (a) In how many ways can the letters be distributed such that none of the addressees gets a letter intended for him or her?

      (b) In how many cases does exactly one (or do exactly two, exactly three) addressees receive the correct letter?

(iii) Two examiners give oral exams to 12 students, each in one subject. The exams take 5 minutes each. How many possible schedules are there such that both examiners meet all 12 students during one hour?

(iv) Find the number of all placements of n rooks on an $n \times n$ chessboard such that no two rooks threaten each other, and none is on the main diagonal.

(v) In how many ways can n married couples sit at a long table with n chairs on each side such that all the men sit on one side and nobody faces their spouse?

(vi) Find the number of all arrangements of $2n$ elements $m_1, m_2, \ldots, m_{2n}$ such that for each odd $i = 1, 3, \ldots, 2n - 1$ the element m_i is not in the ith place?

(vii) Find the number of all permutations of the set $A = \{a_1, a_2, \ldots, a_n\}$ such that the s given elements $a_{i_1}, a_{i_2}, \ldots, a_{i_s}$ $(1 \le s \le n)$ are not fixed points of the permutation (the remaining elements may or may not be fixed points).

6.13 Examples

We will now consider four important problems.

(i) King Arthur invites n pairs of mutually hostile knights for dinner at his round table. In how many ways can the royal butler place the knights around the table such that no two enemies sit next to each other?

SOLUTION.   We first agree that we consider two placements to be different if at least one chair is occupied by different knights in the two placements; we denote the number of all such placements by $w(n)$. (If we were to consider two placements identical if one can be obtained from the other by "rotation," then we would always have to identify $2n$ placements, and for the total number $W(n)$ we would have $W(n) = \frac{w(n)}{2n}$.)

    We now determine the number $w(n)$: Let M_i be the set of all placements where the ith pair of enemies sit next to each other. Then clearly,

$$w(n) = (2n)! - |M_1 \cup M_2 \cup \cdots \cup M_n|.$$

Let $\{j_1, j_2, \ldots, j_r\}$ be a nonempty subset of $\{1, 2, \ldots, n\}$. Then $|M_{j_1} \cap M_{j_2} \cap \cdots \cap M_{j_r}|$ gives the number of all placements in which r fixed pairs of enemies sit next to each other (and the remaining ones sit in an arbitrary manner in the remaining free chairs). Two neighboring seats for the first pair of enemies can be chosen in $2n$ ways (it suffices to consider from among the $2n$ spaces between the chairs the one that divides the two chosen chairs). We combine each of the remaining $(r-1)$ pairs of enemies into one object, and together with the remaining $(2n - 2r)$ knights we consider all arrangements of these $(r - 1 + 2n - 2r) = (2n - r - 1)$ objects, which are $(2n - r - 1)!$ in number. These arrangements do not yet determine the placements of individual knights around the table, since each of the r pairs of enemies had only chosen a pair of neighboring chairs. Since these $2r$ knights can sit in the chosen chairs in 2^r ways, it follows that

$$|M_{j_1} \cap M_{j_2} \cap \cdots \cap M_{j_r}| = 2^r \cdot 2n(2n - r - 1)! = 2^{r+1}n(2n - r - 1)!.$$

Therefore, by (49) we have

$$|M_1 \cup M_2 \cup \cdots \cup M_n| = \sum_{r=1}^{n}(-1)^{r+1}\binom{n}{r}2^{r+1}n(2n - r - 1)!,$$

and thus

$$w(n) = (2n)! - \sum_{r=1}^{n}(-1)^{r+1}\binom{n}{r}2^{r+1}n(2n - r - 1)!$$

$$= \sum_{r=0}^{n}(-1)^{r}\binom{n}{r}2^{r+1}n(2n - r - 1)!,$$

and for $W(n)$ we obtain

$$W(n) = \sum_{r=0}^{n}(-1)^{r}\binom{n}{r}2^{r}(2n - r - 1)!. \qquad \square$$

(ii) In how many ways can n married couples be placed at a round table such that men and women alternate and no couple sits together?

SOLUTION.   As in (i) we first consider as distinct any two seatings in which at least one chair is occupied by different persons, and we denote the total number by $z(n)$.

We divide all $z(n)$ seatings into $2 \cdot n!$ groups according to the placements of the women, and then we determine the number s of seatings in each of these groups, that is, for a fixed distribution of the women, s gives the number of ways in which the men can be placed such that none of them sits next to his wife. For $i = 1, 2, \ldots, n$ we denote by M_i the set of all placements of men such that the ith man sits next to his wife; then we clearly have

$$s = n! - |M_1 \cup M_2 \cup \cdots \cup M_n|.$$

We determine the number $|M_1 \cup M_2 \cup \cdots \cup M_n|$ by way of the inclusion–exclusion principle. It is clear that the identity (49) cannot be used, since for two different r-tuples $\{j_1, j_2, \ldots, j_r\}$, $\{k_1, k_2, \ldots, k_r\}$ the summands $|M_{j_1} \cap M_{j_2} \cap \cdots \cap M_{j_r}|$ and $|M_{k_1} \cap M_{k_2} \cap \cdots \cap M_{k_r}|$ are, in general, different: For instance, if the r chosen women sit close together, then their husbands have considerably fewer choices of seats than in the case where these women are more "spread out"; therefore, the calculation of the individual summands $|M_{j_1} \cap M_{j_2} \cap \cdots \cap M_{j_r}|$ is very complicated. Fortunately, it is quite easy to determine the sum $\sum |M_{j_1} \cap M_{j_2} \cap \cdots \cap M_{j_r}|$ for each fixed $r \in \{1, 2, \ldots, n\}$ (where the sum is taken over all r-element subsets of $\{1, 2, \ldots, n\}$).

Indeed, this sum gives the total number of all placements of all r-tuples of men next to their wives (the remaining $(n-r)$ men can be placed arbitrarily on the remaining $(n-r)$ chairs), and we now determine this number as follows.

There are $2n$ chairs placed around the table, with $2n$ spaces between them. We assume that some r-tuple of men sit next to their wives, and we mark the r spaces between the chairs occupied by the corresponding couples. No two of these spaces can be adjacent to each other, for otherwise, some man would have a wife on both sides, or some woman a husband on both sides.

It is not difficult to find the number of ways in which we can choose r spaces from among $2n$ spaces placed on a circle such that no two of the chosen spaces are adjacent: It follows from the result of Exercise 3.2.(viii) that this number is $\frac{2n}{2n-r}\binom{2n-r}{r}$. In view of the fact that the remaining $(n-r)$ men can be placed in the remaining $(n-r)$ chairs in $(n-r)!$ ways, we obtain for the desired sum

$$\sum |M_{j_1} \cap M_{j_2} \cap \cdots \cap M_{j_r}| = \frac{2n}{2n-r}\binom{2n-r}{r}(n-r)!,$$

and thus

$$s = n! - \sum_{r=1}^{n}(-1)^{r+1}(n-r)!\frac{2n}{2n-r}\binom{2n-r}{r}$$

$$= \sum_{r=0}^{n}(-1)^r(n-r)!\frac{2n}{2n-r}\binom{2n-r}{r}.$$

Altogether, therefore,

$$z(n) = 2n! \cdot s = 2n! \sum_{r=0}^{n}(-1)^r(n-r)!\frac{2n}{2n-r}\binom{2n-r}{r}. \qquad \square$$

(iii) (*Euler's φ-function.*) We consider an arbitrary integer $m > 1$ and its prime decomposition $m = p_1^{\alpha_1} \cdots p_n^{\alpha_n}$, where $p_1 < p_2 < \cdots < p_n$ are primes and $\alpha_1, \ldots, \alpha_n \in \mathbb{N}$. In [5], Chapter 3, Sections 3.6–3.10, we showed by induction that among the integers $1, \ldots, m$ there are exactly $\varphi(m)$ that are relatively prime to m, where

$$\varphi(m) = m\left(1 - \frac{1}{p_1}\right) \cdots \left(1 - \frac{1}{p_n}\right).$$

We will now show that we can easily prove this expression by way of the inclusion–exclusion principle. We set $M = \{1, 2, \ldots, m\}$, and for each $i = 1, \ldots, n$ we set $M_i = \{k \in M; \; p_i \mid k\}$. Then for any nonempty subset $\{j_1, \ldots, j_r\} \subseteq \{1, \ldots, n\}$ we have $M_{j_1} \cap \cdots \cap M_{j_r} = \{k \in M; \; p_{j_1} \cdots p_{j_r} \mid k\}$, and thus $|M_{j_1} \cap \cdots \cap M_{j_r}| = m/(p_{j_1} \cdots p_{j_r})$. Since $M_1 \cup \cdots \cup M_n$ is the set of all integers in M that have at least one prime divisor in common with m, then by (47) the number of integers in M that are relatively prime to m is

$$m - |M_1 \cup \cdots \cup M_n| = m - \sum (-1)^{r+1} |M_{j_1} \cap \cdots \cap M_{j_r}|$$

$$= m \left(1 + \sum \frac{(-1)^r}{p_{j_1} \cdots p_{j_r}} \right), \tag{62}$$

where the sums are taken over all $\emptyset \neq \{j_1, \ldots, j_r\} \subseteq \{1, \ldots, n\}$. Let us now consider the expansion of the expression $(1 - \frac{1}{p_1}) \cdots (1 - \frac{1}{p_n})$. We obtain a sum of 2^n products of the form $c_1 c_2 \cdots c_n$, where either $c_i = 1$, or $c_i = -\frac{1}{p_i}$ for each $i \in \{1, \ldots, n\}$. Each such product can be uniquely written by way of one of the 2^n subsets $L \subseteq \{1, \ldots, n\}$ as follows:

$$c_1 c_2 \cdots c_n \mapsto L = \left\{ i \in \{1, \ldots, n\} : c_i = -\frac{1}{p_i} \right\}.$$

If $L = \emptyset$, then $c_1 c_2 \cdots c_n = 1^n = 1$, while in the case $L \neq \emptyset$ we have

$$c_1 \cdots c_n = \prod_{i \in L} \frac{-1}{p_i},$$

and thus

$$\left(1 - \frac{1}{p_1} \right) \cdots \left(1 - \frac{1}{p_n} \right) = 1 + \sum_{\emptyset \neq L \subseteq \{1,\ldots,n\}} \prod_{i \in L} \frac{-1}{p_i},$$

and with (62) we obtain the desired result.

(iv) (*The Möbius inversion formula.*) The Möbius function $\mu(n)$ is defined for positive integers n as follows: If in the prime decomposition of the integer $n > 1$ no prime appears to a power higher than 1, then we set $\mu(n) = (-1)^v$, where v is the number of primes in the decomposition of n. For all other n we set $\mu(n) = 0$, with the only exception $\mu(1) = 1$. Let $f, g : \mathbb{N} \to \mathbb{R}$ be mappings satisfying the identity

$$g(k) = \sum_{d \mid k} f(d) \tag{63}$$

for each $k \in \mathbb{N}$, where the sum is taken over all divisors $d \in \mathbb{N}$ of k. (Note that the function f is arbitrary, while g is uniquely determined by f according to (63).) Show that for any $m \in \mathbb{N}$ we then have

$$f(m) = \sum_{d \mid m} \mu(d) g \left(\frac{m}{d} \right), \tag{64}$$

where the sum is taken over all divisors $d \in \mathbb{N}$ of m. (This identity is often called the *Möbius inversion formula*.)

SOLUTION.   It is clear that for $m = 1$ the identity (64) holds (it is actually identical with (63) for $k = 1$). We therefore assume that $m > 1$, and we write the integer m in its prime decomposition

$$m = p_1^{\alpha_1} \cdots p_n^{\alpha_n},$$

where $p_1 < \cdots < p_n$ are primes, and $\alpha_1, \ldots, \alpha_n \in \mathbb{N}$. Then we denote $M = \{d \in \mathbb{N}; d \mid m\}$ and for each $i = 1, \ldots, n$ we set $M_i = \{d \in M; p_i^{\alpha_i} \nmid d\}$. Just as we did in 6.2 we construct the function F from f. Then for any nonempty subset $\{j_1, \ldots, j_r\}$ of $\{1, \ldots, n\}$ we have

$$F(M_{j_1} \cap \cdots \cap M_{j_r}) = \sum_{d \in M_{j_1} \cap \cdots \cap M_{j_r}} f(d).$$

We can convince ourselves that the elements of the set $M_{j_1} \cap \cdots \cap M_{j_r}$ are exactly all the divisors $d \in \mathbb{N}$ of m that are not divisible by any of the numbers $p_{j_1}^{\alpha_{j_1}}, \ldots, p_{j_r}^{\alpha_{j_r}}$. With the unique factorization theorem we find that these are exactly the divisors of the number $m/(p_{j_1} \cdots p_{j_r})$, and thus by (63) we have

$$F(M_{j_1} \cap \cdots \cap M_{j_r}) = g\left(\frac{m}{p_{j_1} \cdots p_{j_r}}\right).$$

We have furthermore $M_1 \cup \cdots \cup M_n = M \setminus \{m\}$, and so by (63),

$$F(M_1 \cup \cdots \cup M_n) = \sum_{d \in M, d \neq m} f(d) = g(m) - f(m).$$

Substituting this into (48), we obtain

$$g(m) - f(m) = \sum (-1)^{r+1} g\left(\frac{m}{p_{j_1} \cdots p_{j_r}}\right),$$

that is,

$$f(m) = g(m) + \sum (-1)^r g\left(\frac{m}{p_{j_1} \cdots p_{j_r}}\right),$$

where in both expressions the sum is taken over all nonempty subsets $\{j_1, \ldots, j_r\}$ of $\{1, \ldots, n\}$. By the definition of the Möbius function we obtain (64) directly from this. $\square$

6.14   Exercises

(i) Show that for all $m > 1$ we have $\sum_{d \mid m} \mu(d) = 0$.

(ii) For $m \in \mathbb{N}$ let $\tau(m)$ denote the number of positive divisors of m, $\sigma(m)$ their sum, and $\nu(m)$ their product. Show that

(a) $\sum_{d|m} \mu(d)\tau(\frac{m}{d}) = 1$,

(b) $\sum_{d|m} \mu(d)\sigma(\frac{m}{d}) = m$,

(c) $\prod_{d|m} (\nu(\frac{m}{d}))^{\mu(d)} = m$,

where the sums and the product are taken over all positive divisors d of m.

(iii) Prove the following converse of the Möbius inversion formula: If $f, g : \mathbb{N} \to \mathbb{R}$ are mappings satisfying the identity (64) for all $m \in \mathbb{N}$, then they also satisfy the identity (63) for all $k \in \mathbb{N}$.

(iv) Prove $\sum_{d|m} \varphi(d) = m$ for all $m \in \mathbb{N}$
(a) directly.   (b) by way of (iii) and (64).

*(v) In this exercise we will denote the greatest common divisor of the positive integers $a_1, \ldots, a_k$ by $\gcd(a_1, \ldots, a_k)$, in order to distinguish it from an ordered k-tuple $(a_1, \ldots, a_k)$. For any positive integer m and divisor $d \in \mathbb{N}$ of m we let $H(m, d)$ denote the set of all ordered pairs (a, b), where $a, b \in \{1, \ldots, m\}$ and $\gcd(a, b, m) = d$. Furthermore, let $h(m, d) = |H(m, d)|$. Show that

(a) $\sum_{d|m} h(m, d) = m^2$.

(b) $h(m, d) = h(\frac{m}{d}, 1)$.

(c) if $p_1, \ldots, p_n$ are all the primes dividing m, then

$$h(m, 1) = m^2 \left(1 - \frac{1}{p_1^2}\right) \cdots \left(1 - \frac{1}{p_n^2}\right).$$

(d) Generalize the problem for the number $h_r(m, d)$ of all r-tuples $(a_1, \ldots, a_r)$, where $a_i \in \{1, \ldots, m\}$ for all $i = 1, \ldots, r$ and $\gcd(a_1, \ldots, a_r, m) = d$.

6.15 Generalization of the Inclusion–Exclusion Principle

In 6.13.(iv) we were given two functions $f, g : \mathbb{N} \to \mathbb{R}$, where the values of the function g were determined in a very easy way by the values of the function f. The subsequently derived Möbius inversion formula then described how we could get back the values of the function f with the help of the values of g; this is useful in the case where we know the function g, but do not know f. In what follows we are going to generalize these considerations.

Theorem. *For each $i = 1, \ldots, n$ let $k_i \in \mathbb{N}$ be given. For an arbitrary n-tuple $(x_1, \ldots, x_n)$, where $x_i \in \{0, 1, \ldots, k_i\}$ for all $i = 1, \ldots, n$, suppose we are given real numbers $F(x_1, \ldots, x_n)$, $G(x_1, \ldots, x_n)$. In other words, F and G are mappings from the Cartesian product $\{0, 1, \ldots, k_1\} \times \cdots \times$*

$\{0, 1, \ldots, k_n\}$ to $\mathbb{R}$. Furthermore, for each $i = 1, \ldots, n$ and for all integers r, s such that $0 \le s \le r \le k_i$, let the real numbers $\nu_i(r, s)$ be given. Suppose that for any n-tuple $(x_1, \ldots, x_n)$, where $x_i \in \{0, 1, \ldots, k_i\}$ for all $i = 1, \ldots, n$, we have

$$G(x_1, \ldots, x_n) = \sum_{y_1=0}^{x_1} \cdots \sum_{y_n=0}^{x_n} \left(\prod_{i=1}^{n} \nu_i(x_i, y_i) \right) F(y_1, \ldots, y_n), \qquad (65)$$

and suppose furthermore that for each $i = 1, \ldots, n$ there exists a function $\mu_i : \{0, 1, \ldots, k_i\} \to \mathbb{R}$ such that for each $s \in \{0, 1, \ldots, k_i\}$ we have

$$\sum_{r=s}^{k_i} \mu_i(r)\nu_i(r, s) = \begin{cases} 1 & \text{if } s = k_i, \\ 0 & \text{if } s \ne k_i. \end{cases} \qquad (66)$$

Then

$$F(k_1, \ldots, k_n) = \sum_{x_1=0}^{k_1} \cdots \sum_{x_n=0}^{k_n} \left(\prod_{i=1}^{n} \mu_i(x_i) \right) G(x_1, \ldots, x_n). \qquad (67)$$

PROOF.   By substitution from (65) we see that the right-hand side of (67) is equal to

$$\sum_{x_1=0}^{k_1} \cdots \sum_{x_n=0}^{k_n} \left(\prod_{i=1}^{n} \mu_i(x_i) \right) \sum_{y_1=0}^{x_1} \cdots \sum_{y_n=0}^{x_n} \left(\prod_{i=1}^{n} \nu_i(x_i, y_i) \right) F(y_1, \ldots, y_n). \qquad (68)$$

If we first sum over all $x_i \in \{0, 1, \ldots, k_i\}$ and then over all $y_i \in \{0, 1, \ldots, x_i\}$, then we sum in fact over all pairs (x_i, y_i) satisfying $0 \le y_i \le x_i \le k_i$; thus we obtain the same result if we first sum over all $y_i \in \{0, 1, \ldots, k_i\}$, and then over all $x_i \in \{y_i, y_i + 1, \ldots, k_i\}$. We can therefore transform the expression (68) as follows:

$$\sum_{y_1=0}^{k_1} \cdots \sum_{y_n=0}^{k_n} \sum_{x_1=y_1}^{k_1} \cdots \sum_{x_n=y_n}^{k_n} \left(\prod_{i=1}^{n} \mu_i(x_i)\nu_i(x_i, y_i) \right) F(y_1, \ldots, y_n)$$

$$= \sum_{y_1=0}^{k_1} \cdots \sum_{y_n=0}^{k_n} \left(\prod_{i=1}^{n} \left(\sum_{x_i=y_i}^{k_i} \mu_i(x_i)\nu_i(x_i, y_i) \right) \right) F(y_1, \ldots, y_n), \quad (69)$$

where the last equality follows from the fact that upon expanding the product

$$\left(\sum_{x_1=y_1}^{k_1} \mu_1(x_1)\nu_1(x_1, y_1) \right) \cdots \left(\sum_{x_n=y_n}^{k_n} \mu_n(x_n)\nu_n(x_n, y_n) \right)$$

we obtain a sum in which the summand corresponding to the n-tuple $(x_1, \ldots, x_n)$, where each $x_i \in \{y_i, y_i + 1, \ldots, k_i\}$, is equal to the product

$$(\mu_1(x_1)\nu_1(x_1, y_1)) \cdots (\mu_n(x_n)\nu_n(x_n, y_n)).$$

By (66), however, $\sum_{x_i=y_i}^{k_i} \mu_i(x_i)\nu_i(x_i, y_i)$ is different from zero only when $y_i = k_i$, and thus $\prod_{i=1}^{n}(\sum_{x_i=y_i}^{k_i} \mu_i(x_i)\nu_i(x_i, y_i))$ is nonzero if and only if $y_1 = k_1, \ldots, y_n = k_n$ (in this unique case the expression is equal to 1). This implies that in the sum on the right-hand side of (69) only one summand is nonzero. Hence the right-hand side of (69) is equal to this summand, namely $F(k_1, \ldots, k_n)$, which was to be shown.   □

Remark. The above theorem is useful if the unknown function F is connected with the known function G by the relation (65). Then we can determine the values of μ_i with the help of the conditions (66) and then use (67). A sufficient condition for the existence of the functions μ_i satisfying (66) is that $\nu_i(s, s) \neq 0$ for all $s \in \{0, 1, \ldots, k_i\}$; then one can use (66) to successively determine the values $\mu_i(k_i), \mu_i(k_i - 1), \ldots, \mu_i(0)$.

6.16 Examples

(i) We first show that the inclusion–exclusion principle in 6.1 is a special case of the theorem in 6.15. Let $M_1, \ldots, M_n$ be arbitrary finite sets. We let $M = M_1 \cup \cdots \cup M_n$ and set $k_1 = \cdots = k_n = 1$. For any n-tuple $(x_1, \ldots, x_n)$ with $x_i \in \{0, 1\}$ we consider for each $i = 1, \ldots, n$ the sets

$$P(x_1, \ldots, x_n) = \{a \in M; \; \forall i = 1, \ldots, n \colon (x_i = 0 \Longleftrightarrow a \in M_i)\},$$
$$Q(x_1, \ldots, x_n) = \{a \in M; \; \forall i = 1, \ldots, n \colon (x_i = 0 \Longrightarrow a \in M_i)\}.$$

Then $P(0, \ldots, 0) = \bigcap_{i=1}^{n} M_i$, $P(1, \ldots, 1) = \emptyset$, $Q(1, \ldots, 1) = M$, and in the remaining cases,

$$P(x_1, \ldots, x_n) = \bigcap_{\substack{i=1,\ldots,n \\ x_i=0}} M_i \cap \bigcap_{\substack{i=1,\ldots,n \\ x_i=1}} (M \setminus M_i),$$

$$Q(x_1, \ldots, x_n) = \bigcap_{\substack{i=1,\ldots,n \\ x_i=0}} M_i.$$

We now show that for an arbitrary n-tuple $(x_1, \ldots, x_n)$, where $x_1, \ldots, x_n \in \{0, 1\}$, we have

$$Q(x_1, \ldots, x_n) = \bigcup_{y_1=0}^{x_1} \cdots \bigcup_{y_n=0}^{x_n} P(y_1, \ldots, y_n).$$

Indeed, each set $P(y_1, \ldots, y_n)$ on the right-hand side is a subset of $Q(x_1, \ldots, x_n)$; conversely, an arbitrary element $a \in Q(x_1, \ldots, x_n)$ lies in the set $P(y_1, \ldots, y_n)$ for which $y_i = 0$ if $a \in M_i$, and $y_i = 1$ if $a \notin M_i$. Furthermore, the sets on the right-hand side are disjoint, and therefore, if we set

$$F(x_1, \ldots, x_n) = |P(x_1, \ldots, x_n)|, \quad G(x_1, \ldots, x_n) = |Q(x_1, \ldots, x_n)|,$$

we obtain the identity

$$G(x_1, \ldots, x_n) = \sum_{y_1=0}^{x_1} \cdots \sum_{y_n=0}^{x_n} F(y_1, \ldots, y_n),$$

valid for any $x_1, \ldots, x_n \in \{0, 1\}$. Hence equation (65) is satisfied if we set $\nu_i(r, s) = 1$ for all $i \in \{1, \ldots, n\}$ and all $0 \le s \le r \le 1$. The conditions (66) will be satisfied if $\mu_i(1) = 1$ and $\mu_i(0) + \mu_i(1) = 0$, that is, $\mu_i(0) = -1$. The theorem now implies

$$F(1, \ldots, 1) = \sum_{x_1=0}^{1} \cdots \sum_{x_n=0}^{1} \left(\prod_{i=1}^{n} \mu_i(x_i) \right) G(x_1, \ldots, x_n).$$

Each one of the 2^n summands of this last sum is uniquely determined by a subset L of $\{1, \ldots, n\}$ (we choose $x_i = 0$ if and only if $i \in L$). Thus we have

$$0 = |M| + \sum_{\emptyset \ne L \subseteq \{1, \ldots, n\}} (-1)^{|L|} \left| \bigcap_{i \in L} M_i \right|,$$

from which we obtain the statement of 6.1 through an easy reformulation.

(ii) In how many ways can the digits of the number $1\,112\,223\,456$ be written such that no two identical digits are adjacent?

SOLUTION.   For arbitrary $x_1, x_2 \in \{0, 1, 2\}$ we let $F(x_1, x_2)$ denote the number of integers formed from the digits of $1\,112\,223\,456$ such that for each $i = 1, 2$,

- all three digits i are adjacent when $x_i = 0$;
- two of the digits i are adjacent and the third one is not, when $x_i = 1$;
- no two digits i are adjacent when $x_i = 2$.

Similarly, we let $G(x_1, x_2)$ denote the number of integers formed from the digits of $1\,112\,223\,456$ such that for each $i = 1, 2$,

- a triple of consecutive digits i is underlined when $x_i = 0$;
- a pair of consecutive digits i is underlined when $x_i = 1$;
- no digits i are underlined when $x_i = 2$.

Here we consider two integers of the same value as different if they are underlined differently.

We note that a triple of digits can be marked (or underlined) in a unique way if they are adjacent. Further, a pair of digits i can be marked in a unique way if two of them are adjacent and the third one is separate, while it can be marked in two different ways if all three digits i are adjacent (for instance, $\ldots\underline{111}\ldots$ or $\ldots 1\underline{11}\ldots$). Hence for arbitrary $x_1, x_2 \in \{0, 1, 2\}$ we have

$$G(x_1, x_2) = \sum_{y_1=0}^{x_1} \sum_{y_2=0}^{x_2} \nu_1(x_1, y_1)\nu_2(x_2, y_2)F(y_1, y_2),$$

where for $i = 1, 2$,

$$\nu_i(0,0) = 1; \quad \nu_i(1,1) = 1, \quad \nu_i(1,0) = 2;$$
$$\nu_i(2,2) = 1, \quad \nu_i(2,1) = 1, \quad \nu_i(2,0) = 1.$$

We want to find the value of $F(2,2)$, while the values of $G(x_1, x_2)$ are easy to determine, since the underlined pairs or triples of digits can be considered as single symbols:

$$G(0,0) = P(6) = 6! = 720,$$
$$G(0,1) = G(1,0) = P(7) = 7! = 5040,$$
$$G(1,1) = P(8) = 8! = 40\,320,$$
$$G(0,2) = G(2,0) = P_0(3,1,1,1,1,1) = \frac{8!}{3!} = 6720,$$
$$G(1,2) = G(2,1) = P_0(3,1,1,1,1,1,1) = \frac{9!}{3!} = 60\,480,$$
$$G(2,2) = P_0(3,3,1,1,1,1) = \frac{10!}{3! \cdot 3!} = 100\,800.$$

We now determine the functions $\mu_i \colon \{0,1,2\} \longrightarrow \mathbb{R}$ satisfying (66), that is, such that for each $i = 1, 2$,

$$1 = \mu_i(2)\nu_i(2,2) = \mu_i(2),$$
$$0 = \mu_i(1)\nu_i(1,1) + \mu_i(2)\nu_i(2,1) = \mu_i(1) + \mu_i(2),$$
$$0 = \mu_i(0)\nu_i(0,0) + \mu_i(1)\nu_i(1,0) + \mu_i(2)\nu_i(2,0)$$
$$= \mu_i(0) + 2\mu_i(1) + \mu_i(2),$$

which implies $\mu_i(2) = 1$, $\mu_i(1) = -1$, $\mu_i(0) = 1$. By our theorem we have therefore

$$F(2,2) = \sum_{x_1=0}^{2} \sum_{x_2=0}^{2} \mu_1(x_1)\mu_2(x_2)G(x_1, x_2)$$
$$= G(0,0) - G(0,1) + G(0,2) - G(1,0) + G(1,1) - G(1,2)$$
$$+ G(2,0) - G(2,1) + G(2,2) = 24\,240,$$

which is the desired result.   □

(iii) Nine guests at a round table have ordered dinners. The forgetful waiter, however, remembers only that three portions each of three different meals have been ordered, and that no two adjacent guests ordered the same meal. How many orders satisfy these conditions?

SOLUTION.   We denote the three different meals by the numbers 1, 2, 3. For arbitrary $x_1, x_2, x_3 \in \{0, 1, 2\}$ we let $F(x_1, x_2, x_3)$ be the number of all possible ways of ordering three portions of each meal 1, 2, 3 such that for each $i \in \{1, 2, 3\}$,

- all three guests who ordered meal i sit next to each other when $x_i = 0$;
- two of the guests who ordered meal i sit next to each other, while the third sits elsewhere, when $x_i = 1$;
- no two guests who ordered meal i are adjacent if $x_i = 2$.

Similarly, let $G(x_1, x_2, x_3)$ denote the number of ways to order three portions each of the meals 1, 2, 3 such that for each $i \in \{1, 2, 3\}$,

- all three guests who ordered meal i sit next to each other when $x_i = 0$;
- pairs of meals i ordered by adjacent guests are marked when $x_i = 1$;
- no further condition is required when $x_i = 2$.

Here we will also consider as distinct those orders that differ only in the way they are marked. Our problem now is to evaluate $F(2, 2, 2)$. As in the previous example we verify that for any $x_1, x_2, x_3 \in \{0, 1, 2\}$ we have

$$G(x_1, x_2, x_3) = \sum_{y_1=0}^{x_1} \sum_{y_2=0}^{x_2} \sum_{y_3=0}^{x_3} \nu_1(x_1, y_1)\nu_2(x_2, y_2)\nu_3(x_3, y_3)F(y_1, y_2, y_3)$$

if for each $i \in \{1, 2, 3\}$ we set

$$\nu_i(x_i, y_i) = \begin{cases} 2 & \text{if } x_i = 1, \ y_i = 0, \\ 1 & \text{otherwise.} \end{cases}$$

The functions $\mu_i : \{0, 1, 2\} \to \mathbb{R}$ are then the same as in the previous example, namely, $\mu_i(x_i) = (-1)^{x_i}$. Hence we have

$$F(2, 2, 2) = \sum_{x_1=0}^{2} \sum_{x_2=0}^{2} \sum_{x_3=0}^{2} (-1)^{x_1+x_2+x_3} G(x_1, x_2, x_3). \tag{70}$$

It remains to find the values $G(x_1, x_2, x_3)$. In view of symmetry it suffices to do this for $x_1 \geq x_2 \geq x_3$. Now, $G(0, 0, 0)$ gives the number of possible ways of ordering where each triple of guests who ordered the same meal sit together. The triple of guests who ordered meal 1 can be chosen in 9 ways (it is enough to choose the "middle one"). The remaining guest can then order in only two ways, and thus $G(0, 0, 0) = 18$. Similarly, we have

$$G(1, 0, 0) = 9 \cdot 3! = 54, \quad G(1, 1, 0) = 9 \cdot 4! = 216, \quad G(1, 1, 1) = 9 \cdot 5! = 1080.$$

Next, $G(2,1,0)$ gives the number of possible ways of ordering such that all guests who ordered meal 3 sit together and such that a pair of meals 2, ordered by neighboring guests, is marked. The guest who ordered the unmarked meal 2 can be chosen in nine ways. The number of possible orders by the remaining guests is then the same as the number of permutations of the integers 1, 1, 1, 2, 3, and so

$$G(2,1,0) = 9 \cdot P_0(3,1,1) = 180.$$

Similarly,

$$
\begin{aligned}
G(2,0,0) &= 9 \cdot P_0(3,1) = 36, & G(2,1,1) &= 9 \cdot P_0(3,1,1,1) = 1080, \\
G(2,2,0) &= 9 \cdot P_0(3,3) = 180, & G(2,2,1) &= 9 \cdot P_0(3,3,1) = 1260, \\
G(2,2,2) &= P_0(3,3,3) = 1680,
\end{aligned}
$$

and thus, substituting into (70), we obtain

$$
\begin{aligned}
F(2,2,2) = {}& G(0,0,0) - 3 \cdot G(1,0,0) + 3 \cdot G(1,1,0) - G(1,1,1) \\
& + 3 \cdot G(2,0,0) - 6 \cdot G(2,1,0) + 3 \cdot G(2,1,1) + 3 \cdot G(2,2,0) \\
& - 3 \cdot G(2,2,1) + G(2,2,2) = 132. \qquad \square
\end{aligned}
$$

6.17 Exercises

(i) Show that the assertion in Remark 6.2 is a special case of the theorem in 6.15.

(ii) Show that 6.13.(iv) is a special case of the theorem in 6.15.

(iii) In how many ways can 3 Canadians, 3 Finns, and 3 Turks stand in a row such that no two persons of the same nationality are next to each other?

*(iv) At a camp $3n$ children sleep in n tents, with three to a tent. In how many ways can they be placed in a row such that no two children from the same tent stand next to each other?

*(v) In how many ways can one place four cubes and four balls into bins numbered $1, \ldots, k$ such that no bin contains more than one object, and for no $i = 1, \ldots, k-1$ do the bins numbered i and $i+1$ contain the same object?

*(vi) Solve question (iv) with $4n$ children and n tents for four each.

*(vii) In how many anagrams of the word MISSISSIPPI are no two identical letters next to each other?

7 Basics of Pólya's Theory of Enumeration

In combinatorial problems the evaluation of numbers of objects or configurations is often more complicated when certain configurations of a given type are considered identical for some reasons (for instance, by symmetry). For example, let us see in how many ways the faces of a given cube can be colored with two colors (black and white, say). If we consider the cube as fixed (immovable), then the set X of all colorings has clearly $2^6 = 64$ elements. However, if we consider the cube as homogeneous (that is, there are no properties that let us distinguish one face from another), then in the coloring that has one face black and the others white it is irrelevant which one is chosen to be black. This means that the six corresponding elements in X are considered identical. On the other hand, in the case of two black faces there are two essentially different colorings: The black faces are either opposite (3 elements of X) or adjacent (12 elements of X) to each other. Thus we see that the groups of elements of X giving the same colorings have various cardinalities; that is, they contain various numbers of elements. Therefore, we cannot determine the number p of such groups as we have often done in the preceding sections (namely, $p = |X|/k$, where k is the number of elements in each group). After some reflection we will certainly arrive at a way of describing what it means for two elements of X to be identical (or better, *equivalent*): Two colorings are equivalent if one can be obtained from the other by some rotation of our cube. Each such rotation φ transforms a coloring $a \in X$ into some coloring $b \in X$. Thus we obtain a mapping $\varphi : X \to X$ that is in fact a bijection. You can convince yourself that for solving the whole problem about coloring the cube it is in fact enough to know the set Φ of all such bijections $X \to X$.

Let us now generalize the above problem. We are given a finite nonempty set X and a nonempty set Φ of bijections on X. The set Φ, however, cannot be completely arbitrary if we require that the relation $\sim$ on X defined by the rule

$$\forall a, b \in X : (a \sim b \iff \exists \varphi \in \Phi : \varphi(a) = b) \tag{71}$$

be in fact an equivalence relation, that is, that it determine a partition $X/\!\!\sim$ into classes of pairwise equivalent elements. Before we introduce a simple sufficient condition, we recall a few basic concepts and results relating to bijections. If φ and ψ are two bijections on X, then $\varphi \circ \psi$ ("φ composed with ψ") is the bijection on X defined by the rule $\varphi \circ \psi(x) = \varphi(\psi(x))$ for all $x \in X$; the *identity* on X is the bijection id $: X \to X$ defined by $\mathrm{id}(x) = x$ ($x \in X$); finally, the inverse of a bijection φ on X is the unique mapping $\varphi^{-1} : X \to X$ for which $\varphi \circ \varphi^{-1} = \varphi^{-1} \circ \varphi = \mathrm{id}$. We add to this that the identity $(\varphi \circ \psi) \circ \tau = \varphi \circ (\psi \circ \tau)$ holds for any three bijections φ, ψ, τ on X, and that $\varphi \circ \psi = \varphi \circ \tau$ and $\psi \circ \varphi = \tau \circ \varphi$ hold only when $\psi = \tau$.

7.1   A Sufficient Condition

Lemma. *Let Φ be a nonempty set of bijections on a nonempty finite set X and let $\varphi \circ \psi \in \Phi$ for any φ, $\psi \in \Phi$. Then (71) defines a relation $\sim$ that* *is an equivalence relation on X; furthermore,* $\mathrm{id} \in \Phi$ *and* $\varphi^{-1} \in \Phi$ *for all* $\varphi \in \Phi$.

Before we prove this lemma, we remark that its hypothesis, namely, the *closure of the set* Φ *under the operation of composition* $\circ$, is satisfied in the example of the cube at the beginning of this section (as it will be in a number of other geometric problems): If the bijections φ and ψ describe two rotations of the cube, then the bijection $\varphi \circ \psi$ that describes the composition of these two rotations is another rotation of the cube. The reader who is familiar with group theory will recognize that the closure condition in the lemma means that the set Φ forms a *subgroup of the group of all bijections* on the set of X with the operation $\circ$.

PROOF.   If n is the number of elements of the set X, then we can construct exactly $n!$ different bijections on X (see 2.2.(iv)). Therefore, Φ has no more than $n!$ elements. We choose an arbitrary $\varphi \in \Phi$, and for each $n \in \mathbb{N}$ we define φ_n as follows:

$$\varphi_1 = \varphi \in \Phi, \ \varphi_2 = \varphi \circ \varphi_1 \in \Phi, \ \ldots, \ \varphi_n = \varphi \circ \varphi_{n-1} \in \Phi, \ \ldots.$$

Thus we obtain an infinite sequence of elements of the finite set Φ. By the pigeonhole principle there exist indices $i < j$ such that $\varphi_i = \varphi_j$. If $i > 1$, then $\varphi \circ \varphi_{i-1} = \varphi_i = \varphi_j = \varphi \circ \varphi_{j-1}$, and since φ is a bijection, we obtain from this $\varphi_{i-1} = \varphi_{j-1}$. If $i - 1 > 1$, we similarly obtain $\varphi_{i-2} = \varphi_{j-2}$, etc. After $i - 1$ steps we arrive at the identity $\varphi_1 = \varphi_{j-i+1}$, that is, $\varphi \circ \mathrm{id} = \varphi_1 = \varphi_{j-i+1} = \varphi \circ \varphi_{j-i}$. This means that $\mathrm{id} = \varphi_{j-i} \in \Phi$. If $j - i \leq 2$, then $\varphi \circ \varphi = \mathrm{id}$, and therefore $\varphi^{-1} = \varphi \in \Phi$. If $j - i > 2$, then $\varphi \circ \varphi_{j-i-1} = \varphi_{j-i} = \mathrm{id}$ and also $\varphi \circ \mathrm{id} = \varphi = \mathrm{id} \circ \varphi = (\varphi \circ \varphi_{j-i-1}) \circ \varphi = \varphi \circ (\varphi_{j-i-1} \circ \varphi)$, which means that $\varphi_{j-i-1} \circ \varphi = \mathrm{id}$ and thus $\varphi^{-1} = \varphi_{j-i-1} \in \Phi$.

It remains to show that $\sim$ is an equivalence relation on X. Indeed, it is easy to see that $\sim$ is reflexive (since $\mathrm{id} \in \Phi$), symmetric (if $\varphi(a) = b$ for $\varphi \in \Phi$, then for $\varphi^{-1} \in \Phi$ we have $\varphi^{-1}(b) = (\varphi^{-1} \circ \varphi)(a) = a$), and transitive (if $\varphi(a) = b$, $\psi(b) = c$ for $\varphi, \psi \in \Phi$, then $\psi \circ \varphi \in \Phi$ and $(\psi \circ \varphi)(a) = c$).   □

7.2   Burnside's Lemma

Lemma. *Given the situation of the lemma in 7.1, the number of classes of the partition* $X/\sim$ *is given by the formula*

$$|X/\sim| = \frac{1}{|\Phi|} \sum_{\varphi \in \Phi} v(\varphi), \tag{72}$$

where $v(\varphi)$ *denotes for* $\varphi \in \Phi$ *the number of those elements of X that are mapped onto themselves by* φ, *that is,*

$$v(\varphi) = |\{a \in X; \ \varphi(a) = a\}|.$$

PROOF. For a given $a \in X$ let $u(a)$ denote the number of bijections belonging to the set Φ that map the element a onto itself; that is,

$$u(a) = |\{\varphi \in \Phi; \ \varphi(a) = a\}|.$$

Then we have

$$\sum_{\varphi \in \Phi} v(\varphi) = \sum_{a \in X} u(a),$$

since both sums give the number of pairs $(\varphi, a) \in \Phi \times X$ with the property $\varphi(a) = a$. Instead of the identity (72) it therefore suffices to show that

$$|X/\!\sim| = \frac{1}{|\Phi|} \sum_{a \in X} u(a).$$

This clearly holds if for each class $T \in X/\!\sim$, $T = \{a_1, \ldots, a_s\}$, we have the identity

$$\sum_{a \in T} u(a) = \sum_{j=1}^{s} u(a_j) = |\Phi|.$$

We use an interesting argument to show that the last equality holds: We will show that all s numbers $u(a_1), \ldots, u(a_s)$ are equal to $\frac{|\Phi|}{s}$. By symmetry of the elements $a_1, \ldots, a_s$ it suffices to prove $u(a_1) = \frac{|\Phi|}{s}$. We consider the sets

$$M_i = \{\varphi \in \Phi; \ \varphi(a_1) = a_i\}$$

for $i = 1, \ldots, s$. Since for any $i \in \{1, \ldots, s\}$ we have $a_1 \sim a_i$ (both elements belong to the same class T), then $M_i \neq \emptyset$. Conversely, for any $\varphi \in \Phi$ we have $a_1 \sim \varphi(a_1)$; thus $\varphi(a_1) \in T = \{a_1, \ldots, a_s\}$, and therefore $\varphi \in M_j$ for an appropriate $j \in \{1, \ldots, s\}$. If we could show that all sets $M_1, \ldots, M_s$ have the same number of elements, then we would be finished, since then $u(a_1) = |M_1| = \frac{|\Phi|}{s}$. Hence it is enough if we show that for any $i = 1, \ldots, s$ there is a bijection $f : M_1 \rightarrow M_i$. We choose an arbitrary but fixed $\psi \in M_i$, and for $\varphi \in M_1$ we set $f(\varphi) = \psi \circ \varphi$. Since $\psi \circ \varphi \in \Phi$ and $(\psi \circ \varphi)(a_1) = \psi(\varphi(a_1)) = \psi(a_1) = a_i$, we have $f(\varphi) \in M_i$, and therefore the mapping f is well-defined. Next, f is injective, since if $f(\varphi) = f(\varphi')$ for some $\varphi, \varphi' \in M_1$, then this gives $\psi \circ \varphi = \psi \circ \varphi'$, which, in view of the fact that ψ is a bijection, implies $\varphi = \varphi'$. Finally, it is easy to show that f is surjective: If we choose an arbitrary $\rho \in M_i$, then $\psi^{-1} \circ \rho \in \Phi$ and $(\psi^{-1} \circ \rho)(a_1) = \psi^{-1}(\rho(a_1)) = \psi^{-1}(a_i) = a_1$, since $\psi(a_1) = a_i$. Therefore, $\psi^{-1} \circ \rho \in M_1$ and $f(\psi^{-1} \circ \rho) = \psi \circ (\psi^{-1} \circ \rho) = (\psi \circ \psi^{-1}) \circ \rho = \rho$. Hence f is a bijection, which remained to be shown. □

7.3 Examples

(i) We first consider the problem from the beginning of this section again. Let X denote the set of all colorings with two colors of the faces of an immovable cube, and Φ the set of all bijections on X induced by the rotations of the cube. Since the cube can be rested on any one of six faces, and can then be rotated in four ways such that one of the neighboring faces points forwards, the set Φ has 24 elements. We now determine the value $v(\varphi)$ for each $\varphi \in \Phi$. If you have a cube or a die at hand, you can use it to follow our arguments.

(a) If $\varphi = \mathrm{id}$, then $\varphi(a) = a$ for all $a \in X$, and therefore $v(\varphi) = 2^6 = 64$.

(b) Let φ be a nonidentical rotation (i.e., $\varphi \neq \mathrm{id}$) about an axis going through opposite vertices (there are four pairs of opposite vertices, and for each pair there are two nonidentical rotations, namely, by $120°$ and by $240°$, for a total of 8 such rotations). Then φ leaves exactly those colorings unchanged in which each triple of faces that lie "around" the two vertices have the same color. Hence $v(\varphi) = 2^2 = 4$.

(c) Let φ be a nonidentical rotation about an axis going through the centers of opposite edges (there are 6 such pairs of edges, and for each one there is a unique rotation by $180°$). Then $\varphi(a) = a$ holds for exactly those colorings a in which the two pairs of faces adjacent to the chosen edges have the same color each, and the remaining two faces also have the same color. For such a φ we therefore have $v(\varphi) = 2^3 = 8$.

(d) Let φ be a nonidentical rotation about an axis going through the centers of opposite faces (there are three such pairs of faces, and for each one of them we have three nonidentical rotations, namely, by $90°$, $180°$, and $270°$). We have to distinguish two cases. In the case of a rotation by $90°$ or $270°$ we have $\varphi(a) = a$ if and only if a is a coloring in which all four faces that are not perpendicular to the axis have the same color. For such a φ (of which there are six) we have $v(\varphi) = 2^3 = 8$. In the case of a rotation by $180°$ (of which there are three) $\varphi(a) = a$ holds exactly for those colorings in which the two pairs of opposite faces that are not perpendicular to the axis have the same color each. Thus we have $v(\varphi) = 2^4 = 16$.

In total we have described $1 + 8 + 6 + 9 = 24$ different rotations, and thus all elements of Φ. By Burnside's lemma the desired number of different colorings is therefore

$$\frac{1}{24}(64 + 8 \cdot 4 + 6 \cdot 8 + 6 \cdot 8 + 3 \cdot 16) = 10.$$

We remark in closing that this problem can be also solved also "by hand," by analyzing the different possibilities, ordered according to the number of faces colored white. The reader should verify by this method that the number of different colorings of a cube in two colors is indeed equal to 10.

(ii) A child's bracelet is to have five white, five red, and five black beads. How many different bracelets can be strung together from these beads?

SOLUTION.   The problem can be reformulated as follows: In a regular poly-gon with 15 sides we have to mark 5 corners white, 5 corners red, and 5 corners black, where two markings of the 15-gon are considered identical if one can be obtained from the other by rotation or by flipping the 15-gon about some axis. Apart from the identity we have therefore 14 nonidentical rotations and 15 flips about an axis that goes through the center and one corner of the 15-gon. In order to use Burnside's lemma we have to find how many markings each of the rotations or flips leaves unchanged. A flip leaves one corner in place and switches the locations of seven pairs of corners. Therefore, if the flip is not to change the markings, then these seven pairs must be marked with the same color each, which, however, is impossible in our case. With a similar argument we can verify for nonidentical rotations that markings that are left unchanged under such rotations can exist only for rotations by 72°, 144°, 216°, and 288°. For each of these rotations there exist exactly six markings (each three consecutive corners are marked with a different color). Therefore, it follows from Burnside's lemma that the desired number of all bracelets is

$$\frac{1}{30}\left(\frac{15!}{5!\cdot 5!\cdot 5!}+4\cdot 6\right)=25\,226. \qquad\qquad \square$$

7.4 Exercises

(i) Find the number of ways in which the faces of a cube can be colored with three colors, where two colorings are considered identical if one can be obtained from the other by rotating the cube.

(ii) If we draw one of the diagonals on each of the faces of a cube, how many different cubes can we obtain?

(iii) If we draw on each face of a cube an arrow pointing to one of the corners of this face, how many different cubes can we obtain?

(iv) How does the result of the previous exercise change if we are allowed not to draw arrows on an arbitrary number of faces?

(v) In how many ways can we color the faces of a cube if two have to be white, two black, and two red?

(vi) In how many ways can one color the edges of a cube with two colors?

(vii) In how many ways can one color the corners of a cube with three colors?

*(viii) In how many ways can one color the faces of a dodecahedron (a regular polyhedron with 12 faces) with two colors?

*(ix) In how many ways can one color the edges of a dodecahedron with three colors?

*(x) In how many ways can one color the corners of a dodecahedron if half of them are to be white and half of them black?

*(xi) How many bracelets can be made with three white, three black, and three red beads if no two beads of the same color are to be next to each other?

7.5 Problems on Integers

We now return to some of the problems we have already dealt with in Section 3.5.

(i) Given $n \in \mathbb{N}$, find the number of all solutions in $\mathbb{N}_0$ of the equation

$$x_1 + x_2 + x_3 + x_4 = n$$

under the condition $x_1 \le x_2 \le x_3 \le x_4$.

SOLUTION. We are in fact dealing with the number of decompositions of the given integer n into a sum of four summands, where the order is irrelevant. We will use 7.2 to solve this problem. Let

$$X = \{(x_1, x_2, x_3, x_4); \ x_1, x_2, x_3, x_4 \in \mathbb{N}_0, \ x_1 + x_2 + x_3 + x_4 = n\},$$

and let Φ be the set of bijections on X induced by the individual permutations of the index set $\{1, 2, 3, 4\}$. If $\varphi \in \Phi$ is induced by the identity, then $v(\varphi) = |X| = \binom{n+3}{3}$ by 2.18.(iv). If $\varphi \in \Phi$ is induced by the interchange of one pair of indices (there are $\binom{4}{2} = 6$ such φ), then $v(\varphi)$ is equal to the number of solutions of the equation $2x + y + z = n$ in $\mathbb{N}_0$, and by 3.6.(xvi) we have $v(\varphi) = ([\frac{n}{2}]+1)([\frac{n+1}{2}]+1)$. If $\varphi \in \Phi$ is induced by the interchange of two pairs of indices (there are $\frac{1}{2}\binom{4}{2} = 3$ such φ), then $v(\varphi)$ is equal to the number of solutions of the equation $2x + 2y = n$ in $\mathbb{N}_0$; for odd n this equation has no solutions, while for even n there are $\frac{n}{2} + 1$ solutions, and thus in both cases $v(\varphi) = ([\frac{n}{2}] - [\frac{n-1}{2}])([\frac{n}{2}] + 1)$ solutions. If $\varphi \in \Phi$ is induced by a permutation that leaves exactly one index unchanged (there are $4 \cdot 2 = 8$ such φ), then $v(\varphi)$ is equal to the number of solutions of the equation $3x + y = n$ in $\mathbb{N}_0$, that is, $v(\varphi) = [\frac{n}{3}] + 1$. Finally, if $\varphi \in \Phi$ is induced by a cyclic permutation of all four indices (of which there are $\frac{4!}{4} = 6$), then $v(\varphi)$ is equal to the number of solutions of the equation $4x = n$ in $\mathbb{N}_0$. There is a unique solution when $4 \mid n$, and no solution when $4 \nmid n$, so that $v(\varphi) = [\frac{n}{4}] - [\frac{n-1}{4}]$. In this way we have exhausted all 24 permutations of a 4-element set, and by Burnside's lemma the desired number of solutions is

$$\frac{1}{24}\left(\binom{n+3}{3} + 6 \cdot \left(\left[\frac{n}{2}\right]+1\right)\left(\left[\frac{n+1}{2}\right]+1\right)\right.$$
$$+ 3 \cdot \left(\left[\frac{n}{2}\right] - \left[\frac{n-1}{2}\right]\right)\left(\left[\frac{n}{2}\right]+1\right)$$
$$\left. +8 \cdot \left(\left[\frac{n}{3}\right]+1\right) + 6 \cdot \left(\left[\frac{n}{4}\right] - \left[\frac{n-1}{4}\right]\right)\right). \qquad \square$$

(ii) Given $n \in \mathbb{N}$, find the number of all solutions of the equation

$$x_1 \cdot x_2 \cdot x_3 \cdot x_4 = n \qquad (73)$$

in $\mathbb{N}$ under the condition that $x_1 \leq x_2 \leq x_3 \leq x_4$.

SOLUTION.   We are in fact dealing with the problem of finding the number of ways in which the integer n can be written as a product of four factors if their order is irrelevant (see also 3.5.(vii)). We write n in its prime decomposition: $n = p_1^{\omega_1} \cdot p_2^{\omega_2} \cdots p_k^{\omega_k}$, where $p_1, p_2, \ldots, p_k$ are distinct primes, and $\omega_i \in \mathbb{N}$ for $i = 1, 2, \ldots, k$. Each of the integers $x_1, \ldots, x_4$ is a divisor of n, which means that

$$x_i = p_1^{\alpha_{i1}} \cdot p_2^{\alpha_{i2}} \cdots p_k^{\alpha_{ik}}, \qquad i = 1, 2, 3, 4.$$

The set

$$X = \{(x_1, x_2, x_3, x_4) : \ x_1, x_2, x_3, x_4 \in \mathbb{N}_0, x_1 \cdot x_2 \cdot x_3 \cdot x_4 = n\}$$

has therefore exactly as many elements as there are solutions of the system of k equations

$$\alpha_{11} + \alpha_{21} + \alpha_{31} + \alpha_{41} = \omega_1, \qquad \ldots, \qquad \alpha_{1k} + \alpha_{2k} + \alpha_{3k} + \alpha_{4k} = \omega_k,$$

in the $4k$ unknowns $\alpha_{ik} \in \mathbb{N}_0$. By the multiplication rule we therefore have $|X| = \prod_{j=1}^{k} \binom{\omega_j + 3}{3}$. Once again we let Φ be the set of all bijections on X induced by the individual permutations of the index set $\{1, 2, 3, 4\}$, and we use the approach of (i). For the permutation $\varphi \in \Phi$ induced by the identity we have $v(\varphi) = |X| = \prod_{j=1}^{k} \binom{\omega_j + 3}{3}$. If $\varphi \in \Phi$ is induced by the interchange of one pair of indices, then $v(\varphi) = \prod_{j=1}^{k}([\frac{\omega_j}{2}] + 1)([\frac{\omega_j + 1}{2}] + 1)$. If $\varphi \in \Phi$ is induced by the interchange of two pairs of indices, then we get $v(\varphi) = \prod_{j=1}^{k}([\frac{\omega_j}{2}] - [\frac{\omega_j - 1}{2}])([\frac{\omega_j}{2}] + 1)$. If $\varphi \in \Phi$ is induced by a permutation that leaves exactly one index unchanged, then $v(\varphi) = \prod_{j=1}^{k}([\frac{\omega_j}{3}] + 1)$. Finally, if $\varphi \in \Phi$ is induced by a cyclic permutation of all four indices, then $v(\varphi) = \prod_{j=1}^{k}([\frac{\omega_j}{4}] - [\frac{\omega_j - 1}{4}])$. By Burnside's lemma the desired number of solutions of equation (73) is therefore

$$\frac{1}{24}\left(\prod_{j=1}^{k} \binom{\omega_j + 3}{3} + 6 \cdot \prod_{j=1}^{k}\left(\left[\frac{\omega_j}{2}\right] + 1\right)\left(\left[\frac{\omega_j + 1}{2}\right] + 1\right) \right.$$

$$+ 3 \cdot \prod_{j=1}^{k}\left(\left[\frac{\omega_j}{2}\right] - \left[\frac{\omega_j - 1}{2}\right]\right)\left(\left[\frac{\omega_j}{2}\right] + 1\right)$$

$$\left. + 8 \cdot \prod_{j=1}^{k}\left(\left[\frac{\omega_j}{3}\right] + 1\right) + 6 \cdot \prod_{j=1}^{k}\left(\left[\frac{\omega_j}{4}\right] - \left[\frac{\omega_j - 1}{4}\right]\right)\right). \quad \square$$

7.6   Exercises

For each equation, with a fixed $n \in \mathbb{N}$, find the number of all solutions under the given conditions:

(i) $x_1 \cdot x_2 \cdot x_3 = n$ in $\mathbb{N}$, $x_1 \leq x_2 \leq x_3$.

(ii) $x_1 \cdot x_2 \cdot x_3 \cdot x_4 = n$ in $\mathbb{N}$, $x_1 \mid x_2$, $x_2 \mid x_3$, and $x_3 \mid x_4$.

*(iii) $x_1 + x_2 + x_3 + x_4 + x_5 = n$ in $\mathbb{N}_0$, $x_1 \leq x_2 \leq x_3 \leq x_4 \leq x_5$.

*(iv) $x_1 \cdot x_2 \cdot x_3 \cdot x_4 \cdot x_5 = n$ in $\mathbb{N}$, $x_1 \mid x_2$, $x_2 \mid x_3$, $x_3 \mid x_4$ a $x_4 \mid x_5$.

*(v) $x_1 \cdot x_2 \cdot x_3 \cdot x_4 \cdot x_5 = n$ in $\mathbb{N}$, $x_1 \leq x_2 \leq x_3 \leq x_4 \leq x_5$.

8   Recursive Methods

Let us return to the problem of finding the number d_n of permutations without fixed points of an n-element set. In Section 6.11 we derived for this number the formula (61), namely,

$$d_n = n! \cdot \sum_{r=0}^{n} \frac{(-1)^r}{r!}.$$

At the end of that section we derived an identity that can easily be brought into the form

$$d_n = n! - \sum_{s=0}^{n-1} \binom{n}{s} d_s,$$

and with this last identity, starting with $d_0 = 1$, we were able to consecutively find the values d_1, d_2, ... (see Exercise 6.12.(i)). This step-by-step evaluation is called *recursive* (from Latin *recurrere: to return*). For the sequence of numbers d_n there are, however, other recurrence relations, for example

$$d_n = n! \cdot \left(\frac{(-1)^n}{n!} + \sum_{r=0}^{n-1} \frac{(-1)^r}{r!} \right) = (-1)^n + n \cdot d_{n-1},$$

which is more suitable for the evaluation of the numbers d_n; its use is even easier than the evaluation by way of the explicit formula (61). Furthermore, recursive methods have often proven useful in problems to which explicit solutions cannot be found. Sometimes, an explicit formula is then derived from the recurrence relation; in this way we derived, for instance, the identity (3) in Section 2.1 for the number of all arrangements of an n element set. However, there is no general method for obtaining explicit formulas from recurrence relations; linear recurrence relations with constant coefficients are the only exceptions to this. We will not deal with this topic in the present book; the interested reader may consult, for instance, [2] or [9].

8.1 Examples

We now introduce a number of counting problems in which recurrence relations can be found.

(i) In how many ways can one walk up a flight of stairs with ten steps if one, two, or three steps can be taken at a time?

SOLUTION. Let s_n be the number of possibilities for n steps. Clearly, $s_1 = 1$, $s_2 = 2$, $s_3 = 4$. For $n > 3$ we have the recurrence relation

$$s_n = s_{n-1} + s_{n-2} + s_{n-3},$$

with the help of which we easily obtain $s_{10} = 274$.   □

(ii) Show that the number $S(n, k)$ of all distributions of n different objects over k indistinguishable bins, where no bin remains empty (see 6.7.(iii)), satisfies the recurrence relation

$$S(n + 1, k) = S(n, k - 1) + k \cdot S(n, k) \tag{74}$$

for all $1 < k < n$. This, together with the obvious identities $S(n, 1) = S(n, n) = 1$, can be used for a step-by-step evaluation of the desired numbers $S(n, k)$.

SOLUTION. We divide all $S(n + 1, k)$ distributions of $(n + 1)$ different objects over k indistinguishable bins into two classes T_1 and T_2 according to whether the $(n + 1)$th object is alone in its bin or whether it shares a bin with other objects. In the first case, if we remove the $(n + 1)$th object together with its bin, we are left with an arbitrary distribution of n different objects over $k - 1$ indistinguishable bins, and thus $|T_1| = S(n, k - 1)$. In the second case, if we remove the $(n + 1)$th object from its bin, we are left with an arbitrary distribution of n different objects over k indistinguishable bins. However, each such distribution corresponds to k different elements of the class T_2 that differ only in the location of the $(n+1)$th object. Hence $|T_2| = k \cdot S(n, k)$, and the addition rule implies (74).   □

(iii) Find the number of solutions of the equation

$$x_1 + x_2 + \cdots + x_k = n \tag{75}$$

in $\mathbb{N}_0$ under the condition that n is fixed and $x_1 \le x_2 \le \cdots \le x_k$.

SOLUTION. This is a problem we have already solved for $k = 2, 3, 4, 5$ (see 3.5.(iii),(iv), 7.5.(i) and 7.6(iii)). Let $M_{n,k}$ denote the set of solutions in question, and $p(n, k)$ the desired number of its elements. We easily see that

$$p(0, k) = 1 \text{ for all } k, \ p(n, 1) = 1, \text{ and } p(n, k) = p(n, n) \text{ for all } k \ge n. \tag{76}$$

Therefore, we assume further that for fixed n, k we have $1 < k \leq n$. We subdivide the set $M_{n,k}$ into classes T_i $(i = 0, 1, \ldots, k-1)$ such that T_i contains exactly those solutions of (75) in $\mathbb{N}_0$ that satisfy the condition

$$0 = x_1 = x_2 = \cdots = x_i < x_{i+1} \leq x_{i+2} \leq \cdots \leq x_k.$$

Then the assignment $(x_1, x_2, \ldots, x_k) \mapsto (x_{i+1} - 1, x_{i+2} - 1, \ldots, x_k - 1)$ defines a bijection from T_i to $M_{n-k+i, k-i}$, since $0 \leq x_{i+1} - 1 \leq x_{i+2} - 1 \leq \cdots \leq x_k - 1$, $(x_{i+1} - 1) + (x_{i+2} - 1) + \cdots + (x_k - 1) = n - k + i$, and the inverse function is determined by the assignment

$$(y_1, y_2, \ldots, y_{k-i}) \mapsto (\underbrace{0, 0, \ldots, 0}_{i}, y_1 + 1, y_2 + 1, \ldots, y_{k-i} + 1).$$

This implies that $|T_i| = |M_{n-k+i, k-i}|$, and thus

$$|M_{n,k}| = \sum_{i=0}^{k-1} |T_i| = \sum_{i=0}^{k-1} |M_{n-k+i, k-i}|,$$

which can be written as the recurrence relation

$$p(n, k) = p(n-1, 1) + p(n-2, 2) + \cdots + p(n-k, k)$$

$(1 < k \leq n)$, and together with (76) this enables us to compute, step by step, the desired numbers $p(n, k)$. $\square$

(iv) Show that the number p_n of all sequences of length n consisting of the integers 0, 1 such that no more than two identical integers are adjacent to each other can be determined by the equations $p_1 = 2$, $p_2 = 4$, and $p_{k+2} = p_{k+1} + p_k$ for $k \geq 1$.

SOLUTION.   The identities $p_1 = 2$ and $p_2 = 4$ are clear, since the boundary conditions do not apply to sequences of length 1 and 2. For fixed $k \geq 1$ we subdivide all admissible sequences $x_1, x_2, \ldots, x_{k+2}$ of length $k+2$ into four groups according to the final pair of elements x_{k+1}, x_{k+2}:

(a) If $(x_{k+1}, x_{k+2}) = (0, 0)$, then necessarily $x_k = 1$, and so $x_1, x_2, \ldots, x_k$ is some admissible sequence of length k ending in a 1.
(b) If $(x_{k+1}, x_{k+2}) = (1, 1)$, then necessarily $x_k = 0$, and so $x_1, x_2, \ldots, x_k$ is some admissible sequence of length k ending in a 0.
(c) If $(x_{k+1}, x_{k+2}) = (0, 1)$, then $x_1, x_2, \ldots, x_{k+1}$ is some admissible sequence of length $k+1$ ending in a 0.
(d) If $(x_{k+1}, x_{k+2}) = (1, 0)$, then $x_1, x_2, \ldots, x_{k+1}$ is some admissible sequence of length $k+1$ ending in a 1.

The first two groups together contain exactly p_k sequences, since each admissible sequence of length k occurs exactly once in either (a) or (b). Similarly, by considering all admissible sequences $x_1, x_2, \ldots, x_{k+1}$ we see that the last two groups together contain exactly p_{k+1} sequences. This proves the relation $p_{k+2} = p_{k+1} + p_k$. $\square$

8.2   Exercises

(i) If d_n is the number of all permutations without fixed points of an n-element set, derive the recurrence relation $d_{n+1} = n(d_n + d_{n-1})$
(a) directly from (61) in Section 6.11.
(b) by a combinatorial argument.

(ii) Show that the number $v_0(k, n)$ of all k-element variations from elements of n different types, such that each variation contains at least one element of each type (see 6.5.(ii)), satisfies the recurrence relation

$$v_0(k + 1, n + 1) = (n + 1) \cdot \big(v_0(k, n) + v_0(k, n + 1)\big),$$

which together with the initial values $v_0(k, 1) = 1$ and $v_0(k, n) = 0$ for $k < n$ allows for the step-by-step evaluation of the $v_0(k, n)$.

*(iii) Let c_n be the number of arrangements of an n-element set $A = \{a_1, \ldots, a_n\}$ such that for no $i \in \{1, 2, \ldots, n - 1\}$ does the element a_{i+1} immediately follow the element a_i (see 6.5.(iii)). Show that c_n satisfies the recurrence relation $c_n = (n-1)c_{n-1} + (n-2)c_{n-2}$, which together with the initial conditions $c_1 = 1$ and $c_2 = 1$ allows for the step-by-step evaluation of the numbers c_n.

(iv) Let f_k be the number of all sequences of k ones and k fives in which for each $i = 1, 2, \ldots, 2k$ there are at least as many fives as there are ones among the first i terms of the sequence (see Section 3.7). Show that f_k satisfies the recurrence relation

$$f_k = \sum_{i=1}^{k} f_{i-1} \cdot f_{k-i},$$

which together with the initial value $f_0 = 1$ allows for the step-by-step evaluation of the numbers f_k.

(v) Let two positive integers d and s be given, and let p_n be the number of all sequences of length n that consist of the integers $0, 1, \ldots, d$ such that no sequence has more than s consecutive identical terms. Find equations with which p_n can be determined recursively.

2

Combinatorial Arithmetic

This chapter is devoted to problems on *numerical configurations*. The rich structure associated with numbers (arithmetical operations, order relations, metrics) gives rise to a great variety and diversity of such problems, which makes it difficult to attempt a methodical or an algorithmic approach to their solutions. This is all the more so, since the individual problems are relatively independent "research problems," widely varying in their degrees of difficulty. They range from simple to very difficult, where the solutions require a considerable amount of creative energy, resourcefulness, and also (in the case of initial failure) patience.

The text of this chapter is structured thematically: The problems are divided into sections according to a particular type of configuration. We therefore mention in this brief introduction only some general methods for solving such problems, and describe more specific methods as the need arises throughout the text.

In a number of situations the well-known *principle of mathematical induction* will be useful; it will not be necessary to describe this principle here. Just as often we will use the *pigeonhole principle* (or *Dirichlet's principle*; see Chapter 1, Section 1.7). With the *method of construction* we will mainly prove statements concerning the existence of a configuration $\mathcal{K}$ with given properties; appropriate examples will be explained (and constructed). Since we will often be dealing with configurations $\mathcal{K}_n$ with parameter n (determined by their "size"), we distinguish between *direct* constructions ($\mathcal{K}_n$ given explicitly for general n) and *recursive* constructions ($\mathcal{K}_n$ written in terms of one or several of the preceding configurations $\mathcal{K}_{n-1}, \mathcal{K}_{n-2}, \dots$). In general, the following approach for finding a construction of $\mathcal{K}_n$ is advisable: Find examples $\mathcal{K}_i$ for small values i, and then try to generalize. In practice, this is often done by discovering "common characteristics" of the examples $\mathcal{K}_i$. This approach, however, is not without its problems: If $\mathcal{K}_i$ is not unique (for a given i), an inappropriate choice of the examples $\mathcal{K}_i$ may make a generalization impossible. We therefore add another recommendation: Select those examples that exhibit some kind of "regularity," such as some *symmetry*.

In this connection we note that the construction of examples is an essential part also of *extremal problems*, where, given an estimate $f(\mathcal{K}) \leq c$ (for all configurations of a given type) we want to prove that the value c is *attained*, that is, we want to prove the existence of a configuration $\mathcal{K}_0$ with the property $f(\mathcal{K}_0) = c$. Here the approach is often reversed: First we "guess" the form of $\mathcal{K}_0$, and then we try to justify the inequality $f(\mathcal{K}) \leq f(\mathcal{K}_0)$ for general $\mathcal{K}$.

We conclude this brief overview of some general methods by mentioning a trick that often provides a surprising, but at the same time decisive, first step toward a successful solution. We call this trick the *method of extremes*, since in concrete examples it works, for instance, as in the following situations: In an appropriately chosen set of numbers we select the *smallest* element; in a given sequence we find the term with *largest* index satisfying appropriate conditions; in a given table we determine the row with the *smallest* number of negative entries; from an appropriate set of tables or arrays we choose the one with the *largest* sum of elements.

To facilitate the preparation of classes or seminars that are more method-oriented, we now give an incomplete list of problems arranged according to methods of solution.

Induction principle: 1.3.(i), 1.4.(viii), 1.5.(ii), 2.1.(vi), 2.5.(ii), 2.6.(i), 2.11.(iv), 2.12.(iv), 3.10.(iv), 3.10.(v), 3.11.(ii), 3.12.(ii), 3.12.(iv), 3.13.(iii), 4.1.(vi), 4.6.(i), 4.6.(v), 4.11.(ii), 4.13.(vi), 5.1.(iii), 5.2.(iii), 5.4.(xi).

Pigeonhole principle: 2.3.(i), 2.5.(i), 2.6.(vii), 2.6.(viii), 2.7.(ii), 2.8.(ii), 2.8.(vi), 2.10.(xiv), 3.1.(i), 3.2.(i), 3.6.(i), 3.8.(i), 3.8.(iv), 4.1.(i), 4.1.(ii), 4.1.(vi), 4.2.(i), 4.2.(vii), 4.4.(v), 4.5.(vi), 4.6.(x), 4.6.(xix), 4.14.(viii), 4.14.(xi).

Direct construction: 1.1.(iii), 1.2.(ii), 1.2.(v), 1.8.(ii), 2.12.(ix), 3.1.(iv), 3.1.(v) (2nd solution), 3.2.(ii), 3.2.(v), 3.2.(viii), 3.4.(i), 3.4.(ii), 3.4.(v), 3.8.(ii), 3.8.(v), 3.10.(i), 3.10.(v), 3.13.(iii), 4.1.(iv), 4.1.(v), 4.6.(iv), 4.6.(xiv), 4.6.(xvi), 4.6.(xx), 4.12.(iv), 4.12.(vi), 4.14.(vii), 4.14.(x), 5.7.(iv).

Recursive construction: 1.1.(ii), 2.1.(iii), 2.2.(v), 2.4.(v), 2.6.(vii), 2.7.(iv), 2.10.(xv), 3.1.(iii), 3.1.(v) (1st solution), 3.9.(iii), 3.12.(i), 4.2.(v), 4.6.(xi), 4.6.(xv), 4.6.(xix), 4.8.(ii), 4.8.(iii), 4.8.(v), 4.8.(vii), 4.11.(i), 4.12.(ii), 4.12.(xi).

Method of extremes: 1.3.(v), 1.4.(v), 2.1.(vi), 2.2.(iii), 2.4.(iv), 2.6.(iv), 2.7.(ii), 2.7.(iii), 2.8.(iv), 3.2.(vi), 3.5.(iv), 3.7.(ii), 3.7.(iv), 3.9.(ii), 3.11.(iii),

3.13.(i), 4.5.(v), 4.6.(ix), 4.6.(x), 4.10.(xi), 4.11.(iii), 4.12.(xii), 4.12.(xiii), 4.12.(xiv), 4.14.(v), 5.2.(i).

1   Arrangements

We have dealt with the term *arrangement* of a given set or collection in detail in Chapter 1, Section 2.1; we therefore proceed directly to the solution of problems concerning sequences of numbers.

1.1   Divisibility Problems

(i) Show that for no composite integer $n > 4$ does there exist an arrangement $a_1, a_2, \ldots, a_n$ of the numbers $1, 2, \ldots, n$ such that the numbers

$$a_1, \quad a_1 a_2, \quad a_1 a_2 a_3, \quad \ldots, \quad a_1 a_2 \cdots a_n$$

have distinct remainders upon division by n. Is this also true for $n = 4$?

SOLUTION.   We assume that such an arrangement exists for some $n > 4$. Then only one of the numbers $a_1, \ a_1 a_2, \ \ldots, \ a_1 a_2 \cdots a_n$ is divisible by n; clearly, this number is the last one (equal to $n!$). Therefore, we have $a_n = n$, and thus $a_1 a_2 \cdots a_{n-1} = (n-1)!$. However, for composite $n > 4$ the number $(n-1)!$ is divisible by n (see [5], Chapter 3, Section 2.8.(iii)), and this is a contradiction. For $n = 4$, however, such an arrangement does exist, namely, 1, 3, 2, 4; the numbers $1, 1 \cdot 3, 1 \cdot 3 \cdot 2$, and $1 \cdot 3 \cdot 2 \cdot 4$ give, upon division by 4, the remainders 1, 3, 2, 0, respectively. (For noncomposite n see [7], Problem 4.17.)   □

(ii) Show that for each $n \geq 2$ one can choose n integers $1 < a_1 < a_2 < \cdots < a_n$ with the following property: If $b_1, b_2, \ldots, b_n$ is any arrangement of the numbers $a_1, a_2, \ldots, a_n$, then $1 + b_2 b_3 \cdots b_n$ is an integer multiple of b_1.

SOLUTION.   Let us first analyze the case $n = 2$. If $1 < a_1 < a_2$ are any integers with the required property, then from $a_2 \mid 1 + a_1$ it follows that $a_2 \leq 1 + a_1$, and thus $1 + a_1 < 1 + a_2 \leq 2 + a_1$. Since $1 + a_1, 2 + a_1$ are two consecutive integers, the last inequality implies that $1 + a_2 = 2 + a_1$, that is, $a_2 = 1 + a_1$. Then the condition $a_1 \mid 1 + a_2$ gives $a_1 \mid 2 + a_1$, that is, $a_1 \mid 2$, which is possible only when $a_1 = 2$. For $n = 2$ we therefore have the unique solution $(a_1, a_2) = (2, 3)$.

We now try to proceed to the case $n = 3$ by adding an appropriate number $a_3 > 3$ to the pair $(2, 3)$. The condition $a_3 \mid 1 + 2 \cdot 3$ immediately gives $a_3 = 7$. Since $2 \mid 1 + 3 \cdot 7$ and $3 \mid 1 + 2 \cdot 7$, our attempt was therefore successful. (Note that in the statement of the problem the order of the integers $b_2, b_3, \ldots, b_n$ does not matter, and thus for each a_i $(1 \leq i \leq n)$ we verify the unique condition $a_i \mid 1 + \frac{s}{a_i}$, where $s = a_1 a_2 \cdots a_n$.)

Was our success in the case $n = 3$ a matter of luck, or is there a deeper reason behind it? Let us therefore ask whether the following conclusion is true in general: If some numbers $a_1, a_2, \ldots, a_k$ satisfy the conditions of the problem for $n = k$ and if we set $a_{k+1} = 1 + a_1 a_2 \cdots a_k$, then the numbers $a_1, a_2, \ldots, a_{k+1}$ satisfy the conditions for $n = k + 1$. First of all, it is clear from the choice of a_{k+1} that $a_{k+1} > a_k$, and moreover, the condition $a_{k+1} \mid 1 + a_1 a_2 \ldots a_k$ holds trivially. Therefore, it suffices to establish whether for each $i = 1, 2, \ldots, k$ we have $a_i \mid 1 + s_i a_{k+1}$, where $s_i = a_1 a_2 \cdots a_k / a_i$. From the identities

$$1 + s_i a_{k+1} = 1 + s_i(1 + a_i s_i) = (1 + s_i) + a_i s_i^2$$

it follows that $a_i \mid 1 + s_i a_{k+1}$ holds if and only if $a_i \mid 1 + s_i$, but this is guaranteed by the assumption for each $i = 1, 2, \ldots, k$. We have thus shown the possibility of choosing an appropriate n-tuple $a_1, a_2, \ldots, a_n$ for each $n \geq 2$, and furthermore we have described the recursive construction of examples; for the first few values of n we obtain $(2, 3)$, $(2, 3, 7)$, $(2, 3, 7, 43)$, $(2, 3, 7, 43, 1807)$. $\qquad\square$

(iii) Show that for each $n > 1$ there exists an arrangement $a_1, a_2, \ldots, a_n$ of the integers $1, 2, \ldots, n$ such that a_{j+1} divides the sum $a_1 + a_2 + \cdots + a_j$ for all $j = 1, 2, \ldots, n - 1$.

SOLUTION.   The proof of this assertion would be easy if for each $k > 1$ the number k divided $1 + 2 + \cdots + (k - 1)$, for then it would be enough to set $a_j = j$ for all j. However, the easy calculation

$$\frac{1 + 2 + \cdots + (k - 1)}{k} = \frac{(k - 1)k}{2k} = \frac{k - 1}{2}$$

shows that only odd k have this property. But this information is still useful; it shows that if the assertion holds for some even $n > 1$, then it also holds for $n + 1$: It suffices to adjoin the term $a_{n+1} = n + 1$ to an appropriate arrangement $a_1, a_2, \ldots, a_n$ of the integers $1, 2, \ldots, n$. It is therefore enough to consider the case $n = 2k$. Then we have $1 + 2 + \cdots + n = k(2k + 1)$, and the condition $a_{2k} \mid a_1 + a_2 + \cdots + a_{2k-1}$ is equivalent to a_{2k} dividing the sum $a_1 + a_2 + \cdots + a_{2k} = k(2k + 1)$. We therefore try $a_{2k} = k$; then we would have $a_{2k-1} \mid k(2k + 1) - k = k \cdot 2k$, which suggests that we take $a_{2k-1} = 2k$. Continuing, we get, similarly, $a_{2k-2} = k - 1$, $a_{2k-3} = 2k - 1$, etc. Thus we arrive at the arrangement

$$k + 1, \;\; 1, \;\; k + 2, \;\; 2, \;\; \ldots, \;\; 2k, \;\; k.$$

We now verify that it actually satisfies the conditions of the problem. Indeed, for $1 \leq i \leq k$ we easily compute

$$(k + 1) + 1 + (k + 2) + 2 + \cdots + (k + i) = ik + \frac{i(i + 1)}{2} + \frac{i(i - 1)}{2} = i(k + i)$$

and

$$(k+1) + 1 + (k+2) + 2 + \cdots + (k+i) + i = i(k+i) + i = i(k+i+1).$$

However, the arrangement thus obtained is not the only one that satisfies the condition of the problem for $n = 2k$. The reader should verify that the required condition is also satisfied by the arrangement

$$2k, \ 2, \ k+1, \ 3, \ k+2, \ 4, \ \ldots, \ 2k-2, \ k, \ 2k-1, \ 1. \qquad \square$$

(iv) Suppose that an integer n has the following property: There is an arrangement of the sequence of $2n$ numbers 1, 1, 2, 2, $\ldots$, n, n such that for each $k = 1, 2, \ldots, n$ there are exactly k elements between the two numbers k. Show that $n^2 + n$ is divisible by 4.

SOLUTION.   Let $a_1, a_2, \ldots, a_{2n}$ be an arrangement with the given property. This means that for each $k = 1, 2, \ldots, n$ there exists an index i_k such that $a_{i_k} = a_{i_k+k+1} = k$. The sum of all indices $1 + 2 + \cdots + 2n = n(2n+1)$ is then equal to

$$n(2n+1) = \sum_{k=1}^{n}(i_k + i_k + k + 1) = 2\sum_{k=1}^{n} i_k + \frac{n(n+3)}{2},$$

which implies

$$2\sum_{k=1}^{n} i_k = n(2n+1) - \frac{n(n+3)}{2} = \frac{n(3n-1)}{2}.$$

Since the term on the left is even, the number $n(3n-1)$ is a multiple of 4. Therefore, the assertion of the problem is proven if we can show that the difference $n(3n-1) - (n^2+n)$ is divisible by 4. But this is clear, since this difference is equal to $2n(n-1)$, and $n(n-1)$ is always even. $\square$

1.2   *Exercises*

(i) Show that if n is odd and $a_1, a_2, \ldots, a_n$ is an arbitrary arrangement of $1, 2, \ldots, n$, then $(a_1 + 1)(a_2 + 2) \cdots (a_n + n)$ is even.

(ii) Show that for each even n one can arrange the integers $1, 2, \ldots, n-1$ in such a way that no sum of any number of consecutive elements is an integer multiple of n.

*(iii) For each $n \geq 1$ determine the number p_n that gives the greatest possible number of odd integers among the sums $s_1 = a_1$, $s_2 = a_1 + a_2$, $\ldots$, $s_n = a_1 + a_2 + \cdots + a_n$, where $a_1, a_2, \ldots, a_n$ is an arrangement of $1, 2, \ldots, n$.

*(iv) Returning to Problem 1.1.(ii), show that for a fixed n the n-tuple $a_1, a_2, \ldots, a_n$ is unique if and only if $n \in \{2, 3, 4\}$.

*(v) Show that if $n^2 + n$ is a multiple of 4, then the arrangement in Problem 1.1.(iv) actually exists.

1.3   Problems on Inequalities

(i) Let $x_1, x_2, \ldots, x_n$ and $y_1, y_2, \ldots, y_n$ be two arrangements of the same n-tuple of real numbers such that $x_1 < x_2 < \cdots < x_n$ and $x_1 + y_1 < x_2 + y_2 < \cdots < x_n + y_n$. Show that the arrangements must be identical; that is, $x_i = y_i$ for all $i = 1, 2, \ldots, n$.

SOLUTION.   Assume that $x_1 = y_1$ does not hold; that is, we have $x_1 = y_k$ for some $k > 1$. From the given inequalities it follows that $y_i < (x_k + y_k) - x_i = x_k + (y_k - x_i) = x_k + (x_1 - x_i) \leq x_k$ for all $i = 1, 2, \ldots, k - 1$, and thus $y_i < x_k$ ($1 \leq i \leq k$). This means that each of the k numbers $y_1, y_2, \ldots, y_k$ is equal to one of the $k - 1$ numbers $x_1, x_2, \ldots, x_{k-1}$, which is a contradiction, since $y_1, y_2, \ldots, y_k$ are distinct (as the first k elements of an arrangement of the distinct numbers $x_1 < x_2 < \cdots < x_n$). Therefore, we have $x_1 = y_1$. Now we drop x_1, y_1, and by repeating the argument for the arrangements $x_2, \ldots, x_n$ and $y_2, \ldots, y_n$ we obtain $x_2 = y_2$, etc., and finally $x_n = y_n$.   □

(ii) Suppose that $2n$ distinct real numbers $x_1, y_1, x_2, y_2, \ldots, x_n, y_n$ satisfy $x_k > y_k$ ($1 \leq k \leq n$). Show that if $u_1 < u_2 < \cdots < u_n$ and $v_1 < v_2 < \cdots < v_n$ are increasing arrangements of the n-tuple $x_1, x_2, \ldots, x_n$, respectively $y_1, y_2, \ldots, y_n$, then we also have $u_k > v_k$ ($1 \leq k \leq n$).

SOLUTION.   For a fixed $k \in \{1, 2, \ldots, n\}$ we prove the inequality $u_k > v_k$ by using the observation that the inequality $u_k > y_i$ holds for at least k different values of the index i, for then some of the values y_i would not lie in the $(k - 1)$-element set $\{v_1, v_2, \ldots, v_{k-1}\}$, and we would be done. Now, since $x_i > y_i$ for all i, we also have the inequality $u_k > y_i$ whenever $u_k \geq x_i$. But this last inequality holds for exactly k different values of the index i, given by $x_i = u_1$, $x_i = u_2$, $\ldots$, $x_i = u_k$.   □

(iii) Let $a_1, a_2, \ldots, a_{2n}$ be an arrangement of the numbers $1, 2, \ldots, 2n$ such that the two chains of inequalities

$$a_1 < a_3 < a_5 < \cdots < a_{2n-1}, \tag{1}$$
$$a_2 > a_4 > a_6 > \cdots > a_{2n}$$

hold. (How many such arrangements are there?) Determine the possible values that can be attained by the sum

$$S = |a_1 - a_2| + |a_3 - a_4| + |a_5 - a_6| + \cdots + |a_{2n-1} - a_{2n}|.$$

SOLUTION.   The number of arrangements with the property (1) is the same as the number of n-element subsets $\{a_1, a_3, \ldots, a_{2n-1}\}$ of the set $\{1, 2, \ldots, 2n\}$, namely, $\binom{2n}{n}$. We will now show that the inequalities (1) mean that for each $k = 1, 2, \ldots, n$ we have either $a_{2k} \leq n < a_{2k-1}$ or $a_{2k-1} \leq n < a_{2k}$. Indeed, if $a_{2k-1} > a_{2k}$, then a_{2k-1} is greater than each of the n numbers

$$a_1, \quad a_3, \quad \ldots, \quad a_{2k-3}, \quad a_{2k}, \quad a_{2k+2}, \quad \ldots, \quad a_{2n},$$

which leads to the inequality $a_{2k-1} > n$, while $a_{2k} \le n$ follows from the fact that a_{2k} is smaller than each of the n numbers

$$a_2, \ a_4, \ \ldots, \ a_{2k-2}, \ a_{2k-1}, \ a_{2k+1}, \ \ldots, \ a_{2n-1}.$$

In the opposite case $(a_{2k-1} < a_{2k})$ one similarly derives $a_{2k-1} \le n < a_{2k}$. If we remove the absolute value brackets in the sum S, we obtain

$$S = \pm(a_1 - a_2) \pm (a_3 - a_4) \pm \cdots \pm (a_{2n-1} - a_{2n})$$
$$= \pm 1 \pm 2 \pm 3 \pm \cdots \pm (2n),$$

with appropriate choices of signs (n-times "+" and n-times "−"). However, the inequalities $a_{2k} \le n < a_{2k-1}$ and $a_{2k-1} \le n < a_{2k}$ show that the "+" sign goes exactly with those numbers that are greater than n. Hence the value of S is, *independently* of the choice of arrangement, equal to

$$S = -1 - 2 - \cdots - n + (n+1) + (n+2) + \cdots + (n+n) = n^2. \quad \square$$

(iv) Determine the number of different arrangements $a_1, a_2, \ldots, a_{10}$ of the integers 1, 2, ..., 10 such that $a_i > a_{2i}$ $(1 \le i \le 5)$ and also $a_i > a_{2i+1}$ $(1 \le i \le 4)$.

SOLUTION.

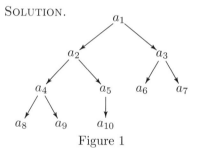

Figure 1

We express the inequalities of this problem by way of the diagram in Figure 1, where the arrows always point from the larger to the smaller number; thus $a_i \to a_j$ means that $a_i > a_j$. Since the relation ">" is transitive ($a > b$ and $b > c$ imply $a > c$), it follows from Figure 1 that a_1 is the largest number; that is, $a_1 = 10$. We now subdivide the remaining numbers $1, 2, \ldots, 9$ into subsets $\{a_2, a_4, a_5, a_8, a_9, a_{10}\}$ and $\{a_3, a_6, a_7\}$ in exactly $\binom{9}{3}$ ways. For each such subdivision the numbers a_2, a_3 are uniquely determined (as largest elements in their subsets). For the choice of a_6, a_7 we then have two possibilities, while for the choice of $\{a_4, a_8, a_9\}$ and $\{a_5, a_{10}\}$ there are $\binom{5}{2}$ possibilities. For each of these choices the numbers a_4, a_5, a_{10} are already determined, and for a_8, a_9 there are two possibilities. The desired number is therefore, by the multiplication rule, equal to

$$\binom{9}{3} \cdot 2 \cdot \binom{5}{2} \cdot 2 = 3360. \quad \square$$

(v) Let $a_1, a_2, \ldots, a_n$ and $b_1, b_2, \ldots, b_n$ be two arrangements of the numbers $1, \frac{1}{2}, \frac{1}{3}, \ldots, \frac{1}{n}$ such that $a_1 + b_1 \ge a_2 + b_2 \ge \cdots \ge a_n + b_n$. Show that the inequality $a_k + b_k \le \frac{4}{k}$ holds for all $k = 1, 2, \ldots, n$.

SOLUTION.   Fix a $k \in \{1, 2, \ldots, n\}$ and suppose that the inequalities $a_j \leq b_j$, respectively $a_j \geq b_j$, hold for α, respectively β, values of the index $j \in \{1, 2, \ldots, k\}$. Since $\alpha + \beta \geq k$, one of the numbers α, β is at least $\frac{k}{2}$; without loss of generality we may assume that this is the case for α. If b_s $(1 \leq s \leq k)$ is the *smallest* of these α numbers b_j $(1 \leq j \leq k)$ for which $a_j \leq b_j$ holds, then clearly $b_s \leq \frac{1}{\alpha}$, since $b_i \geq b_s$ holds for at least α values of the index i. Thus we obtain

$$a_k + b_k \leq a_s + b_s \leq 2b_s \leq \frac{2}{\alpha} \leq \frac{2}{k/2} = \frac{4}{k},$$

and the proof is complete.   □

1.4   Exercises

(i) Show that if $a_1 \leq a_2 \leq \cdots \leq a_n$ is a nondecreasing arrangement of an n-tuple of real numbers $b_1, b_2, \ldots, b_n$, then for each $j = 1, 2, \ldots, n$ we have $a_j \leq \max\{b_1, b_2, \ldots, b_j\}$.

(ii) Let $u_1 \leq u_2 \leq \cdots \leq u_n$, respectively $v_1 \geq v_2 \geq \cdots \geq v_n$, be nondecreasing, respectively nonincreasing, arrangements of an n-tuple of real numbers $x_1, x_2, \ldots, x_n$, respectively $y_1, y_2, \ldots, y_n$. Prove the inequality

$$\max\{x_k + y_k; \ 1 \leq k \leq n\} \geq \max\{u_k + v_k; \ 1 \leq k \leq n\}.$$

(iii) Decide whether there exists an arrangement $a_1, a_2, \ldots, a_{2n+1}$ of the integers $1, 2, \ldots, 2n + 1$ such that for each $k = 1, 2, \ldots, 2n + 1$ we have either $a_{k-j} < a_{k+j}$ $(1 \leq j \leq n)$, or $a_{k-j} > a_{k+j}$ $(1 \leq j \leq n)$, with the convention that $a_p = a_{p+2n+1}$ for all p.

*(iv) Suppose that the sum of the nonnegative numbers $a_1, a_2, \ldots, a_n$ is equal to 1. Show that there exists an arrangement $b_1, b_2, \ldots, b_n$ of these numbers such that

$$b_1 b_2 + b_2 b_3 + \cdots + b_{n-1} b_n + b_n b_1 \leq \frac{1}{n}.$$

(v) Show that the sums $a_k + b_k$ of 1.3.(v) satisfy the lower bound

$$a_k + b_k \geq \frac{4}{n + k + 1} \qquad (1 \leq k \leq n).$$

*(vi) A generalization of 1.3.(iv). Let $Q(n)$ denote the number of different arrangements $a_1, a_2, \ldots, a_n$ of the integers $1, 2, \ldots, n$ that satisfy $a_i > a_{2i}$ $(1 \leq i < 2i \leq n)$ and $a_i > a_{2i+1}$ $(1 \leq i < 2i + 1 \leq n)$. Find a recurrence relation among the numbers $Q(n)$, and use it to compute $Q(100)$. (Leave the result $Q(100)$ in the form of a product of binomial coefficients.)

(vii) Nine diagrams are shown in Figure 2. For each of them, determine
the number of different ways in which the integers $1, 2, \ldots, n$ (where
n is the number of nodes) can be placed on the nodes in such a way
that for any pair of nodes connected by a line the one with the larger
number attached to it is in a higher position.

(Such diagrams are called *Hasse diagrams of partially ordered sets*; see,
e.g., [3], page 373. In this exercise you will actually answer the question:
In how many ways can a given partial ordering be completed to obtain a
linear ordering?

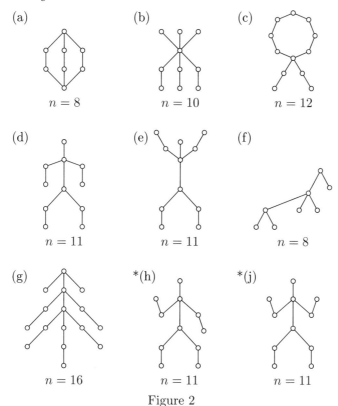

(a)   (b)   (c)

$n = 8$ $n = 10$ $n = 12$

(d)   (e)   (f)

$n = 11$ $n = 11$ $n = 8$

(g)   *(h)   *(j)

$n = 16$ $n = 11$ $n = 11$

Figure 2

(viii) Determine the number of arrangements $a_1, a_2, \ldots, a_n$ of the set
$\{1, 2, \ldots, n\}$ that satisfy the following condition for each $i = 1, 2,$
$3, \ldots, n$: If $a_i \neq 1$, then $a_i > \min\{a_{i-1}, a_{i+1}\}$, with the convention
that a_0 and a_{n+1} are any numbers greater than n.

1.5 Extremal Problems

The examples of the next two subsections can in general be character-
ized as follows: For a given function F in n variables we want to find the

arrangement $x_1, x_2, \ldots, x_n$ of the given numbers $a_1, a_2, \ldots, a_n$ for which the corresponding value $F(x_1, x_2, \ldots, x_n)$ is smallest, respectively largest. Such extremal values of F always exist, since they are the smallest, respectively largest, element of a finite set of numbers that has no more than $n!$ elements.

(i) Find the smallest possible value of the sum

$$S = |x_2 - x_1| + |x_3 - x_2| + \cdots + |x_n - x_{n-1}| + |x_1 - x_n|,$$

where $x_1, x_2, \ldots, x_n$ is an arbitrary arrangement of the integers $1, 2, \ldots, n$.

SOLUTION.   Our approach is based on the inequality $|x| \geq x$, which holds for all $x \in \mathbb{R}$. Since the value S does not change under cyclic permutations of the numbers $x_1, x_2, \ldots, x_n$, we may assume that $x_1 = 1$. Let k $(1 < k \leq n)$ be the index for which $x_k = n$. Adding the inequalities

$$n - 1 = x_k - x_1 = (x_k - x_{k-1}) + (x_{k-1} - x_{k-2}) + \cdots + (x_2 - x_1)$$
$$\leq |x_k - x_{k-1}| + |x_{k-1} - x_{k-2}| + \cdots + |x_2 - x_1|$$

and

$$n - 1 = x_k - x_1 = (x_k - x_{k+1}) + (x_{k+1} - x_{k+2}) + \cdots + (x_n - x_1)$$
$$\leq |x_{k+1} - x_k| + |x_{k+2} - x_{k+1}| + \cdots + |x_1 - x_n|,$$

we obtain the bound $2(n - 1) \leq S$. We have equality $S = 2(n - 1)$ if and only if both of the above inequalities are equalities, that is, if $x_k > x_{k-1} > x_{k-2} > \cdots > x_2 > x_1$ and $x_k > x_{k+1} > x_{k+2} > \cdots > x_n > x_1$ hold. For $k = 2$, for example, the arrangement $1, n, n - 1, n - 2, \ldots, 2$ is of this type. The smallest possible value for S is therefore $2n - 2$.   □

(ii) Find the largest possible value of the expression

$$x_1 x_2 + x_2 x_3 + \cdots + x_{n-1} x_n + x_n x_1$$

for a given $n \geq 3$, where $x_1, x_2, \ldots, x_n$ is an arbitrary arrangement of the integers $1, 2, \ldots, n$.

SOLUTION.   Let $S_n(x_1, x_2, \ldots, x_n)$ denote the above expression, and let M_n be its maximal value. Since $S_3(x_1, x_2, x_3) = 1 \cdot 2 + 2 \cdot 3 + 3 \cdot 1 = 11$ when $\{x_1, x_2, x_3\} = \{1, 2, 3\}$, we have $M_3 = 11$. Since our expression is independent of cyclic permutations, it suffices for $n > 3$ to consider only those arrangements $x_1, x_2, \ldots, x_n$ of $1, 2, \ldots, n$ for which $x_1 = n$. We then obtain

$$S_n(n, x_2, x_3, \ldots, x_n) = S_{n-1}(x_2, x_3, \ldots, x_n) - x_2 x_n + n x_2 + n x_n$$
$$= S_{n-1}(x_2, x_3, \ldots, x_n) + n^2 - (n - x_2)(n - x_n)$$
$$\leq M_{n-1} + n^2 - 1 \cdot 2,$$

with equality when $S_{n-1}(x_2, x_3, \ldots, x_n) = M_{n-1}$ and, at the same time, $\{x_2, x_n\} = \{n-2, n-1\}$. If we make the induction hypothesis T_n that there exists an arrangement $y_1, y_2, \ldots, y_{n-1}$ of $1, 2, \ldots, n-1$ for which $y_1 = n-1$, $y_{n-1} = n-2$, and $S_{n-1}(y_1, y_2, \ldots, y_{n-1}) = M_{n-1}$ (T_n holds for $n = 4$), we obtain the recurrence relation $M_n = M_{n-1} + n^2 - 2$, where $n, y_{n-1}, y_{n-2}, \ldots, y_2, y_1$ is the arrangement of $1, 2, \ldots, n$ belonging to the assumption T_{n+1}. Now we can use any method of summation (see, e.g., [5], Section 1.2) to evaluate

$$M_n = M_3 + \left(4^2 - 2\right) + \left(5^2 - 2\right) + \cdots + \left(n^2 - 2\right)$$

$$= 11 + \left(4^2 + 5^2 + \cdots + n^2\right) - 2(n-3) = \frac{1}{6}\left(2n^3 + 3n^2 - 11n + 18\right).\ \square$$

(iii) For an arrangement $x_1, x_2, \ldots, x_n$ of the integers $1, 2, \ldots, n$ we define $d_1 = |x_2 - x_1|, d_2 = |x_3 - x_2|, \ldots, d_{n-1} = |x_n - x_{n-1}|, d_n = |x_1 - x_n|$ and set $d_* = \min\{d_i : 1 \leq i \leq n\}$. Then the largest possible value of d_* is equal to $\left[\frac{n-1}{2}\right]$. Prove this assertion in the case where n is divisible by 4. (For the remaining cases this is done in Exercise 1.6.(iii).)

SOLUTION. For $n = 4k$ we have $\left[\frac{n-1}{2}\right] = 2k - 1$. Therefore, we have to verify that for any arrangement $x_1, x_2, \ldots, x_{4k}$ the inequality $d_* \leq 2k - 1$ holds, with equality occurring for some arrangements. We base our proof of the inequality on considering those pairs of elements of the arrangement that are placed next to the integer $\frac{n}{2} = 2k$ (the elements x_1, x_n will also be considered neighbors). Since the situation is cyclic, we may assume that $2k = x_1$. Then at least one of the integers $|x_2 - x_1|, |x_1 - x_n|$ is less than $2k$, since the inequality $|2k - x| \geq 2k$ holds only for a unique x from the set $\{1, 2, \ldots, 4k\}$. This means that $d_1 \leq 2k - 1$ or $d_n \leq 2k - 1$, and thus $d_* \leq 2k - 1$. To construct an arrangement with the $d_* = 2k - 1$ we use the following interesting trick: The sequence of integers $b_j = b_0 + (2k-1)j$ is for any $b_0 \in \mathbb{Z}$ an arithmetic sequence with difference $d = 2k - 1$; since $2k - 1$ and $n = 4k$ are relatively prime, the integers $b_1, b_2, \ldots, b_n$ give different remainders upon division by n. Hence there exists an arrangement $x_1, x_2, \ldots, x_n$ of the integers $1, 2, 3, \ldots, n$ such that $x_i \equiv b_i \pmod{n}$, $1 \leq i \leq n$. Since $x_{i+1} - x_i \equiv 2k - 1 \pmod{n}$, each number $d_i = |x_{i+1} - x_i|$ is equal either to $2k - 1$ or to $n - (2k-1) = 2k + 1$, for $1 \leq i \leq n$. This leads to the inequality $d_* \geq 2k - 1$, and thus to the equality $d_* = 2k - 1$. Thus, for example, we obtain for $b_0 = 2k + 1$ the arrangement $\underbrace{4k, 2k - 1}, \underbrace{4k - 2, 2k - 3}, \ldots, \underbrace{2k + 2, 1}, 2k, \underbrace{4k - 1, 2k - 2}, \underbrace{4k - 3, 2k - 4}, \ldots, \underbrace{2k + 3, 2}, 2k + 1$. (For better orientation we have marked those pairs of neighboring integers whose differences are $2k + 1$.) $\square$

1.6 Exercises

(i) Let $x_1, x_2, \ldots, x_n$ and $y_1, y_2, \ldots, y_n$ be arbitrary arrangements of the same n-tuple of integers $1, 2, \ldots, n$. For a given $n \geq 1$ find the smallest possible value of the largest of the products $x_1y_1, x_2y_2, \ldots, x_ny_n$, and the largest possible value of the smallest of these products.

(ii) Find the largest possible value of the sum

$$S = |x_1 - 1| + |x_2 - 2| + \cdots + |x_n - n|,$$

where $x_1, x_2, \ldots, x_n$ is an arbitrary arrangement of $1, 2, \ldots, n$.

(iii) Prove the assertion of 1.5.(iii) in the remaining cases, where n is not divisible by 4.

(iv) In the situation of 1.5.(iii), find for a fixed $n \geq 3$ the smallest possible value of $d^* = \max\{d_i : 1 \leq i \leq n\}$.

(v) Decide whether in the case $n = 4k$ the numbers d_i of 1.5.(iii) can satisfy the system of inequalities $k \leq d_i \leq 2k - 1$, $1 \leq i \leq n$.

*(vi) For which numbers $n \geq 3$ can one choose n different integers $a_1, a_2, \ldots, a_n$ from the set $\{1, 2, \ldots, n + 1\}$ such that $|a_1 - a_2|$, $|a_2 - a_3|, \ldots, |a_{n-1} - a_n|, |a_n - a_1|$ are distinct?

(vii) In the examples 1.5.(i)–(iii) and the exercises 1.6.(i), (ii) the underlying n-tuple was always $1, 2, \ldots, n$. Replace it with a general n-tuple $a_1, a_2, \ldots, a_n$ of real numbers, and decide whether the indicated approach for solving these problems can be easily modified to the general case.

1.7   Miscellaneous Problems

To conclude this section, we solve another triple of interesting problems on arrangements of a given set of numbers.

(i) Let $a < b < c < d$. How many different values does the expression

$$V = (x - y)^2 + (y - z)^2 + (z - t)^2 + (t - x)^2$$

take on if x, y, z, t are arbitrary arrangements of the integers a, b, c, d? Also determine those arrangements for which the value V (for given a, b, c, d) is largest and smallest.

FIRST SOLUTION.   The value of V does not change under a cyclic permutation of the integers x, y, z, t; without loss of generality we may therefore set $x = a$. Changing the quadruple a, y, z, t to a, t, z, y changes only the sign of the four differences occurring in V, but this does not influence the value of V. Therefore, of the six arrangements y, z, t of the integers b, c, d only three can lead to distinct values; these are, for instance, the arrangements (b, c, d), (c, b, d), and (b, d, c). An easy calculation shows that

$V(a, c, b, d) > V(a, b, c, d) > V(a, b, d, c)$. Indeed,

$$V(a, b, c, d) - V(a, b, d, c) = (b - c)^2 + (d - a)^2 - (b - d)^2 - (c - a)^2$$
$$= 2(b - a)(d - c) > 0,$$
$$V(a, c, b, d) - V(a, b, c, d) = (a - c)^2 + (b - d)^2 - (a - b)^2 - (c - d)^2$$
$$= 2(c - b)(d - a) > 0.$$

Thus we have shown that there exist exactly three different values of V, and at the same time we have determined the smallest and the largest of them. □

SECOND SOLUTION.   The expression $W = V(x, y, z, t) + (x - z)^2 + (y - t)^2$ is invariant (that is, it does not depend on the arrangement of x, y, z, t), and it is equal to $V(x, y, z, t) + (x^2 + y^2 + z^2 + t^2) - 2(xz + yt)$. Since the sum $x^2 + y^2 + z^2 + t^2$ is also invariant, the value of V depends only on the expression $U = xz + yt$, which in turn depends on how the set $\{a, b, c, d\}$ is divided into the classes $\{x, z\}$ and $\{y, t\}$. Hence there exist at most three different values $U_1 = ab + cd$, $U_2 = ac + bd$, $U_3 = ad + bc$. We see immediately that $U_1 - U_2 = (d - a)(c - b) > 0$ and $U_2 - U_3 = (b - a)(d - c) > 0$; that is, $U_1 > U_2 > U_3$. It then follows from the identity $V = 2U - (a^2 + b^2 + c^2 + d^2) + W$ that V has exactly three distinct values: $V(a, c, b, d) > V(a, b, c, d) > V(a, b, d, c)$. □

(ii) Find the largest integer A with the following property: If the numbers 1, 2, ..., 100 are written in an arbitrary order, then there are ten consecutive terms whose sum is greater than or equal to A.

SOLUTION.   We divide an arbitrary arrangement $x_1, x_2, \ldots, x_{100}$ of the numbers $1, 2, \ldots, 100$ into ten consecutive "sections" of ten terms each and form the sums $S_1 = x_1 + x_2 + \cdots + x_{10}, \ldots, S_{10} = x_{91} + x_{92} + \cdots + x_{100}$. Since the sum $S_1 + S_2 + \cdots + S_{10}$ is equal to $1 + 2 + \cdots + 100 = 5050$, at least one of the numbers S_i is at least $5050/10 = 505$. We have therefore shown that the number $A = 505$ has the desired property. In order to show that no integer greater than 505 has this property, we consider the arrangement

$$100, 1, 99, 2, 98, 3, \ldots, 51, 50.$$

It is easy to determine the sum S of any ten consecutive integers: If the first number from the left is greater (respectively less) than 50, then $S = 505$ (respectively $S = 500$); this can be verified by dividing the ten integers into five pairs of two neighboring elements. The greatest integer with the desired property is therefore $A = 505$. □

(iii) A sequence $a_1, a_2, \ldots, a_k$, with $k \geq n$, is called n-universal if upon appropriate elimination of $k - n$ terms we can obtain any arrangement of the integers $1, 2, \ldots, n$. Thus, for example, the sequence $1, 2, 3, 1, 2, 1, 3$

is 3-universal, while $1, 2, 3, 2, 1, 3$ is not (the arrangement $3, 1, 2$ cannot be obtained through elimination). Give an example of an n-universal sequence of length $k = n^2$ and one of length $k = n^2 - n + 1$.

SOLUTION.   Even though it would suffice to give an example for $k = n^2 - n + 1$ (and then "fill up" the sequence with $n - 1$ arbitrary terms to obtain length $k = n^2$), we begin with a simple example for $k = n^2$. An appropriate example is the sequence consisting of an n-fold repetition of the same arrangement of the integers $1, 2, \ldots, n$, such as the sequence

$$\underbrace{1, 2, \ldots, n}_{B_1}, \underbrace{1, 2, \ldots, n}_{B_2}, \ldots, \underbrace{1, 2, \ldots, n}_{B_n} . \tag{2}$$

From this sequence of length n^2 we obtain any arrangement $a_1, a_2, \ldots, a_n$ of the integers $1, 2, \ldots, n$ by eliminating all integers in the section B_k with the exception of a_k (for all $k = 1, 2, \ldots, n$).

In order to obtain an example for $k = n^2 - n + 1$, we shorten the sequence (2) by removing the numbers $2, 3, \ldots, n$ from the final section B_n, that is, we leave only the number 1. This new sequence of length $n^2 - n + 1$ is indeed n-universal: If we carry out the choice of the integers a_k from the section B_k as above, we can "save" one of the sections; that is, we can choose two neighboring integers a_j, a_{j+1} from the same section B_j, provided that $a_j < a_{j+1}$. Then we choose a_{j+2} from B_{j+1}, a_{j+3} from B_{j+2}, etc., and a_n from B_{n-1}. This modified approach works for all arrangements $a_1, a_2, \ldots, a_n$ with the exception of the case $a_1 > a_2 > \cdots > a_n$, that is, $a_k = n + 1 - k$ $(1 \le k \le n)$. But in this case we can still use the original approach, since in the section B_n in (2) we have kept exactly the integer 1. $\square$

1.8   Exercises

(i) Find the largest integer A with the following property: If the digits 0, 1, 2, $\ldots$, 9 are placed on the circumference of a circle in an arbitrary order, then there are three neighboring digits whose sum is greater than or equal to A.

*(ii) Show that there exists an n-universal sequence in the sense of 1.7.(iii) with length $k = n^2 - 2n + 4$. (The length of the shortest n-universal sequence for general n is not known to us.)

In Exercises (iii)–(vii) let $A(k, n)$, for $k < n$, denote the largest integer A with the following property: If the numbers $1, 2, \ldots, n$ are written in any order, then there are k consecutive elements whose sum is greater than or equal to A. (In Problem 1.7.(ii) we found the value of $A(10, 100)$.)

(iii) What results about the numbers $A(k, n)$ can be obtained by a direct generalization of the argument in the solution of 1.7.(ii)?

(iv) Find the values of $A(k, n)$ in the cases $k = 2$, $k = n - 1$, and $k = n - 2$.

(v) Prove the inequality $A(k, n) \leq \frac{n(k+1)}{2}$ for all odd $k < n$ (for even k there is a sharper upper bound, derived in the answer to (iii)). Try to improve this bound in the case where k is odd and n is an even integer.

(vi) Prove the identity $A(k, 2k) = k^2 + \left[\frac{k+1}{2}\right]$ for all $k \geq 1$.

*(vii) Show that for $\frac{n}{2} < k < n$ we have

$$A(k, n) = \frac{n+1}{2}(2k - n) + A(n - k, 2n - 2k).$$

Note that from the results of (vi) and (vii) it follows immediately that

$$A(k, n) = \frac{(n + 1)(2k - n)}{2} + (n - k)^2 + \left[\frac{n - k + 1}{2}\right] \quad \left(\frac{n}{2} \leq k < n\right).$$

We do not know exact values for $A(k, n)$ in the case $2 < k < \frac{n}{2}$, except when k is an even divisor of n (see the answer to (iii)). We already consider the problem of finding the value of $A(3, n)$ to be difficult.

(viii) Find those $n \geq 3$ for which there is an arrangement $x_1, x_2, \ldots, x_n$ of the integers $1, 2, \ldots, n$ such that $x_j \neq \frac{x_i + x_k}{2}$ whenever $1 \leq i < j < k \leq n$.

2   Sequences

The majority of problems in this section will be about *finite sequences*. By a finite sequence of *length* n we mean any ordered n-tuple of numbers a_1, $a_2, \ldots, a_n$. If there is a number a_i associated with every value $i \in \mathbb{N}$, we talk about an *infinite sequence* $a_1, a_2, \ldots, a_n, \ldots$. In some situations the length of a given (finite) sequence is irrelevant, or it doesn't even matter whether the sequence is infinite; then we simply write $a_1, a_2, \ldots$. It is also convenient to write certain sums in this fashion: If, for instance, the sequence $a_1, a_2, \ldots, a_n$ is given, then $a_1 + a_3 + a_5 + \cdots$ denotes the sum of all numbers a_i with odd index $i \leq n$ (the last summand is then either a_n or a_{n-1}, depending on the parity of n).

The element a_i of a sequence $a_1, a_2, \ldots$ is called its ith *term*, and the integer i its index. A pair of terms a_i, a_{i+1} of a sequence $a_1, a_2, \ldots$ is called a *pair of consecutive terms*. Note that two different terms of the same sequence (that is, two terms with different indices) may be equal to the same number; for instance, the first and the fourth terms in the sequence $2, 3, 4, 2, 5$ have this property. If we have $a_i = a_j$, then this is an *equality of numbers*; it does not mean in general that $i = j$.

Several problems are such that their formulation does not change if the finite sequence $a_1, a_2, \ldots, a_n$ under consideration is shifted cyclically, that is, if it is changed to the sequence $a_{1+k}, a_{2+k}, \ldots, a_n, a_1, a_2, \ldots, a_k$ for some $k \in \{1, 2, \ldots, n-1\}$. We can picture this as the numbers $a_1, a_2, \ldots, a_n$ being written along the circumference of a circle, and oriented in one of two possible directions. In this case the sequence $a_1, a_2, \ldots, a_n$ is called *cyclic*; for an easier statement of the conditions in such problems we set $a_{kn+i} = a_i$ for all $k \in \mathbb{Z}$ and $i = 1, 2, \ldots, n$. Furthermore, we will not consider the two cyclic sequences $a_1, a_2, \ldots, a_n$ and $a_{j+1}, a_{j+2}, \ldots, a_{j+n}$, with $j \in \mathbb{Z}$, as different.

2.1   Problems on Inequalities

(i) Suppose that each term in the sequence $x_1, x_2, \ldots, x_n$ is either -1, or 0, or 1. Find the smallest possible value of the sum S of all $\binom{n}{2}$ products $x_i x_j$ (where $1 \le i < j \le n$).

SOLUTION.   Let the given sequence $x_1, x_2, \ldots, x_n$ contain exactly p (respectively q, respectively r) terms equal to 1 (respectively -1, respectively 0). Then $p + q + r = n$, and we have

$$S = \binom{p}{2} + \binom{q}{2} - pq = \frac{(p-q)^2}{2} - \frac{p+q}{2} = \frac{(p-q)^2}{2} - \frac{n}{2} + \frac{r}{2}.$$

This implies $S \ge \frac{-n}{2}$, and if n is even, then the equality $S = \frac{-n}{2}$ occurs if and only if $p = q = \frac{n}{2}$ and $r = 0$. If n is odd, then $S \ge \frac{-n}{2}$ implies $S \ge \frac{-n+1}{2}$, since $S \in \mathbb{Z}$; in this case the equality $S = \frac{-n+1}{2}$ occurs in exactly two cases: Either $r = 1$ and $p = q = \frac{n-1}{2}$, or $r = 0$ and $\{p, q\} = \{\frac{n+1}{2}, \frac{n-1}{2}\}$. The final answer for all cases can then be written as $S_{\min} = -\left\lfloor \frac{n}{2} \right\rfloor$.   □

(ii) Suppose that the $2n$ real numbers $x_1, x_2, \ldots, x_n, y_1, y_2, \ldots, y_n$ satisfy the inequality

$$x_1 + x_2 + \cdots + x_n > y_1 + y_2 + \cdots + y_n,$$

but if we switch the numbers x_i and y_i for any index i, then the inequality no longer holds. For which values of n is this possible?

SOLUTION.   We set $X = x_1 + x_2 + \cdots + x_n$ and $Y = y_1 + y_2 + \cdots + y_n$. By hypothesis we have $X - x_i + y_i \le Y - y_i + x_i$, that is, $X - Y \le 2(x_i - y_i)$, for all $i = 1, 2, \ldots, n$. By adding these n inequalities, we obtain $n(X - Y) \le 2(X - Y)$, and upon dividing by the positive number $X - Y$ we get the inequality $n \le 2$. Both values $n = 1$ and $n = 2$ are possible, which is illustrated by the examples $1 > 0$ and $1 + 1 > 0 + 0$.   □

(iii) Let the positive numbers $a_1, a_2, \ldots, a_n$ satisfy $a_k \le a_{k+1} \le 2a_k$ for all $1 \le k \le n-1$. Show that in the sum $S = \pm a_1 \pm a_2 \pm \cdots \pm a_n$ one can choose the signs in such a way that $0 \le S \le a_1$ holds.

SOLUTION.   We use induction on n. If $n = 2$, we can choose $S = -a_1 + a_2$. If the signs in the sum $S_1 = \pm a_2 \pm a_3 \pm \cdots \pm a_n$ are chosen such that $0 \leq S_1 \leq a_2$ holds, then we set either $S = a_1 - S_1$ or $S = -a_1 + S_1$, according to whether $S_1 \leq a_1$ or $S_1 > a_1$. In both cases we have $0 \leq S \leq a_1$.   □

(iv) Suppose that the difference between the largest and the smallest of the n real numbers $x_1, x_2, \ldots, x_n$ is equal to 1. What is the largest possible value for the corresponding difference for the numbers

$$y_1 = x_1, \; y_2 = \frac{x_1 + x_2}{2}, \; \ldots, \; y_n = \frac{x_1 + x_2 + \cdots + x_n}{n} \; ?$$

SOLUTION.   The values of the two differences do not change if we replace the original sequence $x_1, x_2, \ldots, x_n$ by the sequence $x_1 - c, x_2 - c, \ldots, x_n - c$ for an arbitrary $c \in \mathbb{R}$. If we choose $c = \min_k x_k$, then such a replacement gives us a new sequence $x_1, x_2, \ldots, x_n$ with $\min_k x_k = 0$ and $\max_k x_k = 1$. We now choose two indices p, q such that $y_p = \min_k y_k$ and $y_q = \max_k y_k$, and we distinguish the following two cases.

(a) $p < q$. Then

$$y_q - y_p = \frac{x_1 + x_2 + \cdots + x_q}{q} - \frac{x_1 + x_2 + \cdots + x_p}{p}$$

$$= (x_1 + x_2 + \cdots + x_p)\left(\frac{1}{q} - \frac{1}{p}\right) + \frac{x_{p+1} + x_{p+2} + \cdots + x_q}{q}.$$

In view of the fact that $x_1 + x_2 + \cdots + x_p \geq 0$, $\frac{1}{q} - \frac{1}{p} < 0$, and $x_{p+1} + x_{p+2} + \cdots + x_q \leq q - p$, we obtain the estimate

$$y_q - y_p \leq \frac{q - p}{q} = 1 - \frac{p}{q} \leq 1 - \frac{1}{n}.$$

(b) $p > q$. Similarly to the above, we derive

$$y_q - y_p = (x_1 + x_2 + \cdots + x_q)\left(\frac{1}{q} - \frac{1}{p}\right) - \frac{x_{q+1} + x_{q+2} + \cdots + x_p}{p}$$

$$\leq q\left(\frac{1}{q} - \frac{1}{p}\right) = 1 - \frac{q}{p} \leq 1 - \frac{1}{n}.$$

In both cases we therefore have $y_q - y_p \leq 1 - \frac{1}{n}$. An easy argument now shows that equality is actually attained in this inequality: In case (a) this happens exactly when $x_1 = x_2 = \cdots = x_p = 0$, $x_{p+1} = x_{p+2} = \cdots = x_q = 1$, $p = 1$, and $q = n$, while in case (b) it occurs exactly when $x_1 = x_2 = \cdots = x_q = 1$, $x_{q+1} = x_{q+2} = \cdots = x_p = 0$, $q = 1$, and $p = n$. This corresponds to the sequence $0, 1, 1, \ldots, 1$ respectively $1, 0, 0, \ldots, 0$, and the desired largest value is therefore $1 - \frac{1}{n}$.   □

(v) Suppose that the sum of the real numbers $x_1, x_2, \ldots, x_n$ is equal to 0 and that $x_i = 1$ for some index $i \in \{1, 2, \ldots, n\}$. Show that the largest of the numbers

$$|x_1 - x_2|, \ |x_2 - x_3|, \ \ldots, \ |x_{n-1} - x_n|, \ |x_n - x_1| \tag{3}$$

is not less than $\frac{4}{n}$.

SOLUTION.   Let ε denote the largest of the numbers in (3). Since this is a problem concerning a cyclic sequence, we may assume that $x_i = 1$ occurs for $i = 1$, that is, $x_1 = 1$ and $x_2 + x_3 + \cdots + x_n = -1$. To proceed further, we use the following idea: If the number $\varepsilon > 0$ were too small, then by adding we would get the estimates

$$
\begin{aligned}
x_2 &\ge x_1 - \varepsilon, & x_n &\ge x_1 - \varepsilon, \\
x_3 &\ge x_2 - \varepsilon \ge x_1 - 2\varepsilon, & x_{n-1} &\ge x_n - \varepsilon \ge x_1 - 2\varepsilon, \\
x_4 &\ge x_3 - \varepsilon \ge x_1 - 3\varepsilon, & x_{n-2} &\ge x_{n-1} - \varepsilon \ge x_1 - 3\varepsilon, \\
&\ \ \vdots & &\ \ \vdots
\end{aligned}
\tag{4}
$$

and upon setting $x_1 = 1$ it would follow that the number $x_2 + x_3 + \cdots + x_n$ was larger than -1. In view of (4) it will be convenient to distinguish whether n is odd or even. If $n = 2k + 1$, then the last "good" row in (4) is formed by the pair of inequalities

$$x_{k+1} \ge x_1 - k\varepsilon, \quad x_{k+2} \ge x_1 - k\varepsilon;$$

therefore, the sum of the inequalities in the first k rows of (4) can be written as

$$x_2 + x_3 + \cdots + x_n \ge (n-1)x_1 - k(k+1)\varepsilon,$$

or $-1 \ge 2k - k(k+1)\varepsilon$, which implies $\varepsilon \ge \frac{2k+1}{k(k+1)} = \frac{4n}{n^2-1} > \frac{4}{n}$. We leave it to the reader to show in the case $n = 2k$ that (4) implies

$$x_2 + x_3 + \cdots + x_n \ge (n-1)x_1 - k^2\varepsilon,$$

or $-1 \ge (2k - 1) - k^2\varepsilon$, which gives $\varepsilon \ge \frac{2}{k} = \frac{4}{n}$.   □

(vi) Suppose that the positive integers $a_1, a_2, \ldots, a_n$ $(n \ge 3)$ are such that all the quotients

$$p_1 = \frac{a_n + a_2}{a_1}, \quad p_2 = \frac{a_1 + a_3}{a_2}, \quad \ldots, \quad p_n = \frac{a_{n-1} + a_1}{a_n} \tag{5}$$

are integers. Show that the sum $S = p_1 + p_2 + \cdots + p_n$ satisfies the inequality $S \le 3n - 1$.

SOLUTION.   As we will see in a moment, considering the fraction in (5) that has the largest denominator will enable us to use induction on n. We begin with the case $n = 3$. By changing the order of the numbers a_1, a_2, a_3 we may assume that $a_1 \leq a_2 \leq a_3$. In doing this, the sum S does not change, since the numbers p_1, p_2, p_3 only change their orders. The inequality

$$p_3 = \frac{a_2 + a_1}{a_3} \leq \frac{a_3 + a_3}{a_3} = 2$$

now implies that either $p_3 = 2$ (then $a_1 = a_2 = a_3$, which means $S = 6$), or $p_3 = 1$; that is, $a_1 + a_2 = a_3$. In the second case we obtain

$$p_1 = \frac{a_3 + a_2}{a_1} = \frac{a_1 + 2a_2}{a_1} = 1 + \frac{2a_2}{a_1}, \quad p_2 = \frac{a_1 + a_3}{a_2} = \frac{2a_1 + a_2}{a_2} = 1 + \frac{2a_1}{a_2}.$$

Both numbers $\frac{2a_2}{a_1}, \frac{2a_1}{a_2}$ are therefore positive integers, and their product is 4. In view of $a_2 \geq a_1$ this implies $\frac{2a_2}{a_1} \in \{2, 4\}$. Hence in the case $p_3 = 1$ there are two possibilities: Either $a_2 = a_1$ and $a_3 = 2a_1$ (then $S = 7$) or $a_2 = 2a_1$ and $a_3 = 3a_1$ (then $S = 8$). For $n = 3$ we therefore have, in general, $S \leq 8 = 3 \cdot 3 - 1$.

In the case $n > 3$ we need an induction hypothesis for some sequence of length $n-1$, for instance for $a_1, a_2, \ldots, a_{n-1}$; that is, we need the inequality

$$S' = \frac{a_{n-1} + a_2}{a_1} + \frac{a_1 + a_3}{a_2} + \cdots + \frac{a_{n-2} + a_1}{a_{n-1}} \leq 3(n-1) - 1. \quad (6)$$

For this we have to make sure that not only the numbers in (5) are integers, but also the numbers $p_1' = \frac{a_{n-1} + a_2}{a_1}$ and $p_{n-1}' = \frac{a_{n-2} + a_1}{a_{n-1}}$. We will now show that this happens in the case where a_n is the largest of the integers a_{n-1}, a_n, a_1 (this can be achieved by a cyclic shift of the given sequence $a_1, a_2, \ldots, a_n$). Indeed, from the inequalities $a_{n-1} \leq a_n$ and $a_1 \leq a_n$ it follows that

$$p_n = \frac{a_{n-1} + a_1}{a_n} \leq \frac{a_n + a_n}{a_n} = 2;$$

that is, $p_n \in \{1, 2\}$. Now, in the case $p_n = 2$ we have $a_{n-1} = a_n = a_1$; thus $p_1' = p_1$ and $p_{n-1}' = p_{n-1}$, which implies $S = S' + p_n = S' + 2$; this, together with (6), gives the inequality $S \leq 3n - 2$. In the case $p_n = 1$ we have $a_{n-1} + a_1 = a_n$, which implies

$$p_1 = \frac{a_n + a_2}{a_1} = \frac{a_{n-1} + a_1 + a_2}{a_1} = 1 + \frac{a_{n-1} + a_2}{a_1} = 1 + p_1',$$

$$p_{n-1} = \frac{a_{n-2} + a_n}{a_{n-1}} = \frac{a_{n-2} + a_{n-1} + a_1}{a_{n-1}} = 1 + \frac{a_{n-2} + a_1}{a_{n-1}} = 1 + p_{n-1}'.$$

Both numbers p_1', p_{n-1}' are again integers, and furthermore, we have

$$S = p_1 + p_2 + \cdots + p_{n-1} + 1 = (S' - p_1' - p_{n-1}') + p_1 + p_{n-1} + 1 = S' + 3,$$

which, together with (6), gives $S \leq 3n - 1$. This completes the proof by induction. □

2.2   Exercises

(i) Given that all terms of the sequence $x_1, x_2, \ldots, x_n$ lie in the interval $[-1, 1]$, determine the smallest possible value of the sum S of all $\binom{n}{2}$ products $x_i x_j$ $(1 \leq i < j \leq n)$. (This is a generalization of 2.1.(i).)

(ii) Do there exist nonnegative real numbers $a_1, a_2, \ldots, a_7$ such that $a_1 = a_7 = 0$ and at the same time $a_{i+1} + a_{i-1} > a_i \sqrt{3}$ $(2 \leq i \leq 6)$?

(iii) Suppose that the sum of the terms of a cyclic sequence $x_1, x_2, \ldots, x_n$ of real numbers is equal to 0. Show that there is an index k such that x_k, $x_k + x_{k+1}$, $x_k + x_{k+1} + x_{k+2}$, $\ldots$, $x_k + x_{k+1} + \cdots + x_{k+n-1}$ are nonnegative numbers.

(iv) Show that for any sequence $x_1, x_2, \ldots, x_n$ of real numbers there exists an index k $(1 \leq k \leq n)$ for which the inequality

$$\left| \sum_{i=1}^{k} x_i - \sum_{i=k+1}^{n} x_i \right| \leq \max\{|x_i| : 1 \leq i \leq n\}$$

is satisfied.

(v) Show that for any sequence $x_1, x_2, \ldots$ of numbers from the interval $[-1, 1]$ one can choose numbers $\varepsilon_i = \pm 1$ such that for each index n the inequality $|\varepsilon_1 x_1 + \varepsilon_2 x_2 + \cdots + \varepsilon_n x_n| \leq 1$ holds.

(vi) Suppose that the sum of the positive integers $a_1, a_2, \ldots, a_n$ is an even number and that furthermore $a_k \leq k$ for all $k = 1, 2, \ldots, n$. Show that in the expression $V = a_1 \pm a_2 \pm a_3 \pm \cdots \pm a_n$ one can choose the signs such that $V = 0$.

*(vii) Suppose that the sum of the squares of the real numbers $x_1, x_2, \ldots, x_n$ is equal to 1. Determine the largest possible value of the sum S of the absolute values of all 2^n numbers of the form $\pm x_1 \pm x_2 \pm \cdots \pm x_n$ for a fixed value of n.

(viii) Suppose that the real numbers $x_1 \leq x_2 \leq \cdots \leq x_n$ satisfy

$$x_1 + x_2 + \cdots + x_n = 0 \quad \text{and} \quad |x_1| + |x_2| + \cdots + |x_n| = A.$$

Prove the inequality $x_n - x_1 \geq \frac{2A}{n}$.

(ix) For each $n \geq 1$ find the largest possible length N of a sequence of numbers $a_1, a_2, \ldots, a_N$ from the set $\{0, 1, \ldots, n\}$ that satisfy the $N - 2$ inequalities $a_i < \frac{a_{i-1} + a_{i+1}}{2}$ $(1 < i < N)$.

*(x) Find the smallest possible value of the difference considered in Problem 2.1.(iv).

*(xi) In the situation of Problem 2.1.(v), prove that the largest of the numbers

$$\left| x_1 - \frac{x_n + x_2}{2} \right|, \left| x_2 - \frac{x_1 + x_3}{2} \right|, \ldots, \left| x_n - \frac{x_{n-1} + x_1}{2} \right|$$

is not less than $\frac{8}{n^2}$.

2.3   Problems on Subsequences

If we choose a part of the terms of a sequence and preserve the order in which they occur, we obtain a *subsequence* of the original sequence. This can be expressed more exactly as follows: A sequence $B = (b_1, b_2, \dots)$ is a subsequence of the sequence $A = (a_1, a_2, \dots)$ if for each index k of B there exists an index i_k of A such that $b_k = a_{i_k}$, where $i_k > i_{k-1}$ for $k > 1$. Since we defined a subsequence as a sequence with certain properties, it is clear what the terms *infinite subsequence* and (finite) *subsequence of length n* will mean.

(i) Show that from any sequence of 101 distinct integers one can choose an increasing or a decreasing subsequence of length 11, that is, a subsequence $b_1, b_2, \dots, b_{11}$ for which either the inequalities $b_1 < b_2 < \cdots < b_{11}$ or $b_1 > b_2 > \cdots > b_{11}$ hold.

SOLUTION.   We denote the given sequence by $a_1, a_2, \dots, a_{101}$. For each $k = 1, 2, \dots, 101$ we find the largest length $n = n_k$ of all subsequences $b_1, b_2, \dots, b_n$ for which $b_1 < b_2 < \cdots < b_n = a_k$. If $n_k > 10$ for some $k \in \{1, 2, \dots, 101\}$, then the proof is complete. Otherwise, there are 101 integers n_k in the 10-element set $\{1, 2, \dots, 10\}$. Then by the pigeonhole principle there are 11 indices $k_1 < k_2 < \cdots < k_{11}$ such that $n_{k_1} = n_{k_2} = \cdots = n_{k_{11}}$. Note that the numbers n_k have the following property: If $k < k'$ and $a_k < a_{k'}$, then $n_{k'} \geq n_k + 1$. (Indeed, each subsequence $b_1 < b_2 < \cdots < b_n = a_k$ can then be extended by the term $b_{n+1} = a_{k'}$.) For the indices $k_1, k_2, \dots, k_{11}$ this means that none of the inequalities in the chain $a_{k_1} < a_{k_2} < \cdots < a_{k_{11}}$ can hold; that is, we have $a_{k_1} > a_{k_2} > \cdots > a_{k_{11}}$ (recall that the numbers $a_1, a_2, \dots, a_{101}$ are distinct). This completes the proof.   □

(ii) Show that for any three infinite sequences $(a_1, a_2, \dots)$, $(b_1, b_2, \dots)$, and $(c_1, c_2, \dots)$ of positive integers there exist indices $p > q$ such that $a_p \geq a_q$, $b_p \geq b_q$, and $c_p \geq c_q$ hold simultaneously.

SOLUTION.   First we show how we can choose from any infinite sequence $(x_1, x_2, \dots)$ of positive integers an infinite subsequence $(y_1, y_2, \dots)$ that is nondecreasing; that is, it satisfies the condition $y_k \leq y_{k+1}$ for all $k \geq 1$. We use induction and first set $y_1 = x_1$. Now we assume that for some $n \geq 1$ we have chosen the numbers

$$y_1 = x_{i_1} \leq y_2 = x_{i_2} \leq \cdots \leq y_n = x_{i_n}, \quad \text{where} \quad i_1 < i_2 < \cdots < i_n,$$

and suppose that it is *not possible* to choose a number y_{n+1}. This means that $x_i < y_n$, that is, $x_i \in \{1, 2, \dots, y_n - 1\}$ for all $i > i_n$. One of the numbers $1, 2, \dots, y_n - 1$ must therefore be equal to infinitely many terms x_i (with index $i > i_n$); from the sequence $(x_1, x_2, \dots)$ we can therefore choose an infinite nondecreasing subsequence of identical numbers $(y, y, \dots)$.

Now we can easily prove the assertion of the problem. First we choose from the sequence $(a_1, a_2, \dots)$ an infinite nondecreasing subsequence

$(a_{i_1}, a_{i_2}, \dots)$; then we take the sequence $(b_{i_1}, b_{i_2}, \dots)$ and choose an infinite nondecreasing subsequence $(b_{j_1}, b_{j_2}, \dots)$, and finally we choose from the sequence $(c_{j_1}, c_{j_2}, \dots)$ an infinite nondecreasing subsequence $(c_{k_1}, c_{k_2}, \dots)$. Since any subsequence of a nondecreasing sequence is nondecreasing, for the sequence of indices $k_1 < k_2 < \cdots$ just obtained we have

$$a_{k_1} \le a_{k_2} \le a_{k_3} \le \cdots,$$
$$b_{k_1} \le b_{k_2} \le b_{k_3} \le \cdots,$$
$$c_{k_1} \le c_{k_2} \le c_{k_3} \le \cdots,$$

which is a considerably stronger statement than what was to be shown.   □

(iii) Show that from an arbitrary sequence $a_1, a_2, \dots, a_n$ of real numbers one can choose part of the terms such that the following two conditions are satisfied:
(a) From each triple a_i, a_{i+1}, a_{i+2} $(1 \le i \le n - 2)$ either one or two terms are chosen,
(b) the absolute value of the sum of all chosen terms is not less than

$$\tfrac{1}{6}(|a_1| + |a_2| + \cdots + |a_n|).$$

SOLUTION.   We divide the given sequence into three parts,

$$X_1 = (a_1, a_4, a_7, \dots), \quad X_2 = (a_2, a_5, a_8, \dots), \quad X_3 = (a_3, a_6, a_9, \dots),$$

and assume that $a_1 + a_2 + \cdots + a_n \ge 0$ (otherwise, we replace each term a_i by its negative value; this does not affect the problem in any way). We consider the six sequences $X_1 \cup (X_2 \cap \mathbb{R}^+)$, $X_1 \cup (X_3 \cap \mathbb{R}^+)$, $X_2 \cup (X_1 \cap \mathbb{R}^+)$, $X_2 \cup (X_3 \cap \mathbb{R}^+)$, $X_3 \cup (X_1 \cap \mathbb{R}^+)$, and $X_3 \cup (X_2 \cap \mathbb{R}^+)$, where the operations used are to be understood as follows: $X \cap \mathbb{R}^+$ is the subsequence of all positive terms of the sequence X; if $Y = (x_{i_1}, x_{i_2}, \dots)$ and $Z = (x_{j_1}, x_{j_2}, \dots)$ are two subsequences of the same sequence $(x_1, x_2, \dots)$, then we set $Y \cup Z = (x_{k_1}, x_{k_2}, \dots)$, where $k_1 < k_2 < \cdots$ is an increasing arrangement of the set $\{i_1, i_2, \dots\} \cup \{j_1, j_2, \dots\}$. The sequences X_1, X_2, X_3 have been chosen such that each of the six subsequences introduced above has the property (a); we will now show that some of them also have the property (b). For this purpose we denote by s_i^+, respectively s_i^-, the sum of the positive, respectively the negative, terms of the sequence X_i, $i = 1, 2, 3$. We have to show that at least one of the six numbers

$$
\begin{array}{lll}
s_1^+ + s_1^- + s_2^+, & s_1^+ + s_1^- + s_3^+, & s_2^+ + s_2^- + s_1^+, \\
s_2^+ + s_2^- + s_3^+, & s_3^+ + s_3^- + s_1^+, & s_3^+ + s_3^- + s_2^+
\end{array}
\tag{7}
$$

is not less than $\tfrac{1}{6}(s_1^+ - s_1^- + s_2^+ - s_2^- + s_3^+ - s_3^-)$. To do this, it suffices to verify that the sum S of all six numbers in (7) satisfies $S \ge s_1^+ - s_1^- +$

$s_2^+ - s_2^- + s_3^+ - s_3^-$, that is, $S = 4(s_1^+ + s_2^+ + s_3^+) + 2(s_1^- + s_2^- + s_3^-) \geq$
$s_1^+ - s_1^- + s_2^+ - s_2^- + s_3^+ - s_3^-$. But this inequality is equivalent to

$$s_1^+ + s_1^- + s_2^+ + s_2^- + s_3^+ + s_3^- \geq 0,$$

which holds because the left-hand side is the number $a_1 + a_2 + \cdots + a_n$,
which was assumed to be nonnegative.   □

2.4 Exercises

(i) Does the statement of Problem 2.3.(i) also hold for an arbitrary
sequence of length 100?

(ii) Determine all integers $k > 8$ with the following property: If the
product of any eight terms of a sequence $a_1, a_2, \ldots, a_k$ of real
numbers is greater than 1, then $a_1 a_2 \cdots a_k > 1$.

(iii) Suppose that the sequence $x_1, x_2, \ldots, x_n$ ($n \geq 7$) of positive integers
has the property that the sum of any seven terms is less than 15 and
the sum of all n terms is equal to 100. Find the smallest possible
length n of such a sequence.

(iv) Suppose that the numbers $x_1, x_2, \ldots, x_n$ lie in the interval $[0, 1]$
and that the terms of the sequence $x_1, x_2, \ldots, x_n$ cannot be divided
into two sets such that both sums of the numbers in the sets are
greater than 1. Find the largest possible value of the sum $S =$
$x_1 + x_2 + \cdots + x_n$. For which n is this bound actually attained?

*(v) Consider the question of when one can choose k terms from a given
sequence of positive numbers $a_1, a_2, \ldots, a_n$ such that the smallest
chosen number is larger than half the largest chosen number. Show
that for a given k, $k > 1$, such a choice is possible when

$$\max\{a_i : 1 \leq i \leq n\} < \frac{a_1 + a_2 + \cdots + a_n}{2k - 2}.$$

(Such a condition can hold only if $2k - 2 < n$, or $k < \frac{n}{2} + 1$.)

(vi) Suppose that the sequence $a_1, a_2, \ldots, a_n$ of integers satisfies $a_1 = 1$,
$a_i \leq a_{i+1} \leq 2a_i$ ($1 \leq i \leq n - 1$). Show that a subdivision of its
terms into two sets such that the sums of the numbers in both sets
are equal is possible if and only if the number $S = a_1 + a_2 + \cdots + a_n$
is even.

*(vii) How can one generalize Problem 2.3.(iii) concerning the sequence
$a_1, a_2, \ldots, a_n$, when for a given integer p ($3 \leq p \leq n$) one has to
choose from each p-tuple of consecutive terms at least one, but not
all p, terms?

2.5 Choosing Subsequences

The following problems require that we show how one can choose from given sequences a subsequence with prescribed properties (mainly concerning the sum of its terms). To begin with, let us clarify that the phrase " ... some terms a_i can be chosen such that their sum S ... " does not exclude the situation where a *single* term a_i is chosen; it is then clear that as "sum" we should take $S = a_i$.

(i) Suppose that the sum of the positive integers $a_1, a_2, \ldots, a_n$ is equal to $2n$, where the largest of these n integers is different from $n+1$. Show that if n is even, then from the sequence $a_1, a_2, \ldots, a_n$ one can choose some terms whose sum is equal to n.

SOLUTION.   We can list the given numbers as $1 \le a_1 \le a_2 \le \cdots \le a_n$. From the condition $a_n = 2n - (a_1 + a_2 + \cdots + a_{n-1}) \le 2n - (n-1) \le n+1$ it follows in view of $a_n \ne n+1$ that $a_n \le n$. Next, it clearly suffices to consider the case where none of the numbers

$$S_i = a_1 + a_2 + \cdots + a_i \quad (1 \le i \le n-1)$$

is divisible by n, since from $1 \le S_i < 2n$ and $n \mid S_i$ it follows immediately that $S_i = n$. Furthermore, we assume that the number $a_1 - a_n$ is also not divisible by n, since from $-n+1 \le a_1 - a_n \le 0$ and $n \mid (a_1 - a_n)$ it follows that $a_1 - a_n = 0$, that is, $a_1 = a_2 = \cdots = a_n = 2$, and thus $S_i = n$ for $i = \frac{n}{2}$ (recall that n is even). None of the numbers of the n-element set

$$M = \{a_1 - a_n, S_1, S_2, \ldots, S_{n-1}\}$$

is therefore divisible by n. By the pigeonhole principle two numbers in M have the same remainder upon division by n. If those are two numbers S_i, S_j $(i < j)$, then in view of $0 < S_j - S_i < 2n$ we have

$$n = S_j - S_i = a_{i+1} + a_{i+2} + \cdots + a_j;$$

if they are $a_1 - a_n$ and S_i, then either $a_n = n$ (if $i = 1$) and we choose the single integer a_n, or

$$n \mid S_i - (a_1 - a_n) = a_2 + a_3 + \cdots + a_i + a_n \quad (\text{if } 1 < i \le n-1),$$

which means that $a_2 + a_3 + \cdots + a_i + a_n = n$. This completes the proof.   □

(ii) Suppose that the given positive integers $x_1, x_2, \ldots, x_n, y_1, y_2, \ldots, y_m$ are such that the sums $x_1 + x_2 + \cdots + x_n$ and $y_1 + y_2 + \cdots + y_m$ are equal to the same number less than $m \cdot n$. Show that from the equation

$$x_1 + x_2 + \cdots + x_n = y_1 + y_2 + \cdots + y_m \tag{8}$$

one can remove some (but not all) summands such that a valid equation remains.

SOLUTION.   We let s denote the common value of both sides of (8) and proceed by induction on the integer $k = m+n$. Since $\max\{m, n\} \le s < mn$, we have $m \ge 2$ and $n \ge 2$, and thus $k \ge 4$. If $k = 4$, then $m = n = 2$ and $s < 2 \cdot 2 = 4$, which means that from the equation $x_1 + x_2 = y_1 + y_2$ one can remove the number 1 from both sides. If $k > 4$, we carry out the induction step by bringing (8) into the form

$$(x_1 - y_1) + x_2 + \cdots + x_n = y_2 + y_3 + \cdots + y_m, \tag{9}$$

which lowers the value of k by 1. In doing this, we assume that x_1 (respectively y_1) is maximal among the numbers x_i (respectively y_i); this can be achieved by changing their orders. Furthermore, we assume that $x_1 > y_1$ (for if $x_1 = y_1$, then we simply remove the pair x_1, y_1 from both sides of (8); if $x_1 < y_1$, we can reduce it to the case $x_1 > y_1$ by switching the sides of (8)). By the induction hypothesis we can remove appropriate summands from equation (9) as long as the sum $s' = y_2 + y_3 + \cdots + y_m$ satisfies $s' < n(m - 1)$. But this is guaranteed by the fact that y_1 is the largest of the numbers y_i: From $s = y_1 + y_2 + \cdots + y_m$ it follows that $y_1 \ge \frac{s}{m}$, and thus $s' = s - y_1 \le \frac{m-1}{m}s < n(m - 1)$, since $s < mn$.

We finish this solution by remarking that from an admissible deletion of summands in (9) we can return to an appropriate deletion in (8), where the two terms x_1, y_1 are deleted if and only if the term $(x_1 - y_1)$ is among the deleted summands in (9). All other deletions are then the same in (8) and (9). $\square$

(iii) Suppose that the sum of all $2n$ terms of a sequence $x_1, x_2, \ldots, x_{2n}$ of real numbers is equal to a number A, and that none of the numbers $|x_{i+1} - x_i|$, $1 \le i \le 2n - 1$, exceeds a given positive number ε. Show that n terms of the sequence $x_1, x_2, \ldots, x_{2n}$ can be chosen such that their sum S satisfies the inequality $|S - \frac{A}{2}| \le \frac{\varepsilon}{2}$.

SOLUTION.   For each subsequence $c = (x_{i_1}, x_{i_2}, \ldots, x_{i_n})$, where $1 \le i_1 < \cdots < i_n \le 2n$, we set $S(c) = x_{i_1} + x_{i_2} + \cdots + x_{i_n}$, and we denote the set of all such subsequences c of length n by M. We show that $|S(c) - \frac{A}{2}| \le \frac{\varepsilon}{2}$ for an appropriate $c \in M$. Two elements $c = (x_{i_1}, x_{i_2}, \ldots, x_{i_n})$ and $d = (x_{j_1}, x_{j_2}, \ldots, x_{j_n})$ will be called *adjacent* if $|i_1 - j_1| + |i_2 - j_2| + \cdots + |i_n - j_n| = 1$. By hypothesis we have $|S(c) - S(d)| \le \varepsilon$ for any adjacent elements $c, d \in M$. It is easy to write down a sequence $c_1, c_2, \ldots, c_{n^2}$ of elements of M in which any two neighboring terms are adjacent, where $c_1 = (x_1, x_2, \ldots, x_n)$ and $c_{n^2} = (x_{n+1}, x_{n+2}, \ldots, x_{2n})$. This can be done, for instance, as follows: First we change the term x_n in c_1 step by step to $x_{n+1}, x_{n+2}, \ldots, x_{2n}$; then we change x_{n-1} to $x_n, x_{n+1}, \ldots, x_{2n-1}$, etc., until finally x_1 is changed to $x_2, x_3, \ldots, x_{n+1}$. Since $S(c_1) + S(c_{n^2}) = A$, the number $\frac{A}{2}$ lies between $S(c_1)$ and $S(c_{n^2})$, and thus also between $S(c_k)$ and $S(c_{k+1})$ for an appropriate $k \in \{1, 2, \ldots, n^2 - 1\}$. Since for such a k we have

$$\left| S(c_k) - \frac{A}{2} \right| + \left| S(c_{k+1}) - \frac{A}{2} \right| = |S(c_{k+1}) - S(c_k)| \le \varepsilon,$$

at least one of the two numbers $\left| S(c_k) - \frac{A}{2} \right|$, $\left| S(c_{k+1}) - \frac{A}{2} \right|$ is equal to at most $\frac{\varepsilon}{2}$.   □

2.6 Exercises

(i) We say that a sequence $a_1, a_2, \ldots, a_{2n}$ of the numbers $0, 1$ is *parity-balanced* if $a_1 + a_3 + \cdots + a_{2n-1} = a_2 + a_4 + \cdots + a_{2n}$. Show that from an arbitrary sequence of the numbers $0, 1$ of length $2n + 1$ one can choose a parity-balanced subsequence of length $2n$.

(ii) How can 2.5.(i) be formulated in the case of an odd integer n?

*(iii) For each integer $k \ge 1$ find the least real number d_k with the following property: If $x_1, x_2, \ldots, x_n$ are arbitrary numbers in the interval $[0, 1]$ such that $x_1 + x_2 + \cdots + x_n > k$, then for an appropriately chosen sum S of some of them we have the inequalities $0 \le k - S \le d_k$.

*(iv) Solve the analogue of (iii) for the bound $|S - k| \le d_k$.

(v) For each $n \ge 1$ find the least real number d_n that satisfies the following: From an arbitrary sequence of length n of real numbers one can choose some terms such that their sum S satisfies the inequality $|S - k| \le d_n$ for some $k \in \mathbb{Z}$.

(vi) Suppose that the sum of the terms of a sequence $x_1, x_2, \ldots, x_n$ of real numbers is equal to A, and that none of the numbers $|x_{i+1} - x_i|$, $1 \le i \le n - 1$, exceeds a given number $\varepsilon > 0$. Show that for all $k \in \{1, 2, \ldots, n - 1\}$ one can choose k terms from $x_1, x_2, \ldots, x_n$ such that their sum S satisfies the inequality $\left| S - \frac{kA}{n} \right| \le \frac{\varepsilon}{2}$. (This is a generalization of 2.5.(iii), where $k = \frac{n}{2}$.)

*(vii) Suppose that the sum of n real numbers from the interval $[-1, 1]$ is equal to A ($n \ge 5$). How can one select some of these numbers such that their sum S satisfies the inequality $\left| S - \frac{A}{3} \right| \le \frac{1}{3}$?

*(viii) Suppose that the sequence $A = (a_1, a_2, \ldots, a_n)$ of elements from the set $M = \{1, 2, \ldots, m\}$ is such that $a_1 + a_2 + \cdots + a_n = 2S$, where S is an integer divisible by each of the elements of M. Show that from the sequence A one can choose some terms such that their sum is equal to S.

2.7 Groups of Neighboring Terms

In the introduction to this section we mentioned neighboring terms of a sequence. We now generalize this notion as follows: A subsequence $a_i, a_{i+1}, \ldots, a_{i+k-1}$ of a sequence $a_1, a_2, \ldots$ will be called a k-*tuple of neighboring terms* of the given sequence. If we do not wish to specify its length $k \geq 1$, we refer to it as a *group of neighboring* (or *consecutive*) *terms*.

(i) Find the largest $k \leq 10$ for which one can place k different digits on the circumference of a circle such that any pair of neighboring digits forms (in an appropriate order) an integer divisible by 7.

SOLUTION.

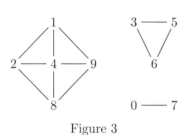

Figure 3

We determine the possible pairs of neighboring digits by writing down all multiples of 7 with at most two digits: 07, 14, 21, ..., 98. This can be represented by the *graph* of Figure 3, where each edge between two digits means that these digits form one of the multiples of 7. As we see, the graph consists of three connected parts (*components*). Since it is clear that we can place only the digits of the same component on the circumference of the circle, we obtain the bound $k \leq 5$. The value $k = 5$ can in fact be achieved, for instance by the sequence $1, 2, 4, 8, 9$ (which can be found by an appropriate closed walk along the largest component).   □

(ii) Suppose that the integer k is greater than the positive integer n, but is not a multiple of n. Show that there exists a sequence of length k of real numbers with the following property: The sum of any n consecutive terms is negative, while the sum of all k terms is positive.

SOLUTION. It is convenient to look for the desired sequence in the form

$$x_1, x_2, \ldots, x_n, x_1, x_2, \ldots, x_n, \ldots,$$

for then all the sums of n consecutive terms are equal to the same number $A = x_1 + x_2 + \cdots + x_n$. In order to express the sum B of all k terms of such a sequence, we divide k by n with remainder: $k = np + q$ $(p, q \in \mathbb{N}, 0 < q < n)$. Then $B = pA + x_1 + x_2 + \cdots + x_q$, and we see that it is very easy to satisfy the conditions $A < 0$, $B > 0$: It suffices to set $x_1 = p + 1$, $x_i = 0$ $(1 < i < n)$, and $x_n = -p - 2$; then $A = -1$ and $B = 1$.   □

(iii) Suppose that the sequences $a_1, a_2, \ldots, a_m$ and $b_1, b_2, \ldots, b_n$ of positive integers satisfy the inequalities $a_i \leq n$ $(1 \leq i \leq m)$ and $b_i \leq m$ $(1 \leq i \leq n)$. Show that from each of the two sequences one can choose a group of neighboring terms such that the sums of the numbers in both groups are the same.

SOLUTION.   We set $A(i) = a_1 + \cdots + a_i$ $(1 \leq i \leq m)$, $B(i) = b_1 + \cdots + b_i$ $(1 \leq i \leq n)$ and assume that $A(m) \geq B(n)$ (in the case where $A(m) < B(n)$ we interchange the two sequences). For each $i = 1, 2, \ldots, n$ there exists a *smallest* index k with the property $A(k) \geq B(i)$; we denote it by k_i. Since $A(k_i) = A(k_i - 1) + a_{k_i} < B(i) + n$, all of the n numbers $d_i = A(k_i) - B(i)$ lie in the set $\{0, 1, \ldots, n-1\}$. If $d_i = 0$ for some i, then we are done $(a_1 + a_2 + \cdots + a_{k_i} = b_1 + b_2 + \cdots + b_i)$; otherwise, by the pigeonhole principle there exist indices i, j $(1 \leq i < j \leq n)$ such that $d_i = d_j$, and therefore $B(j) - B(i) = A(k_j) - A(k_i)$. But this means that $b_{i+1} + b_{i+2} + \cdots + b_j = a_{k_i+1} + a_{k_i+2} + \cdots + a_{k_j}$.   □

(iv) Suppose that a finite sequence of real numbers contains positive and negative numbers. We underline all the positive terms and also every term that is such that its sum with one or more immediately following terms is positive. Show that the sum of all underlined terms is positive.

SOLUTION.   We will base the proof on dividing the underlined numbers of the given sequence $a_1, a_2, \ldots, a_n$ into groups of neighboring terms such that the sum of the terms in each of these groups is positive.

Let a_i be the underlined term with smallest index $(1 \leq i \leq n)$. We choose the *smallest* $k \geq 0$ such that $a_i + a_{i+1} + \cdots + a_{i+k}$ is positive (we set $k = 0$ when $a_i > 0$). If $k > 0$, then all the numbers $a_i, a_i+a_{i+1}, \ldots, a_i+a_{i+1}+\cdots+a_{i+k-1}$ are nonpositive, which together with $a_i+a_{i+1}+\cdots+a_{i+k} > 0$ means that all the numbers $a_{i+1}+a_{i+2}+\cdots+a_{i+k}, a_{i+2}+a_{i+3}+\cdots+a_{i+k}, \ldots, a_{i+k}$ are positive. Therefore, the numbers $a_i, a_{i+1}, \ldots, a_{i+k}$ are all underlined (which is clearly true also in the case $k = 0$). This implies that the sum of all underlined terms in the group $a_1, a_2, \ldots, a_{i+k}$ is positive. If $i + k = n$, we are done. Otherwise, we repeat the whole process for the "shortened" sequence $a_{i+k+1}, a_{i+k+2}, \ldots, a_n$. After a finite number of steps we have then exhausted the whole sequence.   □

(v) We say that the sequence $a_1, a_2, \ldots, a_N$ is of type ABA if for some $k \leq \frac{N}{2}$ the k-tuple of its first terms agrees with the k-tuple of its last terms, that is, if $a_j = a_{N-k+j}$ $(1 \leq j \leq k)$. Show that for each n there exists a finite sequence $a_1, a_2, \ldots, a_N$ such that any "extended" sequence $x, a_1, a_2, \ldots, a_N$, where $x \in \{1, 2, \ldots, n\}$, is of type ABA.

SOLUTION.   It is not difficult to find a suitable sequence P_n for small values of n, for instance $P_1 = (1)$, $P_2 = (2, 1, 2)$, $P_3 = (3, 2, 3, 1, 3, 2, 3)$. By carefully comparing the examples P_2 and P_3, you will certainly notice the trick that will help us with a recursive construction: If the sequence $x, b_1, b_2, \ldots, b_N$ is of type ABA for every x from some set X, then the sequence $x, y, b_1, y, b_2, \ldots, y, b_N, y$ has the same property, with an arbitrary (fixed) element y. Let us therefore carry out the following recursive construction: $P_1 = (1)$, and if $P_k = (a_1, a_2, \ldots, a_{2^k-1})$ for some $k \geq 1$, we put

$$P_{k+1} = (k+1, a_1, k+1, a_2, \ldots, k+1, a_{2^k-1}, k+1),$$

which is a sequence of length $2(2^k - 1) + 1 = 2^{k+1} - 1$. We can verify that the property of these sequences P_k can be made precise (by induction on k): If $P_k = (a_1, a_2, \ldots, a_{2^k-1})$, then for each x $(1 \leq x \leq k)$ the group of the first 2^{k-x} terms in the sequence $x, a_1, a_2, \ldots, a_{2^k-1}$ is the same as the group of the last 2^{k-x} terms.

It is possible that upon examining the examples P_1, P_2, P_3 you thought of a different, "arithmetical," description of the recursive construction: If $P_k = (a_1, a_2, \ldots, a_{2^k-1})$, then

$$P_{k+1} = (a_1 + 1, a_2 + 1, \ldots, a_{2^k-1} + 1, 1, a_1 + 1, \ldots, a_{2^k-1} + 1). \qquad \square$$

2.8 Exercises

(i) Find the largest $k \leq 9$ for which one can place k different nonzero digits on the circumference of a circle such that from any pair of neighboring digits one can form a two-digit prime number.

(ii) Suppose that $n+1$ positive integers, whose sum is equal to $3n$, are written on the circumference of a circle. Show that one can choose a collection of neighboring integers whose sum is equal to $2n$.

(iii) For which $n \geq 3$ can one place n different real numbers on the circumference of a circle such that each one of them is equal to the product of both its neighbors?

(iv) Find all cyclic sequences $x_1, x_2, \ldots, x_{30}$, given that $x_1 + x_2 + \cdots + x_{30} = 20$ and $x_i = |x_{i+2} - x_{i+1}|$ for all i.

(v) The serial numbers on the tram tickets in Brno range from 000000 to 999999. Let us call a ticket with number $ABCDEF$ *unlucky* if $A + B + C \neq D + E + F$. Find the largest possible number of consecutive unlucky tickets.

(vi) Suppose that 80 nonzero digits are written on a strip of paper. We cut the strip into several pieces such that each one contains at least two digits. Then we add all the numbers formed by the digits on the individual pieces. Show that there are at least 11 different ways of cutting the strip for which we always obtain the same sum.

*(vii) Suppose that for the relatively prime integers n, k there exists a sequence of length m with the property that the sum of any n consecutive terms is negative and the sum of any k consecutive terms is positive. Show that $m < n + k - 1$. For $n = 13$ and $k = 7$ show that the case $m = 18$ is possible. Finally, decide how the upper bound for m changes in the case where the integers n, k are not relatively prime.

2.9 Standard Sequences

In practical situations *arithmetic* and *geometric sequences* are sometimes found to be useful, and this is why these sequences occur in most high-school mathematics courses. We recall that a sequence $a_1, a_2, \ldots$ is called an arithmetic sequence with difference d if for each index $i > 1$ we have $a_i - a_{i-1} = d$, and thus $a_i = a_1 + (i - 1)d$. The sequence $b_1, b_2, \ldots$ is called a geometric sequence with quotient q if for each index $i > 1$ we have $b_i = qb_{i-1}$, and thus $b_i = b_1 q^{i-1}$. For finding the sums of the first n terms of both sequences there are well-known formulas, which can be found, for instance, in [5], Chapter 1, Section 2. However, these formulas will not be needed here.

(i) Suppose that the first term a_1 and the difference d of an infinite arithmetic sequence are positive integers. Show that some term of the sequence has the digit 9 in its decimal expansion.

SOLUTION.   The term $a_{i+1} = a_1 + id$ of the given sequence is a $(k+1)$-digit number with first digit 9 if $9 \cdot 10^k \le a_1 + id < 10^{k+1}$, which is equivalent to $i \in [\alpha, \beta)$, where the positive numbers α, β are given by

$$\alpha = \frac{9 \cdot 10^k - a_1}{d} \quad \text{and} \quad \beta = \frac{10^{k+1} - a_1}{d}.$$

Such an index i exists for a chosen k if $\alpha \ge 1$ and $\beta - \alpha > 1$. It is easy to see that these two conditions can be brought into the form

$$a_1 + d \le 9 \cdot 10^k \quad \text{and} \quad d < 10^k.$$

It is clear that we can find an appropriate k, no matter what $a_1, d \in \mathbb{N}$ are (for instance, we can take the number of digits of a_2 as k).   □

(ii) For each $n \ge 1$ choose positive integers $a_1 < a_2 < \cdots < a_{2n+1}$ that form an arithmetic sequence such that the product $a_1 a_2 \cdots a_{2n+1}$ is the square of an integer.

SOLUTION.   We try the choice $A, 2A, 3A, \ldots, (2n+1)A$ with an appropriate $A \in \mathbb{N}$. The product of these numbers is $(2n+1)! A^{2n+1}$, which is a perfect square for $A = (2n+1)!$.   □

(iii) Show that from an infinite arithmetic sequence $a, a + d, a + 2d, \ldots$ (where $a, d \in \mathbb{R}, d \ne 0$) one can choose an infinite geometric sequence if and only if the number $\frac{a}{d}$ is rational.

SOLUTION.   If $a + kd, a + md, a + nd$, where $0 \le k < m < n$, are a triple of neighboring terms of some geometric sequence, then

$$(a + kd)(a + nd) = (a + md)^2,$$

from which upon dividing by d^2 and substituting $t = \frac{a}{d}$ we obtain

$$(t+k)(t+n) = (t+m)^2,$$

or $t(2m-k-n) = kn-m^2$. The number t is therefore rational if $2m-k-n = 0 = kn - m^2$ does not hold. But this is not possible, since it would mean that

$$m = \frac{k+n}{2} = \sqrt{kn},$$

and we know that the arithmetic and the geometric means of the numbers k, n agree only if $k = n$ (see, e.g., [5], Chapter 2, Section 8).

Conversely, let $\frac{a}{d} \in \mathbb{Q}$. We may clearly assume that the number a has the same sign as $d \neq 0$ (otherwise, we simply delete the first few terms $a, a+d, a+2d, \ldots$, that is, change the number a to $a + kd$, with k large enough such that $d(a+kd) > 0$; note that this change does not affect the condition $\frac{a}{d} \in \mathbb{Q}$). We may therefore write $\frac{a}{d} = \frac{p}{q}$, where $p, q \in \mathbb{N}$. By the binomial theorem (see, e.g., [5], Chapter 1, Section 1) each term of the geometric sequence of the numbers $b_m = a(1+q)^m$, where $m = 1, 2, \ldots$, is of the form

$$b_m = a + aq \left[\binom{m}{1} + q \binom{m}{2} + \cdots + q^{m-1} \binom{m}{m} \right] = a + aqk_m = a + pk_m d,$$

where $k_m \in \mathbb{N}$, that is, b_m lies in the sequence $a, a+d, a+2d, \ldots$. Since furthermore $k_m > k_{m-1}$ for all $m > 1$, we have thus found an infinite geometric subsequence. $\qquad \square$

(iv) Suppose that the infinite sequence $x_1, x_2, \ldots$ of real numbers is such that for any indices m, n we have the inequality

$$|x_{m+n} - x_m - x_n| < \frac{1}{m+n}.$$

Show that this sequence is an arithmetic sequence.

SOLUTION.   By the hypothesis we have for any k, n,

$$|(x_{n+1} - x_n) - (x_{k+1} - x_k)|$$
$$= |(x_{n+k+1} - x_n - x_{k+1}) - (x_{n+k+1} - x_{n+1} - x_k)|$$
$$\leq |x_{n+k+1} - x_n - x_{k+1}| + |x_{n+k+1} - x_{n+1} - x_k| \leq \frac{2}{n+k+1}.$$

If we use this inequality twice, we obtain for any m,

$$|(x_{n+1} - x_n) - (x_{k+1} - x_k)|$$
$$\leq |(x_{n+1} - x_n) - (x_{m+1} - x_m)| + |(x_{m+1} - x_m) - (x_{k+1} - x_k)|$$
$$\leq \frac{2}{m+n+1} + \frac{2}{m+k+1} < \frac{2}{m} + \frac{2}{m} = \frac{4}{m}.$$

The system of inequalities

$$|(x_{n+1} - x_n) - (x_{k+1} - x_k)| < \frac{4}{m} \quad (m \in \mathbb{N})$$

can hold only when the nonnegative number on the left (which is independent of m) is zero, that is, when $x_{n+1} - x_n = x_{k+1} - x_k$. But this means that the sequence $x_1, x_2, \ldots$ is arithmetic.

For the sake of interest we add that the inequalities from the statement of the problem are satisfied exactly by those sequences $x_1, x_2, \ldots$ that are of the form $x_n = nd$, where $d \in \mathbb{R}$ is arbitrary.   □

(v) Show that for every $n \geq 3$ one can choose an arithmetic sequence of length n consisting of composite pairwise relatively prime positive integers.

SOLUTION.   For a given $n \geq 3$ we choose a prime number $p > n$ and an integer $N \geq p + (n-1) \cdot n!$. Then the numbers $a_k = N! + p + (k-1) \cdot n!$, $1 \leq k \leq n$, form an arithmetic sequence of length n and are composite (since the number a_k is a multiple of $p + (k-1) \cdot n!$, which is one of the factors in the product $2 \cdot 3 \cdots N = N!$). If we assume that some two numbers a_i, a_j $(1 \leq i < j \leq n)$ have a common prime divisor q, then q divides the difference $a_j - a_i = (j-i) \cdot n!$, which means that $q \leq n$. Therefore, q divides the numbers $N!$ and $n!$, and thus also the number $a_j - N! - (j-1) \cdot n! = p$, which is a contradiction to the choice of p. This means that the numbers $a_1, a_2, \ldots, a_n$ are pairwise relatively prime, and the proof is complete.   □

We add that it is of interest to compare this result with the famous *Dirichlet's theorem on primes in arithmetic progressions*: "If a_1, d are relatively prime positive integers, then among the numbers $a_k = a_1 + (k-1)d$ there are infinitely many primes."

2.10   Exercises

(i) Is there an infinite geometric sequence of positive numbers a_1, $a_2, \ldots, a_n, \ldots$ such that the term a_i is an integer if and only if $1 \leq i \leq 1991$?

(ii) Suppose that the infinite sequence of integers $a_1, a_2, a_3, \ldots$ is geometric with quotient q. Show that if $a_1 \neq 0$, then q is an integer.

(iii) Suppose that a geometric sequence of positive numbers has infinitely many integer terms. Does this imply that its quotient is an integer?

(iv) Can 3, 5, 7 be a triple of (not necessarily neighboring) terms of some geometric sequence?

(v) For which $k \geq 3$ can one choose an arithmetic sequence of length k from the infinite sequence $1, \frac{1}{2}, \frac{1}{3}, \frac{1}{4}, \ldots$?

(vi) Suppose that a given sequence of length $4n$ of positive numbers has the property that from each quadruple of different values of of its terms one can construct a geometric sequence. Show that at least n terms of the given sequence have the same value.

(vii) Show that for infinitely many values n the sequence of binomial coefficients $\binom{n}{0}, \binom{n}{1}, \ldots, \binom{n}{n}$ has a triple of neighboring terms that form an arithmetic sequence. Is there a similar quadruple for some n? Answer the analogous question for geometric triples and quadruples of neighboring terms.

(viii) Find the largest possible number of three-term arithmetic subsequences that can be chosen from the same sequence of numbers $a_1 < a_2 < \cdots < a_n$.

*(ix) Suppose that it is possible to choose a geometric subsequence of length $k \geq 3$ from some arithmetic sequence of length n. Show that if the difference of the sequence is nonzero, then $n \geq 2^{k-1}$.

*(x) For which n is there an arithmetic sequence $a_1, a_2, \ldots, a_{n+1}$ and a geometric sequence $b_1, b_2, \ldots, b_n$ such that

$$a_1 < b_1 < a_2 < b_2 < \cdots < a_n < b_n < a_{n+1}?$$

*(xi) We say that a numerical sequence is *antiarithmetic* if it does not contain any arithmetic subsequence of length 3. Show that for each $k \geq 0$ the sequence $1, 2, 3, \ldots, 5 \cdot 3^k$ contains an antiarithmetic subsequence of length $4 \cdot 2^k$.

*(xii) Show that there exists an antiarithmetic sequence (see (xi)) of positive integers such that each $x \in \mathbb{N}$ that is not a term of the sequence is the arithmetic mean of two of its terms.

(xiii) For each $a \in \mathbb{N}$ denote by $\mathcal{P}(a)$ the set of all prime numbers that divide the integer a (thus, for example, $\mathcal{P}(12) = \{2, 3\}$). Show that from any infinite arithmetic sequence of positive integers $a_1, a_2, \ldots$ one can choose infinitely many a_i with the same sets $\mathcal{P}(a_i)$.

(xiv) Without using Dirichlet's theorem (see 2.9.(v)) prove the following: For any $d \in \mathbb{N}$ there exists an $a_1 \in \mathbb{N}$ such that the arithmetic sequence $a_k = a_1 + (k-1)d$ contains infinitely many prime numbers.

As a *perfect power* we consider in Exercises (xv)–(xix) any integer of the form k^m, where k, m are integers greater than 1.

*(xv) Show that for each $n \geq 3$ there is an arithmetic sequence of length n that consists of n different perfect powers.

(xvi) Explain why there is no infinite arithmetic sequence that consists exclusively of different perfect powers.

(xvii) Show that for every integer $a_1 > 1$ that is not a perfect power there is a $d \in \mathbb{N}$ such that no term of the arithmetic sequence $a_1, a_1 + d, \ldots$ is a perfect power. Is this also true for $a_1 = 1$?

*(xviii) Show that for each n there are n consecutive positive integers among which there is no perfect power.

(xix) In the sequence of triples $(2, 3, 5)$, $(6, 15, 10)$, ... each triple (a, b, c) is followed by (ab, bc, ca). Show that no triple contains a perfect power.

2.11   Miscellaneous Problems

(i) Let $x_1, x_2, \ldots, x_{19}$ be a sequence of positive integers less than 71, and consider the sum of the terms of an arbitrary subsequence. Show that among the sums that occur there are fewer than 1200 different values.

SOLUTION.   If M is the set of values of all the sums that occur, then the integer $S = x_1 + x_2 + \cdots + x_{19}$ is its largest value. We estimate the number of integers from $1, 2, \ldots, S$ that do not lie in M. To do this, we assume that $x_1 \leq x_2 \leq \cdots \leq x_{19}$, which is no loss of generality. Since $S > S - x_1 > x_2 \geq 1$, the sets $M_1 = \{1, 2, \ldots, x_2\}$ and $M_2 = \{S - x_1, S - x_1 + 1, \ldots, S\}$ are disjoint, and each of them contains at most two elements of M (namely, M_1 contains the numbers x_1 and x_2, and M_2 the numbers $S - x_1$ and S). Therefore, M does not contain at least $x_2 - 2$ elements of M_1 and at least $x_1 - 1$ elements of M_2, and thus

$$|M| \leq S - (x_2 - 2) - (x_1 - 1) = x_3 + x_4 + \cdots + x_{19} + 3$$
$$\leq 17 \cdot 70 + 3 = 1193. \qquad \square$$

(ii) We are given 20 integers $1 \leq a_1 < a_2 < \cdots < a_{20} \leq 70$. Show that among the differences $a_j - a_k$ ($1 \leq k < j \leq 20$) at least four values are identical.

SOLUTION.   We assume that among the 19 integers $a_{20} - a_{19}, a_{19} - a_{18}, \ldots, a_2 - a_1$ there are no four identical ones. If we enumerate them according to their sizes $1 \leq b_1 \leq b_2 \leq \cdots \leq b_{19}$, then this means that $1 \leq b_1 \leq b_2 \leq b_3$, $2 \leq b_4 \leq b_5 \leq b_6$, $3 \leq b_7 \leq b_8 \leq b_9$, ..., $6 \leq b_{16} \leq b_{17} \leq b_{18}$, and $7 \leq b_{19}$. Therefore, $b_1 + b_2 + \cdots + b_{19} \geq 3(1 + 2 + \cdots + 6) + 7 = 70$, which is a contradiction, since $b_1 + b_2 + \cdots + b_{19} = a_{20} - a_1 \leq 69$. $\qquad \square$

(iii) Suppose that the cyclic sequence $x_1, x_2, \ldots, x_n$ consisting of the numbers $+1$ and -1 has the property that for each $k = 1, 2, \ldots, n - 1$ the identity

$$x_1 x_{1+k} + x_2 x_{2+k} + \cdots + x_n x_{n+k} = 0$$

holds. Show that the number n is the square of an integer.

SOLUTION.   If $S_k = x_1 x_{1+k} + x_2 x_{2+k} + \cdots + x_n x_{n+k}$, $0 \le k < n$, then

$$(x_1 + x_2 + \cdots + x_n)^2 = S_0 + S_1 + \cdots + S_{n-1}.$$

(The reader should determine the coefficients with which a product $x_i x_j$, where $i, j \in \{1, 2, \ldots, n\}$, occurs on both sides of this equation.) By hypothesis we have $S_k = 0$ $(1 \le k \le n-1)$ and $S_0 = x_1^2 + x_2^2 + \cdots + x_n^2 = n$. Therefore, the above equation takes the form $(x_1 + x_2 + \cdots + x_n)^2 = n$. Hence the number n is the square of the integer $x_1 + x_2 + \cdots + x_n$.   □

(iv) Suppose that the sequence of positive integers $a_1, a_2, \ldots, a_d$ satisfies $a_{j+2} = |a_{j+1} - a_j|$ for each $j = 1, 2, \ldots, d-2$. Show that $d \le \frac{3}{2}(a_2 + 1)$ if $a_1 \le a_2$.

SOLUTION.   We prove the statement by induction on $n = a_2$. A given sequence is uniquely determined by its first two terms a_1, a_2. By a straightforward analysis of all possible choices of the pair (a_1, a_2), $1 \le a_1 \le a_2 = n$, we find that $d \le 2$ for $n = 1$, $d \le 4$ for $n = 2$, $d \le 6$ for $n = 3$, and $d \le 7$ for $n = 4$. We assume now that $n \ge 5$ and that the assertion is true for all pairs (a_1, a_2), where $1 \le a_1 \le a_2 < n$, and we prove the assertion for the sequence

$$a_1, n, n - a_1, a_1, |n - 2a_1|, \ldots, a_d, \tag{10}$$

supposing that $1 \le a_1 < n$ (the case $a_1 = n$, where $d = 2$, can be disregarded). We try to use the induction hypothesis for the shortened sequence $a_3, a_4, \ldots, a_d$, that is, the sequence

$$n - a_1, a_1, |n - 2a_1|, \ldots, a_d. \tag{11}$$

This is possible only when $n - a_1 \le a_1 < n$, that is, when $\frac{n}{2} \le a_1 \le n - 1$; then by the induction hypothesis we have $d - 2 \le \frac{3}{2}(a_1 + 1)$, which implies the required inequality $d \le \frac{3}{2}(n + 1)$ if $3a_1 \le 3n - 4$, that is, if $a_1 \le n - 2$. It therefore remains to consider the cases $1 \le a_1 < \frac{n}{2}$ and $a_1 = n - 1$. To deal with the first case, we put the number $n - 2a_1$ (≥ 1) in front of the sequence (11); we then obtain a sequence of length $d - 1$ satisfying the given conditions. Since $n - 2a_1 < n - a_1 \le n - 1$, we have by the induction hypothesis that

$$d - 1 \le \frac{3(n - a_1 + 1)}{2} \le \frac{3n}{2}, \quad \text{and thus} \quad d \le \frac{3(n + 1)}{2}.$$

In the remaining case $a_1 = n - 1$ the sequence (10) has the form

$$n - 1, n, 1, n - 1, n - 2, 1, n - 3, n - 4, 1, \ldots, 2, 1, 1$$

We can now convince ourselves by induction that for each $n \ge 2$ this sequence has length $\left\lceil \frac{3(n+1)}{2} \right\rceil$ (to do this we replace its first four terms by the single term $n - 3$).   □

2.12 Exercises

(i) Find real numbers $x_1, x_2, \ldots, x_{100}$ such that $x_1 = 1$, $0 \leq x_2 \leq 2x_1$, $0 \leq x_3 \leq 2x_2$, $\ldots$, $0 \leq x_{100} \leq 2x_{99}$, and the sum $S = x_1 - x_2 + x_3 - x_4 + \cdots + x_{99} - x_{100}$ has the largest possible value.

(ii) Given an arbitrary sequence of real numbers $a_1, a_2, \ldots, a_{15}$, we set up a sequence $b_1, b_2, \ldots, b_{15}$, where for each $i = 1, 2, \ldots, 15$ the term b_i is equal to the number of those terms a_j, $1 \leq j \leq 15$, for which $a_j < a_i$ holds. Can such a sequence have the form

$$1, 0, 3, 6, 9, 4, 7, 2, 5, 8, 8, 5, 10, 13, 13?$$

(iii) What interesting property with respect to the construction described in (ii) does the sequence 6, 0, 3, 3, 7, 1, 7, 7, 11, 13, 1, 3, 7, 13, 11 have? Describe all sequences with this property.

(iv) Suppose that the infinite sequence of positive integers $a_1 = 1$, a_2, a_3, $\ldots$ satisfies the condition $a_k \leq 1 + a_1 + a_2 + \cdots + a_{k-1}$ for all $k > 1$. Show that every positive integer is the sum of the terms of one of its subsequences.

*(v) For each of the sets $K = \{0\}$ and $K = \{0, 1\}$ decide whether there exists a sequence of integers $a_1, a_2, \ldots$ such that any $x \in \mathbb{Z}$ can be written in the form

$$x = a_{i_1} + a_{i_2} + \cdots + a_{i_k} \quad (1 \leq i_1 < i_2 < \cdots < i_k) \qquad (12)$$

for some $k \geq 1$ exactly when $x \notin K$, where for each $x \in \mathbb{Z} \setminus K$ the expression (12) is unique.

*(vi) Let $n > k \geq 2$. We call the sequence of numbers $a_1, a_2, \ldots, a_k$ from the interval $(0, 1)$ *exceptional* if for any expansion $n = n_1 + n_2 + \cdots + n_k$ of n into k nonnegative integer summands n_i at least one of the numbers $a_i n_i$ $(1 \leq i \leq k)$ is an integer. For which pairs n, k are there exceptional sequences, and what do they all look like?

*(vii) Suppose that the positive integers x_1 and x_2 are less than $10\,000$, and that the number x_3 is equal to $|x_1 - x_2|$, that x_4 is equal to the smallest of $|x_1 - x_2|$, $|x_2 - x_3|$, and $|x_1 - x_3|$, that x_5 is equal to the smallest of $|x_1 - x_2|$, $|x_1 - x_3|$, $|x_1 - x_4|$, $|x_2 - x_3|$, $|x_2 - x_4|$, and $|x_3 - x_4|$, etc. Show that $x_{21} = 0$.

(viii) Suppose that each term of the sequence $x_1, x_2, \ldots, x_n$ is either $+1$ or -1. Show that the equality

$$x_1 x_2 + x_2 x_3 + \cdots + x_{n-1} x_n + x_n x_1 = 0 \qquad (13)$$

is possible only when n is divisible by 4.

3   Arrays

While in the previous section we considered *linear* configurations, where numbers were assigned to isolated points on a line or a circle, we will now proceed to studying configurations of numbers arranged in a certain scheme in the plane, namely in an *array*. Each array is formed by a grid of objects in the plane, the so-called *fields* (in the great majority of cases these will be congruent squares), that contain numbers. Unless otherwise specified, we suppose that each field of the array contains exactly one number; only in exceptional problems may some fields remain empty. An array formed by mn congruent squares in m rows and n columns will be called an $m \times n$ *array*; the rows and columns of such an array will be numbered downwards, respectively from left to right, by the indices $i = 1, 2, \ldots, m$, respectively $j = 1, 2, \ldots, n$. Finally, in order to denote the number that lies at the intersection of the ith row and the jth column of the array, we use the pair of subscripts i, j. An $m \times n$ array is therefore a configuration of numbers

$$\begin{pmatrix} a_{11} & a_{12} & \cdots & a_{1n} \\ a_{21} & a_{22} & \cdots & a_{2n} \\ \vdots & \vdots & \ddots & \vdots \\ a_{m1} & a_{m2} & \cdots & a_{mn} \end{pmatrix},$$

which in linear algebra and elsewhere is called an $m \times n$ *matrix*.

3.1   Row Sums and Column Sums

For the $m \times n$ array with elements a_{ij} we define the ith *row sum*

$$r_i = a_{i1} + a_{i2} + \cdots + a_{in} \quad (i = 1, 2, \ldots, m)$$

and the jth *column sum*

$$c_j = a_{1j} + a_{2j} + \cdots + a_{mj} \quad (j = 1, 2, \ldots, n).$$

The row and column sums of each $m \times n$ array satisfy the basic relation

$$r_1 + r_2 + \cdots + r_m = c_1 + c_2 + \cdots + c_n, \tag{14}$$

since both sums in this equation represent the sums of all mn numbers of the array. For an $n \times n$ square array we also define the diagonal sums

$$u = a_{11} + a_{22} + \cdots + a_{nn} \quad \text{and} \quad v = a_{1n} + a_{2\,n-1} + a_{3\,n-2} + \cdots + a_{n1}.$$

(i) Is it possible to construct an $n \times n$ square array from the numbers $-1, 0, 1$ such that all the sums, namely, row, column, and both diagonal sums, are distinct?

SOLUTION.   Each of the $2n+2$ sums in question lies in the $(2n+1)$-element set

$$\{-n, -n+1, \ldots, -1, 0, 1, \ldots, n-1, n\};$$

hence by the pigeonhole principle two of these sums have the same value. Therefore, it is not possible to construct an array with the desired properties.   □

(ii) Suppose that an $m \times n$ array has the following property: If we multiply the sum of numbers in an arbitrary row with the sum of numbers in an arbitrary column, we obtain the number in the field at the intersection of this row and this column. Show that either the sum of all numbers in the array is equal to 1, or all numbers in the array are equal to 0.

SOLUTION.   For an element a_{ij}, row sum r_i, and column sum c_j of the array in question we have $a_{ij} = r_i c_j$ $(1 \le i \le m, 1 \le j \le n)$. We then obtain for the sum S of all numbers in the array,

$$S = \sum_{i=1}^{m} \sum_{j=1}^{n} r_i c_j = \sum_{i=1}^{m} r_i \cdot \sum_{j=1}^{n} c_j = S \cdot S = S^2;$$

the equation $S = S^2$ then implies either $S = 1$ or $S = 0$. If $S = 0$, then for each $i = 1, 2, \ldots, m$ we have

$$r_i = \sum_{j=1}^{m} a_{ij} = \sum_{j=1}^{m} r_i c_j = r_i \sum_{j=1}^{m} c_j = r_i \cdot S = 0,$$

and thus $a_{ij} = r_i c_j = 0$, $1 \le i \le m$, $1 \le j \le n$.   □

(iii) Suppose that the $m + n$ positive numbers $r_1, r_2, \ldots, r_m, c_1, c_2, \ldots, c_n$ satisfy the identity (14). Show that for some $k \le m + n - 1$ one can write into k fields of an empty $m \times n$ array k positive numbers such that the sum of these numbers in the ith row (respectively the jth column) is equal to r_i (respectively c_j) for each $i = 1, 2, \ldots, m$ (respectively each $j = 1, 2, \ldots, n$).

SOLUTION.   We cannot enter "too many" numbers in the array. Consider, for instance, entering the number a_{11} as only number in the first row or the first column. If you then continue to reason correctly, you will certainly come up with an inductive approach to fill in the whole $m \times n$ array.

Let us now carry out the proof by induction on $m + n \ge 2$. The assertion is clear in the case $m + n = 2$, as well as in the more general case where $1 \in \{m, n\}$. Therefore, we consider now an $m \times n$ array with $m > 1$ and $n > 1$, under the assumption that the assertion is true for all $m' \times n'$ arrays, with $m' + n' < m + n$. We now distinguish three cases according to the relative sizes of the numbers r_1, c_1:

(a) $r_1 < c_1$. We set $a_{11} = r_1$ and delete the first row of the array. To the remaining $(m - 1) \times n$ array and the numbers $r_2, r_3, \ldots, r_m, c_1 - r_1$, $c_2, c_3, \ldots, c_n$ we now apply the induction hypothesis.

(b) $r_1 = c_1$. We set $a_{11} = r_1$ and delete the first row and the first column of the array. To the remaining $(m - 1) \times (n - 1)$ array and the numbers $r_2, r_3, \ldots, r_m, c_2, c_3, \ldots, c_n$ we apply the induction hypothesis.

(c) $r_1 > c_1$. We set $a_{11} = c_1$ and delete the first column of the array. As before, we apply the induction hypothesis to the remaining $m \times (n-1)$ array and the numbers $r_1 - c_1, r_2, \ldots, r_m, c_2, c_3, \ldots, c_n$.

This completes the proof by induction. We now show that the above solution has a nice geometric interpretation. We divide the line segment CD of length $r_1 + \cdots + r_m = c_1 + \cdots + c_n$ by the points $A_1 = C$, $A_2, \ldots,$ $A_{m+1} = D$ such that $|A_i A_{i+1}| = r_i$ $(1 \le i \le m)$, and then by the points $B_1 = C$, $B_2, \ldots,$ $B_{n+1} = D$ such that $|B_j B_{j+1}| = c_j$ $(1 \le j \le n)$. We enter a number a_{ij} in the array exactly when for a pair i, j of subscripts the intersection $A_i A_{i+1} \cap B_j B_{j+1}$ is nonempty, and in this case we set $a_{ij} = |A_i A_{i+1} \cap B_j B_{j+1}|$. Since the whole line segment CD is subdivided by the points A_i, B_j into no more than $m+n-1$ subintervals, the number k of entries in the array satisfies $k \le m + n - 1$. □

(iv) For given positive integers $k \le n$ write the numbers $1, 2, 3, \ldots, n^2$ into an $n \times n$ square array such that in each row the entries are placed in increasing order (from left to right) and such that the sum c_k of numbers in the kth column is

(a) as small as possible.

(b) as large as possible.

SOLUTION.   (a) We first show that for any distribution of the numbers we have the bound $c_k \ge k + 2k + \cdots + nk = nk(n + 1)/2$. To do this, it suffices to verify the following assertion: If $x_1 < x_2 < \cdots < x_n$ are the numbers of the kth column (written in increasing order), then $x_i \ge ik$ for all $i = 1, 2, \ldots, n$. This inequality (for a fixed i) follows from the fact that the number x_i is the largest of the ik numbers in the array that lie in the first k columns and in those i rows in which the numbers $x_1, x_2, \ldots, x_i$ are located. On the other hand, the equalities $x_i = ki$ $(1 \le i \le n)$ can all be attained at the same time with an appropriate distribution of numbers: We place the numbers $1, 2, \ldots, nk$ in the first k columns of the array according to the rule $a_{ij} = (i - 1)k + j$, $1 \le i \le n$, $1 \le j \le k$, and the remaining numbers $nk + 1, nk + 2, \ldots, n^2$ in the last $n - k$ columns according to the rule $a_{ij} = nk + (i - 1)(n - k) + j - k$, $1 \le i \le n$, $k < j \le n$. Therefore, the smallest possible value of the sum c_k is $nk(n + 1)/2$.

(b) In a similar way as in part (a) one can show that the largest possible value of the sum c_k is equal to

$$\frac{n\left[(n - 1)^2 + k(n + 1)\right]}{2}.$$

However, we can also use a trick that is typical in mathematics, namely, the use of an appropriate transformation to reduce problem (b) to the already settled problem (a). Indeed, the set of all $n \times n$ arrays with the numbers $1, 2, \ldots, n^2$ can be mapped bijectively onto itself by way of the rule $A = (a_{ij}) \mapsto B = (b_{ij})$, where $b_{ij} = n^2 + 1 - a_{i,n+1-j}$. Verify that this mapping is identical to its inverse, and satisfies the required order of the numbers in the rows. In this way the problem of the largest possible kth column sum of the array A is reduced to the problem of the smallest possible $(n + 1 - k)$th column sum of the array B.   □

(v) Show that one can write the squares of mn appropriate different positive integers in the fields of an $m \times n$ array such that all row and column sums of the resulting array are also squares of positive integers.

FIRST SOLUTION.   We use induction on $m + n \geq 2$. In the case of the smallest integer 2, namely $m = n = 1$, we write, for instance, the number 1^2 in the only field of the array. We now show how we can construct an $(m+1) \times n$ array with the help of an appropriate $m \times n$ array consisting of the integers $x_{ij} = a_{ij}^2$ $(1 \leq i \leq m, \ 1 \leq j \leq n)$, where the integers $a_{ij} \in \mathbb{N}$ are distinct, and

$$\sum_{i=1}^{m} a_{ij}^2 = s_j^2, \ \sum_{j=1}^{n} a_{ij}^2 = t_i^2, \ s_j \in \mathbb{N}, t_i \in \mathbb{N} \ (1 \leq i \leq m, \ 1 \leq j \leq n).$$

We further assume that the integers $s_1, s_2, \ldots, s_n$ are distinct, that the $t_1, t_2, \ldots, t_m$ are also distinct, and that the sum of all mn integers a_{ij}^2 is also a perfect square:

$$\sum_{i=1}^{m} \sum_{j=1}^{n} a_{ij}^2 = s_1^2 + s_2^2 + \cdots + s_n^2 = t_1^2 + t_2^2 + \cdots + t_m^2 = S^2, \quad S \in \mathbb{N}.$$

We choose the $(m + 1) \times n$ array as follows:

$$\begin{pmatrix} (2pa_{11})^2 & (2pa_{12})^2 & \cdots & (2pa_{1n})^2 \\ (2pa_{21})^2 & (2pa_{22})^2 & \cdots & (2pa_{2n})^2 \\ \vdots & \vdots & \ddots & \vdots \\ (2pa_{n1})^2 & (2pa_{n2})^2 & \cdots & (2pa_{nn})^2 \\ (p^2 - 1)^2 s_1^2 & (p^2 - 1)^2 s_2^2 & \cdots & (p^2 - 1)^2 s_n^2 \end{pmatrix},$$

where we will fix the value of the parameter $p \in \mathbb{N}$ in a moment. We convince ourselves that the ith row sum of this array is equal to $(2pt_i)^2$ for $i = 1, 2, \ldots, m$, while for $i = m + 1$ it is $(p^2 - 1)^2 S^2$, and that the jth column sum is equal to

$$4p^2 s_j^2 + (p^2 - 1)^2 s_j^2 = (p^2 + 1)^2 s_j^2 \quad \text{for } j = 1, 2, \ldots, n.$$

All these numbers are then perfect squares, as is the sum of all $(m + 1)n$ numbers of this array, which is equal to

$$\left(p^2 + 1\right)^2 \left(s_1^2 + s_2^2 + \cdots + s_n^2\right) = \left(p^2 + 1\right)^2 \cdot S^2.$$

Now we choose the number p such that the finitely many conditions

$$\left(p^2 - 1\right)^2 s_j^2 \neq (2pa_{ik})^2, \quad (2pt_i)^2 \neq \left(p^2 - 1\right)^2 S^2$$

are satisfied for all admissible values of the indices i, j, k (consider why each of these conditions "forbids" the use of at most two positive integers p). Then the array constructed above has all the required properties, which completes the step from an $m \times n$ array to an $(m + 1) \times n$ array; the step to an $m \times (n + 1)$ array is carried out analogously. □

SECOND SOLUTION. We use the following trick: From the positive integers $p_1, p_2, \ldots, p_m$ and $q_1, q_2, \ldots, q_n$ we construct the $m \times n$ array of numbers $x_{ij} = p_i^2 q_j^2$. The row and column sums of these arrays are perfect squares as long as the two numbers

$$S_p = p_1^2 + p_2^2 + \cdots + p_m^2, \quad S_q = q_1^2 + q_2^2 + \cdots + q_n^2$$

are. We now describe an appropriate choice of the integers p_i for an arbitrary $m > 1$. We choose an odd $p_1 \geq 3$ and even numbers $p_2, p_3, \ldots, p_{m-1}$ such that $p_1 < p_2 < \cdots < p_{m-1}$. Then $p_1^2 + p_2^2 + \cdots + p_{m-1}^2 = 2k + 1$ for an appropriate integer $k \geq 4$. If we furthermore choose $p_m = k$ (clearly $p_m > p_{m-1}$), then we will have $S_p = (2k + 1) + k^2 = (k + 1)^2$. Similarly, we choose an odd $q_1 \geq 3$ and even $q_2, \ldots, q_{n-1}$ such that $q_2 > p_m q_1, q_3 > p_m q_2, \ldots, q_{n-1} > p_m q_{n-2}$, and then we set $q_n = r$, where $2r + 1 = q_1^2 + q_2^2 + \cdots + q_{n-1}^2$. Then $S_q = (r + 1)^2$, and the inequalities for the integers p_i, q_j guarantee that

$$p_1 q_1 < p_2 q_1 < \cdots < p_m q_1 < p_1 q_2 < \cdots < p_m q_2 < \cdots$$

$$\cdots < p_1 q_n < p_2 q_n < \cdots < p_m q_n;$$

therefore, the numbers $x_{ij} = p_i^2 q_j^2$ are distinct. □

3.2 Exercises

(i) For which $n \geq 2$ can one construct an $n \times n$ array from the numbers $-1, 1$ such that the row sums and diagonal sums $r_1, r_2, \ldots, r_n, u, v$ are distinct?

(ii) Complete a $2n \times 2n$ array with the numbers $-1, 0, 1$ such that all row and column sums of the array are distinct.

(iii) Write the numbers $1, 2, 3, \ldots, n^2$ into an $n \times n$ array such that all column sums of the array are identical.

(iv) Suppose that an $m \times n$ array is filled with mn numbers such that the elements of each row and column form an arithmetic sequence. Determine the sum of all numbers of the array if you know that the sum of the four numbers in the corner fields is equal to a given number S.

(v) Decide for which $n \geq 3$ the following assertion holds: If an $n \times n$ array consists of integers such that all its row, column, and both diagonal sums are equal, then the sum S of all n^2 numbers of the array is necessarily divisible by n^2.

*(vi) Show that for no $n > 1$ does there exist an $n \times n$ array formed by the numbers $1, 2, \ldots, n^2$ such that each row and column sum is equal to some (not necessarily the same) integer power of 2.

*(vii) Suppose that an $n \times n$ square array is formed by the integers $1, 2, \ldots, n^2$ such that for each $k = 1, 2, \ldots, n^2 - 1$ the integer $k + 1$ lies in the row that has the same index as the column in which the integer k lies. Which value must the difference $c - r$ have, where c is the sum of numbers in the column containing n^2, and r is the sum of numbers in the row containing the integer 1?

*(viii) Determine the numbers $n \geq 2$ for which the integers $1, 2, \ldots, 2n$ can be placed in a $2 \times n$ array such that the row and column sums satisfy the equations $r_1 = r_2$ and $c_1 = c_2 = \cdots = c_n$.

3.3 Problems on Subarrays

If from an $m \times n$ array we select a set of p consecutive rows and a set of q consecutive columns, then the pq fields lying in the intersection of the chosen rows and columns form a $p \times q$ *subarray* of the original $m \times n$ array.

(i) Is it possible to write 25 numbers into a 5×5 array such that the sum of the four numbers in every 2×2 subarray is negative, while the sum of all 25 numbers of the array is positive?

SOLUTION.   A negative answer to this problem would certainly be true if one could divide the entire 5×5 array into 2×2 subarrays without common fields (which, however, is not possible). In fact, an array with the desired property does exist; the reader is encouraged to find one independently. One example is the array

1	1	1	1	1
1	−4	1	−4	1
1	1	1	1	1
1	−4	1	−4	1
1	1	1	1	1

The sum of all 25 numbers is equal to 5, while the sum of numbers in each 2×2 subarray is equal to $1 + 1 + 1 + (-4) = -1$.   □

(ii) Suppose that the numbers $1, 2, \ldots, 81$ are written in some order into a 9×9 array. Show that there is a 2×2 subarray whose numbers have a sum greater than 137.

SOLUTION.   Ordered by size we let $S_1 \leq S_2 \leq \cdots \leq S_{64}$ denote the sums of the numbers in the 2×2 subarrays of the given 9×9 array. We assume that the assertion of the problem does not hold, that is, that the largest of the sums in question satisfies the inequality $S_{64} \leq 137$. Then we have the upper bound
$$S_1 + S_2 + \cdots + S_{64} \leq 64 \cdot 137 = 8768.$$
On the other hand, in the sum $S_1 + S_2 + \cdots + S_{64}$ a number from a field is counted either once, or twice, or four times, according to whether it is a "corner field," a field along one of the sides, or an "interior field." We have therefore the lower bound
$$S_1 + S_2 + \cdots + S_{64} \geq (81 + 80 + 79 + 78) + 2(77 + 76 + \cdots + 50)$$
$$+ 4(49 + 48 + \cdots + 1) = 8774,$$
which is a contradiction to the upper bound.   □

(iii) Suppose that each field of a $(2n+1) \times (2n+1)$ array contains a number from the interval $[-1, 1]$, and that the sum of the four numbers of every 2×2 subarray is 0. What is the largest possible sum of all $(2n+1)^2$ numbers in the array when n is fixed?

SOLUTION.

Figure 4

We begin with the smallest, 3×3 $(n = 1)$, array; if we use the condition for the upper right and the lower left 2×2 subarrays (see Figure 4), we can write the entire sum S of the 9 numbers in the array as follows:

$$S = (a_{12} + a_{13} + a_{22} + a_{23}) + (a_{21} + a_{22} + a_{31} + a_{32})$$
$$+ a_{11} - a_{22} + a_{33} = a_{11} - a_{22} + a_{33}.$$

Using the condition $|a_{ij}| \leq 1$, we obtain the bound $S \leq 3$. Furthermore, the value $S = 3$ is in fact possible: If we set $a_{11} = 1$, $a_{22} = -1$, and $a_{33} = 1$, the remaining 6 numbers can be entered in such a way that we obtain an array with the desired property; it suffices to set $a_{ij} = (-1)^{i+1}$ for each pair of indices $i, j \in \{1, 2, 3\}$.

The above expression for the sum S for 3×3 arrays can be easily used to study $5 \times 5, 7 \times 7, \ldots$ arrays (you should first attempt this yourself). We now use induction to prove the following general hypothesis: The sum S of the numbers in the $(2n + 1) \times (2n + 1)$ array under consideration can be determined with the help of its diagonal elements a_{ii} by the formula

$$S = a_{11} - a_{22} + a_{33} - a_{44} + \cdots + a_{2n+1,2n+1}. \tag{15}$$

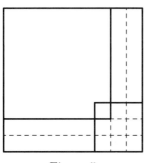

Figure 5

Out of such an array (for a given $n > 1$) we separate a $(2n-1) \times (2n-1)$ subarray with sum of elements S_1 in its upper left corner, and a 3×3 subarray with sum of elements S_2 in its lower right corner; see Figure 5. These two subarrays have exactly one field in common, namely, the field with entry $a_{2n-1,2n-1}$. What remains of the original array is divided into two subarrays, one of size $2 \times (2n-2)$ and the other of size $(2n-2) \times 2$, in which the sums of entries are zero (each of them can be divided into $n-1$ 2×2 subarrays). Hence we have $S = S_1 + S_2 - a_{2n-1,2n-1}$. Since by the first part of the solution we have $S_2 = a_{2n-1,2n-1} - a_{2n,2n} + a_{2n+1,2n+1}$, we can substitute this and the induction hypothesis

$$S_1 = a_{11} - a_{22} + a_{33} - a_{44} + \cdots + a_{2n-1,2n-1}$$

into the expression for S, and obtain (15). This proves the expression (15) by induction; it results in the bound $S \le 2n + 1$, with equality $S = 2n + 1$ obtained, as in the case $n = 1$, for an array with the numbers $a_{ij} = (-1)^{i+1}$, where $i, j \in \{1, 2, \ldots, 2n+1\}$ (note that all the 2×2 subarrays consist of the four numbers $1, 1, -1, -1$). Therefore the largest possible value of the sum S, for a fixed n, is $2n + 1$.   □

(iv) Suppose that each field of a 20×20 array contains one of the numbers $-1, 0, 1$ such that the sum of all 400 numbers in the array is equal to 80. Show that the sum S of all the numbers of at least one of its 5×5 subarrays satisfies $|S| \le 5$.

SOLUTION.   Let $S(C)$ denote the sum of all the numbers of a 5×5 subarray C. If we divide the whole 20×20 array into 16 subarrays of size 5×5, then in view of the equality $80 = 16 \cdot 5$ there are two subarrays C and C' among them that satisfy $S(C) \le 5$ and $S(C') \ge 5$. Suppose that neither of them has the desired property, that is, that $S(C) < -5$ and $S(C') > 5$. Then we "connect" C and C' with a sequence of subarrays

$$C_1 = C, \ C_2, \ C_3, \ \ldots, \ C_k = C'$$

such that the subarray C_{i+1} is obtained by shifting C_i by one position in the direction of the rows or the columns of the array ($1 \le i \le k - 1$). Since two neighboring subarrays C_i, C_{i+1} have exactly 20 fields in common, we have the inequality $|S(C_i) - S(C_{i+1})| \le 10$, $1 \le i \le k - 1$. In view of the fact that $S(C_1) < -5$ and $S(C_k) > 5$, there exists an index i ($1 < i \le k$) for which $S(C_{i-1}) < -5$ and $S(C_i) \ge -5$ hold, and thus

$$-5 \le S(C_i) \le S(C_{i-1}) + 10 < -5 + 10 = 5.$$

Therefore, the 5×5 subarray C_i has the desired property $|S(C_i)| \leq 5$.   □

3.4 Exercises

(i) Is it possible to construct an $n \times n$ array with n^2 appropriate distinct positive integers such that any square subarray has identical products of the numbers in its two diagonals?

(ii) Let two integers $k < n$ be given, where k does not divide n. Show that one can write n^2 real numbers into an $n \times n$ array such that the sum of the k^2 numbers of each of its $k \times k$ subarrays is negative, and at the same time the sum of all the numbers of the original array is positive. (This is a generalization of 3.3.(i).)

(iii) In how many ways can one write the numbers $-1, 0, 1$ in an $n \times n$ array such that the sum of the numbers in any 2×2 subarray is divisible by 3?

*(iv) How many arrays considered in (iii) are such that the sum of the numbers in any 2×2 subarray is zero?

*(v) Suppose that the fields of an $n \times n$ array contain real numbers from the interval $[-1, 1]$, where for a fixed k $(k \leq n)$ the sum of the k^2 numbers of any of its $k \times k$ subarrays is equal to 0. What is the largest possible sum S of all numbers of the array (depending on n and k)?

3.5 Problems on Sets of Fields

In the following problems we will consider *sets of fields* of an $m \times n$ array, which are more general than the *subarrays* defined in the introduction to Section 3.3.

(i) Determine the $n \geq 3$ for which one can write the numbers $1, 2, \ldots, n^2$ into an $n \times n$ array such that the sum of the four numbers of any part of the array that has one of the four shapes shown in Figure 6 is even.

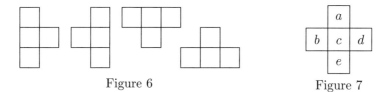

Figure 6   Figure 7

SOLUTION.   We show that if such a distribution exists, then the numbers a, b, c, d, e of any "cross" in the array (see Figure 7) must have the same parity. Indeed, from the expression $a - e = (a + b + c + d) - (e + b + c + d)$ it follows that $a \equiv e$ (all congruences here are modulo 2), which implies $0 \equiv a + b + c + e \equiv b + c$, that is, $c \equiv b$. Since we can rotate the cross by $90°$

and repeat the above argument, we also have $c \equiv a$, $c \equiv d$, and $c \equiv e$, in addition to $c \equiv b$. This proves the assertion for any cross in the $n \times n$ array. From this it immediately follows that the numbers in all the fields of the array, with the possible exception of the corner fields, have the same parity. But this is possible only when $n = 3$, since for $n \geq 4$ there are no $n^2 - 4$ of the numbers $\{1, 2, \ldots, n^2\}$ that have the same parity. In the case $n = 3$ an appropriate distribution does exist: We place the numbers $2, 4, 6, 8$ in the corner fields, and the numbers $1, 3, 5, 7, 9$ in any order in the unique cross inside the 3×3 array.   □

(ii) Suppose that the four corner fields and several other fields of an $n \times n$ array contain real numbers such that for each $k \geq 2$ the following holds: If some straight line goes through the centers of exactly k fields that contain numbers, then the sum of these k numbers is zero. Show that every number in the array must be zero.

SOLUTION.   Let S denote the sum of all numbers written into the array, x_P the number in a field P, and α_P the number of straight lines mentioned in the problem that go through the center of the field P. If for a fixed P we add all α_P equations for the sums of the numbers on the straight lines, we obtain $S + (\alpha_P - 1)x_P = 0$. By the statement of the problem we have $\alpha_P \geq 2$ for each occupied field P. Therefore, if we assume that $S \neq 0$, it follows from the above equation that S and x_P have different signs; this cannot hold for all fields P, since $S = \sum_P x_P$. Hence S is equal to zero; but then it follows from $(\alpha_P - 1)x_P = 0$ that $x_P = 0$ for all occupied fields P.   □

(iii) Suppose that an $m \times n$ array, where $m \geq 2$ and $n \geq 2$, is constructed with the numbers -1 and 1 so that each of the two integers occurs at least twice. Show that one can choose two rows and two columns in the array such that the sum of the numbers in their four intersections is zero.

SOLUTION.   We first show that it suffices to prove the following assertion: In each array with the given properties one can choose two rows such that their union contains each of the numbers $1, -1$ at least twice.

Indeed, if this assertion holds, we use it to choose the two rows from a given $m \times n$ array, and use them as columns to form an $n \times 2$ array. We apply the assertion now to this array and choose two rows (of length 2); then these rows form a 2×2 array with the numbers $1, 1, -1, -1$, which is the intersection of two rows and two columns of the original array.

We now prove the above assertion. Everything is clear in the two cases where

(a) the number of rows that contain both numbers ± 1 is greater than 1;

(b) there is a row that contains only $+1$, and one that contains only -1.

In the case where neither (a) nor (b) holds, we have, because of $m \geq 2$, the following situation: There is a row (say, the ith one) such that all numbers of the array that do not lie in the ith row are equal to the same number $\varepsilon \in \{-1, 1\}$. But then, by hypothesis, the ith row contains the number $-\varepsilon$ at least twice. Thus it suffices to choose the ith row and any other row. □

(iv) Suppose that an $m \times n$ array consists of mn distinct real numbers. For given numbers m_1, n_1 $(1 \leq m_1 \leq m, 1 \leq n_1 \leq n)$ we mark the m_1 largest numbers in each column blue, and the n_1 largest numbers in each row red. Show that at least $m_1 n_1$ numbers in the array are marked with both colors.

SOLUTION.   We prove the assertion by induction on $p = m + n \geq 2$. When $p = 2$, there is nothing to show. If the assertion holds for some $p \geq 2$, then for any $m \times n$ array with $m + n = p + 1$ we distinguish two cases. If every marked entry is marked both blue and red, then the number r of marked entries satisfies $r = mn_1 = m_1 n \geq m_1 n_1$. In the opposite case there are entries in the array that are marked with only one color; let A be the *largest* such entry.

Let us first assume that the number A is marked blue. Then the n_1 red entries $x_1, x_2, \ldots, x_{n_1}$, which lie in the same row as A, are also marked blue (since each of them is greater than A, the largest entry marked with only one color). In the case $m_1 = 1$ we are done, and in the case $m_1 > 1$ we delete from the array the row that contains A. The remaining $(m-1) \times n$ array has the following property: In each of its rows the n_1 largest entries are marked red, and in each of its columns either the largest m_1 or the largest $m_1 - 1$ entries are marked blue. If in each column with m_1 entries that are marked blue we eliminate the blue mark on the smallest of these entries, then the induction hypothesis asserts that in the $(m-1) \times n$ array at least $(m_1 - 1)n_1$ entries are marked with both colors. If we now put the numbers $x_1, x_2, \ldots, x_{n_1}$ back in, we obtain at least $(m_1 - 1)n_1 + n_1 = m_1 n_1$ entries in the original array that are marked with both colors.

Similarly, if the entry A is marked red, we remove from the array the column in which it lies, and use induction for the remaining $m \times (n-1)$ array. □

3.6 Exercises

(i) Given the 10×10 array

$$\begin{pmatrix} 0 & 1 & 2 & \cdots & 9 \\ 10 & 11 & 12 & \cdots & 19 \\ \vdots & \vdots & \vdots & \ddots & \vdots \\ 90 & 91 & 92 & \cdots & 99 \end{pmatrix}, \tag{16}$$

put minus signs in front of any 50 of the entries in such a way that there are exactly five in each row and column. Show that with such a sign distribution the sum of all 100 numbers in the array is equal to zero.

(ii) Choose ten numbers in the array (16), one from each row and column. Which values can the sum of the chosen numbers have?

(iii) A *generalized diagonal* of an $n \times n$ array will be any n-element set of its fields that contains exactly one field from each row and each column. Prove the following assertion: The sums of the numbers in all generalized diagonals of an $n \times n$ array with entries a_{ij} are identical if and only if there are sequences $b_1, b_2, \ldots, b_n$ and $c_1, c_2, \ldots, c_n$ such that $a_{ij} = b_i + c_j$ holds for all $i, j \in \{1, 2, \ldots, n\}$.

*(iv) Suppose that the entries a_{ij} of an $n \times n$ array are such that $a_{ij} + a_{jk} + a_{ki} = 0$ for all $i, j, k \in \{1, 2, \ldots, n\}$. Show that there is a sequence $t_1, t_2, \ldots, t_n$ such that $a_{ij} = t_i - t_j$ for all $i, j \in \{1, 2, \ldots, n\}$.

*(v) Determine the $n \geq 3$ for which the numbers $1, 2, \ldots, n^2$ can be written into an $n \times n$ array such that the sum of the four numbers of any part of the array that has one of the four shapes shown in Figure 8 is divisible by 9.

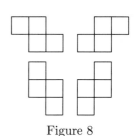

Figure 8

3.7 Problems on Neighboring Fields

Two fields of an $m \times n$ array are called *neighboring* if they form a 2×1 or 1×2 subarray in the sense of Section 3.3. Geometrically, this means that the two fields (squares) have a common side.

(i) Suppose that an $n \times n$ array is filled with integers such that the numbers in two neighboring fields do not differ by more than 1. Show that some integer occurs in the array at least $([\frac{n}{2}] + 1)$ times.

SOLUTION.   First we show that for any integers a, b lying in some $p \times q$ subarray of the given array we have the inequality

$$|a - b| \leq p + q - 2.$$

Formally, one could use induction on $p + q$, but here we prefer a direct geometric description: Any two fields P, Q of the array can be "connected" by a sequence of fields

$$P = P_1, \ P_2, \ P_3, \ \ldots, \ P_k = Q, \tag{17}$$

where P_i, P_{i+1} are neighboring fields of the array for any $i = 1, 2, \ldots, k-1$. (Such a sequence corresponds to the walk of a king on a chessboard from P to Q in the direction of rows and columns). It is now clear that if the fields P, Q belong to some $p \times q$ subarray, one can choose a sequence (17) of length $k \leq p + q - 1$. If a_i is the integer in the field P_i from the sequence (17), then by the hypothesis of the problem we have

$$\begin{aligned} |a_k - a_1| &= |(a_k - a_{k-1}) + (a_{k-1} - a_{k-2}) + \cdots + (a_2 - a_1)| \\ &\leq |a_k - a_{k-1}| + |a_{k-1} - a_{k-2}| + \cdots + |a_2 - a_1| \\ &\leq \underbrace{1 + 1 + \cdots + 1}_{k-1} = k - 1. \end{aligned}$$

This proves the assertion about the inequality $|a - b| \leq p + q - 2$. It implies that for any two integers a, b from the whole $n \times n$ array we have $|a - b| \leq 2n - 2$; therefore among the n^2 integers in the array no more than $2n - 1$ are distinct. Hence by the pigeonhole principle some integer occurs p times in the array, where

$$p \geq \frac{n^2}{2n-1} > \frac{n^2}{2n} = \frac{n}{2}. \qquad \square$$

(ii) For the array in the previous problem (i), show that it contains a number that appears at least n times.

SOLUTION. We will prove a stronger assertion: There is a number that occurs in each row or in each column of the array. If m_k, respectively M_k, is the *smallest*, respectively the *largest*, number in the kth row, then this row clearly contains exactly the numbers $m_k, m_k + 1, \ldots, M_k - 1, M_k$ ($1 \leq k \leq n$). Therefore, our assertion holds in the case where every interval $[m_k, M_k]$, $1 \leq k \leq n$, contains the same integer x. If this is not the case, then there exist indices $\alpha, \beta \in \{1, 2, \ldots, n\}$ such that

$$m_\alpha \leq M_\alpha < m_\beta \leq M_\beta.$$

But this means that in each column of the array we find numbers x, y such that $x \leq M_\alpha$ and $y \geq m_\beta$ (it suffices to take the number x, respectively y, from the row with index α, respectively β). Then, for instance, the number M_α (and more generally, every integer in the interval $[M_\alpha, m_\beta]$) occurs in every column. This completes the proof. $\qquad \square$

(iii) Suppose that in each field of an infinite plane chessboard a different real number is written. Show that some field contains a number that is less than
(a) at least two of the four numbers in neighboring fields;
(b) at least four of the eight surrounding fields (that is, those that have at least one corner in common with the given field).

SOLUTION.   From the infinite array we choose an arbitrary subarray T_2 of size 2×2. Each field in T_2 abuts at least two other fields in T_2, so the field in the array T_2 that contains the least of its nine numbers has property (a). However, the array T_2 (or any other array T_n of size $n \times n$) does not solve part (b), since each of its corner fields is surrounded by only three fields. A way out is to use, for instance, the array T_4 and remove the corner fields (see Figure 9). We thus obtain a finite (12-element) set of fields in which each field is surrounded by at least 4 others from the set. Therefore, the field that has the smallest of the corresponding numbers is the one with the desired property.   □

Figure 9

(iv) Suppose that the numbers $1, 2, \ldots, n^2$, with $n > 1$, are arbitrarily placed in an $n \times n$ array. Show that some two neighboring fields contain numbers x and y such that $|x - y| \geq n$.

SOLUTION.   Before the actual solution we remark that using the trick from Problem (i) with the sequence (17) of neighboring fields we can obtain a weaker conclusion in which the inequality is $|x - y| \geq 1 + \left[\frac{n}{2}\right]$. Indeed, from the inequality $|x - y| \leq \left[\frac{n}{2}\right]$ for the numbers x, y from any two neighboring fields we obtain the bound $|x - y| \leq (2n - 2) \left[\frac{n}{2}\right]$ for any two numbers x, y of the array. If we set $x = n^2$ and $y = 1$, then

$$n^2 - 1 \leq 2(n-1) \left[\frac{n}{2}\right], \quad \text{or} \quad \frac{n+1}{2} \leq \left[\frac{n}{2}\right],$$

and this is a contradiction.

We will now try to do better with the more "refined" approach of Problem (ii): As we did there, we denote by m_k, respectively M_k, the *smallest*, respectively the *largest*, number in the kth row of the array. Since the numbers $m_1, M_1, m_2, M_2, \ldots, m_n, M_n$ are distinct (each of them is equal to a different element of the set $\{1, 2, 3, \ldots, n^2\}$), the numbers

$$m = \max\{m_1, m_2, \ldots, m_n\} \quad \text{and} \quad M = \min\{M_1, M_2, \ldots, M_n\}$$

are distinct as well. We now distinguish two cases: $m < M$ and $m > M$. In the first case we have $m_k \leq m < M_k$ for all $k = 1, 2, \ldots, n$. This means that in any kth row of the array there are numbers a_k, b_k such that $a_k \leq m < b_k$, and certainly we can choose a_k, b_k from two neighboring fields of the kth row. Since the numbers $b_1, b_2, \ldots, b_n$ are greater than m and are distinct, we have for some $k \in \{1, 2, \ldots, n\}$ the bound $b_k \geq m + n$, and thus

$$b_k - a_k \geq (m + n) - m = n.$$

This completes the proof of the assertion in the case $m < M$.

When $m > M$, we first find indices $i, j \in \{1, 2, \ldots, n\}$ such that $m_i > M = M_j$. Then in any kth column of the array $(1 \le k \le n)$ there are numbers a_k, b_k such that $a_k \le M < b_k$ (it suffices to choose a_k respectively b_k from the jth respectively the ith row), where again the numbers a_k, b_k can be chosen from two neighboring fields in the kth column. Just as in the first case, we have the bound $b_k \ge M + n$ for some $k \in \{1, 2, \ldots, n\}$, and thus

$$b_k - a_k \ge (M + n) - M = n.$$

This concludes the proof in the case $m > M$.

We add that this proof remains valid in the more general setting where an $n \times n$ array is formed with n^2 different integers.   □

3.8   Exercises

(i) Suppose that a 100×100 array consists of integers such that the numbers x, y from any two neighboring fields satisfy $|x - y| \le 20$. Show that three fields of the array contain the same number.

(ii) Given the situation of 3.7.(iii), does there always exist a field with a number that is less than at least five of the eight numbers in the surrounding fields?

(iii) Formulate and prove an analogue of 3.7.(iii) for an infinite plane array formed with fields in the shape of equilateral triangles.

(iv) Suppose that each field of an infinite plane chessboard contains a different positive integer. Show that for each $n \ge 1$ there are two neighboring fields whose numbers x, y satisfy $|x - y| > n$.

(v) Find the smallest positive integer N with the following property: The fields of an infinite plane chessboard can be filled with the integers $1, 2, 3, \ldots$ such that each integer k occurs exactly k times, and the integers x, y from any two neighboring fields always satisfy $|x - y| \le N$.

3.9   Transformations of Arrays

We now introduce several problems on the properties an array of numbers may gain if we change its elements according to certain rules. In Section 5 of this chapter we will return to this kind of problem in a more general setting (transformations of arbitrary numerical configurations). There we will also describe general methods for solving such problems, with further problems on transformations that are solved with the techniques presented there.

(i) Suppose that a 3×3 array consists of the numbers 1 and -1. We replace the numbers in all the fields by the products of the numbers in their neighboring fields (in the sense of Section 3.7), and repeat this process successively. Show that after a certain time we obtain an array without the number -1.

SOLUTION.   We denote the original elements of the array by $a, b, c, \ldots, h, i$ and see how they change after the first few transformations (keep in mind that $a^2 = b^2 = \cdots = i^2 = 1$):

a	b	c
d	e	f
g	h	i

$\mapsto$

bd	aec	bf
aeg	$bdfh$	cei
dh	egi	fh

$\mapsto$

cg	bh	ai
df	1	df
ai	bh	cg

$\mapsto$

$\mapsto$

$bdfh$	$acgi$	$bdfh$
$acgi$	1	$acgi$
$bdfh$	$acgi$	$bdfh$

$\mapsto$

1	1	1
1	1	1
1	1	1

.

We see that after at most four transformations we obtain an array consisting only of ones.   □

(ii) In an $m \times n$ array containing real numbers we consider the transformation of replacing all the numbers in some row or in some column by their opposites. Show that by appropriate repetitions of such a transformation we can obtain an array in which each row sum and column sum is nonnegative.

SOLUTION.   From among all the arrays that can be obtained from the original one by repeating the transformation described in the problem, we choose the one that has the *largest* sum S of all mn numbers. Such an array A exists, since the number of ways in which a part of the mn numbers can be changed into their opposites is finite. If there are several arrays with the maximal sum S, we choose an arbitrary one. It is clear that all the row sums and column sums of the array A are nonnegative: If some row sum or column sum had the value $-\varepsilon$ ($\varepsilon > 0$), then by an appropriate transformation of this row, respectively column, we would obtain an array A' with sum of elements $S' = S + 2\varepsilon > S$, which contradicts the way we chose the array A.   □

(iii) Suppose that an $m \times n$ array consists of positive integers, and that we can carry out the following two types of transformations: multiplying all the numbers of some row by 2, or subtracting 1 from all the numbers of some column. Show that after applying a number of these transformations to the original array we can obtain an array whose entries are all zero.

SOLUTION.   It clearly suffices to describe a way of applying the transformations by which all the entries in a fixed column C are "annulled". First we apply the subtraction transformation until the smallest number in C is 1. If all the other numbers in C are also equal to 1, we apply one more subtraction, and we are done. Otherwise, if there are some numbers in C

that are greater than 1, it is enough to use the following transformation, which preserves the entries 1 in the column C, but reduces the other numbers by 1: Multiply by two exactly those rows in the array that have a 1 in C, and then subtract 1 from all the numbers in C. By repeating this transformation we finally obtain only 1's in the column C. □

(iv) Suppose that an 8×8 array consists of integers and that we may choose any 3×3, respectively 4×4, subarray and increase all its 9, respectively 16, numbers by 1. Is it always possible to obtain, by appropriate repetitions of these transformations, an array in which all 64 numbers are divisible by 10?

SOLUTION.   We may clearly replace the numbers in each field of the array by their remainders upon division by 10; such an array (that is, one consisting of the numbers $0, 1, 2, \ldots, 9$) will be called a *residue* array. We will prove that there are at most 10^{61} residue arrays that can be reduced, by repeated application of the transformation in question, to the residue array A_0 consisting only of zeros. Since the number of all residue arrays is 10^{64}, this will imply that there are "bad" original arrays for which it is impossible to reach the desired goal.

Since the transformation in question has an *inverse* (that is, subtraction of 1 from all numbers of an arbitrary 3×3 or 4×4 subarray), it suffices to estimate the number of residue arrays A that can be obtained from A_0 by this inverse. The number of all 3×3 or 4×4 subarrays is equal to

$$(8 - 3 + 1)^2 + (8 - 4 + 1)^2 = 61;$$

we therefore number them $1, 2, \ldots, 61$. The transition from A_0 to A can be described by a 61-tuple $(p_1, p_2, \ldots, p_{61})$, where p_i for each i describes how many times the numbers of the ith subarray were decreased by 1 (you should convince yourself that the resulting transition is independent of the order in which the individual transformations are performed). It is also clear that the numbers p_i can be replaced by their remainders upon division by 10 (if we repeat the transformation of the ith subarray 10 times in a row, we obtain the same residue array as before). Therefore, the number of arrays A is equal to at most the number of different 61-tuples $(p_1, p_2, \ldots, p_{61})$ with $p_i \in \{0, 1, \ldots, 9\}$ for each $i \in \{1, 2, \ldots, 61\}$, that is, 10^{61}. □

3.10   Exercises

(i) Give an example for an initial array in Problem 3.9.(i) where the number of transformations that lead to the final array is 4.

(ii) Suppose that an $n \times n$ array, $n \geq 2$, consists of n^2 real numbers whose sum is positive. Show that the rows of the array can be rearranged (while the orders of elements in the individual rows remain unchanged) in such a way that the diagonal sum

$$u = a_{11} + a_{22} + \cdots + a_{nn}$$

of the new array is positive. Is it always possible to achieve that in addition to the sum u the second diagonal sum

$$v = a_{n1} + a_{n-1,2} + \cdots + a_{1n}$$

is also positive?

(iii) Suppose that each field of a 3×3 array is occupied by an integer. We replace the number in each field by the sum of the numbers in the fields abutting the given field, and we repeat this transformation. Show that after a certain number of steps (depending on n) all the numbers in the array are divisible by 2^n, for any integer $n \geq 1$.

(iv) Show that in (iii) the number 2^n cannot be replaced by an odd prime number.

(v) Suppose that an $n \times n$ array consists of integers and that we can choose any pair of neighboring fields and increase both of their numbers by 1. Given an $n > 1$, is there a $k > 1$ such that an appropriate repetition of the transformation always results in an array in which all the integers are divisible by k?

*(vi) Suppose that in an $n \times n$ array consisting of real numbers the following transformation $P_{j,k}^i$ is admissible for all triples of indices $i, j, k \in \{1, 2, \ldots, n\}$: The entries of the ith row are all added, in their given order, to the entries of the jth column, and at the same time subtracted from the entries of the kth column (row entries are taken from left to right, and column entries from top to bottom). Show that if in the original array all row sums and column sums are equal to zero, then after an appropriate sequence of transformations we can arrive at an array consisting only of zeros.

3.11   Triangular Arrays

The most famous triangular array is *Pascal's triangle*

$$
\begin{array}{ccccccccc}
 & & & & 1 & & & & \\
 & & & 1 & & 1 & & & \\
 & & 1 & & 2 & & 1 & & \\
 & 1 & & 3 & & 3 & & 1 & \\
1 & & 4 & & 6 & & 4 & & 1 \\
\end{array}
\tag{18}
$$

whose nth row consists of the sequence of binomial coefficients $\binom{n-1}{k}$, $k = 0, 1, \ldots, n-1$. Thanks to the basic formula

$$\binom{n}{k} + \binom{n}{k-1} = \binom{n+1}{k}$$

for binomial coefficients (see, e.g., [5], Chapter 1, equation (2)) we can describe a simple rule for constructing the array (18): In every row each number (with the exception of the two end numbers) is equal to the sum of the two closest numbers in the previous row.

(i) Suppose that ten distinct positive integers $a, b, c, \ldots, j$ form the array

$$
\begin{array}{ccccc}
a & \longrightarrow & b & \longrightarrow & c & \longrightarrow & d \\
& & \uparrow & & \uparrow & & \uparrow \\
& & e & \longrightarrow & f & \longrightarrow & g \\
& & & & \uparrow & & \uparrow \\
& & & & h & \longrightarrow & i \\
& & & & & & \uparrow \\
& & & & & & j
\end{array}
\tag{19}
$$

and that we know that each number into which two arrows aim is equal to the sum of the two numbers from which the arrows originate. Find the smallest possible value of the number d.

SOLUTION.   First we prove by contradiction that $d \geq 20$. To do this, let us assume that $d < 20$ holds for some array (19). Since

$$d = c + g = b + f + f + i = (a + e + h + j) + 2f$$

and the number in parentheses is at least $1 + 2 + 3 + 4 = 10$, we obtain from $d < 20$ the estimate $2f < 10$, that is, $f \leq 4$. On the other hand, $f = e + h \geq 1 + 2 = 3$, so altogether, $f \in \{3, 4\}$. We therefore distinguish two cases:

(a) $f = 3$. Then $\{e, h\} = \{1, 2\}$, and from $d = a + j + 3f$ it follows that $a + j = d - 9 < 11$, that is, $\{a, j\} = \{4, 5\}$ or $\{a, j\} = \{4, 6\}$. If we take the first case, that is, $\{a, j\} = \{4, 5\}$, then $b + i \geq 6 + 7 = 13$, then this is in contradiction to $b + i = a + e + h + j = 12$. Hence $\{a, j\} = \{4, 6\}$. From $b + i = a + e + h + j = 13$ it follows that $\{b, i\} = \{5, 8\}$, which implies $\{c, g\} = \{b+3, i+3\} = \{8, 11\}$ and thus $\{b, i\} \cap \{c, g\} \neq \emptyset$; but this is again a contradiction.

(b) $f = 4$. Then $\{e, h\} = \{1, 3\}$, and from $d = a + j + 3f$ it follows that $a + j = d - 12 < 8$, that is, $\{a, j\} = \{2, 5\}$. But then $\{b, i\} \cap \{1, 2, 3, 4, 5\} = \emptyset$, and thus $b + i \geq 6 + 7 = 13$, which is a contradiction, since $b + i = a + e + h + j = 11$.

It remains to give an example to show that the value $d = 20$ is in fact possible:

$$
\begin{array}{ccccccc}
4 & \longrightarrow & 6 & \longrightarrow & 9 & \longrightarrow & 20 \\
 & & \uparrow & & \uparrow & & \uparrow \\
 & & 2 & \longrightarrow & 3 & \longrightarrow & 11 \\
 & & & & \uparrow & & \uparrow \\
 & & & & 1 & \longrightarrow & 8 \\
 & & & & & & \uparrow \\
 & & & & & & 7
\end{array}
$$

$\square$

(ii) The triangular array

$$
\begin{array}{ccccccccc}
 & & & & 1 & & & & \\
 & & & 1 & 1 & 1 & & & \\
 & & 1 & 2 & 3 & 2 & 1 & & \\
 & 1 & 3 & 6 & 7 & 6 & 3 & 1 & \\
1 & 4 & 10 & 16 & 19 & 16 & 10 & 4 & 1
\end{array}
$$

is obtained as follows: The first row contains only the number 1, and if we put two zeros to the left and the right of each row, then each element in the $(n + 1)$th row is equal to the sum of the three closest numbers in the nth row. Show that from the third row on, each row of the array contains at least one even number.

SOLUTION. It clearly suffices to consider the corresponding array consisting of the remainders upon division by 2:

$$
\begin{array}{ccccccccccc}
 & & & & & 1 & & & & & \\
 & & & & 1 & 1 & 1 & & & & \\
 & & & 1 & 0 & 1 & 0 & 1 & & & \\
 & & 1 & 1 & 0 & 1 & 0 & 1 & 1 & & \\
 & 1 & 0 & 0 & 0 & 1 & 0 & 0 & 0 & 1 & \\
1 & 1 & 1 & 0 & 1 & 1 & 1 & 0 & 1 & 1 & 1
\end{array}
$$

Note that the first four numbers in the third row are $(1, 0, 1, 0)$, and verify that if for some k the kth row begins with the four numbers $(1, 0, 1, 0)$, then the rows immediately following begin with $(1, 1, 0, 1)$, $(1, 0, 0, 0)$, $(1, 1, 1, 0)$, $(1, 0, 1, 0)$, respectively, which means that the $(k + 4)$th row begins again with the numbers $(1, 0, 1, 0)$. This implies that for each $k \geq 3$ there is at least one zero among the first four numbers of the kth row. $\square$

(iii) Suppose that a triangular array contains 2^{n-1} numbers in its nth row $(n = 1, 2, \dots)$, and is constructed as follows: The first row contains some integer $a > 1$, and underneath each number k in the array there is the number k^2 to the left, and $k + 1$ to the right. For $a = 2$, for instance, the array looks as follows:

Show that no such array has two identical numbers in the same row.

SOLUTION.   We prove this by contradiction. We assume that some array of the described form has two identical numbers in one or more rows, and we choose the row of this kind with *smallest* index n. Since $t^2 > t + 1$ for all $t \geq 2$, we have $n > 2$. The number x that occurs twice in the nth row is of the form $x = k^2$, $k \in \mathbb{N}$ (otherwise, the number $x - 1$ would occur twice in the $(n-1)$th row). Then the $(n-1)$th row contains both the numbers k and $k^2 - 1$. In view of the fact that the smallest number a of the array occurs only in the first row, we have $2 \leq a < k \leq k^2 - 2k$. Since none of the numbers

$$k^2 - 1, \; k^2 - 2, \; \dots, \; k^2 - 2k + 2$$

are squares of other integers, it follows by induction on $i = 1, 2, \dots, 2k - 1$ that the number $k^2 - i$ lies in the $(n-i)$th row, which makes sense only when $n - (2k - 1) > 1$, that is, $n - 2k > 0$. On the other hand, the number k, as well as every other number in the $(n-1)$th row, satisfies the estimate $k \geq a + n - 2$, which means that

$$n - 2k \leq n - 2(a + n - 2) = -n - 2(a - 2) < 0;$$

but this contradicts the inequality $n - 2k > 0$.   □

3.12   *Exercises*

(i) Show that for each $n > 1$ it is possible to construct the following array with $\frac{n^2+n}{2}$ appropriate distinct numbers $a_{ij} \in \mathbb{N}$, where the arrows have the same meaning as in 3.11.(i):

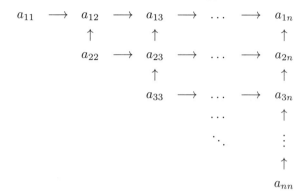

(ii) For a given n, in the triangular array

$$
\begin{array}{ccccccccc}
0 & 1 & 2 & 3 & \ldots & & n-2 & n-1 & n \\
& 1 & 3 & 5 & 7 & \ldots & 2n-5 & 2n-3 & 2n-1 \\
& & 4 & 8 & 12 & \ldots & & 4n-8 & 4n-4
\end{array}
$$

with $n+1$ rows, each number (except the numbers in the first row) is equal to the sum of the two closest numbers in the previous row. For each $k = 1, 2, \ldots, n + 1$, find the sum of the numbers in the kth row of the array.

(iii) Suppose that the first row of a triangular array is formed by a sequence of numbers $a_0, a_1, \ldots, a_n$, and the rest is formed as in Exercise (ii). Find an expression of the (unique) number in the last row of the array in terms of the numbers in the first row.

*(iv) The triangular array

$$
\begin{array}{ccccc}
& & a_{11} & & \\
& a_{21} & & a_{22} & \\
a_{31} & & a_{32} & & a_{33}
\end{array}
$$

of real numbers is constructed as follows: For each n we have $a_{n1} = a_{nn} = 1$, and for $n > 2$ also $a_{ni} \geq a_{n-1,i-1} + a_{n-1,i}$ $(1 < i < n)$. (For instance, the Pascal triangle (18) is of this type.) Show that each number $x > 1$ occurs in the array at most k-times, where the integer k is determined by the relation $2^k < x^2 \leq 2^{k+1}$.

3.13   Miscellaneous Problems

(i) Suppose that an $n \times n$ array of nonnegative integers has the following property: If some field contains the integer 0, then the sum of all the numbers in the row and column of this field is greater than or equal to n. Show that the sum S of all n^2 integers of the array satisfies $S \geq \frac{n^2}{2}$.

SOLUTION. From the row sums $r_1, r_2, \ldots, r_n$ and the column sums $c_1, c_2, \ldots, c_n$ (see Section 3.1) we choose the *smallest* value; we may assume without loss of generality that

$$r_1 = \min\{r_1, r_2, \ldots, r_n, c_1, c_2, \ldots, c_n\}$$

(if this is not the case, we simply change the indexing of the rows and, if necessary, rotate the array by $90°$ to replace rows by columns and vice versa). If $r_1 \geq \frac{n}{2}$, then

$$S = r_1 + r_2 + \cdots + r_n \geq nr_1 \geq \frac{n^2}{2},$$

and therefore we may further assume that $r_1 < \frac{n}{2}$. This means that there are some zeros in the first row of the array; if k denotes their number, then $k \geq n - r_1$. By hypothesis we have $c_i + r_1 \geq n$ for k different indices i of those columns that have a zero in their first field (that is, the field in the first row), and for the remaining $(n-k)$ indices i we have $c_i \geq r_1$, according to our choice. Altogether, we obtain

$$S = c_1 + c_2 + \cdots + c_n \geq \underbrace{(n - r_1) + (n - r_1) + \cdots + (n - r_1)}_{k}$$

$$+ \underbrace{r_1 + r_1 + \cdots + r_1}_{n-k} = k(n - r_1) + (n - k)r_1 = k(n - 2r_1) + nr_1$$

$$\geq (n - r_1)(n - 2r_1) + nr_1 = n^2 - 2nr_1 + 2r_1^2$$

$$= 2\left(r_1 - \frac{n}{2}\right)^2 + \frac{n^2}{2} > \frac{n^2}{2}. \qquad \square$$

(ii) Find the largest integer n with the following property: If a 10×10 array contains each of the numbers $0, 1, 2, \ldots, 9$ exactly ten times, then some row or some column contains at least n different numbers.

SOLUTION. We prove by contradiction that $n = 4$ has the desired property. Suppose to the contrary that in each row and column there are no more than three different numbers. If the number k occurs in exactly α_k rows and β_k columns, then it occurs at most $\alpha_k \cdot \beta_k$ times in the array. Hence for each $k \in \{0, 1, \ldots, 9\}$ we have $\alpha_k \beta_k \geq 10$, which implies $\alpha_k + \beta_k \geq 7$, and thus

$$(\alpha_0 + \beta_0) + (\alpha_1 + \beta_1) + \cdots + (\alpha_9 + \beta_9) \geq 70. \tag{20}$$

Let X denote the set of all pairs (k, j) such that the number k occurs in the jth row. Then X has exactly $\alpha_0 + \alpha_1 + \cdots + \alpha_9$ elements, and since by our assumption there are at most three pairs (k, j) with the same index j

0	0	0	9	9	5	5	5	4	4
0	0	0	9	9	5	5	5	4	4
0	0	1	1	1	5	5	6	6	6
0	0	1	1	1	5	5	6	6	6
7	7	1	1	2	2	2	6	6	7
7	7	1	1	2	2	2	6	6	7
7	8	8	8	2	2	3	3	3	7
7	8	8	8	2	2	3	3	3	7
4	8	8	9	9	9	3	3	4	4
4	8	8	9	9	9	3	3	4	4

Figure 10

in X, where $1 \leq j \leq 10$, we obtain the bound $\alpha_0 + \alpha_1 + \cdots + \alpha_9 \leq 3 \cdot 10 = 30$ for the number of elements in X. Similarly, we also get the inequality $\beta_0 + \beta_1 + \cdots + \beta_9 \leq 30$, and by adding these two, we obtain an inequality that contradicts (20). This proves that $n = 4$ has the desired property.

The array in Figure 10 now shows that no number $n \geq 5$ has the property in question. Indeed, it is easy to check that each row and column has at most four different numbers; to make this easier, the various blocks consisting of identical numbers have been indicated.   □

(iii) Suppose that 1000 real numbers are written in the first row of an array. The second row is formed by writing underneath each number x from the first row the number of occurrences of x in the first row. (If, for instance, all the numbers in the first row are distinct, then the second row consists of 1000 ones.) In the same way the third row is formed from the second, the fourth row from the third, etc.

(a) Show that some two neighboring rows coincide (that is, they consist of the same numbers in the same order).

(b) Show that the 11th row of the array is identical with the 12th row.

(c) Give an example of a first row such that the 10th row of the array is not identical with the 11th row.

SOLUTION.   If in the nth row of the array some number x occupies exactly p places, then in the $(n + 1)$th row these p places are occupied by the number p. Hence for each $n \geq 2$ we have the following: The number of occurrences of an integer x in the nth row is an integer multiple of x, say αx, $\alpha \in \mathbb{N}_0$, where α denotes the number of those (distinct) entries that occur exactly x times in the $(n - 1)$th row; underneath the entry x from the nth row we have therefore the integer $\alpha x \geq x$ in the $(n + 1)$th row. Hence the sequence of numbers in each column of the array (beginning with the second term) is nondecreasing and bounded above by 1000, which means that in each of the 1000 columns of the array from a certain place (depending on the index of the column) the same number is repeated. This proves (a).

Next we use induction on the number $n \geq 2$ to prove the following: If the number of occurrences of an entry x in the nth row is greater than x, then we have $x \geq 2^{n-2}$. This is clear for $n = 2$. If $n > 2$ and the integer x occurs in the nth row more than x times, then in the $(n-1)$th row there are exactly x copies of at least two different integers y, z, which by the above are divisors of x. If, for instance, $y < z$, then $x \geq 2y$, and by the induction hypothesis we have $y \geq 2^{n-3}$, which implies $x \geq 2^{n-2}$, and the proof by induction is complete.

The assertion (b) is now easily proved by contradiction: We assume that the 11th row of some array is not identical with the 12th row. Then there is a number x appearing more than x times in the 11th row. The assertion we just proved shows that $x \geq 2^9$. Thus the number αx of occurrences of x in the 11th row is at least $2x \geq 2^{10} > 1000$, which is a contradiction.

Finally, we consider the array with the first row

$$0, 1, 2, 2, 4, 4, 4, 4, \underbrace{8, \ldots, 8}_{8}, \ldots, \underbrace{256, \ldots, 256}_{256}, \underbrace{488, \ldots, 488}_{488}.$$

Then the next rows are as follows:

$n = 2:$ $1, 1, 2, 2, 4, \ldots, 4, 8, \ldots, 8, \ldots, 256, \ldots, 256, 488, \ldots, 488,$

$n = 3:$ $2, 2, 2, 2, 4, \ldots, 4, 8, \ldots, 8, \ldots, 256, \ldots, 256, 488, \ldots, 488,$

$\ldots$

$n = 10:$ $256, \ldots\ldots\ldots\ldots\ldots\ldots\ldots\ldots\ldots, 256, 488, \ldots, 488,$

$n = 11:$ $512, \ldots\ldots\ldots\ldots\ldots\ldots\ldots\ldots\ldots, 512, 488, \ldots, 488.$

This is an example for Part (c) of the problem.   □

3.14 Exercises

(i) Suppose that an $m \times n$ array consists of real numbers. Let m_i denote the smallest number in the ith row ($1 \leq i \leq m$), and M_j the largest number in the jth column ($1 \leq j \leq n$). Which of the the following two numbers is larger: The largest of the m_i or the smallest of the M_j? When are the two numbers equal?

(ii) Suppose that mn different real numbers are written into an $m \times n$ array in such a way that the numbers in each row are increasing (from left to right). Show that this property is preserved if we simultaneously change the orders of the numbers in all columns such that they are increasing from top to bottom.

(iii) A 50×50 chessboard consists of square fields with side lengths that are identical with those of a usual die with numbers 1 to 6. We roll the die from the lower left to the upper right corner field such that in each step we roll it along an edge onto a neighboring field, always either to the right, or up. Along the way we add the numbers on the faces of the die with which it touches the chessboard; this gives us a sum of 99 numbers. Determine its smallest and largest possible values.

(iv) Suppose that the first row of an array contains arbitrary integers, some of which may be repeated. The second row is then constructed as follows: We examine the first row from left to right, and underneath each integer x we write the integer k if x has occurred for the kth time at that place. In the same way we construct the third row from the second, and the fourth row from the third. In such a way we obtain, for instance, the array

$$\begin{pmatrix} 1 & 3 & 4 & 4 & 3 & 3 & 1 & 1 & 3 & 4 \\ 1 & 1 & 1 & 2 & 2 & 3 & 2 & 3 & 4 & 3 \\ 1 & 2 & 3 & 1 & 2 & 1 & 3 & 2 & 1 & 3 \\ 1 & 1 & 1 & 2 & 2 & 3 & 2 & 3 & 4 & 3 \end{pmatrix},$$

in which the second row is identical with the fourth row. Prove that this is no coincidence: All such arrays have this property.

4   Unordered Configurations

The problems in this section are about configurations of numbers, where the elements are not ordered according to any scheme. Such configurations are therefore *sets* of numbers (in the usual sense), or *multisets* (in the sense of Chapter 1, Section 2.13). Even if the properties we are interested in will not depend on the order in which we write or enumerate the elements, it is sometimes useful to choose the order that corresponds to the usual order relation of the set $\mathbb{R}$ of all real numbers ($a < b \iff b - a \in \mathbb{R}^+$).

4.1   Choosing Subsets

(i) Suppose that for some integer $k \geq 1$ a sum of $2k + 1$ different positive integers is less than $(k+1)(3k+1)$. Show that there are two of the integers whose sum is equal to $2k + 1$.

SOLUTION.   We enumerate the given integers according to size: $x_1 < x_2 < \cdots < x_{2k+1}$. The proof of the assertion is easy if $x_{k+1} \leq 2k$, for then each of the $k + 1$ integers $x_1, x_2, \ldots, x_{k+1}$ lies in one of the k sets $T_1 = \{1, 2k\}$, $T_2 = \{2, 2k - 1\}$, $\ldots$, $T_k = \{k, k + 1\}$; by the pigeonhole principle this means that two integers x_i, x_j ($1 \leq i < j \leq k + 1$) lie in the same set T_s, and thus $x_i + x_j = 2k + 1$.
     If we assume that $x_{k+1} \leq 2k$ does not hold, that is, if $x_{k+1} \geq 2k + 1$, then the sum S of all $2k + 1$ integers satisfies

$$\begin{aligned} S &= (x_1 + x_2 + \cdots + x_k) + (x_{k+1} + x_{k+2} + \cdots + x_{2k+1}) \\ &\geq (1 + 2 + \cdots + k) + ((2k + 1) + (2k + 2) + \cdots + (3k + 1)) \\ &= \frac{k(k + 1)}{2} + \frac{(5k + 2)(k + 1)}{2} = (k + 1)(3k + 1), \end{aligned}$$

which contradicts the hypothesis of the problem.   □

(ii) Show for all $n > 1$ that from any $(n + 2)$-element subset A of the set $\{1, 2, \ldots, 3n\}$ one can choose two numbers whose difference is greater than n but less than $2n$.

SOLUTION.   Since the differences of the elements of the set A do not change if we increase all of its elements by the same number, we may assume that the greatest element of A is equal to $3n$, that is, $3n \in A$. We note that the inequalities $n < 3n - x < 2n$ hold for $x = n+1, n+2, \ldots, 2n-1$; therefore, it suffices to consider only the case where the set A contains none of these integers. Then we have $A = \{3n\} \cup B$, where B is some $(n + 1)$-element subset of the $2n$-element set $\{1, 2, \ldots, n, 2n, 2n+1, \ldots, 3n-1\}$. If we divide the elements of this last set into n subsets

$$T_1 = \{1, 2n\}, \quad T_2 = \{2, 2n + 1\}, \quad \ldots, \quad T_n = \{n, 3n - 1\},$$

then we see by the pigeonhole principle that B contains both elements of some subset T_s. The difference d of these two elements is equal to $2n - 1$, and thus satisfies the required condition $n < d < 2n$.   □

(iii) For all $n > 5$ find the smallest number of integers that one has to delete from the set $\{2, 3, \ldots, n-1, n\}$ so that no remaining integer is equal to the product of two other (distinct) remaining integers.

SOLUTION.   Since $2 \cdot 3 = 6 \leq n$, we have to delete at least one integer for each n. In the case $n < 12$ it suffices to delete just the integer 2, since the product of any two different remaining integers is at least $3 \cdot 4 = 12$. In the case $12 \leq n < 20$ it is enough to delete 2 and 3 (since $4 \cdot 5 = 20 > n$); if we deleted only one of these, then among the remaining integers we would find the triple $\{2, 5, 10\}$ or $\{3, 4, 12\}$. The generalization of this is now easy: For a given n we first determine the number k that satisfies $k(k + 1) \leq n < (k + 1)(k + 2)$. Then it suffices to delete the $k - 1$ integers $2, 3, \ldots, k$ from the set $\{2, 3, \ldots, n - 1, n\}$; if fewer than these $k - 1$ are deleted, then among the remaining integers there is one of the $k - 1$ triples

$$\{2, 2k - 1, 2(2k - 1)\}, \{3, 2k - 2, 3(2k - 2)\}, \ldots, \{k, k + 1, k(k + 1)\}.$$

(The reader should verify that each triple consists of three different positive integers not exceeding n, and that any two triples are disjoint.)   □

(iv) For which $n \geq 3$ can one choose n consecutive positive integers such that the largest one of them is a divisor of the least common multiple of the remaining $n - 1$ chosen integers? For which n is such a choice unique?

SOLUTION.   We begin with the case $n = 3$: Let $(m, m+1, m+2)$ be a triple with the desired property. Since m and $m + 1$ are relatively prime, their least common multiple is $m(m+1)$. However, the number $m(m+1)$ is not a multiple of $m+2$ for any $m \geq 1$, since $(m-1)(m+2) < m(m+1) < m(m+2)$. Hence there is no triple with the desired property.

We will now give an example of an appropriate n-tuple for each $n \geq 4$: For a given number n we first determine a number $k \geq 1$ that satisfies $2^k < n \leq 2^{k+1}$, and we choose the n-tuple $3 \cdot 2^k - n + 1, 3 \cdot 2^k - n + 2, \ldots, 3 \cdot 2^k$ of consecutive positive integers. The largest among them is $3 \cdot 2^k$, and among the remaining $n - 1$ integers there is some multiple of 2^k (since $n - 1 \geq 2^k$) and a multiple of 3 (since $n - 1 \geq 3$). Therefore, each common multiple of these $n - 1$ integers is divisible by $3 \cdot 2^k$.

The above example is not unique for any $n \geq 5$, since for $n \geq 6$ one can give $(5 \cdot 2^k - n + 1, 5 \cdot 2^k - n + 2, \ldots, 5 \cdot 2^k)$ as a second example, and in the case $n = 5$ we also have the admissible quintuple $(2, 3, 4, 5, 6)$, in addition to $(8, 9, 10, 11, 12)$. In the case $n = 4$ the above example of the quadruple $(m, m + 1, m + 2, m + 3)$ is in fact unique, with the value $m = 3$: From the condition $(m + 3) \mid m(m + 1)(m + 2)$ and the identity $m(m+1)(m+2) = (m+3)(m^2+2) - 6$ it follows that $(m+3) \mid 6$; that is, $m = 3$. $\qquad \square$

(v) For each $n \geq 6$ find the largest k for which one can choose k disjoint 3-element subsets $\{a_1, b_1, c_1\}$, $\{a_2, b_2, c_2\}$, $\ldots$, $\{a_k, b_k, c_k\}$ from the set $\{1, 2, \ldots, n\}$ such that the sums $a_1 + b_1 + c_1$, $a_2 + b_2 + c_2$, $\ldots$, $a_k + b_k + c_k$ are k different elements of $\{1, 2, \ldots, n\}$.

SOLUTION. For the sum of all $3k$ chosen integers we have the estimate

$$1 + 2 + 3 + \cdots + 3k \leq S \leq n + (n - 1) + \cdots + (n - k + 1),$$

which implies that $\frac{3}{2}k(3k + 1) \leq \frac{1}{2}(2n - k + 1)k$. From this we obtain the necessary condition $3(3k + 1) \leq 2n - k + 1$, that is, $k \leq \frac{n-1}{5}$. It remains to give an example that confirms that the value $k = \left[\frac{n-1}{5}\right]$ is possible: We write the elements of the individual admissible triples (a_i, b_i, c_i) as columns of the following array of $3k$ different integers, namely,

$$\begin{pmatrix} 1 & 2 & 3 & \cdots & k \\ 2k & 2k - 1 & 2k - 2 & \cdots & k + 1 \\ 2k + 1 & 2k + 2 & 2k + 3 & \cdots & 3k \end{pmatrix}.$$

The numbers $a_i + b_i + c_i = i + (2k + 1 - i) + (2k + i) = 4k + i + 1$ $(1 \leq i \leq k)$ are distinct, and the largest of them has value $5k + 1$, which in the case $k = \left[\frac{n-1}{5}\right]$ does not exceed n. $\qquad \square$

(vi) Show that for each n one can choose three different numbers from an arbitrary $(2n + 1)$-element subset of the set $K_n = \{x \in \mathbb{Z}; |x| \leq 2n - 1\}$ such that their sum is equal to 0.

SOLUTION.   We use induction on n. The assertion is clearly true for $n = 1$, since $K_1 = \{-1, 0, 1\}$ is the unique 3-element subset of K_1, and $(-1) + 0 + 1 = 0$. Now suppose that the assertion holds for some $n \geq 1$ and let X be an arbitrary $(2n + 3)$-element subset of K_{n+1}. If X contains at most two of the numbers $2n$, $2n + 1$, $-2n$, $-2n - 1$, then the set $X' = X \cap K_n$ has at least $2n + 1$ elements; hence in X' (and thus also in X) we find three different numbers whose sum is zero. Otherwise, it suffices to consider the following two cases:

(a) $\{2n, 2n + 1, -2n - 1\} \subseteq X$.   (b) $\{2n, 2n + 1, -2n\} \subseteq X$.

(The remaining cases can be reduced to these by changing all numbers in X to their negatives.) In case (a) we consider the $2n + 1$ pairs of numbers

$$\{1, 2n\}, \{2, 2n - 1\}, \ldots, \{n, n + 1\},$$
$$\{0, -2n - 1\}, \{-1, -2n\}, \ldots, \{-n, -n - 1\},$$

which form a decomposition of the set $K_{n+1} \setminus \{2n + 1\}$. Since the set $X \cap (K_{n+1} \setminus \{2n + 1\})$ has exactly $2n + 2$ elements, by the pigeonhole principle it contains both elements x, y of some of the above pairs. In view of the fact that $x + y \in \{2n + 1, -2n - 1\}$, the triple $\{x, y, -x - y\}$ contains distinct elements of X whose sum is equal to 0. In case (b) it is then enough to restrict ourselves to the case where $-2n - 1 \notin X$ (otherwise, we have (a)) and also $0 \notin X$ (because of the triple $\{0, 2n, -2n\}$). Then in the $2n - 1$ disjoint pairs

$$\{1, 2n - 1\}, \{2, 2n - 2\}, \ldots, \{n - 1, n + 1\},$$
$$\{-1, -2n\}, \{-2, -2n + 1\}, \ldots, \{-n, -n - 1\}$$

there are either $2n$ or $2n + 1$ elements of X (depending on whether or not $n \in X$). Hence by the pigeonhole principle X contains both elements of some of these pairs, and the proof is complete. $\square$

4.2   Exercises

(i) For each $n \geq 1$, find the smallest number k, $3 \leq k \leq 2n + 1$, with the property that from any k-element set $X \subseteq \{1, 2, \ldots, 2n + 1\}$ one can choose three distinct integers a, b, c such that $a + b = c$.

(ii) Show that from any 10-element set $X \subseteq \{1, 2, 3, \ldots, 25\}$ one can choose four distinct integers a, b, c, d such that $a + b = c + d$.

*(iii) Determine for which $n \geq 3$ the following assertion holds: From any n-element set X of positive numbers one can choose two elements a, b $(a < b)$ such that none of the remaining $n - 2$ elements of X is equal to either $a + b$ or to $b - a$.

(iv) For each $n \geq 2$, find the greatest integer k for which it is possible to choose k different elements from the set $\{1, 2, \ldots, n\}$ such that the sum of any two of the chosen integers is not divisible by its difference.

(v) Decide for which $n \geq 2$ there exists an n-element set of positive integers such that the sum of any two distinct elements of this set is divisible by its difference.

(vi) Is there an infinite set of positive integers that has the property required in (v)?

(vii) For each $n \geq 1$ we state the hypothesis H_n: From any $(2n - 1)$-element set of integers one can choose n different elements such that their sum is divisible by n. Show that if for some k, n both the hypotheses H_k and H_n hold, then so does H_{kn}. Therefore, in order to verify H_n, it suffices to prove H_p for all primes p. An elementary (but not easy) proof of H_p can be found as Problem M45 in [49], 1971, no. 3. Prove H_p yourself for at least $p = 2, 3$, and 5.

(viii) For each $n \geq 3$ choose n positive integers such that none of the chosen numbers is divisible by another one of the chosen numbers, but the product of any two different chosen numbers is divisible by each of the remaining $(n - 2)$ numbers.

*(ix) Show that for each $n \geq 2$ one can choose n distinct positive integers such that the chosen numbers are pairwise relatively prime and any sum of several (two or more) chosen numbers is composite.

*(x) For which $n \geq 3$ can one choose n distinct positive integers such that both the sum and the product of any $(n - 1)$ chosen numbers are divisible by the remaining nth number?

*(xi) Show that for any integers k, n ($2 \leq k \leq n$) one can choose n distinct positive integers such that the product of any k chosen numbers is divisible by their sum.

*(xii) Show that there exists an infinite set X of positive integers such that no number in X, nor any sum of distinct numbers from X, is a perfect power (see Section 2.10).

4.3 Sets of Type DS

We will now present several problems about sets that are interesting from the point of view of comparing sums of elements of their subsets with each other. We say that a finite set $X \subseteq \mathbb{N}$ is of type DS if the sums of the elements of any two distinct subsets of X are different. (We introduce this terminology only for the purpose of a more concise formulation of the following text; the letters DS stand for *different sums*.)

(i) Show that for each $n \geq 1$ there is an n-element set of type DS.

SOLUTION.   The set of integers $\{1, 10, 10^2, \ldots, 10^{n-1}\}$ is clearly of type DS, since the decimal representation of the sum of a number of its elements contains only the digits 0 and 1, and the positions occupied by the digit 1 determine uniquely which numbers were added together. Similarly, with the help of binary representations we can give the n-element set $\{1, 2, 2^2, \ldots, 2^{n-1}\}$ as a different example of a set of type DS, which interestingly illustrates the assertion of Exercise 4.4.(ii).   □

(ii) Show that the set $X = \{220, 310, 401, 540, 604, 902\}$ is of type DS.

SOLUTION.   We write the digits of the numbers $x_1, x_2, \ldots, x_6 \in X$ in columns in such a way that the rows always contain digits from the same places (units, tens, hundreds):

$$
\begin{array}{c|cccccc}
 & x_1 & x_2 & x_3 & x_4 & x_5 & x_6 \\
10^2 & 2 & 3 & 4 & 5 & 6 & 9 \\
10^1 & 2 & 1 & 0 & 4 & 0 & 0 \\
10^0 & 0 & 0 & 1 & 0 & 4 & 2
\end{array}
\tag{21}
$$

Let S denote the sum of the elements of an arbitrarily chosen subset $Y \subseteq X$. From the array (21) one can see that the last digit of S is at most 7; if we write 7 in its binary representation, we determine $Y \cap \{x_3, x_5, x_6\}$. Similarly, with the "tens" digit of the number S we determine $Y \cap \{x_1, x_2, x_4\}$. Therefore, the set Y is uniquely determined by the number S. This means that X is of type DS.   □

(iii) Suppose that the numbers $a_1, a_2, \ldots, a_m \in \mathbb{N}$ are given. Show that there is a set X of type DS with fewer than 2^m elements and such that among the sums of the elements of its subsets there are all the numbers $a_1, a_2, \ldots, a_m$.

SOLUTION.   Given the numbers $a_1, a_2, \ldots, a_m$, we first find an integer k such that $a_i < 2^{k+1}$ holds for all $i = 1, 2, \ldots, m$. Then the set $X = \{1, 2, 4, \ldots, 2^k\}$ of type DS has the property that among the sums of the elements of its subsets we can find every integer $x \in \{1, 2, 3, \ldots, 2^{k+1} - 1\}$, and in particular every integer a_i ($1 \leq i \leq m$). We write them in binary representation

$$
a_i = c_k^{(i)} \cdot 2^k + c_{k-1}^{(i)} \cdot 2^{k-1} + \cdots + c_1^{(i)} \cdot 2 + c_0^{(i)},
\tag{22}
$$

where $c_j^{(i)} \in \{0, 1\}$, $0 \leq j \leq k$, $1 \leq i \leq m$ (and in doing this, we pad the representations of the numbers that are less than 2^k with zeros from the left). If $|X| = k + 1 < 2^m$, then the proof is complete. We therefore let $k + 1 \geq 2^m$ in what follows. We now carry out a reduction of the number of elements of the set X, based on the following trick: If $y_1, y_2, \ldots, y_s$ are distinct elements of a set Y of type DS, $s > 1$, then $y_1 + y_2 + \cdots + y_s \notin Y$; if we replace the collection of elements $y_1, y_2, \ldots, y_s$ in Y by the number $y_1 + y_2 + \cdots + y_s$, we obtain again a set of type DS. We therefore need to answer the question: Which collections of elements $2^{j_1}, 2^{j_2}, \ldots, 2^{j_s}$ of our set X can we replace by their sums such that the numbers $a_1, a_2, \ldots, a_m$ remain among the sums of the elements of subsets? They are clearly those collections $2^{j_1}, 2^{j_2}, \ldots, 2^{j_s}$ for which

$$c_{j_1}^{(i)} = c_{j_2}^{(i)} = \cdots = c_{j_s}^{(i)} \quad \text{for each} \quad i = 1, 2, \ldots, m, \tag{23}$$

where the $c_j^{(i)}$ are the digits in the representation (22). The condition (23) is satisfied if and only if all sequences $c_j = \left(c_j^{(1)}, c_j^{(2)}, \ldots, c_j^{(m)} \right)$, $j \in \{j_1, j_2, \ldots, j_s\}$, are the same. We therefore approach the reduction of the set X as follows: First we remove from X those elements 2^j for which $c_j = (0, 0, \ldots, 0)$, since they are not required in forming the sums $a_1, a_2, \ldots, a_m$. We then divide the remaining elements 2^j from X into classes determined by the equivalence relation $2^i \sim 2^j \iff c_i = c_j$. Since each sequence c_j has length m and consists of the digits 0, 1, we obtain at most $2^m - 1$ classes. The sums of the elements in the individual classes then form a set of type DS with the desired properties.   □

4.4   Exercises

    (i) Show that the sum of all n elements of a given set of type DS is at least $2^n - 1$.

   (ii) Decide for which $n \geq 1$ there is an n-element set of type DS whose largest element is less than 2^{n-1}.

* **(iii)** Find all n-element sets $X \subseteq \mathbb{R}^+$ with the property that if we form the sums of the elements of all nonempty subsets of X, then we obtain $2^n - 1$ consecutive positive integers.

   (iv) Show that for any n-element set $X \subseteq \mathbb{R}^+$, among all $2^n - 1$ sums of the elements of its nonempty subsets, at least $\frac{1}{2}n(n+1)$ numbers are distinct. Can the number $\frac{1}{2}n(n+1)$ in general be replaced by some larger integer?

* **(v)** Decide whether there exists a 7-element set of type DS with largest element less than 27.

4.5   Problems on Partitions of Sets

To begin, we briefly recall one of the basic concepts of set theory: A system $\{T_i; i \in I\}$ of nonempty pairwise disjoint sets T_i is called a *partition* of the set X if

$$\bigcup_{i \in I} T_i = X;$$

the sets T_i $(i \in I)$ are called *classes* of this partition.

(i) Decide whether the numbers $1, 2, \ldots, 1997$ can be divided into several groups such that the largest integer in each group is equal to the sum of all its remaining numbers.

SOLUTION.   We assume that such a partition of the set $\{1, 2, \ldots, 1997\}$ exists. Then the sum of the numbers in each of the classes is equal to twice its largest element and is therefore an even number. Hence the sum of all the numbers,

$$1 + 2 + \cdots + 1997 = \frac{1997 \cdot 1998}{2} = 1997 \cdot 999,$$

would have to be even as well, but this is a contradiction, since the product of two odd numbers is odd. Therefore, no such partition exists.   □

(ii) Suppose that an arbitrary partition of the set $\{1, 2, \ldots, 100\}$ into 7 classes is given. Show that in some class there are four distinct numbers a, b, c, d for which $a + b = c + d$, or there are three distinct numbers e, f, g with $e + f = 2g$.

SOLUTION.   Since $14 \cdot 7 = 98 < 100$, by the pigeonhole principle some class T of the given partition has at least 15 elements. We consider all differences $a - b$, for $a, b \in T$, $a > b$. There are at least $\binom{15}{2} = 105$ of them, but their values are among the numbers $1, 2, \ldots, 99$. By the pigeonhole principle there are two different pairs $\{x, y\}$ and $\{u, v\}$ in T such that $x - y = u - v > 0$. If x, y, u, v are four distinct numbers, then we are done $(x + v = y + u)$; otherwise, we have either $x = v$ (then $y + u = 2x$) or $y = u$ (then $x + v = 2y$).   □

(iii) An interesting and deep result is given by *van der Waerden's theorem*:

   For any $k, n \in \mathbb{N}$ there exists a number $A = A(k, n) \in \mathbb{N}$ such that in any partition of the set $\{1, 2, 3, \ldots, A\}$ into k classes there is a class that contains an arithmetic sequence of length n with nonzero difference.

   A sophisticated proof which is, however, completely elementary and uses induction on n, can be found in [6]. Finding the numbers $A(k, n)$ is non-trivial for all $k > 1$ and $n \geq 3$; show at least the existence of $A(2, 3)$ and find its smallest value.

SOLUTION.   We have to find the smallest number A with the property that for any partition of the set $\{1, 2, 3, \ldots, A\}$ into two classes T_1 and T_2, one of the classes T_i contains three numbers $x < y < z$ such that $x + z = 2y$. We clearly need $A \geq 5$, so for $A = 5, 6, 7, \ldots$ we look for partitions that contradict the required property for the number A. Thus, for example, for $A = 8$ such a partition is given by

$$\{1, 2, 3, \ldots, 8\} = \{1, 4, 5, 8\} \cup \{2, 3, 6, 7\};$$

to check this, it suffices to consider the pairs of numbers $x < z$ of the same parity in each class and see whether $\frac{x+z}{2}$ belongs to the same class. If we try this with $A = 9$, we are already unable to find such a partition (you should try it nevertheless; perhaps you will be able to find a reason why it will not work). We therefore assume that there exists a partition $\{1, 2, \ldots, 9\} = T_1 \cup T_2$ such that in none of the two classes T_i is there a triple of the desired form. We distinguish two cases according to whether or not 1 and 9 belong to the same class. Since the particular indexing of the classes is irrelevant, we may distinguish only the two cases $\{1, 9\} \subseteq T_1$, $1 \in T_1$ and $9 \in T_2$.

(a) $\{1, 9\} \subseteq T_1$. Then the number $\frac{1+9}{2} = 5$ cannot lie in T_1, which means that $5 \in T_2$. In view of the fact that $5 = \frac{3+7}{2}$, at least one of the numbers 3, 7 lies in T_1; let $3 \in T_1$ without loss of generality (if $7 \in T_1$, we replace every x by $10 - x$). From $\{1, 3\} \subseteq T_1$ it follows that $2 \in T_2$, $\{3, 9\} \subseteq T_1$ implies $6 \in T_2$, and then $\{5, 6\} \subseteq T_2$ implies $4 \in T_1$ and $7 \in T_1$. We have thus obtained $\{1, 4, 7\} \subseteq T_1$, which is a contradiction.

(b) $1 \in T_1$ and $9 \in T_2$. Without loss of generality let $5 \in T_1$ (otherwise, we carry out the change $x \mapsto 10 - x$, as mentioned above). From $\{1, 5\} \subseteq T_1$ it follows that $3 \in T_2$; then $\{3, 9\} \subseteq T_2$ implies $6 \in T_1$, and then $\{5, 6\} \subseteq T_1$ implies $4 \in T_2$ and $7 \in T_2$. Next, from $\{7, 9\} \subseteq T_2$ it follows that $8 \in T_1$, and $\{3, 4\} \subseteq T_2$ implies $2 \in T_1$. We have thus obtained $\{2, 5, 8\} \subseteq T_1$, which is again a contradiction.

In summary, we have shown that the smallest value of $A(2, 3)$ is equal to 9.   □

(iv) We place 1000 index cards with the numbers $000, 001, \ldots, 999$ into boxes numbered $00, 01, \ldots, 99$. We can place a card with number x only in a box whose number is obtained from x by deleting one of its three digits. Show that all the cards

(a) can be placed in 50 boxes,

(b) cannot be placed into n boxes if $n < 40$.

SOLUTION.   (a) We describe a way in which all cards can be placed into 50 boxes. We divide all the digits $0, 1, \ldots, 9$ into two classes A and B with five elements each, for instance $A = \{0, 1, \ldots, 4\}$ and $B = \{5, 6, \ldots, 9\}$, and choose those boxes whose numbers have both digits in the same class. There are exactly $2 \cdot 5 \cdot 5 = 50$ such boxes. From the number x of any given card we now delete one digit such that the remaining two belong to the same one of the two classes A, B; the number formed by this pair of digits determines the box in which the card will be placed.

(b) For any distribution of the cards, the boxes numbered $00, 11, \ldots, 99$ will always occur, since they are required for the cards with numbers $000, 111, \ldots, 999$. None of these boxes can receive cards with numbers $x = \overline{pqr}$, where $p \neq q \neq r \neq p$. Therefore, it suffices to show that for the placement of these cards, of which there are $10 \cdot 9 \cdot 8 = 720$, we require at least 30 boxes. But this is easy: A box with number $\overline{ab}$, where $a \neq b$, can receive only cards with numbers $\overline{pab}$, $\overline{apb}$, $\overline{abp}$, where p is any digit different from a and b, namely $3 \cdot 8 = 24$ cards. It remains to remark that $\frac{720}{24} = 30$. $\qquad\square$

(v) Show that for any partition of the set $\{1, 2, \ldots, 3n\}$ into three classes with n elements one can choose one number from each of the classes such that one of the chosen numbers is equal to the sum of the other two.

SOLUTION.   We consider a partition and denote its classes by A, B, C in such a way that $1 \in A$, $\{1, 2, \ldots, k - 1\} \subseteq A$, and $k \in B$ (the number k is therefore uniquely determined). We call a triple of numbers $(a, b, c) \in A \times B \times C$ a *good* triple if $a + b = c$, $a + c = b$, or $b + c = a$. We assume that no good triple exists and show that then $c - 1 \in A$ for all $c \in C$, which in view of the fact that $1 \in A$ and $2 \notin C$ would mean that $|A| > |C|$, but this is a contradiction to $|A| = |B| = |C| = n$. Indeed, we assume to the contrary that there exists a $c \in C$ such that $c - 1 \notin A$, and choose the *smallest* such c. Then clearly $c - 1 \notin B$ (otherwise, $(1, c - 1, c)$ would be a good triple), and thus $c - 1 \in C$. By considering the triples $(c - k, k, c)$ and $(k - 1, c - k, c - 1)$, where k is as defined at the beginning, we find that $c - k \notin A$ and $c - k \notin B$, that is, $c - k \in C$. Since $c - k < c$, this means in view of the choice of the number c that $c - k - 1 \in A$. But then $(c - k - 1, k, c - 1)$ is a good triple, which is a contradiction. $\qquad\square$

(vi) For $k \geq 1$ set $n_k = k! \left(1 + \frac{1}{1!} + \frac{1}{2!} + \cdots + \frac{1}{k!}\right)$. Show that for any partition of the set $\{1, 2, \ldots, n_k\}$ into k classes, some class contains numbers x, y, z (where we allow $x = y$) with the property $x + y = z$. (To give a better idea about the size of the numbers n_k we note that n_k is the integer part of the number $e \cdot k!$, where $e \approx 2.718$ is the base of the natural logarithm; see, e.g., [5], Chapter 2, Section 3.8.(iv).)

SOLUTION.   We set $n_0 = 1$. Then for each $t \geq 1$ we have $n_t = t \cdot n_{t-1} + 1$. To prove the assertion by contradiction, we assume that for some k, which we will then fix, there is a partition of $\{1, \ldots, n_k\}$ into k classes such that none of the classes contains numbers x, y, z with $x + y = z$.

We use induction on $i = 1, 2, \ldots, k$ to show that among the classes of the partition there are classes $T_1, \ldots, T_i$ such that from T_i one can choose $n_{k-i} + 1$ numbers $a_0^{(i)} < a_1^{(i)} < \cdots < a_{n_{k-i}}^{(i)}$ with the property that none of the differences $a_r^{(i)} - a_s^{(i)}$ (where $0 \leq s < r \leq n_{k-i}$) lies in any of the classes $T_1, \ldots, T_i$.

Induction beginning. Since $n_k = k \cdot n_{k-1} + 1$, by the pigeonhole principle some class contains at least $n_{k-1} + 1$ numbers. We denote this class by T_1 and arbitrarily choose from it the elements $a_0^{(1)} < a_1^{(1)} < \cdots < a_{n_{k-1}}^{(1)}$. If some of the differences $a_r^{(1)} - a_s^{(1)}$ (where $0 \leq s < r \leq n_{k-1}$) were in T_1, we would obtain a contradiction for $x = a_r^{(1)} - a_s^{(1)}$, $y = a_s^{(1)}$, $z = a_r^{(1)}$.

Induction step. We assume that $i \in \{2, \ldots, k\}$ and that we have classes $T_1, \ldots, T_{i-1}$, and in T_{i-1} we have chosen numbers $a_0^{(i-1)} < a_1^{(i-1)} < \cdots <$

$a_{n_{k-i+1}}^{(i-1)}$ such that none of the differences $a_r^{(i-1)} - a_s^{(i-1)}$ (where $0 \le s < r \le n_{k-i+1}$) lies in any of the classes $T_1, \ldots, T_{i-1}$. Then $n_{k-i+1} = (k - i + 1) \cdot n_{k-i} + 1$ numbers $a_1^{(i-1)} - a_0^{(i-1)}$, $a_2^{(i-1)} - a_0^{(i-1)}$, $\ldots$, $a_{n_{k-i+1}}^{(i-1)} - a_0^{(i-1)} \in \{1, \ldots, n_k\}$ lie in the remaining $k - i + 1$ classes (note that $\{1, \ldots, n_k\}$ is the union of all k classes). By the pigeonhole principle there exists a class, which we denote by T_i, in which there are at least $n_{k-i} + 1$ such numbers. We denote them by $a_0^{(i)} < a_1^{(i)} < \cdots < a_{n_{k-i}}^{(i)}$. In view of the construction of these numbers, the difference of any two of them is equal to $a_r^{(i-1)} - a_s^{(i-1)}$ for some $1 \le s < r \le n_{k-i+1}$, and so they are not in any of the classes $T_1, \ldots, T_{i-1}$. At the same time they are also not in the class T_i (otherwise, we would find $x, y, z \in T_i$ such that $x + y = z$).

This completes the proof by induction; for $i = k$ we have therefore shown that from the class T_k one can choose $n_0 + 1 = 2$ numbers $a_0^{(k)} < a_1^{(k)}$ such that their difference does not lie in any of the classes $T_1, \ldots, T_k$. But this is a contradiction to $0 < a_1^{(k)} - a_0^{(k)} < n_k$ and $T_1 \cup \cdots \cup T_k = \{1, \ldots, n_k\}$. $\qquad \square$

4.6   Exercises

(i) Determine those nonnegative k for which one can partition the set $\{1, 2, \ldots, n\}$ into two classes such that the difference of the sums of the numbers in the individual classes is equal to k.

(ii) Find the smallest number k for which there exists a partition of the 100-element set $X = \{1, 2, \ldots, 100\}$ into k classes such that any two distinct numbers from X
(a) that are not relatively prime
(b) that are relatively prime
lie in different classes.

(iii) Show that for any partition of the set $X = \{1, 2, 3, 4, 5\}$ into two classes there are two numbers from X that belong to the same class as their difference.

(iv) For each of the values $\alpha = 2$ and $\alpha = 3$ determine those n for which the numbers $1, 2, \ldots, n$ can be distributed into some classes such that in each class there is a number whose product with α is equal to the sum of the remaining numbers of the class.

*(v) Suppose that the sum of the integers a, b is not divisible by 3. Show that there is no partition of $\mathbb{Z}$ into three classes such that for every integer t, each of the three numbers $t, t + a, t + b$ lies in a different class.

(vi) Decide for which values of $k, n, d \in \mathbb{N}$ there is a number $A = A(k, n, d) \in \mathbb{N}$ such that for any partition of the set $\{1, 2, \ldots, A\}$ into k classes, one of these classes contains an arithmetic sequence of length n with difference d.

(vii) Use van der Waerden's theorem 4.5.(iii) to prove its analogue for geometric sequences: For any k, n there exists a number $G = G(k, n)$ such that for any partition of the set $\{1, 2, \ldots, G\}$ into k classes, one of these classes contains a geometric sequence of length n with a quotient different from 1.

*(viii) Consider an arbitrary partition of $\mathbb{N} = T_1 \cup T_2$ such that the class T_1 does not contain any three numbers that form an arithmetic sequence. It follows from van der Waerden's theorem that for each n the class T_2 contains an n-tuple of different numbers forming an arithmetic sequence. Decide whether the following stronger assertion holds: The class T_2 contains an infinite arithmetic sequence with a nonzero difference.

*(ix) Find the smallest number of boxes required to distribute the cards of Problem 4.5.(iv).

*(x) For any s and k, $k \geq 3$, consider the generalization of 4.5.(iv) when s^k cards with k-digit numbers formed with s different digits are distributed over s^2 boxes labeled with two-digit numbers from the same s digits. Assume that each card can be placed only in the box with a number that can be obtained by deleting $k - 2$ appropriate digits from the card's number. Show that the smallest number of boxes needed for distributing all cards is equal to the smallest value of the sum $d_1^2 + d_2^2 + \cdots + d_{k-1}^2$, where $d_1, d_2, \ldots, d_{k-1}$ are positive integers satisfying $d_1 + d_2 + \cdots + d_{k-1} = s$.

*(xi) For any integer $k \geq 1$ set $n_k = \frac{1}{2}(3^k - 1)$. Show that the set $\{1, 2, 3, \ldots, n_k\}$ can be partitioned into k classes such that none of the classes contains numbers x, y, z with the property $x + y = z$ (where $x = y$ is allowable).

(xii) Let $\mathbb{N} = T_1 \cup T_2 \cup \cdots \cup T_k$ be a partition such that the elements of each class T_i form an arithmetic sequence. Show that the difference of one of these k sequences divides its first term.

(xiii) Let d_i be the difference of the arithmetic sequence of the elements of T_i in Exercise (xii). Show that $\frac{1}{d_1} + \frac{1}{d_2} + \cdots + \frac{1}{d_k} = 1$.

*(xiv) Does there exist a partition of $\mathbb{N}$ into infinitely many classes $T_1, T_2, \ldots$ such that T_1 is infinite and every class T_k ($k > 1$) is obtained by adding the same number to all the elements of T_1?

*(xv) For each ε ($0 < \varepsilon < 1$) construct a partition of $\mathbb{N}$ into infinitely many classes $T_1, T_2, \ldots$ such that the elements of each class T_i form an infinite arithmetic sequence with difference d_i and such that the inequality $\frac{1}{d_1} + \frac{1}{d_2} + \cdots + \frac{1}{d_n} < \varepsilon$ holds for all $n \geq 1$.

In Exercises (xvi)–(xix) the symbol K_n denotes the 2^n-element set of all n-digit positive integers formed with the digits 1 and 2.

(xvi) Show that for all $n \geq 2$ one can partition K_n into two classes such that the sum of any two different elements of the same class is a number among whose digits there are at least two threes.

(xvii) Decide for which $n \geq 3$ there exists a partition of K_n into three classes such that the sum of any two different numbers in the same class has among its digits at least three threes.

(xviii) Show that every subset of K_n that has the property of the classes described in (xvii) has at most $2^n/(n+1)$ elements.

*(xix) For each $n \geq 1$ find the largest integer p_n with the following property: From the set K_{2^n} one can choose p_n different numbers such that the sum of any two different chosen numbers has among its digits at least 2^{n-1} threes.

*(xx) In a given n-element set of positive numbers we consider all sums of the elements of its nonempty subsets. Show that these $2^n - 1$ numbers can be divided into n classes such that in each class the quotient of the largest and the smallest number does not exceed 2.

4.7 Balanced Partitions

We say that a partition of a set X of numbers into n classes $T_1, T_2, \ldots, T_n$ is *cardinality-balanced* (respectively *sum-balanced*) if all classes T_i have the same number (respectively sum) of elements.

In what follows we will deal with the question for which values n, k greater than 1 the following hypothesis $H(n, k)$ holds: There exists a partition of the set of numbers $\{1, 2, \ldots, nk\}$ into k classes that are both cardinality-balanced and sum-balanced. We divide the solution to this problem into three stages, from which it will follow that hypothesis $H(n, k)$ holds if and only if either the number n is even or both n and k are odd.

(i) Prove hypothesis $H(n, k)$ in the case where n is even; disprove it in the case where n is odd and k is even.

SOLUTION.   The proof of hypothesis $H(2, k)$ is easy: From the numbers $1, 2, \ldots, 2k$ we form the pairs of summands

$$1 + 2k = 2 + (2k - 1) = 3 + (2k - 2) = \cdots = k + (k + 1),$$

so the classes for a required partition are $\{1, 2k\}, \{2, 2k-1\}, \ldots, \{k, k+1\}$. This method can be used also for the proof of $H(n, k)$ for any even $n = 2m$: From the numbers $1, 2, \ldots, 2mk$ we first form the pairs $d_1 = \{1, 2km\}$, $d_2 = \{2, 2km - 1\}, \ldots, d_{mk} = \{mk, mk + 1\}$; then we *arbitrarily* divide these km pairs d_i into k groups with m members each and take unions, for instance as follows: $T_1 = d_1 \cup d_2 \cup \cdots \cup d_m$, $T_2 = d_{m+1} \cup d_{m+2} \cup \cdots \cup d_{2m}$, $\ldots, T_k = d_{m(k-1)+1} \cup d_{m(k-1)+2} \cup \cdots \cup d_{mk}$. Then each class T_i has $2m$ elements, and their sum is $m(2km + 1)$; that is, this partition is both cardinality-balanced and sum-balanced.

We now assume that hypothesis $H(n,k)$ holds for some odd n and even k. Then the sum of the n numbers in each of the k classes of a sum-balanced partition is equal to

$$(1 + 2 + \cdots + nk)\frac{1}{k} = \frac{n(nk+1)}{2},$$

which is not an integer, since both n and $nk + 1$ are odd. We have thus obtained a contradiction, which proves the second part of assertion (i).   □

(ii) Show that hypothesis $H(3,k)$ implies hypothesis $H(n,k)$ for all odd $n > 3$.

SOLUTION.   Let $T_1, T_2, \ldots, T_k$ be the classes of a cardinality-balanced and sum-balanced partition of the set $\{1, 2, \ldots, 3k\}$. If $n > 3$ is odd, then $n - 3$ is an even number, and by (i) there is a cardinality- and sum-balanced partition of the set $\{1, 2, \ldots, (n-3)k\}$ into k classes $X_1, X_2, \ldots, X_k$. We now augment the classes X_i appropriately with the remaining numbers from $\{1, 2, \ldots, nk\}$, that is, the $3k$ numbers $(n-3)k + 1$, $(n-3)k + 2$, $\ldots$, nk. These numbers form a set that can be obtained from $\{1, 2, \ldots, 3k\}$ by adding the same value $(n-3)k$ to all elements, and therefore the k subsets $Y_i = \{x + (n-3)k; \ x \in T_i\}$, $1 \le i \le k$, form a cardinality- and sum-balanced partition of this set. Finally, this is then also true for the partition of the set $\{1, 2, \ldots, 3n\}$ into the k classes $X_1 \cup Y_1, X_2 \cup Y_2, \ldots, X_k \cup Y_k$.   □

(iii) Prove hypothesis $H(3,k)$ for all odd $k \ge 3$.

SOLUTION.   It is useful to begin by looking for appropriate partitions for small values $k = 3, 5, \ldots$ and trying to determine a general rule for their construction. Are you able to see such a rule from the following tables for $k = 3, 5, 7$ (each class is formed by the numbers in one column)?

1	2	3		1	2	3	4	5		1	2	3	4	5	6	7
5	6	4		8	9	10	6	7		11	12	13	14	8	9	10
9	7	8		15	13	11	14	12		21	19	17	15	20	18	16

The general construction for odd $k = 2m + 1$ is

$$T_i = \{i, 3m + i + 1, 6m - 2i + 5\} \quad (1 \le i \le m + 1),$$
$$T_i = \{i, m + i, 8m - 2i + 6\} \quad (m + 2 \le i \le 2m + 1).$$ □

4.8   Exercises

(i) For which nk-element sets of real numbers can one prove the existence of a cardinality- and sum-balanced partition into k classes by using the approach of 4.7?

(ii) For which n does there exist a sum-balanced partition of the set $\{1, 2, \ldots, n\}$ into two classes?

(iii) Answer (ii) with three instead of two classes.

(iv) Show that any 5-element set X of positive real numbers has at most one sum-balanced partition into two classes.

(v) Construct a cardinality- and sum-balanced partition of the set of the thousand perfect squares $\{1^2, 2^2, 3^2, \ldots, 1000^2\}$ into two classes.

(vi) For each n construct a cardinality- and sum-balanced partition of $\{1^2, 2^2, 3^2, \ldots, (27n)^2\}$ into three classes.

*(vii) Given positive integers m, n, k with $1 + 2 + \cdots + n = mk$, show that a sum-balanced partition of the set $\{1, 2, \ldots, n\}$ into k classes exists if and only if $m \geq n$.

4.9   Arithmetic Operations on Sets

A number of arithmetical problems are easier to formulate, and their solutions easier to write down, if we translate arithmetical operations between numbers to operations between sets. If $*$ is some binary operation on a set of numbers K, then for any two nonempty sets $A, B \subseteq K$ we define

$$A * B = \{a * b : \; a \in A \land b \in B\}.$$

Thus, for example, for $A = \{1, 2\}$ and $B = \{3, 4\}$ the set $A + B$ contains exactly the numbers $1 + 3$, $1 + 4$, $2 + 3$, and $2 + 4$; that is, $A + B = \{4, 5, 6\}$. Similarly, we obtain

$$A - B = \{-3, -2, -1\}, \quad A \cdot B = \{3, 4, 6, 8\}, \quad A \div B = \left\{\frac{1}{3}, \frac{1}{4}, \frac{2}{3}, \frac{1}{2}\right\}.$$

(For the product we use the sign "$\cdot$" in order to distinguish between the arithmetic product and the Cartesian product $A \times B$; similarly, we must distinguish between the arithmetic difference $A - B$ and the set-theoretic difference $A \setminus B = \{a : \; a \in A \land a \notin B\}$.) We further simplify the above notation $A * B$ in the case where one of the sets A, B has only one element: Instead of $\{a\} * B$, respectively $A * \{b\}$, we will write $a * B$, respectively $A * b$; finally, $-A$ will mean the same as $(-1) \cdot A$.

We note that, for instance, for $A = \{1, 2\}$ we have $A + A = \{2, 3, 4\}$, while $2 \cdot A = \{2, 4\}$. This means that the equality $A + A = 2 \cdot A$, which in analogy with numbers suggests itself, does not hold in general (but it does hold for some sets with more than one element, for instance for $A = \mathbb{R}^+$). Therefore, when we deal with arithmetical operations with sets, we have to be a little careful, at least at the beginning. The following exercises will serve as a good introductory preparation.

4.10   Exercises

Which relations hold in general between the following pairs of sets? When appropriate, use counterexamples to disprove equality between the sets.

(i) $A + B$ and $B + A$.

(ii) $(A \cdot B) \cdot C$ and $A \cdot (B \cdot C)$.

(iii) $(A \cdot B) \div C$ and $A \cdot (B \div C)$.

(iv) $A + A$ and $2 \cdot A$.

(v) $(A + B) \cdot C$ and $(A \cdot C) + (B \cdot C)$.

(vi) $(A \cap B) \cdot C$ and $(A \cdot C) \cap (B \cdot C)$.

(vii) $(A + B) \cap C$ and $(A \cap C) + (B \cap C)$.

(viii) $(A \cup B) + C$ and $(A + C) \cup (B + C)$.

(ix) $(A \cup B) \cdot C$ and $(A \cdot C) \cup (B \cdot C)$.

(x) A and $(A + B) - B$.

(xi)   Find all nonempty sets $A \subseteq \mathbb{N}$ (respectively $A \subseteq \mathbb{Z}$) for which $A + A = 2 \cdot A$ holds.

(xii)   Show that for every set $A \subseteq \mathbb{N}$ there is a nonempty set $B \subseteq \mathbb{N}$ such that either $B + B \subseteq A$ or $A \cap (B + B) = \emptyset$.

(xiii)   If a set $A \subseteq \mathbb{Z}$ contains 0, then $A \subseteq A + A$. Is the converse also true?

4.11   Problems on Operations

After the introductory exercises in 4.10 we now turn to five less trivial problems.

(i) Show that for every number $c \in \mathbb{Q}^{+}$, $c \neq 1$, there are sets A, B such that $A \cup B = \mathbb{N}$, $c \notin A \div A$, and $c \notin B \div B$.

SOLUTION.   Since the conditions on the sets A, B do not change if we replace the number c by $\frac{1}{c}$, we assume that $0 < c < 1$. We construct the sets A, B inductively. Let $1 \in A$ and let $1, 2, \ldots, n-1$ be already distributed over the sets A, B (each number in exactly one of them). We then place the number n as follows: If $nc \in \mathbb{N}$, then because of $nc < n$ the number nc is already in one of the sets A, B; we then place the number n in the other one; if $nc \notin \mathbb{N}$, we place n in the set A. We thus obtain a partition $\mathbb{N} = A \cup B$; if $c = x \div y$ holds for some $x, y \in \mathbb{N}$, then $x < y$, and according to our rule for placing the number y, we have neither $[x, y] \subseteq A$ nor $\{x, y\} \subseteq B$. $\square$

(ii) Suppose that a given nonempty set $M \subseteq \mathbb{Q}$ satisfies

$$M + M \subseteq M \quad \text{and} \quad M \cdot M \subseteq M, \tag{24}$$

and that for each $c \in \mathbb{Q}$ exactly one of the relationships

$$c \in M, \quad -c \in M, \quad c = 0, \qquad (25)$$

holds. Show that $M = \mathbb{Q}^+$. (This is a characterization of the set of positive elements of the field $(\mathbb{Q}, +, \cdot)$.)

SOLUTION.   In view of (25) it suffices to verify that $\mathbb{Q}^+ \subseteq M$. First of all, we have $1 \in M$ (since otherwise, $-1 \in M$ by (25), which with (24) would imply $1 = (-1) \cdot (-1) \in M$, a contradiction). From $1 \in M$ we get with (24) successively $2 = 1 + 1 \in M$, $3 = 2 + 1 \in M$, etc., and so $\mathbb{N} \subseteq M$. Let us now suppose that $p \div q \notin M$ for some $p, q \in \mathbb{N}$. But then $(-p) \div q \in M$ by (25), which together with $q \in M$ gives, according to (24),

$$-p = \left(\frac{-p}{q} \right) \cdot q \in M.$$

Thus we have found that both numbers p and $-p$ belong to M, which is a contradiction to (25).   □

(iii) Find all nonempty sets $A \subseteq \mathbb{Z}$ for which $A - A \subseteq A$. (From the basics of *group theory* it is well known that the nonempty subsets A of an additive group $(G, +)$ with the property $A - A \subseteq A$ are exactly all its *subgroups*.)

SOLUTION.   For each fixed $m \in \mathbb{N}_0$ the set $A = m \cdot \mathbb{Z}$ has the desired property, since for any $x, y \in \mathbb{Z}$ we have $mx - my = m(x - y)$ and $x - y \in \mathbb{Z}$. We now show that any set A with the desired property is of the above form. First of all, it is clear that $0 \in A$, since a nonempty set A has at least one element, and $0 = a - a \in A - A \subseteq A$. Since $-x = 0 - x$ and $x = 0 - (-x)$, we find that $x \in A$ if and only if $-x \in A$. Therefore, if we set $B = A \cap \mathbb{N}$, then the set A is necessarily of the form $A = B \cup \{0\} \cup (-B)$. If $B = \emptyset$, then $A = \{0\} = 0 \cdot \mathbb{Z}$; therefore, we assume now that B is a nonempty set and m is its *smallest* element. Since $m \in A$ and $(-m) \in A$, we obtain $2m = m - (-m) \in A$, which then implies $3m = 2m - (-m) \in A$, etc., and thus $m \cdot \mathbb{N} \subseteq B$. We show that in fact we have the equality $m \cdot \mathbb{N} = B$. If this were not the case, there would exist a number $x \in B$ for which $m \mid x$ did not hold; we take the *smallest* such number. From the definition of m it follows that $x > m$. However, the positive number $x - m$, as a difference between two elements of A, lies in A and therefore also in B. But at the same time $m \mid (x - m)$ does not hold, which is a contradiction to the choice of x. This shows the equality $m \cdot \mathbb{N} = B$, which finally implies

$$A = m \cdot \mathbb{N} \cup \{0\} \cup (-m \cdot \mathbb{N}) = m \cdot (\mathbb{N} \cup \{0\} \cup (-\mathbb{N})) = m \cdot \mathbb{Z}.$$

We remark in closing that the sets $m \cdot \mathbb{Z}$, $m \in \mathbb{N}$, are clearly infinite and distinct, since they differ in their smallest positive elements. This last fact explains the method of proof in the above solution.   □

(iv) Find all pairs A, B of nonempty subsets of $\mathbb{Z}$ that satisfy the conditions $\mathbb{Z} = A \cup B$, $A + A \subseteq A$, $B + B \subseteq A$, and $A + B \subseteq B$. Find these pairs A, B also in the case where the set $\mathbb{Z}$ in the statement of the problem is replaced by $\mathbb{Q}$.

SOLUTION.   All the conditions are clearly satisfied by the pair of sets $A = B = \mathbb{Z}$. We first show that this is the only such pair of sets that are not disjoint. Let therefore A, B be any pair of sets with the desired properties such that $A \cap B \neq \emptyset$. We choose an arbitrary $p \in A \cap B$. Any $x \in \mathbb{Z}$ can be written in the form $x = p + (x - p)$, from which it easily follows with the conditions of the problem that $x \in A \cap B$. Indeed, we have $x - p \in A$ or $x - p \in B$; in the case $x - p \in A$, for instance, we proceed as follows:

$$\left.\begin{array}{r} p \in A \\ x - p \in A \\ A + A \subseteq A \end{array}\right\} \Rightarrow x \in A, \qquad \left.\begin{array}{r} p \in B \\ x - p \in A \\ A + B \subseteq B \end{array}\right\} \Rightarrow x \in B.$$

The reader should do this for the case $x - p \in B$. This proves the equality $A = B = \mathbb{Z}$ in the case $A \cap B \neq \emptyset$. Therefore, we need to look only for pairs A, B with the property $A \cap B = \emptyset$. It is not difficult to guess such pairs: $A = S$ and $B = L$, where S, respectively L, is the set of all even, respectively odd, integers. We show now that this is the only pair. From the equality $s = \frac{s}{2} + \frac{s}{2}$ it follows that $s \in A$ for each $s \in S$; that is, $S \subseteq A$. The nonempty set B therefore contains exclusively odd numbers; we call one of them ℓ_0. Now, each number $\ell \in L$ can be written in the form $\ell = \ell_0 + (\ell - \ell_0)$, where $\ell - \ell_0 \in S$; from $\ell_0 \in B$ and $\ell - \ell_0 \in A$ it therefore follows that $\ell \in A + B \subseteq B$; that is, $L \subseteq B$. Finally, from $S \subseteq A$, $L \subseteq B$, $A \cap B = \emptyset$, and $A \cup B = S \cup L$ we obtain the desired equalities $A = S$ and $B = L$.

If we replace the set $\mathbb{Z}$ in the statement of the problem by $\mathbb{Q}$, then the equality $A = B = \mathbb{Q}$ in the case $A \cap B \neq \emptyset$ is derived just as before. However, in the case $A \cap B = \emptyset$ the equality $x = \frac{x}{2} + \frac{x}{2}$ implies $x \in A$ for $x \in \mathbb{Q}$; that is, $A = \mathbb{Q}$. This last identity means in view of $B \neq \emptyset$ that $A \cap B \neq \emptyset$, which is a contradiction. Therefore, the pair $A = B = \mathbb{Q}$ is the unique solution of the modified problem. $\square$

(v) For any positive integer n consider the set

$$K_n = \left\{x \in \mathbb{Z} : \ n^2 < x < (n+1)^2\right\}.$$

How many elements does the arithmetic product $K_n \cdot K_n$ have?

SOLUTION.   The set K_n has $(n+1)^2 - n^2 - 1 = 2n$ elements. We show that in the set $K_n \cdot K_n$ we have unique factorization into two factors from K_n: From $xy = uv$ for some $x, y, u, v \in K_n$ it follows that $(x, y) = (u, v)$ or $(x, y) = (v, u)$. This then means that the set $K_n \cdot K_n$ contains exactly $2n$

numbers of the form x^2 ($x \in K_n$) and exactly $\binom{2n}{2}$ numbers of the form xy ($x, y \in K_n$, $x \neq y$); in total the set $K_n \cdot K_n$ has therefore exactly

$$2n + \binom{2n}{2} = n(2n+1)$$

elements. So, let $xy = uv$ for some numbers $x, y, u, v \in K_n$. We can order them such that $x \leq y$, $u \leq v$, and $x \leq u$; then it suffices to show that $x = u$. To do this, we assume that $x \neq u$, that is, $x < u$. From the estimate

$$(x+y)^2 - (y-x)^2 = 4xy = 4uv \leq (u+v)^2$$

it follows that

$$(y-x)^2 \geq (x+y)^2 - (u+v)^2 = (x+y-u-v)(x+y+u+v) > 4n^2,$$

since $x + y + u + v > 4x > 4n^2$, and the integer $x + y - u - v$ is positive, as we will now verify:

$$x + y - u - v = x + \frac{uv}{x} - u - v = x - u + uv\left(\frac{1}{x} - \frac{1}{u}\right) =$$
$$= \frac{(u-x)(v-x)}{x} > 0.$$

On the other hand, $0 < y - x \leq \left((n+1)^2 - 1\right) - (n^2 + 1) = 2n - 1 < 2n$, which implies the inequality $(y-x)^2 < 4n^2$; this contradicts the previously derived inequality $(y-x)^2 > 4n^2$.   □

4.12 Exercises

(i) The sets A, B in the solution of 4.11.(i) are described *recursively*. Give a *direct* rule for deciding whether $n \in A$ or $n \in B$.

(ii) A generalization of 4.11.(i). Let C be a given k-element subset of $\mathbb{Q}^+$, and let $1 \notin C$. Show that there are sets $A_1, A_2, \ldots, A_{k+1}$ with $\mathbb{N} = A_1 \cup A_2 \cup \cdots \cup A_{k+1}$ and $C \cap (A_i \div A_i) = \emptyset$ ($1 \leq i \leq k+1$).

(iii) Determine those positive integers n for which there are sets A, B such that $\mathbb{N} = A \cup B$, and such that neither of the sets $A - A$ and $B - B$ contains n or $n + 1$.

(iv) Show that for any positive integer n there exist sets A, B, C such that $\mathbb{N} = A \cup B \cup C$ and none of the sets $A - A, B - B, C - C$ contains any of the numbers $n, n+1, n+2, \ldots, 2n$.

(v) For a given pair of sets A, B ($\mathbb{N} = A \cup B$) let C denote the set of all those $c \in \mathbb{N}$ that lie neither in $A - A$ nor in $B - B$. Determine the set C if it is given that $2 \in C$.

(vi) For some set $K \subseteq \mathbb{N}$ we have $K * K \subseteq K$, where $*$ denotes the operation $a * b = a + b + ab$. Decide whether K necessarily contains the number 12131, if you know that $\{1, 2\} \subseteq K$.

(vii) Show that the set $\mathbb{N} \setminus (\mathbb{N} * \mathbb{N})$, where $*$ denotes the operation $a * b = a + b + 2ab$, is infinite.

(viii) Suppose that the set $A \subseteq \mathbb{N}_0$ satisfies $A + A = \mathbb{N}_0$. Show that if its elements are written in increasing order $a_0 < a_1 < a_2 < \cdots$, then for every $n \geq 0$ we have $a_n \leq \frac{1}{2}(n^2 + n)$.

(ix) Let $K = \{0, 7, 70, 77, \ldots\}$ be the set of all numbers in $\mathbb{N}_0$ that can be written with the digits 0 and 7. Show that $7 \cdot \mathbb{N}_0 \subseteq K + K + \cdots + K$ (9 summands).

(x) Let $K \subseteq \mathbb{N}$ denote the set of all those numbers x that can be written in the form $x = a_1 + a_2 + \cdots + a_n$, where the a_i are positive (not necessarily distinct) integers such that $\frac{1}{a_1} + \frac{1}{a_2} + \cdots + \frac{1}{a_n} = 1$. Show that each of the sets $\{n^2 : n \in \mathbb{N}\}$, $K \cdot K$, $2 \cdot K + 2$, $2 \cdot K + 8$, $2 \cdot K + 9$, $2 \cdot K + 20$, $3 \cdot K + 6$, $3 \cdot K + 8$, $4 \cdot K + 6$, $4 \cdot K + 13$, and $6 \cdot K + 5$ is a subset of K.

*(xi) Decide whether there exists a set $K \subseteq \mathbb{N}$ with $\mathbb{N} \subseteq K - K$, where for each $x \in \mathbb{N}$ the expression $x = a - b$ $(a, b \in K)$ is unique.

*(xii) Let A_1, A_2, A_3 be nonempty subsets of $\mathbb{Z}$ with the property that if $\{i, j, k\} = \{1, 2, 3\}$, then $A_i + A_j \subseteq A_k$ and $A_i - A_j \subseteq A_k$. Show that at least two of the sets A_1, A_2, A_3 are identical. Can two of these sets be disjoint?

(xiii) Suppose that a set $A \subseteq \mathbb{Z}$, which contains both positive and negative numbers, satisfies $A + A \subseteq A$. Show that $A = m \cdot \mathbb{Z}$ for an appropriate $m \in \mathbb{N}$.

*(xiv) Suppose that a nonempty set $A \subseteq \mathbb{N}$ satisfies $A + A \subseteq A$. Show that there are numbers $m, M \in \mathbb{N}$ such that $A \subseteq m \cdot \mathbb{N}$ and that for every $x \geq M$ we have $x \in A$ if and only if $x \in m \cdot \mathbb{N}$.

*(xv) A *Fibonacci sequence* of numbers is any sequence of integers $a_1, a_2, \ldots$ such that $a_{n+2} = a_{n+1} + a_n$ for all $n \geq 1$. A *Fibonacci sequence of sets* will be any sequence of nonempty sets $A_1, A_2, \ldots$ of integers such that for each $n \geq 1$ we have

$$A_{n+2} = A_{n+1} + A_n \quad \text{and} \quad A_n = A_{n+2} - A_{n+1}.$$

Prove the following assertion: For every Fibonacci sequence of sets $A_1, A_2, \ldots$ there is a number $k \in \mathbb{N}_0$ and a Fibonacci sequence of numbers $a_1, a_2, \ldots$ such that

$$A_n = a_n + k \cdot \mathbb{Z} \quad \text{for all} \quad n \geq 1. \tag{20}$$

Conversely, if $a_1, a_2, \ldots$ is any Fibonacci sequence of numbers, then for each fixed $k \in \mathbb{N}_0$ the sequence of sets $A_1, A_2, \ldots$ defined by (26) is also a Fibonacci sequence.

4.13   Miscellaneous Problems

(i) Peter chose five real numbers $x_1, x_2, \ldots, x_5$ (not necessarily distinct), evaluated the ten sums $x_i + x_j$ $(1 \le i < j \le 5)$, and wrote their values on ten cards, which he handed us after shuffling them. Show that one can determine the numbers chosen by Peter from the numbers $y_1, y_2, \ldots, y_{10}$ on the cards.

SOLUTION.   Since $y_1 + y_2 + \cdots + y_{10} = 4(x_1 + x_2 + \cdots + x_5)$, we can first of all determine the value of the sum $S = x_1 + x_2 + \cdots + x_5$. If $y_1 \ge y_2 \ge \cdots \ge y_{10}$ and $x_1 \ge x_2 \ge \cdots \ge x_5$ (both sets of numbers can be ordered like this in advance), then $y_1 = x_1 + x_2$, $y_2 = x_1 + x_3$, $y_9 = x_3 + x_5$, and $y_{10} = x_4 + x_5$. We can therefore successively determine $x_3 = S - y_1 - y_{10}$, $x_1 = y_2 - x_3$, $x_2 = y_1 - x_1$, $x_5 = y_9 - x_3$, and $x_4 = y_{10} - x_5$.   □

(ii) Decide whether there is a nonempty set $X \subseteq \mathbb{Z}$ such that for any $d \in \mathbb{N}$ and for every $x \in X$ there is exactly one number y in X with the property $|x - y| = d$.

SOLUTION.   We assume that some set X has the property in question, and we choose a fixed $x_0 \in X$. Using the condition for $d = 1$ and $x = x_0$, we find that X contains exactly one of the numbers $x_0 + 1$ and $x_0 - 1$; let, for instance, $x_0 + 1 \in X$ and $x_0 - 1 \notin X$ (otherwise, it suffices to change the number x_0 to $x_0 - 1$). Then we obtain successively

$$x_0 + 2 \notin X \quad (d = 1, \ x = x_0 + 1),$$
$$x_0 - 2 \in X \quad (d = 2, \ x = x_0),$$
$$x_0 + 3 \in X \quad (d = 2, \ x = x_0 + 1),$$
$$x_0 + 4 \in X \quad (d = 1, \ x = x_0 + 3).$$

Since X contains the numbers $x_0 - 2$, $x_0 + 1$, and $x_0 + 4$, the condition of the problem is not satisfied for $d = 3$ and $x = x_0 + 1$. Therefore, no such set X exists.

In view of Exercises 4.14.(vi)–(vii) we note that we used only the conditions for $d \in \{1, 2, 3\}$ in order to obtain the negative answer.   □

(iii) Suppose that the smallest element of a given set $A \subseteq \mathbb{N}$ is 1, and the largest one is 100, and that every element $x \ne 1$ of A is equal to the sum of two (possibly identical) numbers also belonging to A. What is the smallest possible number of elements of the set A?

SOLUTION.   Let the given set A have exactly n elements, which we order according to their size: $1 = x_1 < x_2 < \cdots < x_n = 100$. Since for each $k \in \{1, 2, \ldots, n - 1\}$ we have $x_{k+1} \le x_k + x_k = 2x_k$, we obtain successively $x_2 \le 2, x_3 \le 4, x_4 \le 8, \ldots, x_7 \le 64$, which means that $n \ge 8$. Let us suppose that $n = 8$. From $x_8 = 100$ and the inequalities $x_6 + x_7 \le 32 +$

$64 = 96 < 100$ it follows that $x_8 = 2x_7$, that is, $x_7 = 50$. Similarly, from $x_5 + x_6 \le 16 + 32$ it follows that $x_6 = 25$, and finally $x_4 + x_5 \le 8 + 16$ implies $x_5 = \frac{25}{2} \notin \mathbb{N}$, which is a contradiction. Hence we have the bound $n \ge 9$; since the 9-element set $\{1, 2, 3, 5, 10, 20, 25, 50, 100\}$ has the required properties, $n = 9$ is the smallest value we were looking for.   □

(iv) Show that for all integers $n \ge 5$ the following assertion holds: In any n-element set $X \subseteq \{1, 2, \ldots, 2n\}$ there are two different numbers whose least common multiple does not exceed $3n + 6$.

SOLUTION.   If the given n-element set $X \subseteq \{1, 2, \ldots, 2n\}$ contains a number $a \le n$, then we distinguish the two cases $2a \in X$ and $2a \notin X$. In the first case there is nothing to show, since $2a$ is a common multiple of the pair of numbers $\{a, 2a\} \subseteq X$, and $2a \le 2n < 3n + 6$. In the second case we delete the number a from X and replace it with $2a$; that is, we consider the n-element set

$$X_1 = (X \setminus \{a\}) \cup \{2a\}.$$

It is clear that if X_1 has the desired property, then so does the original set X. If X_1 contains a number $a_1 \le n$, we repeat this approach until we obtain a set X_p that is such that either $\{b, 2b\} \subseteq X_p$ or $X_p = \{n+1, n+2, \ldots, 2n\}$. We show that this last set contains the two distinct numbers $2(k+1)$ and $3(k+1)$, where $k = \left[\frac{n}{2}\right]$. Indeed, from $\frac{n-1}{2} \le k \le \frac{n}{2}$ it follows that

$$2(k+1) \ge 2\left(\frac{n-1}{2} + 1\right) = n + 1 \quad \text{and} \quad 3(k+1) \le 3\left(\frac{n}{2} + 1\right) \le 2n$$

for all $n \ge 6$. (In the case $n = 5$ the numbers 6 and 9 also lie in the appropriate set $\{6, 7, \ldots, 10\}$.) It remains to note that the number $6(k+1)$ is a common multiple of $2(k+1)$ and $3(k+1)$, satisfying the inequality

$$6(k+1) \le 6\left(\frac{n}{2} + 1\right) = 3n + 6.$$ □

(v) For each n-element set $X = \{x_1, x_2, \ldots, x_n\} \subseteq \mathbb{R}$ let $p(X)$ denote the number of different values of the sums $x_i + x_j$ $(1 \le i < j \le n)$. Find the smallest and the largest possible values of $p(X)$ for a fixed $n \ge 2$.

SOLUTION.   If the indices of the elements of X are chosen in such a way that $x_1 < x_2 < \cdots < x_n$, then we have

$$x_1 + x_2 < x_1 + x_3 < \cdots < x_1 + x_n < x_2 + x_n < \cdots < x_{n-1} + x_n, \quad (27)$$

so that $p(X) \ge 2n - 3$. If we want to obtain equality $p(X) = 2n - 3$, we must try to choose the numbers $x_1, x_2, \ldots, x_n$ such that every sum $x_i + x_j$ $(i \ne j)$ is equal to one of the numbers in the chain (27). This is possible if the numbers x_i form an arithmetic sequence:

$$x_2 - x_1 = x_3 - x_2 = \cdots = x_n - x_{n-1} = d > 0,$$

for then we have in the case $i + j \leq n$,

$$x_i + x_j = x_1 + (i-1)d + x_1 + (j-1)d = 2x_1 + (i+j-2)d = x_1 + x_{i+j-1},$$

while in the case $n < i + j \leq 2n - 1$,

$$x_i + x_j = 2x_1 + (i+j-2)d = x_{i+j-n} + x_n.$$

We have thus shown that the smallest value of $p(X)$ is $2n - 3$. Since the number of all sums $x_i + x_j$ ($1 \leq i < j \leq n$) is $\binom{n}{2}$, we have the upper bound $p(X) \leq \frac{n(n-1)}{2}$. We will show that in the last inequality we can actually have equality, for instance for the geometric sequence

$$x_1 = 1, \quad x_2 = q, \quad x_3 = q^2, \quad \ldots, \quad x_n = q^{n-1},$$

with integer quotient $q \geq 2$. Let us assume that among the sums $x_i + x_j$ there are two identical numbers; that is, $q^i + q^j = q^r + q^s$ for some nonnegative integers $i < j$ and $r < s$. From the equality $q^i (1 + q^{j-i}) = q^r (1 + q^{s-r})$ it follows that $q^i \mid q^r$ and $q^r \mid q^i$, and thus $i = r$, since both integers $1 + q^{j-i}$, $1 + q^{s-r}$ are relatively prime to q. Now, $i = r$ means that $q^j = q^s$, and therefore also $j = s$. We have thus shown that $p(X) = \frac{1}{2}n(n-1)$ is the largest possible value. $\square$

(vi) Suppose that the set $X \subseteq \mathbb{R}$ contains the numbers 0 and 1 and that for all $n \geq 2$ it has the property that if n different numbers lie in X, then their arithmetic mean also lies in X. Show that X contains all rational numbers from the interval $(0, 1)$.

SOLUTION.   We first note that the conditions of the problem imply

$$\frac{1}{2} = \frac{1}{2}(0 + 1) \in X, \quad \frac{1}{4} = \frac{1}{2}\left(0 + \frac{1}{2}\right) \in X, \quad \frac{3}{4} = \frac{1}{2}\left(\frac{1}{2} + 1\right) \in X, \quad \text{etc.}$$

It is therefore clear that one can verify by induction on q that $\frac{p}{2^q} \in X$ for any $p, q \in \mathbb{N}_0$ with the property $p \leq 2^q$: If $\frac{p}{2^q}$ is such a fraction with odd numerator p, that is, $p = 2s - 1$, $1 \leq s \leq 2^{q-1}$, then it suffices to use the expression

$$\frac{p}{2^q} = \frac{2s - 1}{2^q} = \frac{1}{2}\left(\frac{s}{2^{q-1}} + \frac{s-1}{2^{q-1}}\right)$$

and the induction hypothesis that both fractions with denominator 2^{q-1} belong to X.

In the second stage of the solution we show by induction that $\frac{k}{n} \in X$ when $0 < k \leq n$. For $n = 1$ this is true, and as induction step from $n - 1$ to n (for all $n \geq 2$) we use the expression $\frac{k}{n} = \frac{k}{n-1} \cdot \frac{n-1}{n}$ and show that $\frac{n-1}{n} \cdot t \in X$ for all $t \in X$. If we choose and fix such a $t \neq 0$ and if in the first stage of the solution we replace the pair of numbers 0, 1 by the pair 0, t, then we deduce that the set X contains all numbers of the form $\frac{pt}{2^q}$, where $p, q \in \mathbb{N}_0$ and $p \le 2^q$. Next we use the identity $1+3+2^2+2^3+\cdots+2^{n-1} = 2^n$ $(n \ge 2)$, which implies the expression

$$\frac{n-1}{n} \cdot t = \frac{1}{n}\left(\frac{2^n-1}{2^n} \cdot t + \frac{2^n-3}{2^n} \cdot t + \sum_{k=2}^{n-1} \frac{(2^n-2^k)}{2^n} \cdot t\right).$$

On the right-hand side there is the arithmetic mean of the n distinct numbers $\frac{p_k}{2^q} \cdot t$, which by the above argument lie in X. Therefore, the number $\frac{n-1}{n} \cdot t$ is also in X, and the proof is complete.   □

4.14 Exercises

(i) Decide for which $n \ge 3$ the following assertion holds: The elements of every n-element set X with the property $p(X) = 2n - 3$ (see 4.13.(v)) form an arithmetic sequence of length n.

(ii) Are there integers $a_1, a_2, \ldots, a_9$ such that the set of values of the sums $a_i + a_j$ $(1 \le i < j \le n)$ consists of 36 consecutive integers?

(iii) A generalization of 4.13.(v). For every n-element set $X \subseteq \mathbb{R}$ and every number k $(1 \le k \le n)$ we let $p_k(X)$ denote the number of distinct values of all sums of k different elements of X. Find the smallest and largest possible values of $p_k(X)$ for fixed k and n.

(iv) Let A be a set of positive integers such that for any two distinct elements x and y the inequality $|x-y| \ge \frac{xy}{25}$ is satisfied. Show that A contains at most 9 elements. Does there exist such a 9-element set A?

*(v) Construct a finite set $X \subseteq \mathbb{R}$ that has at least 4 elements and the property that $ab + cd \in X$ for any distinct elements $a, b, c, d \in X$. Can all elements of X be positive?

(vi) Let $K = \{2^n : n \in \mathbb{N}\}$. Decide whether there is a nonempty set $X \subseteq \mathbb{Z}$ such that for every $d \in K$ and every $x \in X$ there is exactly one integer $y \in X$ with the property $|x - y| = d$.

(vii) Show that a set X with the properties as described in (vi) exists in another case, namely, when K consists of an arbitrary sequence of positive integers $k_1, k_2, \ldots$ that for every $n > 1$ satisfy the condition

$$k_n - k_{n-1} > k_1 + k_2 + \cdots + k_{n-1}. \tag{28}$$

(viii) Show that in any $(n+1)$-element set $X \subseteq \{1, 2, \ldots, 2n\}$ there are two different numbers, one of which is an integer multiple of the other. Is this true for every n-element set $X \subseteq \{1, 2, \ldots, 2n\}$?

(ix) Show that if we choose an integer less than 16 and another 99 integers from the numbers $1, 2, 3, \ldots, 200$, then among these 100 integers there are two different ones whose quotient is an integer.

(x) Can the number 16 in the statement of (ix) be replaced by 17?

*__(xi)__ Suppose that n different numbers are chosen from the set $\{1, 2, \ldots, 2n - 1\}$ such that every chosen number is not divisible by any other one of the chosen numbers. Show that none of these numbers can be less than 2^k, where the integer k is uniquely determined by the condition $3^k < 2n < 3^{k+1}$.

*__(xii)__ A generalization of 4.13.(i). Given the collection of n real numbers $x_1, x_2, \ldots, x_n$, form the $N = \binom{n}{2}$ sums $x_i + x_j$ $(1 \le i < j \le n)$, and denote them by $y_1, y_2, \ldots, y_N$ (in any order). Show that if $n \ne 2^k$, $k \in \mathbb{N}$, then the original numbers $x_1, x_2, \ldots, x_n$ are uniquely determined by the collection of numbers $y_1, y_2, \ldots, y_N$. Furthermore, show that this conclusion is not valid for any number $n = 2^k$.

5   Iterations

By the term *iteration* we usually understand in mathematics the result of some repetition of the same mathematical operation, algorithm, rule, etc. In Section 3.9 of this chapter we have already solved several problems concerning iterations of certain operations on arrays. Before turning to further problems on numerical configurations, we present a description of such problems in a general setting. This will enable us in the following sections to formulate in general some possible approaches to their solutions.

The letter X will denote the set of all configurations to be considered in a given iterative problem. This may be, for instance, the set of all 4×4 integer arrays, or the set of all real sequences of a given length. The given "rule" for the individual elements $x \in X$ that we will iterate in the problem then introduces a certain *relation* Ω on X, that is, a (nonempty) subset of the Cartesian product $X \times X$ $(\Omega \subseteq X \times X)$. In all problems the relation Ω can be described as follows: We have a rule that, given an element $x \in X$, allows us to form all elements of the set

$$\Omega(x) \stackrel{\mathrm{def}}{=} \{y \in X : (x, y) \in \Omega\}$$

(if the set $\Omega(x)$ has only one element for all $x \in X$, then the relation Ω is a *mapping* $\Omega \colon X \to X$). If the set X does not have too many elements, then it is convenient to visualize the relation Ω by way of a *directed graph* (see Example 5.1.(i)).

A finite or infinite sequence $x_1, x_2, \ldots$ of elements of the set X will be called *iterative* (with the relation Ω) if for any two neighboring terms x_i, x_{i+1} we have $x_{i+1} \in \Omega(x_i)$. We now state the three most frequent questions that occur in the study of iterative sequences.

(a) The *attainability problem*. Given two sets $A \subseteq X$ and $B \subseteq X$, decide whether there is an iterative sequence $x_1, x_2, \ldots, x_n$ such that $x_1 \in A$ and $x_n \in B$. Of special importance is the case where one or both of the sets A, B have only one element. We say that the element b is *attainable* from the element a if there is an iterative sequence $a = x_1, x_2, \ldots, x_n = b$; if at the same time a is also attainable from b, then we call a, b a pair of mutually attainable elements.

(b) The *finiteness problem*. Decide whether there are infinite iterative sequences; if the answer is negative, establish or estimate an upper bound for the maximal length of these sequences. (In this formulation we have to exclude from Ω all pairs of the form (x, x); otherwise, we would have to talk about the *stabilization problem* of infinite iterative sequences.)

(c) The *periodicity problem*. Decide whether there exist periodical iterative sequences; possibly describe all of them or find their periods.

5.1 Introductory Examples

We begin by illustrating the problems surrounding iterative sequences, which we just sketched, with a few examples.

(i) On a table there are 6 pebbles, divided into several piles. From each pile we take one pebble and form a new pile with them. We keep repeating this operation. Decide how many piles there will be on the table after 30 steps (the initial distribution of the pebbles is not known).

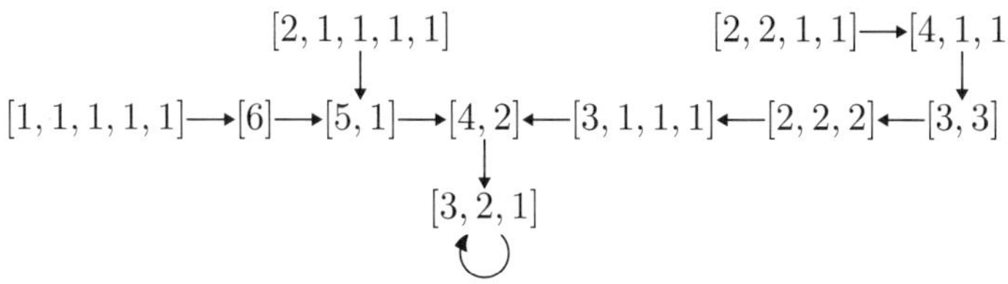

$$[2, 1, 1, 1, 1] \qquad\qquad [2, 2, 1, 1] \longrightarrow [4, 1, 1]$$
$$\downarrow \qquad\qquad\qquad\qquad\qquad \downarrow$$
$$[1, 1, 1, 1, 1] \longrightarrow [6] \longrightarrow [5, 1] \longrightarrow [4, 2] \longleftarrow [3, 1, 1, 1] \longleftarrow [2, 2, 2] \longleftarrow [3, 3]$$
$$\downarrow$$
$$[3, 2, 1]$$
$$\circlearrowleft$$

Figure 11

SOLUTION.   We describe each distribution of the pebbles into piles by a collection of numbers, each giving the number of pebbles in a pile. Since the order clearly does not matter, each distribution is described by one of the collections [6], [5,1], [4,2], [3,3], [4,1,1], [3,2,1], [2,2,2], [3,1,1,1], [2,2,1,1], [2,1,1,1,1], [1,1,1,1,1,1]. The course of the operations can then be visualized by the directed graph of Figure 11, in which the arrows represent the change of the composition of the piles in one step. We easily see from the graph that after at most 6 steps we obtain the distribution [3,2,1], which will not change any further (and thus after 30 steps there will be three piles on the table).   □

We note that the final sections, 5.11 and 5.12, will be devoted to a generalization of Problem (i) to the case of an arbitrary number of pebbles.

(ii) From a quadruple of positive numbers (a, b, c, d) we form a new quadruple (ab, bc, cd, da), and keep repeating this operation. Show that in the resulting iterative sequence of quadruples the original quadruple (a, b, c, d) does not occur again, with the exception of the case where $a = b = c = d = 1$.

SOLUTION. Suppose that after several steps we obtain the original quadruple, and we set $s = abcd$. Since $(ab)(bc)(cd)(da) = s^2$, we can easily deduce by induction that after k steps we obtain a quadruple of numbers whose product is s^{2^k}. Therefore, it follows from our assumption that $s^{2^k} = s$ for some $k \geq 1$, and thus $s = 1$. Let us now see what in the case $abcd = 1$ the fourth quadruple looks like:

$$(a, b, c, d) \to (ab, bc, cd, da) \to \left(ab^2c, bc^2d, cd^2a, da^2b\right) \to$$
$$\to \left(ab^3c^3d, bc^3d^3a, cd^3a^3b, da^3b^3c\right) = \left(b^2c^2, c^2d^2, d^2a^2, a^2b^2\right).$$

Thus the fourth quadruple comes from the second quadruple by squaring the elements and then changing their orders; in a similar way the sixth quadruple comes from the fourth, the eighth from the sixth, etc. Therefore, the largest number in the $2k$th quadruple is equal to $t^{2^{k-1}}$, where t is the largest one of the numbers ab, bc, cd, da. Since by assumption the iterative sequence is periodic, the sequence of numbers $t, t^2, t^4, t^8, \ldots$ can have only finitely many different terms; this is possible only when $t = 1$. On the other hand, we have

$$1 = a^2b^2c^2d^2 = (ab)(bc)(cd)(da) \leq t \cdot t \cdot t \cdot t = t^4.$$

From this it follows that $ab = bc = cd = da = 1$. The second quadruple is then $(1, 1, 1, 1)$; therefore, all following quadruples have the same form, and finally the original one does as well. $\square$

(iii) From the n-tuple of numbers $x_1, x_2, \ldots, x_n$ consisting of $+1$ and -1 we form the new n-tuple $(x_1x_2, x_2x_3, \ldots, x_nx_1)$, and keep repeating this operation. Show that if $n = 2^k$ for some integer $k \geq 1$, then after a certain number of steps we obtain the n-tuple $(1, 1, \ldots, 1)$.

SOLUTION. We prove the assertion, which is clear for $k = 1$, by induction on k. We assume that it is true for some $k \geq 1$ and consider an arbitrary sequence $(x_1, x_2, \ldots, x_n)$ of the numbers ± 1 of length $n = 2^{k+1}$. Since $x_i^2 = 1$ for all i, the second iteration

$$\left(x_1x_2^2x_3, x_2x_3^2x_4, \ldots, x_{n-1}x_n^2x_1, x_nx_1^2x_2\right)$$

can be written as the n-tuple $(x_1x_3, x_2x_4, \ldots, x_{n-1}x_1, x_nx_2)$, which arises as a regular interlacing of the terms of the two 2^k-tuples

$$(x_1x_3, x_3x_5, x_5x_7, \ldots, x_{n-1}x_1) \quad \text{and} \quad (x_2x_4, x_4x_6, \ldots, x_nx_2). \qquad (29)$$

The same rule can also be used for obtaining the fourth iteration from the second, the sixth from the fourth, etc. Therefore, after $2j$ steps ($j \geq 2$) we obtain from the original n-tuple one in which the terms of the $(j - 1)$th iterations of both sequences in (29) are regularly interlaced. But by the induction hypothesis these iterations consist only of ones for sufficiently large j. This completes the proof by induction. □

(iv) In a triple of positive integers we replace one of them by the sum of the remaining two, decreased by 1, where this transformation is considered as an iterative step only when the original triple becomes in fact a different one. Show that in any finite iterative sequence

$$[a_0, b_0, c_0] \to [a_1, b_1, c_1] \to \cdots \to [a_n, b_n, c_n] \tag{30}$$

one can determine from the final iteration $[a_n, b_n, c_n]$ all preceding ones with the exception of the original triple $[a_0, b_0, c_0]$. (The order of the numbers in a triple is irrelevant.)

SOLUTION.   We note that in contrast to the previous three examples the rule under consideration does not determine a mapping, since, for example,

$$\Omega([1, 2, 3]) = \{[4, 2, 3], [1, 3, 3], [1, 2, 2]\}.$$

Because the triple $[1, 2, 2]$ is at the same time the first iteration of all triples of the form $[1, 2, n]$, we don't even have the assertion that each triple is determined by its first iteration. Nevertheless, we will show that in any iterative sequence (30) and for every index $k > 1$ the triple $[a_k, b_k, c_k]$, where $a_k \leq b_k \leq c_k$, is necessarily preceded by a triple of the form $[a_k, b_k, b_k - a_k + 1]$. Indeed, from the rule $[a_{k-1}, b_{k-1}, c_{k-1}] \to [a_k, b_k, c_k]$ it follows that one of the equalities

$$a_k = b_k + c_k - 1, \quad b_k = a_k + c_k - 1, \quad c_k = a_k + b_k - 1$$

must hold. We show that only the third one can hold: Since $a_k \leq b_k \leq c_k$, the first (respectively second) equality is possible only if $a_k = b_k = c_k = 1$ (respectively $a_k = 1$ and $b_k = c_k$), which in both cases is a contradiction to $k > 1$. This means that we have $[a_{k-1}, b_{k-1}, c_{k-1}] = [a_k, b_k, x]$ for some $x \in \mathbb{N}$. To determine the number x, we repeat the previous consideration: Since $k > 1$, one of the equalities

$$a_k = b_k + x - 1, \quad b_k = a_k + x - 1, \quad x = a_k + b_k - 1 \tag{31}$$

is true, where the third one means that $x = c_k$, and thus $[a_{k-1}, b_{k-1}, c_{k-1}] = [a_k, b_k, c_k]$, which is a contradiction. From $a_k \leq b_k \leq b_k + x - 1$ it follows that the first equality in (31) is possible only when $a_k = b_k$; in any case we therefore have the middle equality, from which we obtain $x = b_k - a_k + 1$. This completes the proof. □

5.2   Exercises

(i) Suppose that k even and $k+1$ odd numbers are written on the circumference of a circle in some order. Between any two neighboring numbers we write their sum, then remove the original numbers, and repeat the process with the new $2k+1$ numbers. Show that after any number of steps at least one of the $2k+1$ numbers will be odd.

(ii) Continue to study the operation described in (i), when in the starting position there are 25 numbers on the circumference of the circle: 12-times the number $+1$ and 13-times -1. Show that after 100 steps one of the 25 numbers will be less than -10^{28}.

(iii) Return to the situation of 5.1.(iii), with a general n. Show that for $n = 2^k \cdot \ell$, where $k \geq 0$ and $\ell \geq 3$ is odd, after a finite number of steps you obtain the n-tuple $(1, 1, \ldots, 1)$ if and only if the elements of the original n-tuple $(x_1, x_2, \ldots, x_n)$ satisfy $x_i = x_{i+2^k}$ for all $i = 1, 2, \ldots, n - 2^k$.

5.3   The Method of Invariants

We now describe an important concept that is often useful in the solution of iterative problems. An *invariant* of a given relation Ω on X with values in K is any mapping $I \colon X \to K$ that is nonconstant (that is, $I(x) \neq I(y)$ for some distinct elements $x, y \in X$) and that has the property that $I(x) = I(y)$ for any pair $(x, y) \in \Omega$. In our problems K will always be some set of numbers. Since the value of the invariant does not change on the elements of an arbitrary iterative sequence, we conclude that the element y is not attainable from the element x if there exists an invariant I such that $I(x) \neq I(y)$. We will illustrate this *nonattainability rule* by way of four examples.

(i) On the circumference of a circle there are 2 ones and 48 zeros in the order $1, 0, 1, 0, \ldots, 0$. It is allowed to change any pair of neighboring numbers x, y by the pair $x+1, y+1$ (in this order). Show that by repeating this operation we cannot end up with all 50 numbers being identical.

SOLUTION.   We denote the numbers on the circumference of the circle by $x_1, x_2, \ldots, x_{50}$, counting in a certain direction. Since we have the identity $(x + 1) - (y + 1) = x - y$, an invariant of the operation in question is an expression that for every pair of neighboring terms x_i, x_{i+1} depends only on the difference $x_i - x_{i+1}$. It is not difficult to guess that the expression

$$I = x_1 - x_2 + x_3 - x_4 + \cdots + x_{49} - x_{50}$$

is of this kind (in checking this, you should not forget that x_{50}, x_1 are also a pair of neighboring terms). For the original sequence we have $I = 1 - 0 + 1 =$

2, while any sequence of 50 identical numbers gives $I = 0$. This completes the proof of nonattainability.   □

(ii) Suppose that an $n \times n$ array consists of the signs $+$ and $-$. It is allowable to change all the signs that lie on the fields of the same row, or the same column, or on the fields of a "slanted row" parallel to one of the two diagonals (such a "row" is also formed by each of the corner fields of the array). For each of the three arrays in Figure 12, where $n = 4$, 5, respectively 6, decide whether upon repeating the operations described above one can transform them into arrays consisting of n^2 copies of the $+$ sign.

+	+	−	+
+	+	+	+
+	+	+	+
+	+	+	+

+	−	−	−	+
+	+	+	+	+
−	+	+	+	−
+	+	−	+	+
+	−	+	+	+

+	+	+	+	+	+
+	+	+	+	+	+
+	+	+	−	−	+
+	+	+	+	+	+
+	+	+	+	+	−
+	+	+	+	+	+

Figure 12

SOLUTION.   We change the signs in the arrays to the numbers $+1$ and -1 and denote them in the obvious way by a_{ij}, $i, j \in \{1, 2, \ldots, n\}$. Then the product of some factors a_{ij} is an invariant of the transformation if and only if any row, column, or slanted row contains an even number of the factors in question. Try to find such a product for $n = 4$; you will notice that it is unique and has the form

$$I = a_{12}a_{13}a_{21}a_{24}a_{31}a_{34}a_{42}a_{43}.$$

Since for the first array in Figure 12 we have $I = -1$, it cannot be transformed into an array with the value $I = 1$. For the other two arrays it is not necessary to look for further invariants; it suffices to consider the same invariant constructed for their 4×4 subarrays placed at their lower right corners. Since in both cases we have again $I = -1$, it is not possible to transform these two to the desired arrays without minus signs.   □

(iii) In the set X_n of all sequences $a = (a_1, a_2, \ldots, a_n)$ consisting of the numbers $0, 1$, the following transformation is allowed: In each sequence $(a_1, \ldots, a_n)$ we may interchange any two neighboring triples of elements (a_i, a_{i+1}, a_{i+2}) and $(a_{i+3}, a_{i+4}, a_{i+5})$, where $1 \leq i \leq n-5$, that is, transform it to the sequence

$$a' = (a_1, \ldots, a_{i-1}, a_{i+3}, a_{i+4}, a_{i+5}, a_i, a_{i+1}, a_{i+2}, a_{i+6}, \ldots, a_n).$$

Show that from the set X_n one can choose at least p sequences that are pairwise nonattainable by the transformation described, where the number

p is given by

$$p = \begin{cases} (k+1)^3 & \text{if } n = 3k, \\ (k+2)(k+1)^2 & \text{if } n = 3k+1, \\ (k+2)^2(k+1) & \text{if } n = 3k+2. \end{cases}$$

SOLUTION.   For each $a = (a_1, a_2, \ldots, a_n) \in X_n$ we set

$$S_1(a) = a_1 + a_4 + a_7 + \cdots,$$
$$S_2(a) = a_2 + a_5 + a_8 + \cdots,$$
$$S_3(a) = a_3 + a_6 + a_9 + \cdots.$$

The sums S_1, S_2, S_3 are clearly invariants, since any allowable interchange of triples causes only the switching of two neighboring summands in each of the sums S_i. In the case $n = 3k$ for any one of $p = (k+1)^3$ triples of numbers chosen from $\{0, 1, \ldots, k\}$ we can easily find a sequence $a \in X_n$ such that $S_i(a) = \alpha_i$ $(i = 1, 2, 3)$, so these p sequences are pairwise nonattainable. Similarly, the appropriate number p of sequences can also be found in the cases $n = 3k+1$ and $n = 3k+2$, where the first, respectively both, of the numbers α_1, α_2 can also take on the value $k+1$.   □

(iv) Suppose that the number -1 is written on the corner A_1 of the regular 12-gon $A_1 A_2 \ldots A_{12}$, and the other corners have the number $+1$. It is allowed to simultaneously change the signs of the numbers at six arbitrary neighboring corners. Show that no repetition of this rule can make the corner A_2 to be -1 and the remaining corners $+1$. Also prove the same assertion for the case where not six but four signs of the numbers at any four neighboring corners can be changed.

SOLUTION.   Let a_k be the number written on the corner A_k $(1 \le k \le 12)$. In the case of changing the signs of the numbers on six neighboring corners the sign of each of the products $a_1 a_7$, $a_2 a_8$, $a_3 a_9$, $\ldots$, $a_6 a_{12}$ will change. Therefore, $I = (a_2 a_8)(a_3 a_9)$ is an invariant that in the starting position has the value $I = 1$. Hence no number of operations can lead to $a_2 = -1$ and $a_3 = a_8 = a_9 = 1$, since this would mean that $I = -1$. In the case of changing the signs on four neighboring corners, the signs of all of the products $a_1 a_5 a_9$, $a_2 a_6 a_{10}$, $a_3 a_7 a_{11}$, $a_4 a_8 a_{12}$ will change, and the desired assertion follows from considering the invariant $I = (a_1 a_5 a_9)(a_3 a_7 a_{11})$.   □

5.4 Exercises

(i) Suppose that in a 4×4 array consisting of the signs $+$ and $-$ it is allowed to change all the signs in any row or in any column. Determine whether the arrays in Figure 13 are mutually attainable.

Figure 13

(ii) Does the assertion from 5.3.(iv) hold if the allowable operation consists of changing signs of the numbers in any triple of neighboring corners?

(iii) Answer the question from (ii) for changing signs in the corners that form an isosceles but not a right triangle.

(iv) Let M be an arbitrary finite subset of $\mathbb{R} \times \mathbb{R}$. If $(x, y) \in M$ is any pair such that $(x + 1, y) \notin M$ and $(x, y + 1) \notin M$, then we may exchange the pair (x, y) in the set M with the two pairs $(x + 1, y)$ and $(x, y + 1)$. By repeating these operations, can an initial set $M = \{(1, 1)\}$ be changed into a set M' that is such that $x + y > 4$ whenever $(x, y) \in M'$?

*(v) A generalization of 5.3.(iv). Suppose that the numbers $a_1, a_2, \ldots,$ a_n, where $a_k \in \{-1, 1\}$, $1 \le k \le n$, are written next to each other on the corners of a regular n-gon. For a fixed integer p $(1 \le p \le n)$ it is allowed to simultaneously change the signs of the numbers a_k in any p neighboring corners. To the n-tuple of numbers $(a_1, a_2, \ldots, a_n)$ we associate the d-tuple $(s_1, s_2, \ldots, s_d)$, where d is the greatest common divisor of the integers n and p, and where $s_k = a_k a_{k+d} a_{k+2d} \cdots a_{k+n-d}$ $(1 \le k \le d)$. Show that two n-tuples $(a_1, a_2, \ldots, a_n)$ and $(a'_1, a'_2, \ldots, a'_n)$ are mutually attainable through repeated use of the operation described above if and only if either $s_k = s'_k$ $(1 \le k \le d)$, or $s_k = -s'_k$ $(1 \le k \le d)$ and the integer $\frac{p}{d}$ is odd. (The numbers s'_k are formed from the a'_k in the same way as the numbers s_k are formed from the a_k.)

5.5 Invariants in Residue Classes

For solving problems concerning operations on integer configurations it is often useful to find invariants with values in the set $\{0, 1, 2, \ldots, m - 1\}$ of residue classes with an appropriately chosen modulus m. We illustrate this with the following five examples.

(i) Suppose that a 10×10 array consists of integers, and we are allowed to choose any 3×3 or 4×4 subarray and increase every number in it by 1. Is it always possible to appropriately repeat this operation such that we obtain a new array all of whose numbers are divisible by 3?

SOLUTION. If $r_1, r_2, \ldots, r_{10}$ are the row sums of the array, then none of the allowed operations with a 3×3 subarray changes the remainder that the sums r_i leave upon division by 3; an operation with a 4×4 subarray changes exactly four of the sums r_i, which are increased by 4. Among four sums changed in this way there is exactly one of the sums r_4, r_8, and therefore exactly three of the other sums. Hence as an invariant we can take the remainder upon division by 3 of the sum

$$I = r_1 + r_2 + r_3 + r_5 + r_6 + r_7 + r_9 + r_{10}.$$

If we choose an initial array such that, for example, $r_1 = 1$ and $r_i = 0$ ($1 < i \leq 10$), then it is not possible to repeat the allowable operations in such a way that the sum I is an integer multiple of 3, which means that not every number in the final array can be a multiple of 3.

We note that in solving this problem it was impossible to use the approach of 3.9.(iv), since the number of subarrays that can be changed is

$$(10 - 3 + 1)^2 + (10 - 4 + 1)^2 = 64 + 49 = 113 > 10^2. \qquad \square$$

(ii) In the four-element set $M = \{(0,0), (1,1), (-3,0), (2,-1)\}$ it is allowed to replace any pair (a, b) by the pair $(a + 2c, b + 2d)$, if (c, d) also belongs to M. Decide whether it is possible to obtain the four-element set $M' = \{(-1, 2), (2, -1), (4, 0), (1, 1)\}$ through an appropriate sequence of such operations.

SOLUTION.   We note that $3 \mid (x - y)$ for each pair (x, y) in the original set M. This property is an invariant, since the number

$$(a + 2c) - (b + 2d) = (a - b) + 2(c - d)$$

is a multiple of 3 if both numbers $a - b$ and $c - d$ are. Therefore, no sequence of operations will lead to the new pair $(4, 0)$. $\qquad \square$

(iii) Suppose that several ones, twos, and threes are written on a blackboard. It is allowed to erase any two different digits and adjoin the remaining third digit (thus the number of digits on the blackboard is decreased by 1). Show that if after a number of such operations one single digit remains on the board, then this digit is determined by the original situation, that is, it does not depend on the particular sequence of the allowable operations.

SOLUTION.   The situation in which there are exactly p ones, q twos, and r threes written on the blackboard will be denoted by the triple (p, q, r). An allowable operation is then the change of (p, q, r) to one of the triples $(p - 1, q - 1, r + 1)$, $(p - 1, q + 1, r - 1)$, $(p + 1, q - 1, r - 1)$. We note that in each of these operations the parities of all three numbers p, q, r change. Therefore, the parity of each of the sums

$$s_1 = p + q, \quad s_2 = p + r, \quad \text{and} \quad s_3 = q + r$$

is invariant. Let us write down the values s_i for the states where one digit remains on the blackboard:

(p,q,r)	s_1	s_2	s_3
$(1,0,0)$	1	1	0
$(0,1,0)$	1	0	1
$(0,0,1)$	0	1	1

Any two of these three states differ in the parity of two of the numbers s_i, and can therefore not be the result of the same starting position.   □

(iv) Along the circumference of a circular park there are n linden trees, on each of which there is one lark. From time to time two of the larks fly simultaneously to a neighboring tree, but in opposite directions. Decide whether it is possible that at some time all the larks end up in one tree.

SOLUTION.   We number the trees consecutively, in one direction, by $1, 2, \ldots, n$. If n is odd, $n = 2k + 1$, then the desired situation can occur, for instance, when successively the pairs from trees 2 and $2k+1$, 3 and $2k$, 4 and $2k - 1$, $\ldots$, $k + 1$ and $k + 2$ all fly to linden tree 1.

We use the method of invariants to show that for even n such a situation can never occur. Suppose that at some moment exactly p_j larks sit on tree j $(1 \le j \le n)$. Then we consider the sum

$$S = 1 \cdot p_1 + 2 \cdot p_2 + \cdots + n \cdot p_n.$$

When a lark flies to the neighboring tree in the direction of the numbering, then the value of S either increases by 1 or decreases by $n - 1$; upon flying in the opposite direction, the value of S either decreases by 1 or increases by $n - 1$. Therefore, when a pair of larks fly to their new trees in an allowable fashion, then the change of S is equal to 0, n, or $-n$. Therefore, the remainder of the number S upon division by n is an invariant. In the starting position the value

$$S = 1 \cdot 1 + 2 \cdot 1 + \cdots + n \cdot 1 = n \cdot \frac{n + 1}{2}$$

is not divisible by n (the number $\frac{n+1}{2}$ is not an integer, since n is even); on the other hand, in the position where all larks sit on linden tree j we have $S = n \cdot j$. This completes the proof.   □

(v) In the sequence $1, 0, 1, 0, 1, 0, 3, 5, 0, \ldots$, each term (beginning with the seventh) is equal to the last digit of the sum of the preceding six terms. Show that in this infinite sequence the numbers $0, 1, 0, 1, 0, 1$ will never occur in this order.

SOLUTION.   Consider the mapping that associates to the sextuple $(x_1, x_2, \ldots, x_6)$ the sextuple $(x_2, x_3, \ldots, x_6, x_7)$, where x_7 is the final digit of the sum $x_1 + x_2 + \cdots + x_6$. An invariant of this mapping is the final digit of the expression

$$I(x_1, x_2, \ldots, x_6) = 2x_1 + 4x_2 + 6x_3 + 8x_4 + 10x_5 + 12x_6,$$

since the difference

$$I(x_2, x_3, \ldots, x_7) - I(x_1, x_2, \ldots, x_6) = 10x_7 + 2(x_7 - (x_1 + x_2 + \cdots + x_6))$$

is, by definition of the digit x_7, divisible by 10. It remains to add that

$$I(1, 0, 1, 0, 1, 0) = 18 \quad \text{and} \quad I(0, 1, 0, 1, 0, 1) = 24;$$

the first sextuple is the starting one, and the second is the one whose nonattainability we wanted to show.   □

5.6 Exercises

(i) On a magic tree there are 25 lemons and 30 oranges. The gardener picks two fruits every day, but the following night one new fruit grows on the tree: An orange (respectively a lemon) if the fruits picked during the day were the same (respectively different). What fruit is the last one to grow on the tree?

(ii) Peter tears a sheet of paper into 10 pieces, then he tears some of the pieces into 10 smaller pieces, etc. Is it possible to get exactly 1991 pieces in this way?

(iii) In the decimal representation of the number 2^{1991} we remove the first digit (on the left) and add it to the remaining number. We keep repeating this operation until we obtain a ten-digit number A. Show that at least two of the digits of A are the same.

(iv) A generalization of 5.5.(iv). We consider the more general initial situation where $p_1, p_2, \ldots, p_n$ larks, in this order, sit in the linden trees, where $p_1 + p_2 + \cdots + p_n = N > 1$. Show that if the same rules as before apply to the pairs of larks, then the state $(p_1, p_2, \ldots, p_n)$ can be changed to $(p'_1, p'_2, \ldots, p'_n)$ if and only if $p'_1 + p'_2 + \cdots + p'_n = N$ and $p_1 + 2p_2 + \cdots + np_n \equiv p'_1 + 2p'_2 + \cdots + np'_n \pmod{n}$.

(v) We return to Example 5.5.(iii), using the notation from its solution. Show that if we can carry out at least one operation on the original position (p, q, r), then upon repetition we can always get exactly one of the four positions $(1, 0, 0)$, $(0, 1, 0)$, $(0, 0, 1)$, $(2, 0, 0)$.

*(vi) Suppose that the numbers $1, 2, \ldots, n$ are written on a blackboard, each number exactly once. We may choose any two of these numbers a, b, erase them, and write the number $|a - b|$ on the blackboard. If we repeat this operation $(n - 1)$ times, a single number will remain on the blackboard. What can its value be?

*(vii) Given $k, m \geq 1$, on the set M_m^k of all k-tuples $(x_1, x_2, \ldots, x_k)$ of numbers chosen from $\{0, 1, \ldots, m - 1\}$ we consider the transformation $(x_1, x_2, \ldots, x_k) \mapsto (x_2, x_3, \ldots, x_k, x_{k+1})$, where $x_{k+1} \equiv x_1 + x_2 + \cdots + x_k \pmod{m}$. Show that an invariant of the form

$$I(x_1, x_2, \ldots, x_k) \equiv c_1 x_1 + c_2 x_2 + \cdots + c_k x_k \pmod{m}$$

with constants $c_1, c_2, \ldots, c_k \in \{0, 1, \ldots, m-1\}$ exists if and only if the numbers $k-1$ and m are not relatively prime. (Recall that every invariant is a nonconstant mapping, and therefore $c_i \neq 0$ for some $i \in \{1, 2, \ldots, k\}$.)

5.7   The Method of Valuations

For studying iterative sequences of a relation Ω on X we will see that in addition to invariants (see 5.3) a more general mapping $J \colon X \to K$ will be useful, where the values upon iterating the elements of X change monotonically. Thus we call the mapping $J \colon X \to K$ a *nonincreasing* (respectively *decreasing*) *valuation* of the relation Ω if for any pair $(x, y) \in \Omega$, $x \neq y$, we have $J(x) \geq J(y)$ (respectively $J(x) > J(y)$). Similarly, we define a *nondecreasing*, respectively *increasing*, valuation; in all cases K is one of the sets of numbers $\mathbb{R}, \mathbb{N}_0, \mathbb{Z}$ with the usual ordering.

If we notice, for instance, that a certain relation has a nonincreasing valuation J, then for any iterative sequence $x_1, x_2, \ldots$ we have

$$J(x_1) \geq J(x_2) \geq \cdots ,$$

which often makes it possible to solve the periodicity problem (the condition $J(x_1) = J(x_2)$ usually leads to a description of all possible elements x_1). If furthermore J is a decreasing valuation with values in $\mathbb{N}_0$, then in each infinite iterative sequence $x_1, x_2, \ldots$ there exists an index n such that $J(x_n) = J(x_{n+1}) = \cdots$, that is, $x_n = x_{n+1} = \cdots$, since each nonempty subset of $\mathbb{N}_0$ contains its smallest element; this would solve the finiteness problem, respectively the stabilization problem (see the introduction to Section 5).

The problems solved here with the method of monotonic valuations will be divided between the two subsections 5.7 and 5.9.

(i) From the quadruple of real numbers (a, b, c, d) we form the new quadruple

$$(a - b, b - c, c - d, d - a),$$

and keep repeating this transformation. Show that, as long as the original quadruple does not satisfy $a = b = c = d$, after a certain number of steps we obtain a quadruple that contains at least one number larger than 10^6.

SOLUTION.   We consider the nonnegative valuation

$$J(a, b, c, d) = a^2 + b^2 + c^2 + d^2$$

and denote by (a_n, b_n, c_n, d_n) the quadruple that is obtained from the original quadruple after n steps. It is easy to see that $a_n + b_n + c_n + d_n = 0$ for all $n \geq 1$, so for each such n we can write

$$
\begin{aligned}
J(a_{n+1}, & b_{n+1}, c_{n+1}, d_{n+1}) \\
&= (a_n - b_n)^2 + (b_n - c_n)^2 + (c_n - d_n)^2 + (d_n - a_n)^2 \\
&= 2J(a_n, b_n, c_n, d_n) - 2(a_n b_n + b_n c_n + c_n d_n + d_n a_n) \\
&= 2J(a_n, b_n, c_n, d_n) - 2(a_n + c_n)(b_n + d_n) \\
&= 2J(a_n, b_n, c_n, d_n) + 2(a_n + c_n)^2 \geq 2J(a_n, b_n, c_n, d_n).
\end{aligned}
$$

This gives the estimate $J(a_n, b_n, c_n, d_n) \geq 2^{n-1} J(a_1, b_1, c_1, d_1)$ for all $n \geq 1$. Therefore, unless $J(a_1, b_1, c_1, d_1) = 0$ (which occurs only when $a = b = c = d$), it means that $J(a_n, b_n, c_n, d_n) = a_n^2 + b_n^2 + c_n^2 + d_n^2 > 36 \cdot 10^{12}$ for sufficiently large n; but then at least one of the numbers $|a_n|, |b_n|, |c_n|, |d_n|$ must be greater than $3 \cdot 10^6$. However, since $a_n + b_n + c_n + d_n = 0$, we would obtain from the assumption $\max\{a_n, b_n, c_n, d_n\} \leq 10^6$ the bound $\min\{a_n, b_n, c_n, d_n\} \geq -3 \cdot 10^6$, and thus $\max\{|a_n|, |b_n|, |c_n|, |d_n|\} \leq 3 \cdot 10^6$, which is a contradiction.

It is worth remarking that the defining transformation is a linear operator $\mathbb{R}^4 \to \mathbb{R}^4$, so the problem could also be solved in a standard way by finding the eigenvalues and eigenvectors of the corresponding 4×4 matrix.   □

(ii) Let a sequence of integers $x_1, x_2, \ldots, x_n$ be given. If i, j are arbitrary indices such that $x_i - x_j = 1$, then we may replace the terms x_i, x_j by the numbers $x_i + 1, x_j - 1$ (in this order). Show that only a finite number of repetitions of this transformation is possible.

SOLUTION.   The integer valuation $J = x_1^2 + x_2^2 + \cdots + x_n^2$ increases by 4 with each transformation, since $(x_i + 1)^2 + (x_j - 1)^2 - (x_i^2 + x_j^2) = 2(x_i - x_j) + 2 = 4$. If we set $m = \min\{x_1, x_2, \ldots, x_n\}$ and $M = \max\{x_1, x_2, \ldots, x_n\}$, it suffices to show that for any attainable n-tuple $(y_1, \ldots, y_n)$ we have the bounds $m - 3n < y_i < M + 3n$ for each $i = 1, 2, \ldots, n$. (These bounds imply that the valuation J is bounded on any iterative sequence.) Our approach is based on the following observation. For each $k \in \mathbb{Z}$ we have that if $\{x_1, \ldots, x_n\} \cap \{k-1, k, k+1\} \neq \emptyset$, then also $\{y_1, \ldots, y_n\} \cap \{k-1, k, k+1\} \neq \emptyset$ for every n-tuple $(y_1, \ldots, y_n)$ that is attainable from $(x_1, \ldots, x_n)$. Therefore, if we suppose that the n-tuple $(y_1, \ldots, y_n)$, which is attainable from $(x_1, \ldots, x_n)$, satisfies $y_i \geq M + 3n$ for some $i \in \{1, 2, \ldots, n\}$, then every integer a, $M \leq a \leq y_i$, has to be an element of some n-tuple occurring in the iteration. Hence the set $\{y_1, \ldots, y_n\}$ must have a nonempty intersection with each of the $(n + 1)$ disjoint sets $\{M - 1, M, M + 1\}$, $\{M + 2, M + 3, M + 4\}$, $\ldots$, $\{M + 3n - 1, M + 3n, M + 3n + 1\}$, which is a contradiction. Similarly, one also excludes the case $y_i \leq m - 3n$ for some $i \in \{1, 2, \ldots, n\}$. This completes the proof.   □

(iii) Suppose that n real numbers, $n \geq 4$, are written on the circumference of a circle. If four adjacent numbers a, b, c, d satisfy $(a - d)(b - c) < 0$, then we may interchange the places of the neighboring numbers b, c. Show that only a finite number of such transformations can be carried out.

SOLUTION. We first remark that even though there are only finitely many arrangements of n numbers along the circumference of a circle, it is not so clear why a sequence of the transformations described cannot be infinite (e.g., periodic). We note that we have

$$(a - d)(b - c) = (ab + cd) - (ac + bd),$$

where within the first parentheses on the right we have the products of the pairs situated at both ends of the quadruple (a, b, c, d), and within the second pair of parentheses we have the same products for the quadruple (a, c, b, d), which is obtained from the original quadruple by switching the places of the numbers b, c; furthermore, this center pair is the same in both quadruples (up to order). It is therefore convenient to denote the numbers, consecutively in one direction, by $x_1, x_2, \ldots, x_n$ and to consider the valuation

$$J = x_1 x_2 + x_2 x_3 + \cdots + x_{n-1} x_n + x_n x_1,$$

which is increasing for the allowable operations. Since for a given n-tuple there can be only finitely many values of J, after a certain number of steps J will attain a maximal value, and further transformations are not possible.   □

(iv) Given a triple a, b, c of integers, we form a new triple

$$|a - b|, \quad |b - c|, \quad |c - a|,$$

and keep repeating this transformation. Show that after a certain number of steps we reach a triple containing the number 0. Does this assertion also hold in the case where the initial numbers a, b, c are real?

SOLUTION. We assume that there is an initial triple of numbers $a, b, c \in \mathbb{Z}$, from which after any number of steps we always obtain triples of nonzero (and thus positive) integers. We may clearly assume that the integers a, b, c are also positive (otherwise, we delete the first triple). Let us set

$$J = \max\{a, b, c\} \quad \text{and} \quad J' = \max\{|a - b|, |b - c|, |c - a|\}.$$

For any positive numbers x, y we have

$$|x - y| = \max\{x, y\} - \min\{x, y\} < \max\{x, y\},$$

so we immediately obtain $J' < J$, that is, $J' \le J - 1$, since the values of J lie in $\mathbb{N}_0$. Hence after n steps we obtain from the triple $[a, b, c]$ the triple of positive numbers $[a_n, b_n, c_n]$ for which

$$0 < \max\{a_n, b_n, c_n\} \le \max\{a, b, c\} - n$$

holds, but this is a contradiction for $n > \max\{a, b, c\}$. This proves the assertion in the case of integers a, b, c.

In the case of real numbers a, b, c we base the construction of a counterexample on the following consideration. If there exists a triple of positive numbers $[a, b, c]$ that is transformed into the triple $[pa, pb, pc]$ for some $p > 0$, then after an arbitrary number n of transformations this triple turns into the triple of positive numbers $[p^n a, p^n b, p^n c]$. Hence we look for numbers $0 < a < b < c$ such that

$$\frac{b-a}{a} = \frac{c-b}{b} = \frac{c-a}{c} = p > 0.$$

The reader should verify that these equalities hold if and only if $b = (p+1)a$, $c = (p+1)^2 a$, where p satisfies the equation $p^2 + p - 1 = 0$. This equation has the positive root $p = \frac{-1+\sqrt{5}}{2}$; if we choose $a = 1$, then we obtain a triple of the form $\left[1, \frac{1+\sqrt{5}}{2}, \frac{3+\sqrt{5}}{2}\right]$. Now verify by direct computation of the iterations that this triple has the desired property: No number of steps will give a triple that contains a 0. □

5.8 Exercises

(i) Suppose that n positive integers are written on the circumference of a circle. Between all pairs of neighboring integers we write their greatest common divisor, then we delete the original numbers and repeat the process with the new n-tuple of integers. Show that after a certain number of steps we obtain an n-tuple of identical integers.

(ii) Suppose that 100 numbers are written on a blackboard: 10 zeros, 10 ones, ..., 10 nines. We may choose any two of these numbers and replace both of them with their arithmetic mean (for example, the pair $[3, 6]$ may be changed to $\left[\frac{9}{2}, \frac{9}{2}\right]$). Determine the smallest positive number that can appear on the blackboard after a sequence of such operations.

(iii) A sequence $x_1, x_2, \ldots, x_n$ is formed with nonzero real numbers. We are allowed to choose any two of its terms x_i, x_j $(i \neq j)$ and replace them by $x_i + \frac{x_j}{2}, x_j - \frac{x_i}{2}$, in this order. Show that by iterating this transformation we can never obtain either the original sequence or a sequence that differs from the original one only by the order of its terms.

(iv) From a quadruple of real numbers (a, b, c, d) we form the quadruple $(a+b, b+c, c+d, d+a)$, and keep repeating this process. Show that if at two different times we obtain the same quadruple of numbers (possibly in a different order), then the initial quadruple was of the form $(a, -a, a, -a)$ for an appropriate $a \in \mathbb{R}$.

(v) A generalization of (iv). From an n-tuple of real numbers $a_1, a_2, \ldots,$ a_n we form the n-tuple $(a_1 + a_2, a_2 + a_3, \ldots, a_n + a_1)$. Does a conclusion similar to that in (iv) hold for every even $n \geq 4$?

*(vi) A generalization of 5.7.(i). For a fixed $n \geq 6$ we consider iterations of n-tuples from $\mathbb{R}^n$ under the transformation

$$(x_1, x_2, \ldots, x_n) \mapsto (x_1 - x_2, x_2 - x_3, \ldots, x_n - x_1).$$

We assume that the initial n-tuple of real numbers does not satisfy $x_1 = x_2 = \cdots = x_n$. Do we always reach, after a certain number of steps, an n-tuple containing at least one number exceeding 10^6?

(vii) Find all triples of real numbers that have the property that after a certain number of the transformations described in 5.7.(iv) the initial triple is obtained again.

*(viii) For a fixed $m \geq 3$ we consider the mapping

$$(a_1, a_2, \ldots, a_m) \mapsto (|a_2 - a_1|, |a_3 - a_2|, \ldots, |a_m - a_{m-1}|, |a_1 - a_m|)$$

defined on the set $\mathbb{R}^m$. Show that an initial m-tuple can be chosen in such a way that in the sequence of its iterations no m-tuple contains the number 0.

*(ix) Suppose that in the initial m-tuple from (viii) all the numbers $a_1, a_2, \ldots, a_m$ are integers. Show that if m is a power of 2, then the iterative sequence contains an m-tuple consisting only of zeros.

(x) Embedded into a horizontal straight line there is a finite number of arrows (as, for example, in Figure 14), some of which point to the left, and the remaining ones to the right. We choose any two neighboring arrows that point to each other (in Figure 14, for example, the fourth and the fifth from the left) and change the directions of both. Show that this transformation can be repeated only several times, and that the final state (where further changes are no longer possible) is determined by the initial situation; that is, it does not depend on the order in which the changes were carried out. Furthermore, explain why the total number of changes also does not depend on the order of carrying them out.

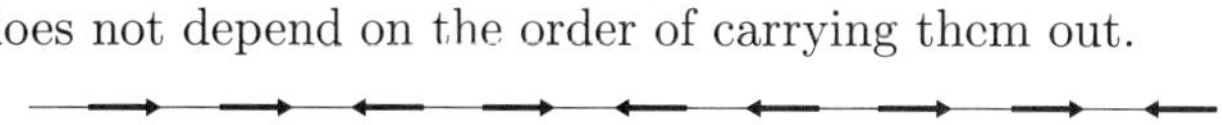

Figure 14

*(xi) Suppose that n piles with $a_1, a_2, \ldots, a_n$ pebbles lie on a table. We may choose any two piles and from one of them move as many pebbles to the other as there were already in this second pile (thus the number of its pebbles is doubled). Find a necessary and sufficient condition on the numbers $a_1, a_2, \ldots, a_n$ that makes it possible to get all pebbles onto one pile after a finite number of steps. (Hint: Consider the change of the greatest common divisor d of all numbers $a_1, a_2, \ldots, a_n$.)

*(xii) Suppose that an arrangement $a_1, a_2, \ldots, a_n$ of the numbers $1, 2, \ldots, n$ is such that $a_k \neq k$ for some $k \in \{1, 2, \ldots, n\}$. We choose any such k and move the number a_k to the a_kth position in the arrangement $a_1, a_2, \ldots, a_n$. To be more exact: If $a_k = c$, then the new arrangement in the case $c < k$ has the form

$$a_1, a_2, \ldots, a_{c-1}, c, a_c, a_{c+1}, \ldots, a_{k-1}, a_{k+1}, a_{k+2}, \ldots, a_n, \quad (32)$$

while in the case $c > k$ it has the form

$$a_1, a_2, \ldots, a_{k-1}, a_{k+1}, a_{k+2}, \ldots, a_c, c, a_{c+1}, a_{c+2}, \ldots, a_n. \quad (33)$$

Show that after a finite sequence of such changes we obtain the arrangement $1, 2, \ldots, n$ (when no further changes are possible), no matter how the numbers a_k to be moved are chosen.

5.9 The Method of Valuations – Continuation

We now consider three further, and somewhat more difficult, problems that can successfully be solved with the method of monotonic valuations, as introduced in 5.7.

(i) Suppose that a sequence of $2n + 1$ integers has the following property: If we remove any of its terms, then the remaining terms can be divided into two n-element sets such that the sums of the numbers in both of them are the same. Show that every such sequence is formed of $2n + 1$ identical numbers.

SOLUTION.   Each term of the sequence under consideration differs from the sum of all $2n + 1$ terms by an even number. This means that the whole sequence consists either exclusively of even or exclusively of odd numbers. Thus we have either a sequence of the form $2a_1, 2a_2, \ldots, 2a_{2n+1}$, or a sequence of the form $2a_1 - 1, 2a_2 - 1, \ldots, 2a_{2n+1} - 1$; in both cases we may instead consider the sequence of integers $a_1, a_2, \ldots, a_{2n+1}$. This "reduction" preserves the original property of the sequence, since for any two n-element sets of indices $I, J \subseteq \{1, 2, \ldots, 2n+1\}$ we have the identities

$$\sum_{i \in I} 2a_i = \sum_{j \in J} 2a_j, \qquad \sum_{i \in I}(2a_i - 1) = \sum_{j \in J}(2a_j - 1),$$

if and only if

$$\sum_{i \in I} a_i = \sum_{j \in J} a_j.$$

Since the inequalities $|x| \leq |2x|$ and $|x| \leq |2x - 1|$ hold for all $x \in \mathbb{Z}$, the above reduction does not increase the sum S of absolute values of all $2n + 1$ terms of the sequence. Furthermore, the value S is a nonnegative integer, and therefore we obtain after a certain number of steps a sequence of integers $b_1, b_2, \ldots, b_{2n+1}$, such that the sum $S = |b_1| + |b_2| + \cdots + |b_{2n+1}|$ does not change upon further reductions. Then for all $i \in \{1, 2, \ldots, 2n+1\}$ we have either $|b_i| = |2b_i|$ (that is, $b_i = 0$) or $|b_i| = |2b_i - 1|$ (that is, $b_i = 1$). However, the numbers $b_1, b_2, \ldots, b_{2n+1}$ have the same parity, as we have already seen, and therefore we have either $b_i = 0$ ($1 \leq i \leq 2n + 1$) or $b_i = 1$ ($1 \leq i \leq 2n + 1$). Now it is easy to verify by induction that all the preceding $(2n + 1)$-tuples also consist of identical numbers.   □

(ii) From an arbitrary n-tuple of integers $a_1, a_2, \ldots, a_n$ we form a new n-tuple

$$\left(\frac{a_1 + a_2}{2}, \frac{a_2 + a_3}{2}, \ldots, \frac{a_{n-1} + a_n}{2}, \frac{a_n + a_1}{2} \right),$$

and keep repeating this transformation. Find all possible initial n-tuples if you know that all n-tuples that occur successively are formed exclusively of integers.

SOLUTION.   If $a_1 = a_2 = \cdots = a_n$, then clearly, after each transformation we obtain the same n-tuple of identical integers. However, let us not jump to the premature conclusion that these are all the desired n-tuples. For instance, in the case where n is even, $(c, d, c, d, \ldots, c, d)$ also has the desired property, where c and d are any pair of integers with the same parity. Let now $(a_1, a_2, \ldots, a_n)$ be any n-tuple of integers with the desired property, and let

$$M = \max\{a_1, a_2, \ldots, a_n\} \quad \text{and} \quad m = \min\{a_1, a_2, \ldots, a_n\}.$$

Since M (respectively m) is a nonincreasing (respectively nondecreasing) integer valuation of the given transformation and since $M \geq m$, this means that the values M and m on our infinite iterative sequence do not change from a certain place on (say, starting with the iteration $(b_1, b_2, \ldots, b_n)$). But we have an equality

$$\max \left\{ \frac{b_1 + b_2}{2}, \frac{b_2 + b_3}{2}, \ldots, \frac{b_n + b_1}{2} \right\} = \max\{b_1, b_2, \ldots, b_n\} \ (= M')$$

if and only if in the sequence $b_1, b_2, \ldots, b_n, b_1$ the number M' occurs in some two neighboring places. If we repeat the same argument for the following iterations, we obtain by induction on $k \geq 1$ the following assertion: In the infinite periodic sequence

$$b_1, b_2, \ldots, b_n, b_1, b_2, \ldots, b_n, \ldots$$

the number M' occurs in some $k + 1$ neighboring places. This is possible for all $k \geq 1$ only if $b_1 = b_2 = \cdots = b_n$. If the n-tuple $(b_1, b_2, \ldots, b_n)$ is not the initial one, then the preceding iteration $(c_1, c_2, \ldots, c_n)$ must satisfy

$$\frac{c_1 + c_2}{2} = \frac{c_2 + c_3}{2} = \cdots = \frac{c_n + c_1}{2},$$

that is, $c_1 = c_3 = c_5 = \cdots$ and $c_2 = c_4 = \cdots$. From this it follows in the case where n is odd that $c_1 = c_2 = \cdots = c_n$ (and thus clearly the initial n-tuple consists of n identical numbers). For even n we then have

$$(c_1, c_2, \ldots, c_n) = (c, d, c, d, \ldots, c, d),$$

where c and d are integers of the same parity. Finally, we prove that in the case $c \neq d$ the n-tuple $(c_1, c_2, \ldots, c_n)$ is the initial one; that is, there do not exist integers $e_1, e_2, \ldots, e_n$ simultaneously satisfying the identities

$$\frac{e_1 + e_2}{2} = \frac{e_3 + c_4}{2} = \cdots = \frac{e_{n-1} + e_n}{2} = c$$

and

$$\frac{e_2 + e_3}{2} = \frac{e_4 + e_5}{2} = \cdots = \frac{e_n + e_1}{2} = d.$$

Indeed, these identities imply

$$nc = e_1 + e_2 + e_3 + e_4 + \cdots + e_{n-1} + e_n = nd,$$

and thus $c = d$, which is a contradiction.   □

(iii) Suppose that the integers $x_1, x_2, \ldots, x_n$ are successively written on the corners of a regular n-gon, where

$$S = x_1 + x_2 + \cdots + x_n > 0.$$

If for some $k \in \{1, 2, \ldots, n\}$ we have $x_k < 0$, we may carry out the following transformation: The numbers x_{k-1}, x_k, x_{k+1} are changed to $x_{k-1} + x_k$, $-x_k$, $x_{k+1} + x_k$, in this order; here we set $x_0 = x_n$ and $x_{n+1} = x_1$. We may repeat this transformation with the new n-tuple, etc. Show that in the cases $n = 3$ and $n = 5$ only a finite number of transformations can be carried out, that is, after a certain number of steps we obtain an n-tuple of nonnegative integers. (The case of general n and real values $x_1, x_2, \ldots, x_n$ is discussed in Section 5.10.)

SOLUTION.   First we note that

$$(x_{k-1} + x_k) + (-x_k) + (x_{k+1} + x_k) = x_{k-1} + x_k + x_{k+1},$$

so the sum S is an invariant of the transformation in question. In the case $n = 3$ it is not difficult to come up with the appropriate valuation

$$J(x_1, x_2, x_3) = x_1^2 + x_2^2 + x_3^2; \tag{34}$$

in view of symmetry it suffices to consider a change of J for $k = 2$. Let therefore $x_2 < 0$; the transformation gives

$$\begin{aligned}
J(x_1 + x_2, -x_2, x_3 + x_2) &= (x_1 + x_2)^2 + (-x_2)^2 + (x_3 + x_2)^2 \\
&= (x_1^2 + x_2^2 + x_3^2) + 2x_2(x_1 + x_2 + x_3) \\
&= J(x_1, x_2, x_3) + 2x_2 S < J(x_1, x_2, x_3).
\end{aligned}$$

Hence J is a decreasing valuation; since by (34) the value of J is a non-negative integer, this completes the proof of the finiteness assertion in the case $n = 3$.

For $n = 5$ the situation is more complicated; we can convince ourselves that a direct analogue with (34) would not work. Guessing an appropriate valuation $J(x_1, x_2, \ldots, x_5)$ would require considerable imagination; however, if we assume that J will be a quadratic form, it is possible to use the method of undetermined coefficients, namely, to search for a J of the form

$$
\begin{aligned}
J(x_1, x_2, x_3, x_4, x_5) = {} & p\left(x_1^2 + x_2^2 + x_3^2 + x_4^2 + x_5^2\right) \\
& + q(x_1 x_2 + x_2 x_3 + x_3 x_4 + x_4 x_5 + x_5 x_1) \quad (35) \\
& + r(x_1 x_3 + x_2 x_4 + x_3 x_5 + x_4 x_1 + x_5 x_2)
\end{aligned}
$$

with unknown constants p, q, r. The coefficients in (35) are chosen such that J is an invariant with respect to those permutations of the quintuple $x_1, x_2, \ldots, x_5$ that do not change the situation of the problem; furthermore, this reduces the study of the change to J again to the case $k = 2$. A routine calculation now gives

$$
\begin{aligned}
& J(x_1 + x_2, -x_2, x_3 + x_2, x_4, x_5) - J(x_1, x_2, x_3, x_4, x_5) \\
& = 2p\left(x_1 x_2 + x_2^2 + x_2 x_3\right) + q\left(-2x_1 x_2 - 2x_2^2 - 2x_2 x_3 + x_2 x_4 + x_2 x_5\right) \\
& \quad + r\left(x_1 x_2 + x_2^2 + x_2 x_3 - x_2 x_4 - x_2 x_5\right) \\
& = x_2[(2p - 2q + r)(x_1 + x_2 + x_3) + (q - r)(x_4 + x_5)].
\end{aligned}
$$

It is clearly convenient to require that $2p - 2q + r = q - r = c > 0$; the term in brackets will then be equal to

$$
c(x_1 + x_2 + x_3 + x_4 + x_5) = cS > 0,
$$

which together with the condition $x_2 < 0$ shows that J is a decreasing valuation. The assumptions on the numbers p, q, r can be rewritten as

$$
p = \frac{r + 3c}{2} \quad \text{and} \quad q = r + c,
$$

where the numbers r and c are still arbitrary. By substituting into (35) and rearranging, we get

$$
\begin{aligned}
J(x_1, x_2, x_3, x_4, x_5) = {} & \frac{r + c}{2}(x_1 + x_2 + x_3 + x_4 + x_5)^2 \\
& + \frac{c}{2}\left[(x_1 - x_3)^2 + (x_2 - x_4)^2 + (x_3 - x_5)^2 + (x_4 - x_1)^2 + (x_5 - x_2)^2\right].
\end{aligned}
$$

From this we see that the value J is a nonnegative integer if both numbers are even, $c > 0$, and $r \geq -c$. The easiest valuation is obtained for $c = 2$ and $r = -2$:

$$
J = (x_1 - x_3)^2 + (x_2 - x_4)^2 + (x_3 - x_5)^2 + (x_4 - x_1)^2 + (x_5 - x_2)^2.
$$

This completes the proof for the case $n = 5$. $\qquad\square$

5.10 Exercises

(i) We now weaken the property of the sequence in 5.9.(i); namely, we no longer require the two sets with equal sums of elements to have the same number of elements. Does the conclusion still hold that all such sequences consist of identical numbers?

The following exercises continue the study of the transformation from 5.9.(iii) under the initial condition $S > 0$.

(ii) Find a decreasing valuation with values in $\mathbb{N}_0$ in the case $n = 4$.

*(iii) Show that if we disallow the transformation of the triple x_{k-1}, x_k, x_{k+1} for a fixed k, for instance $k = 1$, then only a finite number of transformations can be carried out, even if the initial numbers $x_1, x_2, \ldots, x_n$ are real (and not necessarily integers).

*(iv) Find a decreasing valuation with values in $\mathbb{N}_0$ in the case $n \geq 6$.

*(v) With the help of (iii) and the valuation from the solution to (iv), prove the assertion concerning finiteness of the number of transformations in the general case $n \geq 6$, $x_1, x_2, \ldots, x_n \in \mathbb{R}$.

(vi) From the initial numbers $x_1, x_2, \ldots, x_n \in \mathbb{R}$ set up a formula for some constant K with the following properties: The absolute value of each of the n numbers that occur in the corners of the n-gon after an arbitrary number of transformations does not exceed the number K.

5.11 Problems on Piles of Pebbles

To finish this section, we return to a problem that we used in 5.1 to introduce the general idea of iterations. We will now consider the general initial situation where n pebbles, divided into several piles, lie on a table. Let us recall the individual transformations: From every pile we take one pebble each, and form a new pile with these pebbles.

It is important to note that in view of the finite number of all possible distributions of n pebbles into piles, after a certain number of steps the situation will necessarily become periodic. However, it is not clear how long the period will be, or whether any initial state will eventually lead to the same "universal" state. We will now proceed to solve these problems.

Let the positive integers $a_1 \geq a_2 \geq \cdots \geq a_m$ denote the numbers of pebbles in the individual piles, so clearly, $1 \leq m \leq n$ and $a_1 + a_2 + \cdots + a_m = n$. We note that the number m of piles in general changes in the course of the transformations. Nevertheless, we consider the valuation

$$J = \sum_{k=1}^{m} \left(a_k^2 + c_k a_k \right), \tag{36}$$

where $c_1, c_2, \ldots$ are constants that we will choose in a moment. After carrying out one transformation, the distribution $(a_1, a_2, \ldots, a_m)$ changes to $(m, a_1 - 1, a_2 - 1, \ldots, a_m - 1)$, and if we substitute the new numbers into J in exactly this order, we obtain the new value

$$J' = m^2 + c_1 m + \sum_{k=1}^{m} \left[(a_k - 1)^2 + c_{k+1}(a_k - 1) \right].$$

We note that if $a_k = 1$ for some $k \in \{1, 2, \ldots, m\}$, then the corresponding pile disappears; but this clearly has no influence on the calculation of the value of J'. We choose the numbers c_k in such a way that the equality $J = J'$ holds identically in the variables $a_1, a_2, \ldots, a_m$. We can rewrite this as

$$\sum_{k=1}^{m} (c_k + 2 - c_{k+1}) a_k = m^2 + m - \sum_{k=1}^{m} (c_{k+1} - c_1), \tag{37}$$

where the left-hand side does not depend on $a_1, a_2, \ldots, a_m$ if $c_k + 2 - c_{k+1} = 0$, that is, if $c_k = c_1 + 2(k - 1)$ for all $k \geq 1$. But then the right-hand side of (37) is equal to zero, as the reader should verify, which then guarantees that $J = J'$. If, for instance, we choose $c_1 = 2$, then we obtain $c_k = 2k$ for all k, and the definition (36) takes the form

$$J = \sum_{k=1}^{m} \left(a_k^2 + 2k a_k \right). \tag{38}$$

We have agreed that for the computation of J the numbers of pebbles are substituted into J in their nonincreasing order. However, the implication

$$a_1 \geq a_2 \geq \cdots \geq a_m \implies m \geq a_1 - 1 \geq \cdots \geq a_m - 1$$

does not hold in general. If we have to change the arrangement of $m, a_1 - 1$, $\ldots, a_m - 1$ into the nonincreasing one, then the value of J decreases because of the choice of the coefficients c_k. Therefore, J is not an invariant, but only a nonincreasing valuation. However, each value of J is a positive integer, and therefore only a finite number of changes is possible. Hence after a certain number of transformations the numbers $a_1, a_2, \ldots, a_m$ of pebbles in the piles must all satisfy the inequalities $m \geq a_k - 1$ $(1 \leq k \leq m)$. We have thus proved the following assertion:

After a certain number of transformations the number of pebbles in any pile is always at most one more than the number of piles at a given moment.

As an illustration we show in Figure 15 an infinite iterative sequence in the case $n = 8$ with initial numbers of pebbles $[5, 1, 1, 1]$. If we compare Figure 15 with Figure 11, we see that in the case $n = 8$ the situation is different from $n = 6$: There is no "final" state that remains unchanged by further transformations and is reached after a certain number of steps from any initial situation.

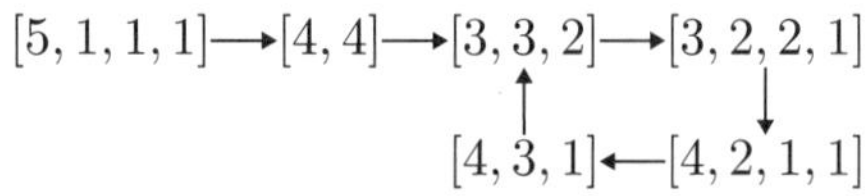

$$[5,1,1,1] \longrightarrow [4,4] \longrightarrow [3,3,2] \longrightarrow [3,2,2,1]$$

$$[4,3,1] \longleftarrow [4,2,1,1]$$

Figure 15

In closing, we show that such a final state exists not only for $n = 6$, but also for every n of the form $n = \frac{d(d+1)}{2}$, where $d \in \mathbb{N}$, and that this state has the form $[1,2,\ldots,d]$. (An analysis of the situations for the remaining values of n is described in Section 5.12.)

So, let $n = 1 + 2 + \cdots + d$. We choose an arbitrary initial situation and let $m_0, m_1, m_2, \ldots$ denote the numbers of piles that successively appear on the table; we then write the numbers of pebbles in the piles, always in a nonincreasing order:

$$
\begin{aligned}
a_0(1) &\geq a_0(2) \geq \cdots \geq a_0(m_0), \\
a_1(1) &\geq a_1(2) \geq \cdots \geq a_1(m_1), \\
a_2(1) &\geq a_2(2) \geq \cdots \geq a_2(m_2),
\end{aligned}
\tag{39}
$$

$$\vdots$$

According to the preceding result we may ignore the first few iterations and therefore assume that we have

$$a_{j+1}(1) = m_j \geq a_j(1) - 1, \ a_{j+1}(k) = a_j(k-1) - 1, \tag{40}$$

where $2 \leq k \leq m_{j+1}$ and $j \in \mathbb{N}_0$. Here the number m_j is determined for each $j \geq 1$ by the relation

$$m_j = 1 + \max\{k \in \{1,2,\ldots,m_{j-1}\} : a_{j-1}(k) > 1\}.$$

Since for a fixed n there is only a finite number of distributions of n pebbles, in the system (39) there is only a finite number of different chains that are repeated from a certain row on. We may therefore assume that this repetition begins with the first row and that the number m_0 is the largest of the numbers $m_0, m_1, m_2, \ldots$. By (40) we have $a_1(1) = m_0$, $a_2(2) = m_0 - 1$, $a_3(3) = m_0 - 2$, $\ldots$, $a_{m_0}(m_0) = 1$, which implies $a_{m_0+1}(1) = m_0$. By induction we easily obtain

$$a_{pm_0+1}(1) = m_0 \quad (p \in \mathbb{N}_0). \tag{41}$$

The right-hand part of the estimate

$$m_0 - 1 \leq a_j(1) \leq m_0 \quad (j \in \mathbb{N}_0) \tag{42}$$

follows from (40) and from the fact that $m_j \leq m_0$ for all $j \in \mathbb{N}_0$. Let us now prove the left-hand part of (42) by contradiction. If we assume that for some $j > 1$ we have $a_j(1) < m_0 - 1$, then for every $q \in \mathbb{N}_0$ we must also have

$$a_{q(m_0-1)+j}(1) < m_0 - 1. \tag{43}$$

Indeed, let (43) be false for some $q \in \mathbb{N}$; we take the smallest such q. Then from $a_{q(m_0-1)+j}(1) \geq m_0 - 1$ and (40) we get successively

$$a_{q(m_0-1)+j-1}(m_0 - 1) \geq 1, \ a_{q(m_0-1)+j-2}(m_0 - 2) \geq 2, \ \ldots$$
$$\ldots, \ a_{(q-1)(m_0-1)+j}(1) \geq m_0 - 1,$$

and this is a contradiction to the choice of the number q. Hence (43) holds for all $q \in \mathbb{N}_0$. However, the relations (41) and (43) with $p = q = j - 1$ lead to a contradiction, since then $pm_0 + 1 = q(m_0 - 1) + j$. This proves the left-hand part of (42).

From the estimates (42) and the relations (40) we now easily derive for each $j \in \mathbb{N}_0$,

$$m_0 - 2 \leq a_{j+1}(2) \leq m_0 - 1, \ \ m_0 - 3 \leq a_{j+2}(3) \leq m_0 - 2, \ \ \text{etc.,}$$

and since the sequence of chains in (39) repeats from the beginning, this implies

$$m_0 - k \leq a_j(k) \leq m_0 - k + 1 \quad (1 \leq k \leq m_j, \ j \in \mathbb{N}_0). \tag{44}$$

If we add these inequalities for $j = 0$ and $k = 1, 2, \ldots, m_0$, we get

$$\frac{(m_0 - 1)m_0}{2} < a_0(1) + a_0(2) + \cdots + a_0(m_0) \leq \frac{m_0(m_0 + 1)}{2}.$$

(The left inequality is strict, since $a_0(m_0) > 0$.) In view of the fact that

$$a_0(1) + a_0(2) + \cdots + a_0(m_0) = n = \frac{d(d + 1)}{2},$$

we obtain from this $m_0 = d$ and $a_0(k) = d - k + 1$ $(1 \leq k \leq d)$. The proof is now complete.

5.12   Exercises

(Continuation of the problem on piles of pebbles from 5.11.)

(i) For each $n \geq 1$ find all "final" distributions of pebbles that do not change under the transformation.

*(ii) Show that in the case $\frac{(d-1)d}{2} < n < \frac{d(d+1)}{2}$ $(d \in \mathbb{N})$ after a certain number of transformations the iteration becomes periodic with a period $p > 1$ that is a divisor of d such that the number $\frac{d}{p}$ divides the difference $n - \frac{(d-1)d}{2}$. Describe those distributions that repeat periodically.

*(iii) In the situation of (ii), find the number of distributions of the pebbles that are periodically repeated in the iterations with a given period p.

3
Combinatorial Geometry

The development of geometry, as inspired by the deep results of Bernhard Riemann in the second half of the nineteenth century, has meant that scientific work in this field moved quite far from the "naive" or elementary geometry practiced by the Greek mathematicians of around the beginning of our era, and their numerous successors in later times. Classically, the main focus of geometry has been on the *proofs* or *constructions* connected with properties of basic geometrical objects (points, straight lines, circles, triangles, half-planes, tetrahedra, etc.), that is, problems that can be visualized. On the other hand, a paradoxical characteristic of contemporary scientific works in "pure" geometry is the fact that the majority of them are completely devoid of pictures; or the fact that "geometrical intuition" is more often required by specialists in mathematical analysis or algebra than by mathematicians who consider themselves geometers. The apparent dissatisfaction of a number of mathematicians with this situation has led to new directions of research in geometry; this is mainly in response to modern problems in *optimization*. As examples of such geometrical optimization problems one can consider the problem of *filling* a plane or a space (or some of their parts) with some system of geometrical objects, or *covering* parts of a plane or a space with the smallest possible number of copies of a geometrical object. These and several other reasons have led, especially in the past half century, to the rapid development of several nontraditional branches of geometry (or areas closely related to geometry), among them *combinatorial geometry*.

It is rather difficult to define the contents of combinatorial geometry precisely, since this discipline is closely connected with a number of related ones, such as *discrete geometry, combinatorial topology*, and especially the *theory of convex sets*. According to the Swiss mathematician *Hugo Hadwiger* (who coined the name of this discipline) the object of combinatorial geometry is the study of problems concerned with finding "optimal" configurations, normally formed with a *finite number* of points or other geometrical objects. A characteristic property of this kind of problems is their connection with determining some *integers*, namely the numbers of points or geometrical objects that satisfy the conditions associated with the particular configuration to be studied. One of the simplest and most frequent such conditions is the convexity of the objects under consideration.

The circle of problems connected with coverings and fillings are closely related to the subject matter of *discrete geometry*, where, however, a significant role is also played by the study of *infinite* systems of covering or filling objects (for instance, *tiling* problems). In contrast to *combinatorial topology*, an essential part of combinatorial geometry is the frequent use of some "measurements" of the objects studied; often it is their area, surface area, or volume.

A characteristic trait of the problems studied in combinatorial geometry is their simplicity of formulation and the possibility of their visualization; we are dealing here with the simplest geometrical notions and their basic properties. However, the solutions of these problems are often not at all trivial, and some of them have to this day withstood the efforts of mathematicians. This unclear boundary between easy and difficult problems is very attractive not only to research mathematicians, but also to students; this might explain the increasing number of such problems included in various student competitions.

In closing, we note that the text of this chapter is only an introduction to this circle of problems and is subject to a number of limitations, especially in scope. It is important to point out that here we will be dealing almost exclusively with *combinatorial planimetry*, concerned with objects in a plane.

Finally, we give a list of problems arranged according to the method of solution used; for a more detailed discussion, see the introduction to Chapter 2.

Induction principle: 1.1.(ii), 1.3.(ii), 1.10.(ii), 1.14.(ii), 1.15.(i), 1.15.(ii), 2.1.(i), 2.1.(iii), 2.2.(i), 2.2.(ii), 2.2.(v), 2.2.(vi), 2.3.(i), 2.4.(iii), 2.5.(ii) (1st solution), 2.5.(iii), 2.5.(iv), 2.6.(i), 2.6.(ii), 2.6.(iv), 2.6.(ix), 2.7.(i), 2.7.(ii), 2.8.(i), 2.8.(ii), 2.8.(iv), 4.4.(iii), 4.5.(iv), 4.6.(ix).

Pigeonhole principle: 1.3.(iv), 1.13.(iii), 2.5.(iv), 2.6.(viii), 2.6.(ix), 2.6.(x), 3.3.(iv), 3.3.(v), 3.4.(iii), 3.4.(iv), 3.4.(v), 3.4.(vi), 3.4.(viii), 3.5.(i), 3.5.(iii), 3.6.(v), 3.10.(iv), 3.11.(i), 3.11.(ii), 3.12.(i), 3.12.(ii), 3.12.(iii), 3.12.(vi), 4.1.(i), 4.5.(iii), 4.6.(vi), 4.6.(vii), 4.8.(vi).

Direct construction: 1.3.(iii), 1.4.(i), 1.5.(v), 1.6.(vi) (2nd solution), 1.12.(ii), 1.13.(ii), 1.14.(ii), 1.15.(iii), 2.3.(i), 2.3.(ii), 2.4.(i), 2.4.(ii), 2.4.(iii),

1 Systems of Points and Curves

In this section we will study *finite* systems of points, line segments, straight lines, and circles in the plane, with regard to their combinatorial properties. This is a rather substantial topic, and it is well suited to illustrate both poles of our preceding combinatorial considerations: On the one hand to determine the *number* of certain geometric configurations, and on the other hand problems of *existence* of geometric configurations with required properties. Problems of the former type will mainly occur in the first few subsections; they will to a large extent be similar to examples from Chapter 1, with boundary conditions given by geometrical properties of the objects under consideration. Problems of the latter type will dominate the closing subsections; there we will mainly use the induction principle and the method of extremes.

In many considerations the concept of a convex polygon will play a key role; we therefore begin with an account of several of its properties.

1.1 Remarks on Convex Bodies

A *convex body* in the plane (on the line, in space) is a set $\mathcal{U}$ of points in the plane (on the line, in space) with the following property: For any two distinct points X, Y in $\mathcal{U}$, each point of the line segment XY is also in $\mathcal{U}$.

Convex sets on the line have a simple structure: Except for the whole line, the empty set, and singletons, they are exactly all the half lines and line segments, with or without boundary points.

The classification of convex bodies in the plane is obviously more difficult, but a significant aid in the construction of convex sets is the fact that the intersection of convex sets is a convex set. An important notion is that of the *convex hull* $\mathcal{K}$ of a plane body $\mathcal{U}$; it is the intersection of all convex sets $\mathcal{M}$ satisfying $\mathcal{U} \subseteq \mathcal{M}$ (thus it is, with respect to inclusion, the *smallest* convex set containing $\mathcal{U}$).

In the majority of problems we will work with *convex polygons*, which we now define. Let $\mathcal{K} = \{A_1, A_2, \ldots, A_n\}$ be an n-element set of points in the plane ($n \geq 3$). The set $\mathcal{M}$ is called a *convex n-gon* $A_1 A_2 \ldots A_n$ if $\mathcal{M}$ is the intersection of n half-planes $\varrho_1, \varrho_2, \ldots, \varrho_n$ with the property that for each

$i = 1, 2, \ldots, n$ the line $A_i A_{i+1}$ (where $A_{n+1} = A_1$) is the boundary of the half plane ϱ_i and each point $A_j \in \mathcal{K} \setminus \{A_i, A_{i+1}\}$ lies inside ϱ_i. The polygon $\mathcal{M}$ is then the convex hull of the set $\mathcal{K}$, and its elements $A_1, A_2, \ldots, A_n$ will be called the *vertices* of the polygon $\mathcal{M}$.

Conversely, if $\mathcal{U}$ is a finite set of $n \geq 3$ points in the plane that are not *collinear* (that is, they do not lie on the same straight line), then the convex hull of $\mathcal{U}$ is a convex polygon whose vertices are some (and possibly all) elements of the set $\mathcal{U}$.

We now derive two basic properties of convex polygons that will be used frequently in solutions throughout this chapter.

(i) The number u_n of diagonals in a convex n-gon $\mathcal{M}$ satisfies

$$u_n = \frac{n(n-3)}{2}. \tag{1}$$

Indeed, the n vertices of the polygon are connected by $\binom{n}{2} = \frac{n(n-1)}{2}$ line segments, of which exactly n form the sides of the polygon $\mathcal{M}$. Hence we have $u_n = \frac{n(n-1)}{2} - n = \frac{n(n-3)}{2}$.

The formula (1) can also be obtained in a different way: Each vertex of the polygon $\mathcal{M}$ is an endpoint of $n - 3$ diagonals. If we add this number over all n vertices of $\mathcal{M}$ and take into account the fact that each diagonal has been counted twice, we obtain again equality (1).

(ii) The sum S_n of interior angles of a convex n-gon satisfies

$$S_n = (n-2) \cdot 180°. \tag{2}$$

First we prove this by mathematical induction. For $n = 3$ it is the well-known formula for the sum of interior angles of a triangle. We assume now that the assertion holds for $i = n$; we want to show that it is true also for $i = n+1$. We subdivide the convex $(n+1)$-gon $A_1 A_2 \ldots A_{n+1}$ by the diagonal $A_1 A_3$ into the triangle $A_1 A_2 A_3$ and the n-gon $A_1 A_3 \ldots A_{n+1}$, whose sum S_n of interior angles is, by the induction hypothesis, $S_n = (n-2) \cdot 180°$. Since the sum S_{n+1} of interior angles of the polygon $A_1 A_2 \ldots A_{n+1}$ is equal to the sum of interior angles of the triangle $A_1 A_2 A_3$, plus that of the n-gon $A_1 A_3 \ldots A_{n+1}$, we obtain

$$S_{n+1} = 180° + (n-2) \cdot 180° = (n-1) \cdot 180°,$$

so the formula (2) holds for all $n \geq 3$.

A direct proof of (2) follows from the observation that the polygon $\mathcal{M} = A_1 A_2 \ldots A_n$ can be divided by the $n-3$ diagonals $A_1 A_k$ ($k = 3, 4, \ldots, n-1$) into $n - 2$ triangles, and the sum of their interior angles is equal to S_n.

To conclude this overview, we state without proof (which would require a considerable effort) another important result from the theory of convex sets, namely the so-called *theorem of Helly* (see [4], where a proof as well as numerous applications can be found).

(iii) Let S be an arbitrary (finite or infinite) system of convex subsets of a given plane such that any three sets in S have at least one common point. Then all sets in S have a point in common.

1.2   Examples

We will now present a first collection of three problems on convex polygons. More of them can be found in the following subsections and also in later sections of this chapter.

(i) Show that for each $n \geq 3$ there exists a convex n-gon $\mathcal{M}$ such that no three of its diagonals intersect in one interior point of $\mathcal{M}$.

SOLUTION.   We show by induction on n that there even exists an n-gon inscribed in a circle and having the desired property. For $n \leq 5$ the assertion is trivially satisfied (consider a regular n-gon). Let us now assume that some n-gon $\mathcal{M}_n = A_1 A_2 \ldots A_n$ inscribed in the circle c has the desired property. We consider the collection of all lines connecting the intersections of all the diagonals of $\mathcal{M}_n$. There are only finitely many such lines, and therefore only finitely many intersections of these lines with the circle c. Hence there exists a point A_{n+1} on the circle c that does not lie on any of the lines of our collection. Furthermore, A_{n+1} can be chosen to lie on the arc that is bounded by A_1, A_n and does not contain $A_2, A_3, \ldots, A_{n-1}$. It remains to convince ourselves that $A_1 A_2 \ldots A_{n+1}$ is indeed a convex $(n + 1)$-gon and that no three of its diagonals intersect in the same point; this is left to the reader.   □

(ii) Suppose that we are given n points $(n \geq 4)$ in the plane such that no three lie on a straight line, and any four are the vertices of a convex quadrilateral. Show that the given points are the vertices of a convex n-gon.

SOLUTION.   We denote the set of the given points by S and construct the convex hull of S; this will be a k-gon, which we denote by $\mathcal{M}$. We show that each point $X \in S$ is a vertex of $\mathcal{M}$. Let us assume to the contrary that this is not the case; then some point $Y \in S$ lies in the interior of $\mathcal{M}$. From an arbitrary vertex Z of $\mathcal{M}$ we draw all diagonals, and thus we subdivide $\mathcal{M}$ into $k - 2$ triangles. The point Y then lies in the interior of one of these triangles, say UVZ. But then the quadruple U, V, Y, Z does not form the vertices of a convex quadrilateral, and this is a contradiction.   □

(iii) Choose a point P in the interior of a convex polygon $\mathcal{M}$, and construct orthogonal projections from P to all straight lines that contain sides of $\mathcal{M}$. Show that at least one of these projections lies on a side of $\mathcal{M}$ (and not on its extension).

SOLUTION.   From among the extended sides of $\mathcal{M}$ we choose one that has the *smallest distance* from the point P, and call it l. (If there are several such lines, we choose an arbitrary one of them.) Let the side AB lie on l.

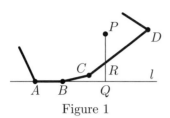

We will show that the orthogonal projection Q of P on the line l lies on the line segment AB. Let us assume that $Q \in l \setminus AB$ (Figure 1). Then the segment PQ connects an interior and an exterior point of the polygon $\mathcal{M}$, and therefore it cuts some side CD in a point R. Since $|PR| < |PQ|$, the distance of P from the line segment CD is less than the distance from the line l, which is a contradiction to the choice of l. (Here and throughout this chapter the notation $|AB|$ stands for the length of the line segment AB.)   □

Figure 1

1.3 Exercises

   (i) Determine the largest number of acute angles in a convex n-gon.

   (ii) Some diagonals are constructed in a convex n-gon in such a way that no two of them intersect in an interior point of the n-gon. Show that there are at least two vertices of the n-gon from which none of the diagonals in question emanates.

   (iii) Find the smallest number of points that are needed in the interior of a convex n-gon to make sure that any triangle, the vertices of which are among those of the n-gon, contains at least one of the points.

   (iv) A point P is placed inside a convex $2n$-gon $A_1 A_2 \ldots A_{2n}$, and all the straight lines PA_i $(i = 1, 2, \ldots, 2n)$ are constructed. Show that at least one side of the polygon is not intersected in its interior by any of the lines in question.

1.4 Counting Intersections

In this subsection and in the following exercises we present several problems on determining the *number of intersections* in systems of line segments, straight lines, and circles. Even though these are all simple problems, we want to mention at this point that the result of Problem (ii) will be used several times throughout this chapter.

We also recall here the notions of *line bundles* of the 1st and 2nd kinds, which will occur in Exercises 1.5.(i) and (ii) and in later subsections. A set of straight lines in the plane is called

   (a) a *line bundle of the 1st kind* if all lines of the set are parallel to each other, that is, they have a common direction;

   (b) a *line bundle of the 2nd kind* if all lines of the set have a point in common, the so-called *center of the line bundle*.

(i) A closed broken line consists of $2n + 1$ segments $(n \geq 1)$. Determine the largest number of points in which it can intersect itself.

SOLUTION.   We consider an arbitrary segment AB of the broken line. It can contain at most $2n - 2$ points of intersection, since AB will not be cut by at least two segments, namely, its immediate neighbors. This gives an estimate for the number k of such points of intersection: $k \leq \frac{(2n+1)(2n-2)}{2} = (2n + 1)(n - 1)$. It is possible to construct a broken line with exactly $(2n + 1)(n - 1)$ points of intersection: Take a polygon $A_1 A_2 \ldots A_{2n+1}$ in which no three diagonals intersect in the same interior point (the existence of such a polygon is guaranteed by 1.2.(i)), and construct the closed broken line $A_1 A_{n+1} A_{2n+1} A_n A_{2n} A_{n-1} \ldots A_2 A_{n+2} A_1$, which has the desired property.   □

(ii) In how many points inside a convex n-gon $A_1 A_2 \ldots A_n$ do its diagonals intersect when no three of them have a common interior point?

FIRST SOLUTION.   We consider the fixed diagonal $A_1 A_k$ $(k = 3, 4, \ldots, n - 1)$ and begin by determining the number of intersection points on this diagonal. In one of the half-planes determined by the line $A_1 A_k$ there are $k - 2$, and in the other one $n - k$ of the remaining $n - 2$ vertices of the polygon $A_1 A_2 \ldots A_n$ (Figure 2). Therefore, the diagonal $A_1 A_k$ intersects with the other diagonals in $(n - k)(k - 2)$ points, and the number $a_1(n)$ of intersection points on all diagonals emanating from A_1 satisfies

$$a_1(n) = \sum_{k=3}^{n-1}(n - k)(k - 2) = \sum_{k=3}^{n-1}\left[(n + 2)k - k^2 - 2n\right]$$
$$= \frac{(n - 1)(n - 2)(n - 3)}{6}$$

(see, for example, (8) and (13) in [5], Chapter 1, Section 2). If we add these numbers for all vertices $A_1, \ldots, A_n$ of the polygon, then we have counted each intersection point four times. Hence for the total number $A(n)$ of all diagonals we obtain

$$A(n) = \frac{n}{4}a_1(n) = \frac{n(n - 1)(n - 2)(n - 3)}{24} = \binom{n}{4}. \tag{3}$$

□

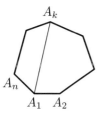

Figure 2

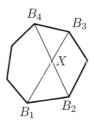

Figure 3

SECOND SOLUTION.   We prove formula (3) by a more direct consideration. Any intersection point X of two diagonals corresponds to a quadruple of vertices of the n-gon, namely the endpoints of the two diagonals that intersect in X. We denote these four points by B_1, B_2, B_3, B_4 in such a way that the point X corresponds to the intersection of the diagonals $B_1 B_3$ and $B_2 B_4$ (Figure 3), while no other two segments connecting these points have any interior point in common. Hence we have exactly as many of the intersection points as there are ways of choosing four of the vertices of the n-gon, namely $\binom{n}{4}$.   □

(iii) Suppose that 5 points are given in the plane, and that among all the straight lines connecting any two of these points, no two are either parallel or perpendicular to each other. From each of the given points construct perpendiculars to all the lines connecting the other four points. Show that the number of intersection points in this system of perpendiculars does not exceed 310.

SOLUTION.   The number of lines determined by the pairs of the given points is $\binom{5}{2} = 10$, and 4 of them go through any given point. Hence from each of the points there are 6 lines perpendicular to the lines connecting the other points. Let us first consider any two of the given points, say A, B, and count the intersection of the perpendiculars from A with those from B. Denote the remaining points by C, D, E. We divide the 6 perpendiculars from A into two groups: The lines p_1, p_2, p_3 will be perpendicular to the lines BC, BD, BE; they intersect all 6 perpendiculars $r_1, r_2, \ldots, r_6$ from the point B in $3 \cdot 6 = 18$ points. Each of the remaining three lines q_1, q_2, q_3 from the point A, perpendicular to one of the remaining three lines CD, CE, DE that do not go through either A or B, intersects 5 of the perpendiculars from B, while it does not meet the sixth one (they are two distinct perpendiculars to the same line; see Figure 4). This adds another $3 \cdot 5 = 15$ intersection points. Hence the perpendiculars emanating from the points A and B intersect in $18 + 15 = 33$ points. One can choose two points from among the given five in $\binom{5}{2} = 10$ ways, so the number p of intersection points cannot exceed $10 \cdot 33 = 330$. We further note that the given points form the vertices of $\binom{5}{3} = 10$ triangles, in each of which all three intersection points of the altitudes coincide. Thus, finally, $p \leq 330 - 2 \cdot 10 = 310$.

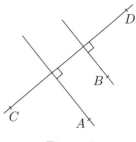

Figure 4   □

1.5   Exercises

(i) Two line bundles of the 2nd kind lie in a plane, with centers A, B and consisting of p, respectively q, lines. Suppose that none of these

$(p + q)$ lines passes through both points A and B, and that no two lines are parallel. Find the number of all intersections of these lines (not counting the points A and B).

(ii) Three line bundles of the 2nd kind lie in a plane, with centers A, B, C, and each consisting of m lines. Suppose that no three lines from different bundles pass through the same point, no line passes through two of the centers, and no two lines are parallel. Find the number of intersections of these lines (not counting the points A, B, C).

(iii) Suppose that n points in a plane are given, of which no three lie on the same line and no four on the same circle. A straight line is drawn through every pair of these points, and a circle through every triple of the points. Show that the number of points in the plane through which one of the lines and also one of the circles passes is not greater than $n + \binom{n}{3}(n - 1)(n - 3)$.

*(iv) Suppose that n points in a plane are given, of which no four lie on the same circle. A circle is drawn through every triple of these points. Show that the possible number of intersections in this system of circles is not greater than $n + \frac{5}{3} \cdot \binom{n}{5} \cdot (2n - 1)$.

(v) Suppose that any two neighboring segments of a closed broken line with 14 segments are perpendicular to each other, while no two segments lie on the same straight line. Find the largest number of points in which this broken line can intersect itself.

(vi) Suppose that no two diagonals in a convex n-gon $\mathcal{M}$ are parallel, and no three diagonals that do not emanate from the same vertex have a point in common. Consider the straight lines containing the diagonals of $\mathcal{M}$ and find the maximal number of all their intersections that can lie outside the polygon $\mathcal{M}$.

*(vii) Suppose that $n \geq 3$ points in a plane are given. Among the straight lines that connect any two of these points, no two are either parallel or perpendicular to each other. Through each of these points we draw the perpendicular to all the lines connecting the remaining $(n - 1)$ points. Show that the number of intersections in the system of these perpendiculars is not greater than $\binom{n}{3} \cdot \frac{3n^3 - 12n^2 + 9n + 4}{4}$.

1.6 Counting Triangles

One of the classical problems of combinatorial geometry is the determination of the number of triangles whose vertices or sides (as lines) lie in a given n-element set (of points or lines, respectively). It is clear that in both cases this number does not exceed $\binom{n}{3}$. In this subsection we study situations where there are special requirements either on the configurations of the possible vertices of the triangles (for instance, in (ii) these are the vertices of a regular n-gon) or on the configurations of the possible sides of the triangles (such as the diagonals of a convex polygon in (iii)).

(i) In the interior of each side of a square we choose n different points. Find the number of all triangles whose vertices are these points.

SOLUTION.   From the given $4n$ points one can choose a total of $\binom{4n}{3}$ triples of points, where in $4 \cdot \binom{n}{3}$ cases the chosen triple lies on one side of the square, so that no triangle is formed. The desired number of triangles is therefore $\binom{4n}{3} - 4 \cdot \binom{n}{3} = 2n^2(5n - 3)$.

The same result can also be obtained in a different way. Each of the triangles is of one of the following two types: Either every vertex of the triangle lies on a different side of the square, or two points of the triangle lie on the same side of the square, and the third one on a different side. There are $4n^3$ triangles of the first type, and $4 \cdot 3 \cdot \binom{n}{2} \cdot n$ of the second type. The number of all triangles is then $4n^3 + 12n\binom{n}{2} = 2n^2(5n - 3)$.   □

(ii) In how many ways can one choose three of the vertices of a regular n-gon $A_1 A_2 \ldots A_n$ $(n \geq 4)$ such that they form the vertices of an obtuse triangle?

FIRST SOLUTION.   Let $P(n)$ denote the desired number of obtuse triangles and let $P_1(n)$ be the number of obtuse angles $A_1 A_i A_j$, where $1 < i < j \leq n$. Clearly, $P(n) = n \cdot P_1(n)$. Any of the considered angles $A_1 A_i A_j$ is obtuse if and only if $j \leq \frac{n+1}{2}$ (see Figure 5); hence $P_1(n)$ equals the number of all two-element subsets of $\{2, 3, \ldots, [\frac{n+1}{2}]\}$. Thus

$$P(n) = n \cdot P_1(n) = n \cdot \frac{1}{2} \cdot \left(\left[\frac{n+1}{2}\right] - 1\right) \cdot \left(\left[\frac{n+1}{2}\right] - 2\right)$$

$$= \frac{n}{2} \cdot \left[\frac{n-1}{2}\right] \cdot \left[\frac{n-3}{2}\right].$$ □

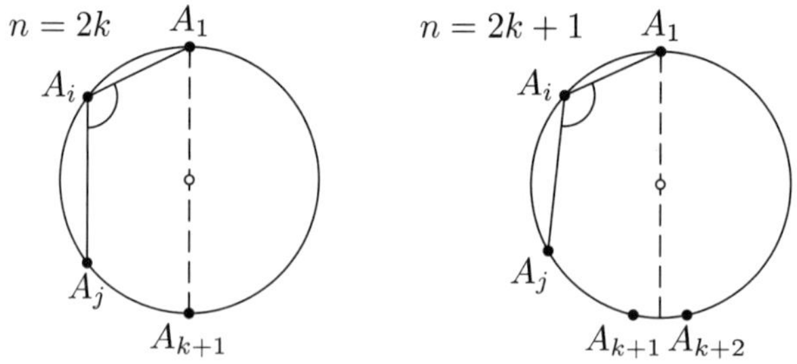

Figure 5

SECOND SOLUTION.   Let $l = \overset{\frown}{A_i A_j}$ be the shortest arc of the circumcircle that contains the vertices of an obtuse triangle under consideration, say $\triangle A_i A_k A_j$. Then the size of l is less than $180°$, and the number of all obtuse triangles $A_i A_k A_j$ with the common longest side $A_i A_j$ is identical to the number $v(l)$ of those vertices A_k of the n-gon that are inner points of the arc l. Clearly, $v(l) \in \left\{1, 2, \ldots, \left[\frac{n-3}{2}\right]\right\}$. Since there are exactly n arcs l with the same value $v(l)$, the desired number is $n \cdot \left(1 + 2 + \cdots + \left[\frac{n-3}{2}\right]\right) = \frac{n}{2} \cdot \left[\frac{n-3}{2}\right] \cdot \left[\frac{n-1}{2}\right]$. $\qquad\square$

(iii) Let the convex n-gon $\mathcal{M}$ ($n \geq 6$) be given. Find the number of all triangles with vertices among those of $\mathcal{M}$ and with sides lying on diagonals of $\mathcal{M}$.

FIRST SOLUTION. From the number of all triangles with vertices among those of $\mathcal{M}$ we subtract the number of those that have either one or two sides lying on the sides of the polygon $\mathcal{M}$. There are $n(n-4)$ triangles of the first kind, since for each of the n sides of $\mathcal{M}$ (and thus pairs of vertices) one can choose the third vertex in $(n-4)$ ways; there are n triangles of the second kind, since each such triangle consists of two neighboring sides of $\mathcal{M}$ and is therefore uniquely determined by their common vertex. The number N of desired triangles is therefore

$$N = \binom{n}{3} - n(n-4) - n = \frac{n(n-4)(n-5)}{6}. \qquad\square$$

SECOND SOLUTION. We first determine the number of all triangles with a vertex of the point A_1 of the polygon $\mathcal{M} = A_1 A_2 \ldots A_n$. Since no allowable triangle contains a pair of neighboring vertices of $\mathcal{M}$, we have to choose from $A_3, A_4, \ldots, A_{n-1}$ a pair of nonneighbors; this can be done in $\binom{n-4}{2}$ ways (see Exercise 3.2.(xvi) in Chapter 1). If we add these numbers for all n vertices of $\mathcal{M}$, then each one of the desired triangles is counted three times; hence we have

$$N = \frac{n}{3} \cdot \binom{n-4}{2} = \frac{n(n-4)(n-5)}{6}. \qquad\square$$

(iv) Suppose that no three diagonals of a convex n-gon $\mathcal{M}$ have a common interior point. Find the number of triangles whose sides lie on the sides or diagonals of the given polygon $\mathcal{M}$.

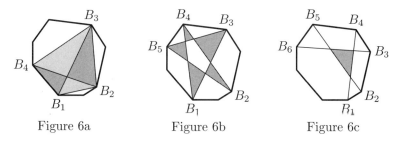

Figure 6a   Figure 6b   Figure 6c

SOLUTION. We divide all $P(n)$ triangles with the required properties into classes S_0, S_1, S_2, S_3 such that S_i contains those triangles that have exactly

i vertices among the vertices of the polygon $\mathcal{M}$ ($i = 0, 1, 2, 3$). It is easy to see that $|S_3| = \binom{n}{3}$. Each triangle in the class S_2 has two vertices, say B_1, B_2, among the vertices of $\mathcal{M}$, while the third vertex is the intersection of some diagonals $B_1 B_3$ and $B_2 B_4$. If we choose any four B_1, B_2, B_3, B_4 from the vertices of $\mathcal{M}$ (see Figure 6a), then we obtain exactly four triangles from the class S_2; hence $|S_2| = 4 \cdot \binom{n}{4}$. Using similar arguments, we obtain $|S_1| = 5 \cdot \binom{n}{5}$ and $|S_0| = \binom{n}{6}$ (see Figures 6b and 6c). Therefore,

$$P(n) = \sum_{i=0}^{3} |S_i| = \binom{n}{3} + 4 \cdot \binom{n}{4} + 5 \cdot \binom{n}{5} + \binom{n}{6}$$

$$= \frac{n(n-1)(n-2)}{6} \cdot \frac{n^3 + 18n^2 - 43n + 60}{120}. \qquad \square$$

(v) We are given a finite set $\mathcal{S}$ of n different points in the plane ($n \geq 3$), of which no three lie on a line. Show that there are at least $\frac{n(n-2)}{3}$ triangles whose vertices are among the points of $\mathcal{S}$ and such that they do not contain any other points of $\mathcal{S}$.

SOLUTION. From the set $\mathcal{S}$ we choose any two points and denote them by X, Y. Given these points, consider a half-plane ϱ with the boundary line XY that contains at least one more point of $\mathcal{S}$. In the interior of ϱ we determine the point $Z \in \mathcal{S}$ that has the *smallest distance* from the line XY (if there are several such points, we choose an arbitrary one). Then the triangle XYZ is "empty"; that is, it does not contain any of the remaining ($n - 3$) points of $\mathcal{S}$. To prove this, assume to the contrary that a point $A \in \mathcal{S}$ is an interior point of $\triangle XYZ$. If d_A, respectively d_Z, denotes the distance of the point A, respectively Z, from the line XY, then the fact that the area of $\triangle AXY$ is less than the area of $\triangle ZXY$ implies that $d_A < d_Z$ (see Figure 7), which is a contradiction to the choice of the point Z. We note that there exist at most n pairs X, Y of points in $\mathcal{S}$ such that the whole set $\mathcal{S}$ lies in one of the two half-planes with boundary line XY (these are exactly the pairs X, Y of neighboring vertices of the convex hull of $\mathcal{S}$). For each such pair X, Y we have shown that at least one

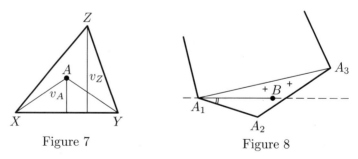

Figure 7   Figure 8

of the triangles XYZ, $Z \in \mathcal{S}$, is "empty." For each of the other pairs X, Y (the number of which is at least $\binom{n}{2} - n$) there are at least two "empty" triangles XYZ, $Z \in \mathcal{S}$. If we sum these numbers over all pairs X, Y, we get at least $n + 2(\binom{n}{2} - n) = n(n-2)$ "empty" triangles, where each of them is counted at most three times. □

(vi) Suppose that in a plane there are $3n$ points ($n \geq 1$), of which no three lie on a straight line. Show that these points form the vertices of n mutually disjoint triangles.

FIRST SOLUTION.   We use mathematical induction. For $n = 1$ the assertion is clearly true; we now assume that it also holds for $n = k$, and we will prove it for $n = k + 1$. We consider $3k + 3$ points in the plane, no three of which lie on the same line. Let the convex hull of these points be the polygon $A_1 A_2 \ldots A_s$ ($3 \leq s \leq 3k + 3$). We now distinguish two cases:
(a) The triangle $A_1 A_2 A_3$ contains none of the remaining $3k$ points. Then all these points lie in the half-plane opposite the half-plane bounded by the line $A_1 A_3$ and containing the point A_2. By the induction hypothesis, these $3k$ points form k disjoint triangles. Together with $\triangle A_1 A_2 A_3$ we then obtain the desired $(k + 1)$ triangles.
(b) The triangle $A_1 A_2 A_3$ contains some of the remaining $3k$ points. Choose a point B from among them such that the angle $B A_1 A_2$ is *minimal*; such a point B is uniquely determined (Figure 8). Then the half-plane opposite the half-plane $A_1 B A_2$ contains $3k$ of the given points, which according to the induction hypothesis are the vertices of k disjoint triangles. If we add to these $\triangle A_1 A_2 B$, then we have $(k + 1)$ disjoint triangles.
We note that the proof clearly gives an approach to dividing the given $3n$ points into n triples of vertices of mutually disjoint triangles. □

SECOND SOLUTION.   An even more intuitive construction of the desired n disjoint triangles follows from a different argument. We draw all lines determined by pairs of the given $3n$ points, and choose any direction that is different from all the directions of the lines already constructed. Through each of the given points we draw a straight line with this direction; we thus obtain a system of $3n$ *distinct* parallel lines. We subdivide these parallel lines (in a unique way) into n triples of neighboring lines; the given points on the lines belonging to the same triple then form the vertices of one of the desired triangles (Figure 9).

We note that this "method of parallels" can be also used to solve the following problem: Suppose that the sum of the positive integers $k_1, k_2, \ldots, k_n$ is equal to N. Show that every N-element set of points in the plane can be divided into n classes with $k_1, k_2, \ldots, k_n$ elements such that the convex hulls of the individual classes are disjoint.

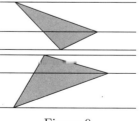

Figure 9   □

1.7 Exercises

(i) Find the number of triangles with vertices in the set $\mathcal{S}$ of n points, of which p $(3 \le p < n)$ lie on the same straight line, while any other straight line contains at most two points of $\mathcal{S}$.

(ii) Find the number of triangles with vertices in the set $\mathcal{S}$ of $(p+q)$ points, of which p lie on the line a, q lie on the line b, where a and b are parallel.

(iii) Suppose that n different interior points are given on each edge of a die. Find the number of all triangles with these points as vertices. How many of these triangles lie on the surfaces of the die?

In (iv)–(vi) $\mathcal{V}_n$ denotes the set of vertices of a given regular n-gon.

(iv) Find the number of all right triangles with vertices in $\mathcal{V}_n$.

(v) Find the number of all isosceles triangles with vertices in $\mathcal{V}_n$.

(vi) Find the number of all obtuse isosceles triangles with vertices in $\mathcal{V}_n$.

*(vii) Let $\mathcal{V}$ denote the set of all vertices of a convex n-gon $\mathcal{M}$ inscribed in a circle k. Find the number of all triangles with vertices in $\mathcal{V}$ and with the following property for a given s: The vertices of the triangle divide k into three arcs such that on each of them there are at least s vertices from $\mathcal{V}$, where the endpoints of the arcs do not count.

1.8 Counting Polygons

In the problems of this subsection and in the subsequent exercises we generalize the considerations from 1.6 and 1.7: In a configuration of n points in the plane we will determine the number of k-gons $(4 \le k \le n)$ with vertices among the given points. Particularly worth mentioning is the *Cayley problem* on the number of convex k-gons with vertices among those of a convex n-gon and sides belonging to the diagonals of the n-gon. This problem will be solved in (i); we already dealt with the special case $k = 3$ in 1.6.(iii). Also note that Exercise 1.9.(ix) generalizes the arguments from 1.7.(vii).

(i) Let $\mathcal{M}$ be a convex n-gon. Find the number of convex k-gons with a given k, $3 \le k \le \frac{n}{2}$, all of whose vertices are among those of $\mathcal{M}$ and whose sides are diagonals of $\mathcal{M}$.

SOLUTION.   Let $\mathcal{M}$ be the n-gon $A_1 A_2 \ldots A_n$ and let us determine how many k-gons have the point A_{n-1} as one of their vertices. We then denote the remaining vertices of one such k-gon by $A_{i_1}, A_{i_2}, \ldots, A_{i_{k-1}}$, where the indices $i_1, i_2, \ldots, i_{k-1}$ satisfy the inequalities $1 \le i_1 < i_2 < \cdots < i_{k-1} \le n - 3$; $i_2 - i_1 \ge 2$, $i_3 - i_2 \ge 2$, $\ldots$, $i_{k-1} - i_{k-2} \ge 2$. By the result of

Exercise 3.2.(xvi) in Chapter 1 such $(k-1)$ indices can be chosen in $\binom{n-k-1}{k-1}$ ways, and therefore the number of k-gons with the desired property and with vertex A_{n-1} is equal to $\binom{n-k-1}{k-1}$. Adding this number for all vertices of $\mathcal{M}$, and in view of the fact that each k-gon is then counted k times, we obtain for the number $c(k,n)$ of allowable k-gons the formula

$$c(k,n) = \frac{n}{k} \cdot \binom{n-k-1}{k-1} = \frac{n(n-k-1)!}{k!(n-2k)!}. \tag{4}$$

We note that for $k = 3$ the identity (4) becomes

$$c(3,n) = \frac{n(n-4)!}{3!(n-6)!} = \frac{n(n-4)(n-5)}{6},$$

which we already derived in 1.6.(iii). $\qquad\square$

(ii) Suppose that five points are given in a plane such that no three lie on a straight line. Find the number of ways in which four of the points can be chosen to form a convex quadrilateral.

SOLUTION.   We denote by $\mathcal{S}$ the set of the given points, and distinguish three cases according to the shape of the convex hull $\mathcal{K}$ of $\mathcal{S}$.

(a) $\mathcal{K}$ is a pentagon with vertices in $\mathcal{S}$. Then each quadruple of points forms the vertices of a convex quadrilateral, and there are $\binom{5}{4} = 5$ of them.

(b) $\mathcal{K}$ is a quadrilateral with vertices A_1, A_2, A_3, $A_4 \in \mathcal{S}$, while the point $A_5 \in \mathcal{S}$ lies in the interior of $\mathcal{K} = A_1 A_2 A_3 A_4$ (Figure 10). Let P be the intersection of the diagonals of this quadrilateral. The point A_5 is an interior point of one of the four triangles $A_1 A_2 P, \ldots, A_4 A_1 P$, say $A_5 \in \triangle A_1 A_2 P$. Then there are exactly three of the desired quadrilaterals: $A_1 A_2 A_3 A_4$, $A_1 A_5 A_3 A_4$, and $A_2 A_3 A_4 A_5$.

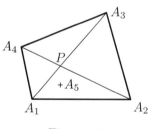

Figure 10

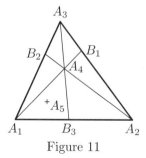

Figure 11

(c) $\mathcal{K}$ is a triangle with vertices $A_1, A_2, A_3 \in \mathcal{S}$, while $A_4, A_5 \in \mathcal{S}$ are interior points of $\mathcal{K}$ (Figure 11). The straight lines $A_4 A_i$ $(i = 1, 2, 3)$ cut the opposite sides of $\triangle A_1 A_2 A_3$ in the points B_i and divide this triangle into six triangles, with the point A_5 lying in the interior of one of them. Without loss of generality we may assume that $A_5 \in \triangle A_1 B_3 A_4$; then among the given points, A_1, A_2, A_3, A_4 (in this order) form vertices of the unique convex quadrilateral. $\qquad\square$

1.9 Exercises

(i) We are given n different points in the interior of each side of a square. Find the number of all quadrilaterals whose vertices are among these points.

(ii) Find the number of all squares whose vertices are among those of a regular n-gon.

In (iii) and (iv) below assume that on a piece of paper there are $(n+1)$ horizontal and $(n+1)$ vertical lines that form a system of n^2 unit squares.

(iii) Find the number of all rectangles whose sides are on the lines.

(iv) Find the number of all squares whose sides are on the lines.

(v) Let n different points be chosen on a circle. Find the number of all convex polygons whose vertices are among these points.

(vi) Suppose that $n \geq 4$ different points on a circle are given, among them the point A. Let m_1, respectively m_2, denote the number of all convex polygons with vertices among the given points that do not have (respectively do have) the vertex A. Show that $m_1 > m_2$, and determine $m_1 - m_2$.

(vii) Let a square $A_1 A_2 A_3 A_4$ be given, and inside this square a quadrilateral $A_5 A_6 A_7 A_8$ with an interior point A_9, such that no three of these nine points lie on a straight line. Show that five of these points can be chosen such that they form a convex pentagon.

(viii) Suppose that $n \geq 5$ points in a plane are given such that no three lie on a straight line. Show that there are at least N convex quadrilaterals whose vertices are among these points, where $N = \frac{1}{5} \cdot \binom{n}{4}$.

*(ix) Let $\mathcal{V}$ denote the set of all vertices of a convex n-gon inscribed in a circle l. Find the number of all convex k-gons with vertices in $\mathcal{V}$ with the following property for a given s: The vertices of the k-gon divide the circle l into k arcs such that each of them contains at least s vertices from $\mathcal{V}$, where the endpoints of the arcs are not counted.

1.10 Problems on Systems of Points and Lines

In this subsection we will study situations that are closely related to *configurations* of finite sets of *points* and *lines* in the plane. We introduce three problems that mainly require the *method of extremes* for their solution.

(i) Let $\mathcal{S}$ be a finite set of points in a plane not all of which lie on one line. Show that there is a line on which exactly two points of $\mathcal{S}$ lie. (This problem is known in the literature as *Sylvester's problem*.)

SOLUTION.   Let $\mathcal{L}$ be the (finite) set of all lines going through at least two points of S. From among all the distances of the points in S to the lines in $\mathcal{L}$ we choose the *smallest positive* one; let it be the distance from the point $A \in S$ to the line $l \in \mathcal{L}$. The orthogonal projection B of the point A onto l determines on this line two half-lines with starting point B; we show that on each of them there is at most one point of S. Indeed, assume to the contrary that one of them contains two different points $C_1, C_2 \in S$ such that $0 \leq |BC_1| < |BC_2|$ (Figure 12). Then the triangle AC_1C_2 is obtuse (or a right triangle, which happens when $C_1 = B$) with the largest angle at C_1. Therefore, $|C_1C_2| < |AC_2|$, and by expressing the area of $\triangle AC_1C_2$ in two different ways we derive that the distance between the point C_1 and the line AC_2 is less than the distance between A and l, which is a contradiction. This means that there are exactly two points from S on the line l. $\square$

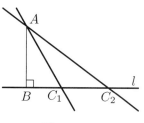

Figure 12

(ii) A finite set S of $n \geq 3$ different noncollinear points in a plane is given. Show that among the connecting lines of pairs of points in S there are at least n distinct lines.

SOLUTION.   We use induction on n. For $n = 3$ the assertion is clear. We assume now that it holds also for $k = n - 1$, and try to prove it for $k = n$. By Sylvester's problem (i), among the connecting lines of the points in S there is one that contains exactly two points $A, B \in S$. Let us choose, for instance, the point A; then we set $S' = S \setminus \{A\}$ and distinguish two cases: If all points of the $(n-1)$-element set S' lie on one line l, then we obtain a system of n distinct lines $\{AX : X \in S'\} \cup \{l\}$; in the opposite case there are, by the induction hypothesis, at least $(n-1)$ different lines among the connecting lines in S', and none of them is identical to the line $l = AB$, since $l \cap S' = \{B\}$. $\square$

(iii) In a plane we are given four points that do not all lie on a line. Show that one can choose three of them as vertices of a triangle, one of whose angles does not exceed $45°$. Show further that there is a configuration of four such points in which none of the triangles with the given points as vertices has even one angle less than $45°$.

SOLUTION.   We begin by proving the first assertion. To do this, we distinguish two cases according to the shape of the convex hull of the given quadruple of points. If the convex hull is the triangle $A_1A_2A_3$, then the point A_4 lies inside the triangle (Figure 13a). Then the line segments A_iA_4 $(i = 1, 2, 3)$ divide the interior angles of $\triangle A_1A_2A_3$ into six angles, at least one of which is not greater than $\frac{180°}{6} = 30° < 45°$. If, on the other hand, the convex hull is the convex quadrilateral $A_1 A_2 A_3 A_4$ (Figure 13b), then its interior angles are divided by its two diagonals into eight angles, and at least one of them is not greater than $\frac{360°}{8} = 45°$. This proves the assertion. An example of an appropriate configuration for the second part of the problem is a quadruple of points that form the vertices of an arbitrary square.   □

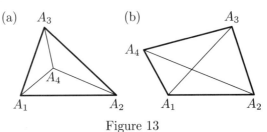

Figure 13

1.11   Exercises

In each of the exercises below, the points or lines all lie in a given plane.

(i) We are given a set $\mathcal{L}$ of $n \geq 3$ pairwise nonparallel lines that do not all go through one point. Show that there are two lines from $\mathcal{L}$ whose intersection is not traversed by another line from this set.

(ii) Decide whether Sylvester's theorem of 1.10.(i) holds also in the case where the set $\mathcal{S}$ is *countable*.

(iii) Suppose that from each quadruple of n given points ($n > 4$) we can choose one point such that the remaining three are collinear. Show that we can then choose one point from the given ones such that the remaining $(n-1)$ points are collinear.

(iv) Given four points that do not all lie on a line, show that one can choose three of them forming the vertices of a right or an obtuse triangle. Show further that there is a configuration of such four points where none of the triangles formed with the given points as vertices is obtuse. (Compare with 1.10.(iii).)

(v) We are given n lines, no two of which are parallel. Show that two of them can be found such that the angle between them is no greater than $\frac{180°}{n}$.

(vi) A country has 100 airports such that the distances between any two are distinct. From each one a plane takes off and then lands at the closest airport. Show that there is no airport at which more than five planes will land.

(vii) Decide whether there is a finite (at least two-element) point set $\mathcal{S}$ with the following property: For any two points $A, B \in \mathcal{S}$ ($A \neq B$) there are points $C, D \in \mathcal{S}$ ($C \neq D$) such that the lines AB and CD

are distinct and parallel. In the affirmative case find the smallest possible number of elements of the set S.

1.12 Problems on Systems of Line Segments

The subject of our investigations in this subsection will be certain properties of finite systems of segments on a line, of chords of a given circle, and of line segments with a given set of endpoints.

(i) We are given 50 segments on a line. Show that at least one of the following assertions holds:
(a) There exist 8 given segments with a common point.
(b) There exist 8 given segments of which any two are disjoint.

SOLUTION.   We denote the given segments by $I_1, I_2, \ldots, I_{50}$ such that their numbering agrees with the ordering of their left endpoints on a (horizontal) line from left to right. We assume that the first assertion does not hold, that is, that no 8 segments have a point in common. We first consider the segments $I_1, I_2, \ldots, I_8$. Among the segments $I_1, I_2, \ldots, I_7$ there is one, which we denote by U_1, that has no point in common with I_8 (otherwise, the left endpoint of I_8 would lie in each of the segments $I_1, \ldots, I_7$, which according to the assumption is not possible). The segment U_1, however, is also disjoint from all of the segments I_j, where $j > 8$. With a similar argument for the group of segments $I_8, I_9, \ldots, I_{14}$ we can prove the existence of the segment $I_k = U_2$ ($8 \leq k \leq 14$), which is disjoint from each of the segments I_j, where $j \geq 15$. Repeating this construction five more times, we obtain eight disjoint segments $U_1, U_2, \ldots, U_7, I_{50}$.   □

(ii) Suppose that on a line we are given $n \geq 3$ segments with the property that if we choose any three of the segments, then at least two of them have a point in common. Show that one can choose two points on the line such that each of the given segments contains at least one of the two chosen points.

FIRST SOLUTION.   We will show that the "rightmost left endpoint" and the "leftmost right endpoint" are two points with the required property. To make it easier to explain the following arguments, we identify the given line with the number line, as well as all segments I_k with intervals $[\alpha_k, \beta_k]$, where $\alpha_k < \beta_k$ ($k = 1, 2, \ldots, n$). We set $\alpha = \max \alpha_k$ (then α is that "rightmost left endpoint"), and similarly, $\beta = \min \beta_k$. We now assume that some segment does not contain either of these points, that is, that there is an index $i \in \{1, 2, \ldots, n\}$ such that $\alpha \notin [\alpha_i, \beta_i]$ and $\beta \notin [\alpha_i, \beta_i]$. Since $\alpha \geq \alpha_i$, we have $\alpha > \beta_i$, and similarly, $\beta < \alpha_i$. Now we find the indices p, q such that $\alpha = \alpha_p$ and $\beta = \beta_q$. Then we have $\alpha_q < \beta_q = \beta < \alpha_i < \beta_i < \alpha = \alpha_p < \beta_p$, that is, I_q, I_i, and I_p are three disjoint segments, which is impossible.   □

SECOND SOLUTION.   Here we prove that we can also take the points A, B as our required points, where A is again the "leftmost right endpoint" of the segments, but B is the "leftmost of the right endpoints" of those segments (if any) that lie to the right of A. We use the same notation as above, including $\beta = \min \beta_k$; then all the segments I_k, $k = 1, 2, \ldots, n$, can be divided into two sets M_1, M_2: $M_1 = \{I_i = [\alpha_i, \beta_i] : \alpha_i \leq \beta\}$, $M_2 = \{I_i = [\alpha_i, \beta_i] : \alpha_i > \beta\}$. It is easy to verify that all segments in M_1 contain the common point A with coordinate β. If $M_2 = \emptyset$, then the proof is complete; the point A is then common to *all* segments, and the second point can be chosen completely arbitrarily. Therefore, we assume now that $M_2 \neq \emptyset$. We will show that all segments in M_2 have the point B with coordinate $\gamma = \min\{\beta_j : [\alpha_j, \beta_j] \in M_2\}$ in common. Indeed, if this were not the case, then there would exist a segment $I_k = [\alpha_k, \beta_k]$ with the property $\beta < \gamma < \alpha_k < \beta_k$. If we take the indices p, q such that $\beta = \beta_p$, $I_q \in M_2$, and $\gamma = \beta_q$, then the segments $I_p \in M_1$, I_q, and I_k would be a triple of pairwise disjoint segments, which is not possible.   □

(iii) Suppose that several chords are drawn in a circle l of radius 1. Show that if every diameter of l crosses at most k chords, then the sum of the lengths of all chords is less than $k\pi$.

SOLUTION.   Let us assume that the sum s of the lengths of all chords satisfies $s \geq k\pi$; we will show that then some diameter crosses at least $(k+1)$ chords. Since the length of a circular arc belonging to a given chord is greater than the length of this chord, the sum of the lengths of the arcs belonging to all chords is greater than $k\pi$. To each arc o we associate an arc o' symmetric to o about the center S of the circle l (Figure 14). The sum of all lengths of this system of both the original and the associated arcs is greater than $2k\pi$, so some point X of the circle l lies on at least $(k+1)$ of the arcs in question. The diameter going through the point X then crosses at least $(k+1)$ chords.   □

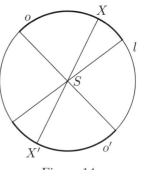

Figure 14

(iv) A point set $\mathcal{S} = \{X_1, X_2, \ldots, X_n\}$, where $n \geq 3$, is given in the plane and is such that the distances between any two elements are distinct. For each $i = 1, 2, \ldots, n$ we draw a segment U_i connecting the point X_i with the closest point $X_j \in \mathcal{S} \setminus \{X_i\}$. Decide whether some of these segments U_i can form a closed broken line.

SOLUTION.   We assume that a closed broken line does occur, and choose its *longest* segment $X_p X_q$, which is neighboring on the segments $X_p X_r$

and $X_q X_s$. Then $|X_p X_r| < |X_p X_q|$; that is, X_q is not the closest point to X_p. Similarly, $|X_q X_s| < |X_q X_p|$; then X_p is not closest to X_q. Hence the segment $X_p X_q$ cannot be drawn, and this is a contradiction.   □

1.13 Exercises

(i) We are given $(n^2 + 1)$ segments on a line. Show that at least one of the following two assertions holds:

(a) There are $(n + 1)$ of the given segments that have a point in common.

(b) There are $(n + 1)$ of the given segments such that no two of them have a point in common.

(ii) We are given $n \geq 3$ segments on a line, where any two of them have a common point. Show that all n segments have a point in common. (This is a one-dimensional variant of Helly's theorem, which was introduced above, in 1.1.(iii).)

(iii) On a circle k we are given some arcs of size less than $\alpha = 180°$ such that any three of them have a point in common. Show that then all the given arcs have a common point. Does this assertion also hold for some $\alpha > 180°$?

(iv) On a circle k we are given some arcs of size less than $120°$ such that any two of them have a point in common. Show that then all the given arcs have a common point.

*(v) On a circle k we are given a finite system of arcs, any two of which have a point in common. Show that one can choose two *opposite* points on k (that is, endpoints of some diameter) such that on each arc of the system there is at least one of the two chosen points.

*(vi) Suppose that in a plane there are several arbitrarily chosen (convex or nonconvex) polygons, any two of which have a point in common.

(a) Show that through every point A of the plane one can draw a straight line that has a point in common with each one of the given polygons.

(b) Show that for every point A of the plane there is a circle with center A that has a point in common with each one of the given polygons.

(vii) On a line we are given $m \geq 3$ segments that satisfy for some n $(1 < n \leq m)$ the following condition: From any n segments one can choose at least two that have a point in common. Show that there are $(n - 1)$ points on the line such that each segment contains at least one of these points.

(viii) Decide whether it is possible to place a finite number of line segments in a plane such that each of their endpoints is an interior point of another segment.

(ix) Suppose that we are given the set $\mathcal{S} = \{A_1, A_2, \ldots, A_n, B_1, B_2, \ldots, B_n\}$ of $2n$ different points in a plane, no three of which lie on the same line. Show that it is possible to construct n segments A_iB_j ($i, j \in \{1, 2, \ldots, n\}$) such that every point from $\mathcal{S}$ is the endpoint of exactly one segment, and such that no two of these segments intersect each other.

*(x) Find the largest N for which the following assertion holds: If $\mathcal{S}$ is any 21-element set of points on a circle c, then there exist N arcs of c whose endpoints lie in $\mathcal{S}$ and whose sizes do not exceed $120°$.

1.14 Problems on Nonconvex Polygons

In concluding this section, we will deal with several interesting properties of nonconvex polygons. In textbooks such polygons are usually mentioned only in passing, with illustrations of some possible shapes (three examples can be seen in Figure 15, even though they spell the Czech word for "darkness"; if the letter A had its usual hole, it would no longer be a polygon).

Figure 15

We therefore begin by defining more precisely the term of a plane polygon, doing this in a way that covers both convex and nonconvex polygons. For this purpose we start with the concept of a *closed broken line*, assuming from the beginning that the broken line does not intersect itself.

The notion of a cyclic sequence was explained in the introduction of Section 2 in Chapter 2; we recall that in a cyclic sequence of length n we consider as neighbors the beginning and the end terms, and the indices of the terms outside of the usual index set $\{1, 2, \ldots, n\}$ are taken modulo n.

We say that a cyclic sequence $A_1, A_2, \ldots, A_n$ of points in a plane forms the vertices of closed broken line $\mathcal{L}$ (Figure 16) if no three neighboring terms (or points) lie on the same straight line, and if in the cyclic sequence of segments

$$A_1A_2, A_2A_3, \ldots, A_{n-1}A_n, A_nA_1$$

only neighboring terms (or segments) have a point in common; this common point can obviously be only an endpoint. A closed broken line $\mathcal{L}$ can then be understood as the union of these n segments, and we write $\mathcal{L} = A_1A_2 \ldots A_nA_1$.

Without proof we now give the statement of Jordan's theorem:

Every closed broken line $\mathcal{L} = A_1A_2 \ldots A_nA_1$ divides the plane into two connected parts, exactly one of which is bounded (that is, it lies inside some circular disk).

This bounded part of the plane, together with the points on the closed broken line $\mathcal{L}$, will be called a *polygon* (or more exactly an n-gon) $\mathcal{M}$, which we will denote by $\mathcal{M} = A_1A_2 \ldots A_n$. The broken line $\mathcal{L}$ will be called the *boundary* of the polygon $\mathcal{M}$, the individual segments A_iA_{i+1} its *sides*, and the points A_i its *vertices*; finally, those points of $\mathcal{M}$ that do not lie on the boundary will be called *interior points* of $\mathcal{M}$. If the vertices A_i and A_j are not neighbors, the segment A_iA_j is called a *diagonal* of the polygon $\mathcal{M}$. (Note that some diagonals may contain points that do not belong to $\mathcal{M}$. In Figure 16 there are 5 such diagonals in total.) According to whether or not the set $\mathcal{M}$ is convex, we talk about a *convex*, respectively *nonconvex*, polygon $\mathcal{M}$.

We will now explain more exactly which one of the two angles $A_{i-1}A_iA_{i+1}$ at the vertex A_i should be called the interior angle of the polygon $\mathcal{M} = A_1A_2\ldots A_n$. To do this, we draw a circle k_i around the point A_i with a radius small enough so that it does not cut the broken line $A_{i+1}A_{i+2}\ldots A_nA_1\ldots A_{i-1}$, that is, such that the circle has only the two points $X_i \in A_{i-1}A_i$ and $Y_i \in A_iA_{i+1}$ in common with the boundary of $\mathcal{M}$ (Figure 17). The points X_i and Y_i divide the circle k_i into two arcs, both of which are usually drawn to mark both the angles $A_{i-1}A_iA_{i+1}$. Exactly one of these arcs lies in $\mathcal{M}$; this one then corresponds to the angle that we will call the *interior angle* of $\mathcal{M}$ at the vertex A_i. If $\mathcal{M}$ is a convex set, then it must contain the whole segment X_iY_i. From this it follows immediately that *all interior angles of a convex polygon are less than* $180°$; *that is, they are convex.* As we will see in Problem (ii), the nonconvex polygons are exactly those that have at least one nonconvex interior angle, that is, one greater than $180°$.

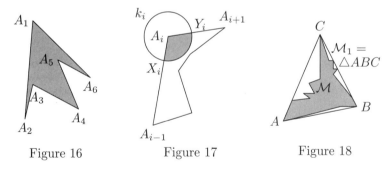

Figure 16   Figure 17   Figure 18

(i) Show that at least three interior angles of every nonconvex polygon are convex.

SOLUTION.   Given a nonconvex n-gon $\mathcal{M}$ we construct its convex hull (Figure 18). This is some convex m-gon $\mathcal{M}_1$, $3 \leq m < n$, where each vertex of $\mathcal{M}_1$ is a certain vertex of this original polygon $\mathcal{M}$. The interior angle of $\mathcal{M}$ at each such vertex is a (convex!) subset of the interior angle of $\mathcal{M}_1$ at the same vertex (since $\mathcal{M} \subseteq \mathcal{M}_1$), and so it is also convex. Hence at least m interior angles of $\mathcal{M}$ are convex, and the proof is complete.   □

(ii) Show that each nonconvex polygon has at least one nonconvex interior angle.

SOLUTION. In the given nonconvex polygon $\mathcal{M}$ we choose two points X and Y such that some points of the segment XY do not lie in $\mathcal{M}$. If necessary, we shift the points X, Y toward each other such that they are the *only* two points of the segment XY that lie in $\mathcal{M}$. Then X and Y necessarily lie on the boundary of $\mathcal{M}$, which they divide into two (nonclosed) broken lines $\mathcal{L}_1 = XA_1A_2\ldots A_mY$ and $\mathcal{L}_2 = XB_1B_2\ldots B_nY$ (Figure 19). The segment XY and the broken lines $\mathcal{L}_1$, $\mathcal{L}_2$ are three lines connecting the points X and Y, and no two of the lines have any other points in common. Hence one of these three lines must lie in the polygon bounded by the other two; however, by the choice of the points X, Y this cannot be the segment XY. Hence let, for instance, the broken line $\mathcal{L}_1$ divide the polygon $\mathcal{M}_2 = XB_1B_2\ldots B_nY$ into two parts: the original polygon $\mathcal{M}$, and the polygon $\mathcal{M}_1 = XA_1A_2\ldots A_mY$. The sum of the interior angles at each vertex A_i of $\mathcal{M}$ and $\mathcal{M}_1$ is $360°$. By Problem (i), however, the polygon $\mathcal{M}_1$ (whether or not it is convex) has at least three convex angles; at least one of them is at a vertex different from X and Y, and thus at some vertex A_i. At this vertex the interior angle of $\mathcal{M}$ is then nonconvex. □

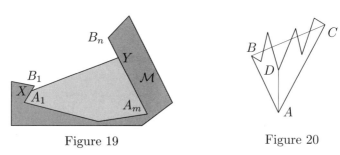

Figure 19   Figure 20

(iii) Show that any nonconvex polygon has a diagonal that lies completely inside the polygon, and a diagonal that contains points outside the polygon.

SOLUTION. We first explain why on some diagonal of an arbitrary nonconvex polygon $\mathcal{M}$ there are points that do not belong to $\mathcal{M}$. It clearly suffices to find points on the boundary of its convex hull $\mathcal{M}_1$ and not belonging to $\mathcal{M}$, since the boundary of $\mathcal{M}_1$ is a broken line consisting of segments, each of which is either a side or a diagonal of $\mathcal{M}$ or a union of such sides and diagonals (see the segment BC in Figure 18). If we assume that this closed line is entirely in $\mathcal{M}$, then the entire polygon defined by it also lies there, and so does the polygon $\mathcal{M}_1$. But then we would have $\mathcal{M}_1 = \mathcal{M}$, so $\mathcal{M}$ would be a convex set.

We now approach the solution of the given problem itself by using the result of Problem (i) to choose a vertex A at which the interior angle of the polygon $\mathcal{M}$ is convex. Let B and C be the vertices that are neighbors of A. If A, B, C are the only vertices of $\mathcal{M}$ that lie inside $\triangle ABC$, then BC is the desired diagonal that lies entirely in $\mathcal{M}$. If this triangle contains also other vertices of $\mathcal{M}$ (Figure 20), we choose and denote by D the one that has the maximal distance from the line BC (this distance may be zero; if there is more than one candidate for D, choose an arbitrary one among them). If the segment AD had a point other than A and D in common with some side EF of the polygon $\mathcal{M}$, then at least one of the vertices E, F would have to lie in $\triangle ABC$ and have a greater distance from the line BC than D. This contradiction means that the points A and D are the only points of the boundary of $\mathcal{M}$ that lie on the segment AD. This segment is therefore the desired diagonal.   □

(iv) Find the smallest possible number of diagonals of an n-gon $\mathcal{M}$ that lie completely inside $\mathcal{M}$.

SOLUTION.   If the n-gon $\mathcal{M}$ is convex, then each of its u_n diagonals has the desired property, where according to (1) we have $u_n = \frac{n(n-3)}{2} > n-3$ for $n \geq 4$. Figure 21 illustrates the construction of a nonconvex n-gon with exactly $(n-3)$ diagonals with the desired property. We now prove by induction that in each n-gon $\mathcal{M}$ there are at least $(n-3)$ such "interior" diagonals. For $n = 4$ the assertion is clear; we now assume that it holds for all k-gons with $k < n$, and prove it also for $k = n$. We divide a nonconvex n-gon with one of the "interior" diagonals into two polygons, namely, a $(k+1)$-gon and an $(n-k+1)$-gon $(2 \leq k \leq n-2)$. By the induction hypothesis they have at least $(k-2)$ and $(n-k-2)$ "interior" diagonals. Therefore, the number of such diagonals of $\mathcal{M}$ is no less than $1 + (k-2) + (n-k-2) = n-3$.

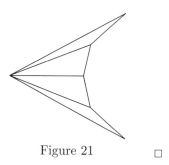

Figure 21   □

1.15 *Exercises*

In the exercises below let $\mathcal{M}$ be an arbitrary nonconvex n-gon $(n \geq 4)$.

    (i) Show that $\mathcal{M}$ can be divided into triangles with vertices among those of $\mathcal{M}$.

  (ii) Show that the sum of all interior angles of $\mathcal{M}$ is $(n-2) \cdot 180°$.

 (iii) Find the largest possible number of acute interior angles of $\mathcal{M}$.

 (iv) Decide for which $n \geq 4$ there is a polygon $\mathcal{M}$ such that some straight line cuts each of its sides in an interior point.

2 Systems of Curves and Regions

In this section we will study problems related to the following situation: In a plane we are given a geometrical object and a system of curves that divide it into several parts of different "shapes." We will be interested in the question concerning their number, and other combinatorial problems. Thus, in Figure 22, the whole plane is divided into eight parts—so-called regions—by two straight lines and one circle. We point out that we are not concerned with the decomposition of the plane in the usual set-theoretic sense. We agree on the convention that the *boundary points* of the individual parts (namely, the points of the dividing curves) remain "undistributed"; that is, they belong to none of the regions formed. The reason for this is that if all the regions were to include their boundary points, then the parts of the subdivision would no longer be disjoint. Even though the terms *regions* and their *boundaries* are intuitively clear, we will describe them more accurately, in order to avoid possible misunderstandings and to show how these intuitive ideas can be put on a firm mathematical foundation.

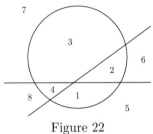

Figure 22

A point X is called an *interior point* of a (plane) point set $\mathcal{A}$ if $\mathcal{A}$ contains some circular disk with center X (and positive radius); the set $\mathcal{A}$ is called *open* if each point $X \in \mathcal{A}$ is an interior point of $\mathcal{A}$. We say that X is a *boundary point* of $\mathcal{A}$ if X is neither an interior point of $\mathcal{A}$ nor an interior point of its complement (in the plane); the set of all boundary points of the given set $\mathcal{A}$ form its *boundary*. If for any two points $X, Y \in \mathcal{A}$ there is a broken line $X Z_1 Z_2 \ldots Z_n Y$ lying in $\mathcal{A}$, we say that the set $\mathcal{A}$ is *connected*. Now we are in a position to define a *region* as any nonempty set of points that is both open and connected. Finally, we say that the region $\mathcal{A}$ is divided into the regions $\mathcal{D}_1, \mathcal{D}_2, \ldots, \mathcal{D}_n$ by the system of curves $\mathcal{L}$ if the following two conditions are satisfied:

(i) Each point of the region $\mathcal{A}$ that does not lie on any curve $l \in \mathcal{L}$, belongs to exactly one of the regions $\mathcal{D}_1, \mathcal{D}_2, \ldots, \mathcal{D}_n$.

(ii) Each point of any region $\mathcal{D}_i$ $(1 \le i \le n)$ is a point of the region $\mathcal{A}$ that does not lie on any of the curves $l \in \mathcal{L}$.

In some of the problems we will consider only *bounded* (respectively *unbounded*) regions; these are regions that lie inside some disk (respectively do not lie inside any disk) with finite radius. Thus, for example, in Figure 22 the bounded regions are exactly those numbered 1, 2, 3, 4.

Finally, we note that this introduction lacks a definition of the term *curve*; however, in our examples the only curves we will be dealing with are straight lines or their parts (half-lines, line segments) and circles or their arcs.

2.1 Dividing the Plane with a System of Lines

We assume that $\mathcal{A}$ is the set of all points in a plane, and we are given a finite set $\mathcal{L}$ of (distinct) straight lines in the plane. Each region of the plane defined by the lines in $\mathcal{L}$ is either *bounded*, in which case it is formed by the interior points of some *convex polygon* (for instance, the regions 5 and 8 in Figure 23), or *unbounded*. The range of "shapes" of the unbounded regions is more varied: Such a region may be the interior of some *angle* (regions 1, 3, 6, 12 in Figure 23) or the interior of an *unbounded "polygon"* with two "infinite sides," namely, half-lines (regions 2, 4, 7, 9, 10, and 13 in Figure 23). This figure, however, does not contain two other possible "shapes" of unbounded regions, namely, the interiors of a *half-plane* and of a *parallel strip*; these occur when the plane is divided by a line bundle of the first kind (see 1.4). The reader should show (by induction) that we have listed all possible "shapes" of regions.

We will now limit our attention to the number of such plane regions; it is clear that this number depends on the relative positions of the "dividing" lines.

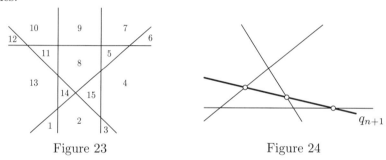

Figure 23   Figure 24

(i) Suppose that n lines are given in a plane. Find the greatest possible number $p(n)$ of regions into which the plane can be divided by these lines. Also, describe the possible distributions of the n lines when this number is attained.

SOLUTION.   Clearly, we have $p(1) = 2$; let us now assume that we know the largest possible number $p(n)$ of regions into which the plane can be divided by the n lines $q_1, q_2, \ldots, q_n$. By adding the line q_{n+1}, the number of parts of the plane increases by as much as the number of parts into which the line q_{n+1} is divided by intersections with the lines $q_1, q_2, \ldots, q_n$ (if no such intersection exists, then the number of parts of the plane increases by 1). There are at most n such intersections on the line q_{n+1}, so it is divided into at most $(n + 1)$ parts, two of which are half-lines, and the remaining ones line segments; see Figure 24. Hence we have the bound

$$p(n + 1) \leq p(n) + (n + 1).$$

To determine an explicit formula for $p(n)$ as a function of n, we use the following trick which will occur frequently in the remainder of this chapter but which we will not explain in detail (but see, for instance, [5], Chapter 1, Section 2.16). We write down the following system of $(n-1)$ inequalities and one equality and sum everything:

$$p(1) = 2,$$
$$p(2) \leq p(1) + 2,$$
$$p(3) \leq p(2) + 3,$$

$$\cdots$$

$$p(n-1) \leq p(n-2) + n - 1,$$
$$p(n) \leq p(n-1) + n.$$

From this we obtain

$$p(n) = \sum_{i=1}^{n} p(i) - \sum_{i=1}^{n-1} p(i)$$

$$\leq 1 + (1 + 2 + \cdots + n) = 1 + \frac{n(n+1)}{2} = \frac{n^2 + n + 2}{2}. \tag{5}$$

The values $p(n) = \frac{1}{2}(n^2 + n + 2)$ can be attained if and only if for each $i = 2, 3, \ldots, n$ the line q_i has exactly $(i-1)$ intersections with the preceding lines $q_1, q_2, \ldots, q_{i-1}$. This occurs exactly when in the given system $\mathcal{L}$ of n lines no two are parallel and no three pass through the same point. If both these conditions are satisfied, we say that the lines in $\mathcal{L}$ are in *general position*. In Exercise 2.2.(iii) you will be asked to show that in this case among the $p(n) = \frac{1}{2}(n^2 + n + 2)$ regions there are exactly $2n$ unbounded and $\frac{1}{2}(n^2 - 3n + 2)$ bounded regions.   □

(ii) Suppose that $n \geq 4$ lines in general position are given in a plane. Show that among all the bounded regions into which the plane is divided by these lines there are at least $\frac{2}{3}(n-1)$ triangles.

SOLUTION.   We denote the set of all the given lines by $\mathcal{L}$, and the set of their intersections by $\mathcal{S}$. First of all, we divide the set $\mathcal{L}$ into two disjoint sets $\mathcal{L}_1$ and $\mathcal{L}_2$: A line $p \in \mathcal{L}$ will belong to $\mathcal{L}_1$ if and only if all the points in $\mathcal{S}$ lie in one half-plane with boundary p; otherwise, $p \in \mathcal{L}_2$. We will show that $|\mathcal{L}_1| \leq 2$, that is, $|\mathcal{L}_2| \geq n-2$. Indeed, if we assume that $|\mathcal{L}_1| \geq 3$, then it can be seen that some three lines $p_1, p_2, p_3 \in \mathcal{L}_1$ describe a triangle $\mathcal{T} = \triangle ABC$ (with vertices denoted as in Figure 25) such that each point $X \in \mathcal{S}$ belongs to the triangle $\mathcal{T}$. But this is impossible, since

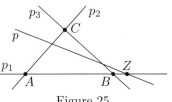

Figure 25

any line $p \in \mathcal{L} \setminus \{p_1, p_2, p_3\}$ cuts at most two of the segments AB, AC, BC, and therefore does not cut one of them, say AB. But then the intersection Z of the lines p and p_1 does not lie in $\triangle ABC$.

We now estimate the number of triangular regions: Each line $p \in \mathcal{L}_1$ determines at least one such region, where two vertices of the triangle lie on p, and the third one is the point $Y \in \mathcal{S}$ that has the *smallest distance* from the line p (compare with 1.6.(v)). Similarly, each line $p \in \mathcal{L}_2$ determines at least two triangular regions, namely, at least one in each of the two half-planes formed by this line. If we sum the numbers of triangular regions for all lines in $\mathcal{L}$, then each region in question is counted at most three times, and therefore their number N satisfies the bound

$$N \geq \frac{1}{3}[k + 2(n - k)],$$

where $k = |\mathcal{L}_1| \in \{0, 1, 2\}$, and thus $N \geq \frac{1}{3}(2n - 2)$.   □

(iii) Consider $n \geq 2$ lines in the plane, forming two bundles of the second kind with centers in the distinct points P and Q, where p (respectively q) lines go through P (respectively Q), with $p, q \geq 1$, $p + q = n$. Assume that none of the lines goes through the points P, Q, and that no two of the lines are parallel. For a fixed n determine those numbers p and q for which the number of regions into which the plane is divided by these n lines is maximal.

SOLUTION.   Let $N(p, q)$ denote the number of regions into which the plane is divided by the two line bundles. The lines of the bundle with center P divide the plane into $N(p, 0) = 2p$ regions (for details, see Exercise 2.2.(ii)). If one line goes through the point Q, then it is divided into $(p+1)$ parts by the lines of the first bundle, and therefore $N(p, 1) = N(p, 0) + p + 1$. Each further line from the bundle with center Q is divided into $(p+2)$ parts; hence for $q \geq 2$ we obtain the recurrence relation $N(p, q) = N(p, q - 1) + p + 2$. In a similar way as in (i) we find that

$$N(p, q) = pq + 2(p + q) - 1. \tag{6}$$

In view of the fact that for positive numbers p, q we have the inequality $pq \leq \left(\frac{p+q}{2}\right)^2$, with equality when $p = q$, we get

$$N(p, q) \leq \left(\frac{p + q}{2}\right)^2 + 2(p + q) - 1 = \frac{n^2}{4} + 2n - 1 = \frac{n^2 + 8n - 4}{4}.$$

Here we obtain equality in the case where n is even and $p = q = \frac{n}{2}$. For the case where n is odd, $n = 2k + 1$, the reader should show that the number of regions is maximal exactly when $\{p, q\} = \{k, k + 1\}$ (set, for instance, $p = \frac{n-1}{2} + r$, $q = \frac{n+1}{2} - r$) and

$$N(k, k + 1) = N(k + 1, k) = k^2 + 5k + 1 = \frac{n^2 + 8n - 5}{4}. \qquad □$$

2.2 Exercises

(i) Find the number of parts into which the plane is divided by a bundle of n lines of the first kind.

(ii) Solve the problem analogous to (i) for a bundle of the second kind.

In (iii) and (iv) assume that the plane is divided into regions by n lines in general position.

(iii) Find the numbers N_1 of bounded and N_2 of unbounded regions.

(iv) Show that if $n = 2k$, $k \geq 2$, then among the unbounded parts there are at most $(2k - 1)$ angular regions.

*(v) In the plane there are two nonempty line bundles $\mathcal{M}$ and $\mathcal{N}$ with m and n lines, respectively. Suppose that the plane is divided into 1987 regions by these lines. Determine all possible values of m and n in each of the following cases:

  (a) $\mathcal{M}$ and $\mathcal{N}$ are line bundles of the second kind with different centers A and B, no two of the lines are parallel, and none of them go through both points A and B.

  (b) $\mathcal{M}$ is a line bundle of the first kind and $\mathcal{N}$ is a bundle of the second kind with center B, where no line of $\mathcal{N}$ has the direction of the bundle $\mathcal{M}$, and the point B does not lie on any of the lines of $\mathcal{M}$.

*(vi) A triple of noncollinear points A, B, C is given in the plane, as well as a set $\mathcal{M}$ of 18 mutually nonparallel lines, each of which goes through exactly one of the points A, B, C. Show that the number of regions into which the plane is divided by the lines from the set $\mathcal{M}$ is at most 142.

2.3 Dividing the Plane with Closed Curves

The contents of this subsection are closely related to the material in 2.1 and 2.2; once again we are interested in the number of regions into which the plane is divided. However, instead of straight lines as "dividing lines" we will now use closed curves, namely, *circles* and *boundaries of* (convex) *polygons*. It is clear that in this connection questions about the number of unbounded regions have lost their meaning, since there is always only one unique such region.

(i) Find the largest possible number $k(n)$ of regions into which a plane can be divided by n circles.

SOLUTION.   Clearly, $k(1) = 2$; now assume as known the largest possible number $k(n)$ of regions into which the plane can be divided by n circles. Add an $(n + 1)$st circle; it has at most $2n$ intersections with the previous circles, and these intersections determine at most $2n$ arcs on the new circle, and each of these arcs divides one of the original regions into two new ones. Hence $k(n + 1) \leq k(n) + 2n$. From this we derive, in a similar way as in 2.1.(i), the estimate

$$k(n) \leq 2 + 2 + 4 + \cdots + 2(n - 1) = n^2 - n + 2. \tag{7}$$

In this inequality we get equality if there is a system of n circles such that any two cut each other in two points, and no three of them go through the same point. It is not difficult to construct such a configuration of circles: We choose an arbitrary line segment AB of length d and consider a system of n congruent circles with radius d and whose centers are different interior points of the segment AB. Any two circles of this system have exactly two different points in common, since the distance of their centers is less than d. If some three circles were to go through one point X, then their centers would form a triple of points on the same line segment and with the same distance from X, which is impossible. Hence, in summary, the desired value of $k(n)$ is equal to $n^2 - n + 2$ for every $n \geq 1$.   □

(ii) Find the largest number $N(n)$ of regions into which a plane can be divided by the boundaries of two convex n-gons.

SOLUTION.   Any region that occurs in the subdivision of the plane by two polygons belongs to one of the following three types:

(a) the "inner" region lying inside both polygons; there is at most one such region (it is the intersection of the interiors of the polygons, which are convex);

(b) the "outer" region lying outside both polygons; this is a unique unbounded region;

(c) regions that lie outside one and inside the other polygon; we will now estimate their number $N'(n)$.

Each part of type (c) is a polygon (with at least three sides) whose boundary is formed by line segments that are either some sides of the given polygons or parts of these sides. The reader should justify why each such segment forms the boundary of a *unique* region of type (c). Now it suffices to note that each side of one of the n-gons can be divided by the sides of the other polygon into at most three such segments; therefore, we have

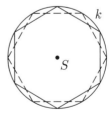

Figure 26

$$N'(n) \leq \frac{1}{3}(3n + 3n) = 2n,$$

and thus $N(n) \leq 1 + 1 + 2n = 2n + 2$. If we consider two regular n-gons with common center S, one of which is obtained from the other by rotation by an angle $\alpha = \frac{\pi}{n}$ about the center S (Figure 26 shows this for $n = 6$), then it is clear that the number of parts in the division can be $2n + 2$. Hence $N(n) = 2n + 2$. □

2.4   Exercises

  (i) Into how many regions can the plane be divided by a circle and the boundary of a square?

  (ii) Find the largest number $M(n)$ of regions into which the plane can be divided by a circle and the boundary of a convex n-gon.

*(iii) Find the largest possible number of regions into which the plane can be divided by a system of k circles and p straight lines.

2.5   Dividing a Convex Polygon

(i) For a given $n \geq 4$ find the greatest number k with the following property: Some convex n-gon $\mathcal{M}$ is divided by the system of all its diagonals into regions among which there is at least one k-gon.

SOLUTION.   We assume that one of the regions into which $\mathcal{M}$ is divided is a (convex) k-gon $\mathcal{P}$. Through each vertex of the polygon $\mathcal{M}$ at most two sides or diagonals that contain sides of $\mathcal{P}$ will pass. Each side of $\mathcal{P}$ is counted at two vertices of $\mathcal{M}$, and therefore we have $k \leq \frac{2n}{2} = n$. If n is odd, it is easy to see that the value $k = n$ can be attained: Consider a regular n-gon inscribed in a circle with center S. None of the diagonals go through S; this point lies in the region $\mathcal{P}$, which is a polygon with exactly n sides, since upon rotation about S by an angle of $\frac{2\pi}{n}$ the polygon $\mathcal{P}$ is mapped into itself (see Figure 27 for $n = 5$).

Now let n be even. We show by contradiction that in this case the value $n = k$ cannot be attained. Indeed, let us assume that one of the regions into which the n-gon $\mathcal{M}$ is divided by its diagonals is itself an n-gon $\mathcal{P}$. Then each vertex X of $\mathcal{M}$ is traversed by exactly two sides or diagonals XY, XZ, that contain sides of $\mathcal{P}$, and the vertices Y, Z are neighbors (consider the diagonal that connects X with the vertex that would otherwise lie between them).

Figure 27

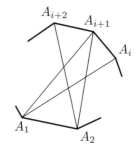

Figure 28

We set $\mathcal{M} = A_1 A_2 \ldots A_n$ and let two sides of $\mathcal{P}$ lie on $A_1 A_i$ and $A_1 A_{i+1}$ (see Figure 28). Since $\mathcal{P}$ lies in the half-plane $A_1 A_{i+1} A_i$, the second side or diagonal that originates in A_{i+1} and contains a side of $\mathcal{P}$ is $A_{i+1} A_2$; hence $\mathcal{P}$ lies in the half-plane $A_2 A_{i+1} A_1$. The second diagonal originating in A_2 and containing a side of $\mathcal{P}$ is then $A_2 A_{i+2}$. Repeating this argument, we obtain the following assertion: For any $k = 1, 2, \ldots, n$ the sides of $\mathcal{P}$ lie on the diagonals $A_{i+k-1} A_k$ and $A_k A_{i+k}$ (where we set $A_{n+j} = A_j$ for all j). But some sides of $\mathcal{P}$ lie on the diagonals $A_{i+1} A_1$ and $A_{i+1} A_2$. Comparing for $k = i + 1$ we obtain $A_1 = A_{2i}$, that is, $i = \frac{1}{2}$ or $i = \frac{n+1}{2}$, which is a contradiction.

In the case of even n, therefore, we cannot have more than an $(n-1)$-gon among the regions into which our n-gon is divided by its diagonals. It remains to show the existence of a suitable n-gon with an $(n-1)$-gon as one of its regions. In the case $n = 4$ such an example is given by a square; so let $n = 2m \geq 6$. Consider the regular $(2m-1)$-gon $A_1 A_2 \ldots A_{2m-1}$ inscribed in a circle with center S. This point S lies in the interior of the $(2m-1)$-gon $\mathcal{P}$. We now choose a point B in the interior of the segment $A_1 A_2$ and a point C in the interior of $A_1 A_{2m-1}$, both sufficiently close to A_1 such that the segments $A_m C$ and $A_{m+1} B$ intersect in the point X outside of $\mathcal{P}$ (see Figure 29). Then the $2m$-gon $\mathcal{M}$ with vertices $B, A_2, A_3, \ldots, A_{2m-1}, C$ (which is obtained from $A_1 A_2 \ldots A_{2m-1}$ by "cutting off" the triangle $A_1 B C$) has the desired property; we have only to consider the region containing the point S. □

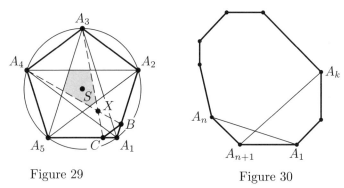

Figure 29   Figure 30

(ii) Suppose that no three diagonals of a convex n-gon $\mathcal{M}$ $(n \geq 4)$ meet in the same interior point. Find the number $f(n)$ of regions into which the polygon $\mathcal{M}$ is divided by its diagonals.

FIRST SOLUTION.   Assuming that a convex n-gon $\mathcal{M}_n$ is divided by its diagonals (no three of which meet inside $\mathcal{M}_n$) into $f(n)$ regions, we determine the number $f(n + 1)$ of regions into which an allowable $(n + 1)$-gon $\mathcal{M}_{n+1}$ is divided by its diagonals. If we do not consider the diagonals that go through the point A_{n+1}, then $\mathcal{M}_{n+1}$ is divided by the remaining diagonals into $f(n) + 1$ regions, since in addition to the $f(n)$ regions into which $A_1 A_2 \ldots A_n$ is divided, there is the triangle $A_1 A_n A_{n+1}$. Each diagonal $A_{n+1} A_k$ $(k = 2, 3, \ldots, n-1)$ adds to the original regions as many additional regions as there are segments into which it is divided by the other diagonals, and this number is clearly $(k-1) \cdot (n-k) + 1$ (see Figure 30; the diagonals in question are connections between the sets of $(k-1)$ and $(n-k)$ vertices). Hence we have the recurrence relation

$$f(n+1) = f(n) + 1 + \sum_{k=2}^{n-1} \big((k-1)(n-k) + 1\big)$$

$$= f(n) + n - 1 + \sum_{k=2}^{n-1} n \cdot (k-1) - \sum_{k=2}^{n-1} (k-1)k$$

$$= f(n) + n - 1 + \frac{n(n-1)(n-2)}{2} - \frac{n(n-1)(n-2)}{3}$$

$$= f(n) + (n-1) + \binom{n}{3},$$

where we have used some standard addition formulas (see, for instance, [5], Chapter 1, Section 2). To simplify further manipulations, it is convenient to set $f(3) = 1$; then $f(4) = 4 = f(3) + (3-1) + \binom{3}{3}$. Hence the expression

$$f(n) = (1 + 2 + \cdots + (n-2)) + \binom{3}{3} + \binom{4}{3} + \cdots + \binom{n-1}{3}$$

can be simplified to

$$f(n) = \frac{(n-1)(n-2)}{2} + \binom{n}{4} = \frac{(n-1)(n-2)(n^2 - 3n + 12)}{24}; \qquad (8)$$

see, for instance, equation (11) in [5], page 10. □

SECOND SOLUTION.   The diagonals divide the polygon $\mathcal{M}$ into regions, each of which is again a convex k-gon for some $k \leq n$. Let r_k (≥ 0) denote the number of regions that are k-gons $(k = 3, 4, \ldots, n)$. Then clearly,

$$f(n) = r_3 + r_4 + \cdots + r_n.$$

We now use two arguments: On the one hand, we determine the sum V of the number of vertices of all polygons into which $\mathcal{M}$ is divided by its diagonals, and on the other hand, we establish the sum U of all interior angles of these polygons. For V we clearly have

$$V = 3r_3 + 4r_4 + \cdots + nr_n,$$

but this number can also be expressed in a different way: Each intersection of diagonals is a vertex of 4 regions (according to (3) there are exactly $\binom{n}{4}$)

such intersections), and each vertex of the polygon $\mathcal{M}$ is a vertex of exactly $(n-2)$ triangular regions. This implies $V = 4 \cdot \binom{n}{4} + n(n-2)$, which means that

$$3r_3 + 4r_4 + \cdots + n \cdot r_n = 4 \cdot \binom{n}{4} + n(n-2). \tag{9}$$

Now we determine the sum U of interior angles of all regions. By 1.1.(ii), the sum of all interior angles of a convex k-gon satisfies $S_k = (k-2) \cdot 180°$, and therefore,

$$U = [(3-2) \cdot r_3 + (4-2) \cdot r_4 + \cdots + (n-2) \cdot r_n] \cdot 180°.$$

However, this sum U can also be determined in a different way: At each of the $\binom{n}{4}$ intersections of the diagonals, the interior angles of the regions that "meet" there form a full angle of $360°$, and the sum of the remaining interior angles is the same as the sum of the interior angles of the polygon $\mathcal{M}$ (see Figure 31). Hence we have

$$U = \binom{n}{4} \cdot 360° + (n-2) \cdot 180°.$$

By comparing both expressions for U we obtain

$$(3-2)r_3 + (4-2)r_4 + \cdots + (n-2)r_n = 2 \cdot \binom{n}{4} + (n-2), \tag{10}$$

and subtracting (10) from (9),

$$2(r_3 + r_4 + \cdots + r_n) = 2f(n) = 2 \cdot \binom{n}{4} + (n-1)(n-2);$$

finally, by expanding the binomial coefficient we once again obtain (8).   □

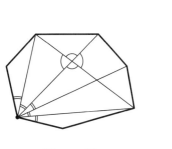

Figure 31

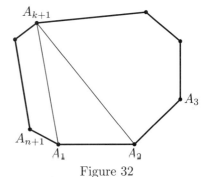

Figure 32

(iii) Suppose that a convex n-gon $\mathcal{M}$ ($n \geq 4$) is divided into triangles by some of its diagonals that do not cross each other inside $\mathcal{M}$.

(a) Show that the number of triangles $t(n)$ and the number $w(n)$ of diagonals used do not depend on the particular decomposition, and determine both numbers.

(b) Determine the number $T(n)$ of ways in which the desired decomposition can be carried out.

SOLUTION.   (a) Using a similar argument as in (ii) concerning the interior angles of the triangular regions, we derive the identity $180° \cdot t(n) = (n-2) \cdot 180°$, or $t(n) = n - 2$. We further note that each one of the n sides of the polygon $\mathcal{M}$ is a side of exactly one of the $t(n)$ triangles, and each of the $w(n)$ chosen diagonals is a side of exactly two of these triangles. Therefore, we have $n + 2 \cdot w(n) = 3 \cdot t(n) = 3 \cdot (n-2)$; that is, $w(n) = n - 3$.

(b) We first derive a recurrence relation for the number $T(n+1)$ of allowable decompositions of an $(n+1)$-gon $\mathcal{M}_{n+1} = A_1 A_2 \ldots A_{n+1}$ under the assumption that the number $T(k)$ of decompositions of a k-gon is already known for $k = 3, 4, \ldots, n$. We choose any two fixed neighboring vertices of the polygon $\mathcal{M}_{n+1}$ (without loss of generality we may take A_1, A_2). Then the following two cases for the decomposition of $\mathcal{M}_{n+1}$ can occur:

(1) The decomposition contains the triangle $A_1 A_2 A_3$ (respectively $A_1 A_2 A_{n+1}$); then the remaining n-gon $A_1 A_3 \ldots A_{n+1}$ (respectively $A_2 A_3 \ldots A_{n+1}$) can always be decomposed into triangles in $T(n)$ ways.

(2) The decomposition contains the triangle $A_1 A_2 A_{k+1}$ ($4 \le k+1 \le n$, that is, $3 \le k \le n-1$). The segments $A_1 A_{k+1}$ and $A_2 A_{k+1}$ divide $\mathcal{M}_{n+1}$ into the k-gon $A_2 A_3 \ldots A_{k+1}$ and the $(n-k+2)$-gon $A_1 A_{k+1} \ldots A_{n+1}$, see Figure 32, and by the induction hypothesis both of them can be subdivided into triangles in $T(k)$, respectively $T(n-k+2)$, ways. By the multiplication rule the number of decompositions in which $\triangle A_1 A_2 A_{k+1}$ occurs is equal to $T(k) \cdot T(n-k+2)$. For the number $T(n+1)$ of all decompositions of an $(n+1)$-gon into triangles we therefore obtain the recurrence relation

$$T(n+1) = 2 \cdot T(n) + \sum_{k=3}^{n-1} T(k) \cdot T(n-k+2), \tag{11}$$

which, together with the initial value $T(3) = 1$, allows for a recursive computation of all the numbers $T(n)$.

We now derive an explicit expression for $T(n)$ in terms of the number of sides n. To do this we determine, for the polygon $\mathcal{M}_n$, the number of decompositions that use a previously chosen diagonal, say $A_1 A_k$ ($k = 3, 4, \ldots, n-1$). This diagonal divides $\mathcal{M}_n$ into a k-gon and an $(n-k+2)$-gon, and therefore the number of such decompositions is $T(k) \cdot T(n-k+2)$. If we add these numbers over all diagonals of $\mathcal{M}_n$, we obtain the sum

$$S = \frac{n}{2} \cdot \sum_{k=3}^{n-1} T(k) \cdot T(n-k+2).$$

In this sum, however, each decomposition of $\mathcal{M}_n$ is counted as many times as the number of diagonals used. By (a) there are $(n-3)$ diagonals of $\mathcal{M}_n$

in each decomposition, so we have $S = (n-3) \cdot T(n)$, and thus

$$(n-3)T(n) = \frac{n}{2} \cdot \sum_{k=3}^{n-1} T(k) \cdot T(n-k+2). \tag{12}$$

By the formula (11) we have $\sum_{k=3}^{n-1} T(k) \cdot T(n-k+2) = T(n+1) - 2T(n)$, and by substituting into (12) we get

$$T(n+1) = \frac{2(2n-3)}{n}T(n). \tag{13}$$

Starting with $T(3) = 1$ and iterating the identity (13) we obtain

$$T(n+1) = 2^{n-1} \cdot \frac{(2n-3)(2n-5)\cdots 3 \cdot 1}{n!},$$

which implies the desired expression

$$\begin{aligned}
T(n) &= 2^{n-2} \cdot \frac{(2n-5) \cdot (2n-7) \cdots 3 \cdot 1}{(n-1)!} \\
&= 2^{n-2} \cdot \frac{(2n-4)!}{(n-2)! \cdot 2^{n-2} \cdot (n-1)!} \\
&= \frac{1}{n-2} \cdot \binom{2n-4}{n-3} = \frac{1}{2n-3} \cdot \binom{2n-3}{n-2}.
\end{aligned} \tag{14}$$

$\square$

(iv) Show that any convex 22-gon cannot be divided by some of its diagonals into 7 pentagons.

SOLUTION.   We will prove in two different ways the more general assertion that no $(3k+1)$-gon can be divided by diagonals into k pentagons, namely, by considering the sum of the interior angles of these polygons and by induction. We assume that such a decomposition of a $(3k+1)$-gon $\mathcal{M}_{3k+1}$ is possible. Since the conditions of the problem do not exclude the possibility that some diagonals that contain sides of the pentagons can cross at interior points of $\mathcal{M}_{3k+1}$, we assume that there are p such points ($p \geq 0$). Then from considering the sums of the interior angles of all pentagons it follows that $k \cdot 3 \cdot 180° = p \cdot 360° + (3k-1) \cdot 180°$, that is, $p = \frac{1}{2}$, which is impossible; hence the assertion holds.

Now, as promised, we present the second proof (by induction), where we will also use the pigeonhole principle. For $k = 1$ the assertion is clearly true; we now assume that it holds for some $k \geq 1$ and suppose that there is a convex $(3k+4)$-gon $\mathcal{M}_{3k+4}$ that can be divided by diagonals into $(k+1)$ pentagons. In view of the inequality $3(k+1) < 3k+4$ we see by the pigeonhole principle that some of these pentagons must have four sides that lie on the sides of the polygon $\mathcal{M}_{3k+4}$. If we separate this pentagon with its fifth side (a diagonal of $\mathcal{M}_{3k+4}$) from the polygon $\mathcal{M}_{3k+4}$, then we obtain a $(3k+1)$-gon that can be divided by diagonals into k pentagons; however, by the induction hypothesis this is not possible.   □

2.6   Exercises

(i) Suppose that n distinct points are chosen from the interior of the side AC of the triangle ABC, and another n distinct points from the interior of the side AB. Into how many regions is the triangle ABC divided by the $2n$ lines connecting the points B and C with the chosen points on the opposite sides?

(ii) Each one of the vertices A, B, C of the triangle ABC is connected by line segments with n distinct interior points on the opposite sides. Into how many regions is the triangle divided by these segments if no three of them have a common interior point?

(iii) Find the largest number of regions into which a circular disk $\mathcal{D}$ can be divided by chords connecting $n \geq 4$ distinct points chosen on the boundary circle of $\mathcal{D}$.

(iv) Suppose that a chocolate bar consists of $m \cdot n$ pieces. How many times do we have to break the chocolate in order to obtain all the individual pieces? (In each move we may break only one connected block of the bar.)

*(v) Find the largest number of triangles into which a given triangle can be divided by line segments with the following property: Each vertex of the triangles obtained is an endpoint of the same number of segments, and no segment meets another segment in its interior, where among the segments there are also the three sides of the given triangle.

(vi) Suppose that k distinct points are given in the interior of a square. Show that the square can be divided into $(2k+2)$ triangles with vertices in the given k points and the vertices of the square.

(vii) In the interior of a square we are given two points, each of which is connected by four line segments with the vertices of the square. Decide whether the square can be divided by these segments into 9 regions with equal areas.

(viii) Suppose that each of nine straight lines divides a given square into two quadrilaterals whose areas are in proportion $2 : 3$. Show that some three of these lines go through one point.

(ix) Show that any convex $(2k+1)$-gon cannot be divided into k quadrilaterals by some of its diagonals.

*(x) On an 8×8 chessboard the centers of all fields are marked. Decide whether it is possible to divide the chessboard by 13 lines (that do

not pass through any of the marked points) into parts such that inside each of them there is at most one of the marked points.

2.7 Dividing Space

In closing this section we generalize several considerations from 2.1 and 2.3 to the case of *three-dimensional* analogues of the problems that we solved there. In doing so we go beyond the restrictions set in the introduction to this chapter. We will consider divisions of space by various systems of several surfaces (planes or spherical surfaces) into three-dimensional regions whose exact definition is comparable with the definition of plane areas introduced at the beginning of this section. The only difference lies in the definition of an open set, where instead of a (plane) disk we have a (three-dimensional) ball.

(i) Find the greatest possible number of parts into which space can be divided by n planes.

SOLUTION.   We denote the desired greatest possible number by $r(n)$; then clearly $r(1) = 2$. We assume that we know the number $r(n - 1)$ that gives the greatest possible number of parts into which space can be divided by planes $\alpha_1, \alpha_2, \ldots, \alpha_{n-1}$ for some $n \geq 2$. A new plane α_n increases this number of parts by exactly as much as there are regions on α_n into which this plane is divided by the intersections with the planes $\alpha_1, \alpha_2, \ldots, \alpha_{n-1}$. The number of these is, by 2.1.(i), at most $p(n - 1)$, since in the plane α_n the intersections $\alpha_i \cap \alpha_n$ $(i = 1, 2, \ldots, n - 1)$ form a system of at most $(n - 1)$ lines. Hence we have the inequality $r(n) \leq r(n - 1) + p(n - 1)$. If we substitute here the expression for $p(n - 1)$ from (5) and use a similar trick as in 2.1.(i), then we derive the estimate

$$r(n) \leq \frac{1}{6} \left(n^3 + 5n + 6 \right), \tag{15}$$

where we have used appropriate summation formulas (see, for instance, equations (8) and (13) in [5], Chapter 1, Section 2). In the inequality (15) we have equality if there is a placement of the planes $\alpha_1, \alpha_2, \ldots, \alpha_n$, such that each α_k is divided by the planes $\alpha_1, \alpha_2, \ldots, \alpha_{k-1}$ into the greatest possible number, namely into $\frac{1}{2} \left(k^2 - k + 2 \right)$ parts for $k = 2, 3, \ldots, n$. This case can occur, and it is easy to see that it does so when the planes are in *general position*: No three are parallel to the same line, and no four go through the same point.   □

(ii) Determine the greatest possible number $K(n)$ of parts into which space can be divided by n spherical surfaces.

SOLUTION.   We proceed as in (i): The nth spherical surface adds to the already existing regions the same number of regions as there are (surface) regions into which this spherical surface itself is divided by the original $(n-1)$ surfaces. The last surface is divided into parts that are bounded by circles, and it is not difficult to justify that there are at most $k(n-1)$ of them, where $k(n-1) = n^2 - 3n + 4$, by the methods of 2.3.(i). Hence we have $K(n) \leq K(n-1) + k(n-1)$, from which in view of $K(1) = 2$ it follows that

$$K(n) \leq \frac{n}{3} \left(n^2 - 3n + 8 \right). \tag{16}$$

In (16) we can, in fact, have equality, since there is a distribution of n spherical surfaces such that any three of them intersect at two points, and no four go through the same point. To construct such a configuration, we choose n points $A_1, A_2, \ldots, A_n$ in a plane such that no three of them lie on a line and no four of them lie on a circle. The desired spherical surfaces share their centers in $A_1, A_2, \ldots, A_n$ and the same radius R, where R is greater than the circumradius of any triangle $A_i A_j A_k$, $1 \leq i < j < k \leq n$. We leave it to the reader to verify the two intersection conditions mentioned before.   □

2.8 *Exercises*

(i) Find the number of regions into which space can be divided by n different planes if all of them have a common point, and any three planes have no other point in common.

(ii) Solve the problem analogous to (i) for the case where all the planes contain the same straight line.

(iii) Find the number of regions into which space can be divided by the planes of the faces of a tetrahedron.

(iv) Solve the problem analogous to (iii) for the faces of a cube.

*(v) We are given $n \geq 5$ planes in general position in space (see 2.7.(i)). Show that among the bounded regions into which space is divided by the given planes there are at least $\frac{n-1}{2}$ tetrahedra (compare with 2.1.(ii)).

3   Coverings and Packings

The problems of this section belong to the most highly developed area of combinatorial geometry. This area involves the study of situations where a certain geometrical object contains some further objects in its interior. We are then interested in whether some of the "interior" objects must overlap, or whether it is possible to optimally distribute these objects such that they cover the entire original object, or perhaps the largest possible part of it. Such problems occur in a number of practical situations (for instance,

cutting material with the least possible amount of waste, an optimal placement of parts of machinery and installations, covering a given territory with
the smallest number of transmitters of a given range, etc.); their successful
solutions can have considerable economic consequences.

To make it easier to express ourselves concisely, we will now explain the
usage of a few words of everyday language (namely, *overlapping, covering,
packing, tiling*). But let us first agree that all geometrical objects will be
considered as *closed*, that is, containing all their boundary points (see the
introduction to Section 2).

We say that two objects $\mathcal{M}_1$ and $\mathcal{M}_2$ *overlap* if some interior point of
$\mathcal{M}_1$ is also an interior point of $\mathcal{M}_2$; otherwise we say that $\mathcal{M}_1$ and $\mathcal{M}_2$
are *nonoverlapping*. If for a system $\mathcal{S} = \{\mathcal{M}_1, \mathcal{M}_2, \ldots, \mathcal{M}_n\}$ of objects we
have $\mathcal{M}_0 \subseteq \bigcup_{i=1}^{n} \mathcal{M}_i$, then we say that the system $\mathcal{S}$ *covers* the object $\mathcal{M}_0$
(or $\mathcal{M}_0$ is *covered* by the system $\mathcal{S}$). If, in particular, $\mathcal{S} = \{\mathcal{M}_1\}$, we talk
more concisely about the covering of $\mathcal{M}_0$ by the object $\mathcal{M}_1$.

If any two objects $\mathcal{M}_i, \mathcal{M}_j$ ($i \neq j$) of a system $\mathcal{S} = \{\mathcal{M}_1, \mathcal{M}_2, \ldots, \mathcal{M}_n\}$
are nonoverlapping, and if furthermore we have $\bigcup_{i=1}^{n} \mathcal{M}_i \subseteq \mathcal{M}_0$, then we
say that the system $\mathcal{S}$ is *packed* in the object $\mathcal{M}_0$. If in the above inclusion
we even have equality, that is, if the object $\mathcal{M}_0$ in which the system $\mathcal{S}$ is
packed, is also covered by this system, then we say that the object $\mathcal{M}_0$ is
tiled by the system $\mathcal{S}$.

3.1 Overlapping Objects

In the problems of the first two parts of this section we will study systems
of objects that lie in some bounded part of the plane and that overlap due
to the sizes of their areas. We will be especially interested in how large an
intersection of two (or more) objects of such a system has to be. The area
of a plane geometrical object $\mathcal{M}$ will be denoted by $|\mathcal{M}|$. We begin with
the following simple problem.

(i) On a floor with an area of 3 m^2 we have placed two carpets $\mathcal{K}_1, \mathcal{K}_2$,
each with an area of 2 m^2. Show that the carpets overlap in an area of at
least 1 m^2.

SOLUTION.   From the diagram in Figure 33 it
follows that

$$|\mathcal{K}_1 \cap \mathcal{K}_2| = q = (p+q) + (q+r) - (p+q+r)$$
$$= |\mathcal{K}_1| + |\mathcal{K}_2| - |\mathcal{K}_1 \cup \mathcal{K}_2|,$$

and in view of $|\mathcal{K}_1| = |\mathcal{K}_2| = 2$ and $|\mathcal{K}_1 \cup \mathcal{K}_2| \leq 3$
we obtain the estimate

$$|\mathcal{K}_1 \cap \mathcal{K}_2| \geq 2 + 2 - 3 = 1. \qquad \square$$

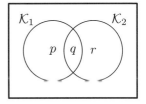

Figure 33

We will now generalize the formula derived with the help of Figure 33 to the case of $n \geq 2$ objects $\mathcal{M}_1, \mathcal{M}_2, \ldots, \mathcal{M}_n$. We are going to show that

$$|\mathcal{M}_1 \cup \mathcal{M}_2 \cup \cdots \cup \mathcal{M}_n| = \sum (-1)^{r+1} |\mathcal{M}_{j_1} \cap \mathcal{M}_{j_2} \cap \cdots \cap \mathcal{M}_{j_r}|, \quad (17)$$

where the sum on the right-hand side is taken over all nonempty subsets $\{j_1, j_2, \ldots, j_r\}$ of $\{1, 2, \ldots, n\}$. (Note the formal similarity between (17) and the identity (47) of Chapter 1, which expresses the inclusion–exclusion principle.)

To prove (17), we divide the object $\mathcal{M}_0 = \mathcal{M}_1 \cup \mathcal{M}_2 \cup \cdots \cup \mathcal{M}_n$ into 2^n disjoint parts of the type $\mathcal{X}_1 \cap \mathcal{X}_2 \cap \cdots \cap \mathcal{X}_n$, where $\mathcal{X}_i = \mathcal{M}_i$ or $\mathcal{X}_i = \mathcal{M}_0 \backslash \mathcal{M}_i$ for all $i = 1, \ldots, n$. (Some of these parts may have an area equal to zero, or may even be the empty set.) If we have $\mathcal{X}_i = \mathcal{M}_i$ for exactly $r \geq 1$ different values of the index $i \in \{1, 2, \ldots, n\}$, then the area of the part $\mathcal{X}_1 \cap \cdots \cap \mathcal{X}_n$ is counted exactly once both on the left-hand side of (17) (which is clear), and also on the right-hand side, since by the binomial theorem we have

$$\binom{r}{1} - \binom{r}{2} + \cdots + (-1)^{r+1} \cdot \binom{r}{r} = 1 - (1-1)^r = 1$$

for every integer $r \geq 1$. This completes the proof of (17).

(ii) Suppose that the polygon $\mathcal{M}_0$ with area 6 cm^2 covers three polygons $\mathcal{M}_1, \mathcal{M}_2, \mathcal{M}_3$ with the same area 3 cm^2. Show that two of these polygons $\mathcal{M}_1, \mathcal{M}_2, \mathcal{M}_3$ overlap in a region with an area of at least 1 cm^2.

SOLUTION.   By (17) with $n = 3$ we obtain (leaving out cm^2)

$$\begin{aligned}
6 &\geq |\mathcal{M}_1 \cup \mathcal{M}_2 \cup \mathcal{M}_3| \\
&= |\mathcal{M}_1| + |\mathcal{M}_2| + |\mathcal{M}_3| \\
&\quad - |\mathcal{M}_1 \cap \mathcal{M}_2| - |\mathcal{M}_1 \cap \mathcal{M}_3| - |\mathcal{M}_2 \cap \mathcal{M}_3| + |\mathcal{M}_1 \cap \mathcal{M}_2 \cap \mathcal{M}_3| \\
&\geq 3 + 3 + 3 - |\mathcal{M}_1 \cap \mathcal{M}_2| - |\mathcal{M}_1 \cap \mathcal{M}_3| - |\mathcal{M}_2 \cap \mathcal{M}_3|.
\end{aligned}$$

Therefore, the sum $|\mathcal{M}_1 \cap \mathcal{M}_2| + |\mathcal{M}_1 \cap \mathcal{M}_3| + |\mathcal{M}_2 \cap \mathcal{M}_3|$ is at least 3, and thus at least one of the three summands is at least 1. □

We note that the inequality obtained from (17) can be generalized in an interesting way. If $i \in \{1, 2, \ldots, n\}$ is odd, then

$$|\mathcal{M}_1 \cup \mathcal{M}_2 \cup \cdots \cup \mathcal{M}_n| \leq \sum_{r=1}^{i} (-1)^{r+1} |\mathcal{M}_{j_1} \cap \mathcal{M}_{j_2} \cap \cdots \cap \mathcal{M}_{j_r}|, \quad (18)$$

while for all even $i \in \{1, 2, \ldots, n\}$ we have

$$|\mathcal{M}_1 \cup \mathcal{M}_2 \cup \cdots \cup \mathcal{M}_n| \geq \sum_{r=1}^{i} (-1)^{r+1} |\mathcal{M}_{j_1} \cap \mathcal{M}_{j_2} \cap \cdots \cap \mathcal{M}_{j_r}|, \quad (19)$$

where on both right-hand sides of (18) and (19) the sums are taken over all nonempty subsets $\{j_1, j_2, \ldots, j_r\}$ of $\{1, 2, \ldots, n\}$ of at most i elements. (In the solution of (ii) we have used (19) for $n = 3$ and $i = 2$.) The proofs of (18) and (19) are left to the reader as Exercise 3.2.(i).

(iii) Suppose that the polygon $\mathcal{M}_0$ with an area of 10 cm^2 covers four polygons $\mathcal{M}_1, \mathcal{M}_2, \mathcal{M}_3, \mathcal{M}_4$ with identical areas of 6 cm^2. Show that the area of the common part of some three of the polygons $\mathcal{M}_1, \mathcal{M}_2, \mathcal{M}_3, \mathcal{M}_4$ is at least 1 cm^2.

SOLUTION.   It is not difficult to see that the desired assertion follows directly from the inequality

$$
\begin{aligned}
|\mathcal{M}_1 \cap \mathcal{M}_2 \cap \mathcal{M}_3| + |\mathcal{M}_1 \cap \mathcal{M}_2 \cap \mathcal{M}_4| & \\
+ |\mathcal{M}_1 \cap \mathcal{M}_3 \cap \mathcal{M}_4| + |\mathcal{M}_2 \cap \mathcal{M}_3 \cap \mathcal{M}_4| & \quad (20) \\
\geq |\mathcal{M}_1| + |\mathcal{M}_2| + |\mathcal{M}_3| + |\mathcal{M}_4| - 2 \cdot |\mathcal{M}_1 \cup \mathcal{M}_2 \cup \mathcal{M}_3 \cup \mathcal{M}_4|. &
\end{aligned}
$$

This inequality (20) (which may surprise you somewhat) can be verified by using a diagram with four sets; however, we prefer an approach used above in the proof of (17). We will directly prove the more general inequality

$$
\sum_{\{i,j,k\}} |\mathcal{M}_i \cap \mathcal{M}_j \cap \mathcal{M}_k| \geq \sum_{i=1}^{n} |\mathcal{M}_i| - 2 \cdot \left| \bigcup_{i=1}^{n} \mathcal{M}_i \right|, \quad (21)
$$

where on the left-hand side the sum is taken over all three-element subsets $\{i, j, k\}$ of $\{1, 2, \ldots, n\}$. As we did in the proof of (17), we divide the set $\mathcal{M}_0 = \mathcal{M}_1 \cup \mathcal{M}_2 \cup \cdots \cup \mathcal{M}_n$ into 2^n disjoint parts $\mathcal{X}_1 \cap \mathcal{X}_2 \cap \cdots \cap \mathcal{X}_n$, where $\mathcal{X}_i = \mathcal{M}_i$ or $\mathcal{X}_i = \mathcal{M}_0 \setminus \mathcal{M}_i$ $(1 \leq i \leq n)$. If we have $\mathcal{X}_i = \mathcal{M}_i$ for exactly r values of $i \in \{1, 2, \ldots, n\}$, where $r \geq 3$, then the area of the part $\mathcal{X}_1 \cap \mathcal{X}_2 \cap \cdots \cap \mathcal{X}_n$ is counted on the left-hand side of (21) exactly $\binom{r}{3}$ times, while on the right-hand side it is counted exactly $(r - 2)$ times. For all $r \geq 3$ the inequality $\binom{r}{3} \geq r - 2$ can be verified directly by writing out the binomial coefficient. For $r = 1$ and $r = 2$ we use $0 > 1 - 2$ and $0 = 2 - 2$, respectively.   □

3.2   Exercises

  (i) Prove the inequalities (18) and (19).

  (ii) Suppose that the polygon $\mathcal{M}_0$ with area 6 cm^2 covers the polygons $\mathcal{M}_1, \mathcal{M}_2, \mathcal{M}_3$ with areas 2, 3, and 4 cm^2, respectively. Show that two of them overlap in a region with an area of at least 1 cm^2

  (iii) Suppose that the polygon $\mathcal{M}_0$ with area 12 cm^2 covers four polygons $\mathcal{M}_1, \ldots, \mathcal{M}_4$ with identical areas of 4 cm^2. Show that two of them overlap in a region with an area of at least $\frac{2}{3}$ cm^2.

(iv) Inside a square with a side length of 2 cm there are 7 polygons with identical areas of 1 cm². Show that two of them overlap in a region with an area of at least $\frac{1}{7}$ cm².

(v) Inside a rectangle with an area of 6 cm² there are three rectangles with areas 3 cm², 5 cm², and 5 cm². Show that there is a place where all three rectangles overlap.

*(vi) Inside a disk with an area of 1 cm² there are five polygons $\mathcal{M}_1, \ldots, \mathcal{M}_5$ with identical areas of 0.5 cm². Show that two of these five polygons overlap in a region with an area of at least 0.2 cm².

(vii) Suppose that the object $\mathcal{M}$ is covered by the system $\mathcal{S}_{2n}$ of objects $\mathcal{U}_1, \mathcal{U}_2, \ldots, \mathcal{U}_{2n}$. Show that there is an n-element subsystem $\mathcal{S}_n$ of $\mathcal{S}_{2n}$ that covers at least half of the area of the object $\mathcal{M}$.

*(viii) Suppose that the object $\mathcal{M}$ is covered by the n-element system $\mathcal{S}_n$ of objects $\mathcal{U}_1, \mathcal{U}_2, \ldots, \mathcal{U}_n$. Show that for each $k \leq n$ there is a k-element subsystem $\mathcal{S}_k$ of the system $\mathcal{S}_n$ that covers a part of the object $\mathcal{M}$ whose area is at least $\frac{k}{n} \cdot |\mathcal{M}|$.

3.3   Coverings with Systems of Congruent Disks

We assume that $\mathcal{S} = \{\mathcal{D}_1, \mathcal{D}_2, \ldots, \mathcal{D}_n\}$, with $n \geq 1$, is a system of disks with the same radius R, and that $\mathcal{M}_0$ is a bounded plane object. Our aim in the following problems will be either to find, for a fixed n, the smallest R for which the given set $\mathcal{M}_0$ can be covered by the system $\mathcal{S}$, or to decide for a given value of R whether such a covering exists at all. The problem can be interesting even in the case $n = 1$ if we have insufficient information about the object $\mathcal{M}_0$ (for instance, in (ii) we know only the length of the perimeter of a polygon $\mathcal{M}_0$). An important result in this direction is *Jung's theorem*:

If the distance between any two points of a plane object $\mathcal{M}_0$ is at most 1, then $\mathcal{M}_0$ can be covered by a disk with radius $\frac{1}{\sqrt{3}}$ (see 3.4.(ii)).

If $n \geq 2$, then the covering of $\mathcal{M}_0$ by a system $\mathcal{S}$ of congruent disks means, in fact, that it is possible to divide the set $\mathcal{M}_0$ into at most n nonoverlapping parts $\mathcal{M}_1, \mathcal{M}_2, \ldots, \mathcal{M}_n$ with the following common property: The distance of the two "furthest" points $X_i, Y_i \in \mathcal{M}_i$ for each $i = 1, 2, \ldots, n$ does not exceed $2R$. (Note that this condition is not sufficient for the existence of such a covering.)

(i) Let $\mathcal{M}$ be an arbitrary (finite or infinite) set of points in the plane with the property that every triple of points from $\mathcal{M}$ can be covered by a disk with radius R. Show that all the points of the set $\mathcal{M}$ can be covered by one disk with radius R.

SOLUTION.   We consider the system $\mathcal{S}$ of disks $\mathcal{D}(X, R)$ of radius R and center X, where $X \in \mathcal{M}$. Since it is possible to cover any triple of points from M with a disk of radius R, this means that any three disks in S have a point in common. By Helly's theorem (see 1.1) this implies that all the disks in S have a common point A. But then the disk $\mathcal{D}(A, R)$ covers the whole set M.   □

(ii) Show that any polygon M with circumference $2a$ can be covered by a disk with radius $R = \frac{a}{2}$. Also show that if $0 < d < a$, then there is a polygon M with circumference $2a$ that cannot be covered by a disk with radius $\frac{d}{2}$.

SOLUTION.   On the boundary of the polygon M we choose points A, B such that in both directions along the boundary they have distance a from each other. Then $|AB| < a$, and for each point M on the boundary of M we have $|AM| + |BM| \le a$. We let S denote the midpoint of the line segment AB, and we show now that the disk $\mathcal{D}(S, \frac{a}{2})$ covers the whole boundary of M (and thus also the whole polygon M). To do this, we choose an arbitrary point M on the boundary of M, and construct the point M' symmetric to M with respect to the midpoint S (Figure 34). From the triangle inequality it follows that $|MM'| \le |MA| + |AM'|$, that is, $|SM| \le \frac{1}{2}(|MA| + |AM'|)$. Since $|AM'| = |BM|$ and $|AM| + |BM| \le a$, we obtain $|SM| \le \frac{1}{2}(|AM| + |BM|) \le \frac{1}{2}a$. This means that $M \in \mathcal{D}$.

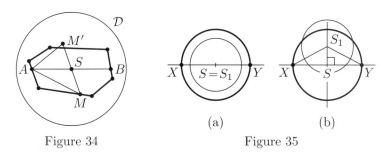

Figure 34   Figure 35

An example of a polygon satisfying the conditions of the second part of the problem is an isosceles triangle with base $(a - d)$ and the lateral sides equal to $\frac{1}{2}(a+d)$. It is clear that such a triangle exists for all $0 < d < a$, and has the desired property, since $\frac{1}{2}(a+d) > \frac{1}{2}(d+d) = d$; this means that the lateral side of the triangle cannot be covered by a disk of radius $\frac{d}{2}$.   □

(iii) Show that a disk of radius R cannot be covered by two disks with radii less than R.

SOLUTION.   We prove this by contradiction. Suppose that the disk $\mathcal{D}(S, R)$ is covered by two disks $\mathcal{D}_1(S_1, r_1)$, $\mathcal{D}_2(S_2, r_2)$, where $\max\{r_1, r_2\} < R$. Let us first assume that $S_1 = S$ (Figure 35a). Then clearly, neither of the endpoints X, Y of some diameter of $\mathcal{D}$ lies in $\mathcal{D}_1$, and thus $X, Y \in \mathcal{D}_2$; this means that $2r_2 \ge |XY| = 2R$, which is a contradiction. In the case $S_1 \ne S$

we have $|S_1 X| > R > r_1$ (see Figure 35b); then $X \notin \mathcal{D}_1$ and $X \in \mathcal{D}_2$, and similarly we obtain $Y \in \mathcal{D}_2$. So altogether, $X, Y \in \mathcal{D}_2$, which is impossible, since $|XY| = 2R > 2r_2$.   □

(iv) Cover a given triangle $\mathcal{T}$ with two congruent disks with smallest possible radius R. Consider only the cases where $\mathcal{T}$ is
  (a) equilateral. (b) right.

SOLUTION. (a) Let $\mathcal{T}$ be an equilateral triangle with side length a. If this triangle is covered by two disks with radii R, then at least one of them must contain two of the triangle's three vertices A, B, C, and therefore $2R \geq a$, that is, $R \geq \frac{a}{2}$. From Figure 36 it is clear how $\mathcal{T}$ can be covered by two congruent disks, each with radius $R = \frac{a}{2}$.

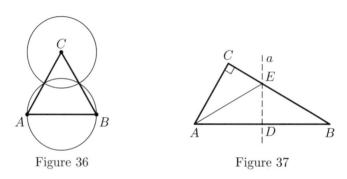

Figure 36   Figure 37

  (b) Let $\mathcal{T}$ be a right triangle ABC with hypotenuse AB, the midpoint of which will be denoted by D (Figure 37). Without loss of generality we may assume that $|AC| \leq |BC|$. Then the axis a perpendicular to the hypotenuse AB and passing through the point D cuts the side BC in the point E (if $|AC| = |BC|$, then $E = C$). We now show that $R = \frac{|AE|}{2} = \frac{|BE|}{2}$. First of all, we have $R \leq \frac{|AE|}{2}$, since the congruent disks on the diameters AE and BE cover the whole of $\triangle ABC$ (the first one circumscribes the quadrilateral $ADEC$, and the second one the right triangle EBD). On the other hand, if two disks cover the triangle ABC, then at least one of them must contain two of the three points $A, E,$ and B. Therefore, we have

$$R \geq \min \left\{ \frac{|AE|}{2}, \frac{|AB|}{2}, \frac{|BE|}{2} \right\} = \frac{|AE|}{2}. \qquad □$$

(v) Find the smallest value of R for which a given square with side length a can be covered by three congruent disks with radius R.

SOLUTION. We show that the smallest R is equal to $\frac{a\sqrt{65}}{16}$. First we prove that $R \geq \frac{a\sqrt{65}}{16}$. To obtain a contradiction, we assume that the square $ABCD$ is covered with disks $\mathcal{D}_1, \mathcal{D}_2, \mathcal{D}_3$, each with radius $R < \frac{a\sqrt{65}}{16}$. Two of the vertices of the square will lie on the same disk $\mathcal{D}_i$, and since $2R < a\sqrt{2}$, they will be neighboring vertices, say $A \in \mathcal{D}_1$ and $B \in \mathcal{D}_1$. Next we consider the points $E, F \in BC$, $G \in CD$, and $H \in DA$ such that $|BE| = |FC| = |HA| = \frac{a}{8}$, $|CG| = \frac{a}{2}$ (Figure 38). It is easy to calculate that $|AE| = |BH| = |GE| = |GH| = |DF| = \frac{a\sqrt{65}}{8} > 2R$. There-

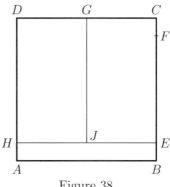

Figure 38

fore, no two of the points A, E, G lie on the same disk $\mathcal{D}_i$; since $A \in \mathcal{D}_1$, we can choose a numbering of the disks such that $E \in \mathcal{D}_2$ and $G \in \mathcal{D}_3$.

With a similar argument for the three points B, G, H we find that $H \in \mathcal{D}_2$. Next we have $D \in \mathcal{D}_3$, since $B \in \mathcal{D}_1$, $|BD| > 2R$, $E \in \mathcal{D}_2$, and $|ED| > 2R$. Since $A \in \mathcal{D}_1$, $H \in \mathcal{D}_2$, and $D \in \mathcal{D}_3$, the inequalities $|AF| > 2R$, $|HF| > 2R$, and $|DF| > 2R$ show that the point F belongs to none of the disks $\mathcal{D}_1, \mathcal{D}_2, \mathcal{D}_3$, which is a contradiction. Thus we have the inequality $R \geq \frac{a\sqrt{65}}{16}$; equality occurs if we cover the square $ABCD$ by three disks that circumscribe, respectively, the rectangles $ABEH$, $ECGJ$, and $JGDH$, where J is the midpoint of the line segment EH.   □

3.4 Exercises

  (i) Cover a given triangle with a disk with smallest possible radius.

 (ii) Let $\mathcal{M}$ be an arbitrary (finite or infinite) set of points in the plane such that the distance between any two points of $\mathcal{M}$ is not greater than 1. Show that all points of $\mathcal{M}$ can be covered by a disk with radius $R = \frac{1}{\sqrt{3}}$.

In (iii)–(vi) the letter R stands for the smallest radius such that the given polygon $\mathcal{M}$ can be covered by n congruent disks of radius R.

 (iii) Find R if $n = 3$ and $\mathcal{M}$ is an equilateral triangle with side length a.

*(iv) Find R if $n = 3$ and $\mathcal{M}$ is a right triangle ABC with hypotenuse AB of length c and with acute angle α at the vertex A.

  (v) Find R if $n = 2$ and $\mathcal{M}$ is a square with side length a.

 (vi) Find R if $n = 4$ and $\mathcal{M}$ is a square with side length a.

(vii) Decide whether a square with side length a can be covered by five congruent disks with radius $R < \frac{a\sqrt{2}}{4}$.

(viii) Find the radius R of the largest disk that can be covered by three congruent disks with radius r.

*(ix) Suppose that three congruent disks with radius R have their centers in the vertices of a given triangle which they cover. Find the smallest value of R for a given triangle.

(x) Let m be the smallest number of disks of radius 1 with which it is possible to cover a given convex polygon $\mathcal{M}$, and let n be the largest number of disks with diameter 1 that are nonoverlapping and whose centers belong to the polygon $\mathcal{M}$. Which of the numbers m and n is larger?

3.5   Further Problems on Coverings

(i) Suppose that 25 points are given in the plane such that from any triple of these points one can choose two whose distance is less than 1. Show that there is a disk $\mathcal{D}$ with radius 1 that covers at least 13 of these points.

SOLUTION.   Let A be one of the given points. If all the remaining points lie on the disk $\mathcal{D}_1(A, 1)$, then we are done. If not, then for some given point B we have $|AB| > 1$. We now consider the disk $\mathcal{D}_2(B, 1)$. Each of the remaining 23 points lies in one of the disks $\mathcal{D}_1, \mathcal{D}_2$, for if we assume the contrary, namely, that one of the remaining points, say C, lies in neither of the two disks, then $|AC| > 1$ and $|BC| > 1$, and thus the triple A, B, C does not satisfy the condition of the problem. Now, by the pigeonhole principle, one of the disks $\mathcal{D}_1, \mathcal{D}_2$ contains at least 13 of the given points. □

(ii) Show that every triangle of area 1 cm^2 can be covered by a rectangle of width $\sqrt{2}$ cm. Furthermore, find the smallest width of a rectangle with which every triangle of area 1 cm^2 can be covered.

SOLUTION.   An equilateral triangle with area 1 cm^2 has side length $a = \frac{2}{\sqrt[4]{3}}$ cm and altitude $v = \sqrt[4]{3}$ cm. We show now that the longest side of *any* triangle with area $S = 1$ cm^2 has length at least $\frac{2}{\sqrt[4]{3}}$ cm, and thus its shortest altitude is at most $\sqrt[4]{3}$ cm. (This last statement means that such a triangle can be covered by a rectangle of width $\sqrt{2}$ cm, since $\sqrt{2} > \sqrt[4]{3}$.) Indeed, if $a \geq b \geq c$ are the sides of some triangle, then with the usual notation we have $\gamma \leq 60°$, and thus

$$S = \frac{1}{2} ab \sin \gamma \leq \frac{a^2}{2} \sin 60° = \frac{a^2 \sqrt{3}}{4},$$

which implies $a \geq \frac{2\sqrt{S}}{\sqrt[4]{3}}$.

For the second part we show that the desired smallest possible width of a covering rectangle is $s = \sqrt[4]{3}$ cm. To do this, it suffices to show that an equilateral triangle ABC of altitude v cannot be covered by a rectangle of width $d < v$. We assume that on each of two opposite sides a, b of the covering rectangle there is at least one vertex of $\triangle ABC$

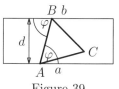

Figure 39

(otherwise we can make the rectangle "narrower"), say $A \in a$, $B \in b$ (Figure 39). Let φ denote the angle of the line AB and the side a; then $\varphi \geq |{\sphericalangle}CAB| = 60°$, so $|AB| = \frac{d}{\sin\varphi} \leq \frac{2d}{\sqrt{3}}$, and thus $v = |AB| \cdot \frac{\sqrt{3}}{2} \leq d$.   □

(iii) Decide whether it is possible to cover a square $\mathcal{S}$ of side length 7 cm with eight smaller squares of side length 3 cm. In the affirmative case, decide whether there exists a covering where the sides of all the smaller squares are parallel to the sides of the square $\mathcal{S}$.

SOLUTION.

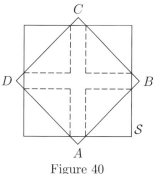

Figure 40

We place four of the smaller squares on the vertices of the square $\mathcal{S}$; the remaining uncovered part of $\mathcal{S}$ can then be covered by the square $ABCD$ (Figure 40). It suffices to show that the side length of the square $ABCD$ is less than 6 cm, for then $ABCD$ can be covered by four squares of side length 3 cm. With a simple calculation we find that $|AB| = 4\sqrt{2} < 6$.

The answer to the second question is negative, which can be seen by applying the pigeonhole principle to the following nine points: the four vertices of the square $\mathcal{S}$, the four midpoints of its sides, and the center of the square $\mathcal{S}$.   □

(iv) Show that each nonequilateral triangle ABC can be covered by two smaller triangles that are similar to ABC.

SOLUTION.   We assume that for the triangle ABC we have $|AB| > |AC|$. We "overlap" the triangle ABC by the triangle $AB'C'$ that is similar to $\triangle ABC$ and such that $|AC| < |AB'| < |AB|$, with $\triangle ABC$ as shown in Figure 41. We now construct a line $DE\|AC$ through the point E that is the intersection between the sides BC and $B'C'$. Then the triangles $AB'C'$ and DBE satisfy the conditions of the problem.   □

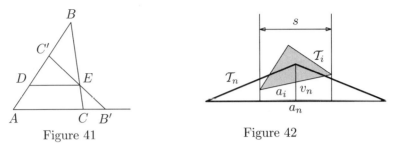

Figure 41   Figure 42

(v) We are given 100 triangles. Decide whether it is always possible to cover one of them with the remaining 99.

SOLUTION.   It is not always possible. We will show that there even exists an infinite sequence of triangles $\mathcal{T}_1, \mathcal{T}_2, \ldots$ of which not even one can be covered by the remaining ones. We first explain the idea of the proof, before giving it in detail. We begin by determining the sequence $\{a_n\}$ of lengths of the longest sides of $\triangle \mathcal{T}_n$ such that for any placement of the triangles $\mathcal{T}_1, \mathcal{T}_2, \ldots, \mathcal{T}_{n-1}$ there remains a certain part (which is independent of n) of the area of the triangle $\mathcal{T}_n$ that is not covered (for instance, one-third of its area). This can be achieved by a choice of side lengths a_n that are "sufficiently" large compared to $a_1, a_2, \ldots, a_{n-1}$. Then we have to choose a sequence $\{S_n\}$ of areas of the triangles $\mathcal{T}_n$ such that the sum of the areas of the triangles $\mathcal{T}_{n+1}, \mathcal{T}_{n+2}, \ldots$ is less than the area of the uncovered part of the triangle $\mathcal{T}_n$; this will occur if the sequence $\{S_n\}$ decreases "sufficiently fast."

We now present an example of one such sequence of triangles: Let $\mathcal{T}_1$ be an equilateral triangle with side length $a_1 = 1$, and for $n > 1$ let $\mathcal{T}_n$ be an isosceles triangle with base $a_n = 4^{n-1}$ and area $S_n = \frac{S_1}{4^{n-1}}$, where $S_1 = \frac{\sqrt{3}}{4}$ is the area of $\triangle \mathcal{T}_1$ (the altitude of $\triangle \mathcal{T}_n$ is $v_n = \frac{2S_n}{a_n} = \frac{2\sqrt{3}}{4^{2n-1}}$). Let us consider the triangles $\mathcal{T}_1, \mathcal{T}_2, \ldots, \mathcal{T}_n$. Each of the triangles $\mathcal{T}_i$ ($i = 1, 2, \ldots, n-1$) can be placed in a strip of width $s \leq a_i$ that is perpendicular on the base of $\triangle \mathcal{T}_n$ (Figure 42). The area covered by this strip in $\triangle \mathcal{T}_n$ is less than $a_i v_n$, and therefore the region covered by the triangles $\mathcal{T}_1, \mathcal{T}_2, \ldots, \mathcal{T}_{n-1}$ in $\mathcal{T}_n$ has area less than $(a_1 + a_2 + \cdots + a_{n-1}) \cdot v_n$. Since

$$\sum_{i=1}^{n-1} a_i = \frac{1}{3}\left(4^{n-1} - 1\right) < \frac{1}{3}a_n,$$

the covered area of $\triangle \mathcal{T}_n$ is less than $\frac{a_n v_n}{3} = \frac{2S_n}{3}$, which means that at least one third of the area of $\triangle \mathcal{T}_n$ remains uncovered. At the same time we have

$$S_{n+1} + S_{n+2} + \cdots = S_1 \left(\frac{1}{4^n} + \frac{1}{4^{n+1}} + \cdots\right) = \frac{S_1}{4^n\left(1 - \frac{1}{4}\right)} = \frac{1}{3} \cdot \frac{S_1}{4^{n-1}} = \frac{1}{3}S_n.$$

We have therefore shown that the triangles $\mathcal{T}_1, \mathcal{T}_2, \ldots, \mathcal{T}_{n-1}, \mathcal{T}_{n+1}, \ldots$ cannot cover the whole triangle $\mathcal{T}_n$ for any n.   □

3.6   Exercises

(i) Suppose that a point A is covered by each of six congruent disks lying in a plane. Show that at least one of these disks contains the center of some of the remaining five disks.

(ii) Suppose that a square with area S_1 is covered by a triangle with area S_2. Show that $2 \cdot S_1 \le S_2$.

(iii) Given a square of side length a, cover it with a triangle of smallest possible area.

(iv) Decide whether it is possible to cover a square of side length 5 cm with three congruent squares of side length 4 cm.

*(v) Find the side length b of the largest equilateral triangle that can be covered by three congruent equilateral triangles with a given side length a.

*(vi) Given 100 rectangles, decide whether it is always possible to cover one of them with the remaining 99.

(vii) Suppose that any three points of some finite set $\mathcal{M}$ of points in the plane are vertices of a triangle with area at most 1 m^2. Show that the set $\mathcal{M}$ can be covered by one triangle with an area of 4 m^2.

3.7   Problems on Packings

In what follows we present some problems on the properties of a system $\mathcal{S}$ of geometrical objects that are packed in a polygon $\mathcal{M}$.

(i) In a rectangle of size 20×25 we have 120 unit squares. Show that we can also place a disk $\mathcal{D}$ with diameter 1 in the rectangle such that it does not overlap with any of the given squares.

SOLUTION.   As indicated in Figure 43, we "shrink" the given rectangle $\mathcal{R}$ by a frame of width $x = \frac{1}{2}$, and we "enlarge" each of the 120 squares by the set of all those points that have a distance of at most $\frac{1}{2}$ from the square. Each of these "enlarged" squares has an area of $S = 1 + 4 \cdot 1 \cdot \frac{1}{2} + \pi \cdot \left(\frac{1}{2}\right)^2 = 3 + \frac{\pi}{4}$, and therefore the sum S' of the areas of all these "enlarged" squares is $120 \cdot \left(3 + \frac{\pi}{4}\right)$; at the same time, the area of the shrunken rectangle is $19 \cdot 24 = 456$. In view of the

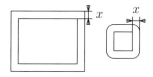

Figure 43

inequality $120 \cdot \left(3 + \frac{\pi}{4}\right) < 456$ we see that some point X of the "shrunken" rectangle does not lie in any of the "enlarged" squares. The disk $\mathcal{D}\left(X, \frac{1}{2}\right)$ then has the desired property.   □

(ii) On a round table with radius R there are $n > 1$ nonoverlapping congruent coins of radius r such that no further congruent coin can be placed on the table without overlapping some of the n coins. Show that we have the inequalities

$$\frac{1}{2}\left(\frac{R}{r}-1\right) < \sqrt{n} < \frac{R}{r}. \tag{22}$$

SOLUTION.   Since the coins do not overlap, the sum of their areas is less than the area of the table; we have therefore $n \cdot \pi r^2 < \pi R^2$, which implies the right-hand inequality in (22). It remains to prove the left-hand one. Denote the given coins by $\mathcal{D}_i(S_i, r)$, $i \in I = \{1, 2, \ldots, n\}$, and the table by $\mathcal{D}(S, R)$. If we increase the radius of all the coins from r to $2r$, we obtain new "coins" $\mathcal{D}_i^*(S_i, 2r)$, and if we decrease the radius of the table from R to $R - r$, we get a new "table" $\mathcal{D}^*(S, R - r)$. The system of disks $\mathcal{D}_i^*$ has to cover the disk $\mathcal{D}^*$, since otherwise there would exist some point $X \in \mathcal{D}^*$ with the property $X \notin \mathcal{D}_i^*$ ($i \in I$), that is, $|XS| \le R - r$ and $|XS_i| > 2r$ ($i \in I$); we would then be able to add a new coin $\mathcal{D}_{n+1}(X, r)$ to the original coins, with the property $\mathcal{D}_{n+1} \subseteq \mathcal{D}$ and $\mathcal{D}_i \cap \mathcal{D}_{n+1} = \emptyset$ ($i \in I$), which by the hypothesis of the problem is not possible. Therefore, since the n disks $\mathcal{D}_i^*$ cover the disk $\mathcal{D}^*$, the sum of their areas satisfies $n \cdot \pi \cdot (2r)^2 \ge \pi \cdot (R - r)^2$, and it is clear that equality cannot hold. An easy transformation of the strict inequality now gives the left-hand one in (22).   □

(iii) Suppose that a disk $\mathcal{D}$ of radius n contains $4n$ line segments of unit length, and that furthermore a straight line l is given. Show that there exists a straight line l' that is either parallel or perpendicular to l, and that cuts at least two of the given line segments.

SOLUTION.   We construct the line segments A_1A_2 and B_1B_2 as diameters of the disk $\mathcal{D}$ such that $A_1A_2 \| l$ and $B_1B_2 \perp l$. We denote the orthogonal projections of the given segments $\mathcal{U}_i$ ($1 \le i \le 4n$) onto the lines A_1A_2, B_1B_2 by $\mathcal{V}_i$, respectively $\mathcal{W}_i$. Each of the sets $\mathcal{V}_i, \mathcal{W}_i$ is a line segment (or possibly a point), and their lengths $a_i = |\mathcal{V}_i|$, $b_i = |\mathcal{W}_i|$ clearly satisfy $a_i + b_i \ge |\mathcal{U}_i| = 1$ ($1 \le i \le 4n$), and thus

$$(a_1 + a_2 + \cdots + a_{4n}) + (b_1 + b_2 + \cdots + b_{4n}) \ge 4n.$$

We first assume that $a_1 + a_2 + \cdots + a_{4n} \ge b_1 + b_2 + \cdots + b_{4n}$; then

$$|\mathcal{V}_1| + |\mathcal{V}_2| + \cdots + |\mathcal{V}_{4n}| \ge 2n.$$

In view of the fact that each of the sets $\mathcal{V}_1, \mathcal{V}_2, \ldots, \mathcal{V}_{4n}$ is a subset of the line segment A_1A_2 of length $2n$, the last inequality means that two segments $\mathcal{V}_i, \mathcal{V}_j$ ($i \ne j$) have a common point C. Then the line perpendicular to the line l and passing through C cuts both segments $\mathcal{U}_i, \mathcal{U}_j$. Similarly, in the case $a_1 + a_2 + \cdots + a_{4n} < b_1 + b_2 + \cdots + b_{4n}$ there is an appropriate parallel to the line l.   □

3.8 Exercises

(i) Suppose that 20 squares of side length 1 cm are placed in a square $\mathcal{K}$ of side length 15 cm. Show that we can still place a disk of radius 1 inside $\mathcal{K}$ such that it does not overlap any of the 20 squares.

(ii) Suppose that several disjoint circles, whose lengths add up to 10 cm, are placed in a square of side length 1 cm. Show that there is a line that has a point in common with at least four of the circles.

(iii) A convex polygon $\mathcal{M}$, whose area is greater than 0.5 cm^2, is placed in a square of side length 1 cm. Show that a line segment of length 0.5 cm that is parallel to one of the sides of the square can be placed inside the polygon $\mathcal{M}$.

(iv) We are given two triangles $\mathcal{T}_1, \mathcal{T}_2$, each of them with an area greater than 1 cm^2. Show that $\mathcal{T}_1, \mathcal{T}_2$ cannot be packed in a disk $\mathcal{D}$ of radius 1 cm without overlapping.

3.9 Problems on Tilings

Problems on tiling a geometrical object $\mathcal{M}_0$ with a system $\mathcal{S} = \{\mathcal{M}_1, \mathcal{M}_2, \ldots, \mathcal{M}_n\}$ can often be found as popular "brain teasers" in the columns of newspapers or magazines; they are usually formulated as follows: From given "parts" $\mathcal{M}_1, \mathcal{M}_2, \ldots, \mathcal{M}_n$ construct a "whole" $\mathcal{M}_0$. To describe a general method for solving such problems is practically impossible; in the majority of cases the solver's imagination and intuition are deciding factors. We will therefore restrict our attention to problems of a *discrete* type. To be more exact, we will assume that the object $\mathcal{M}_0$ is formed of a grid of congruent fields (squares) and that each of the tiling objects consists of several whole fields of those that make up $\mathcal{M}_0$. It is clear that such a *discretization* is meaningful also in a more general setting: If we form in the plane a sufficiently "fine" grid of fields, we can with its help get a good "approximation" of more complicated plane figures.

A square of k^2 fields ($k = 1, 2, \ldots$) will be called a $k \times k$ *chessboard* (or *checkerboard*); apart from these we also consider shapes as shown in Figure 44, which we will refer to as a *domino* (a), *tromino* (b), *tetromino* (c), piece of *type* Γ (d), and piece of *type* T (e).

We note that problems on these shapes will also be solved later, in Section 4.5; there, as a rule, we will use an appropriate *coloring* to show that a desired tiling of a chessboard does not exist. To those who are interested in further studying these problems we recommend the excellent book [1].

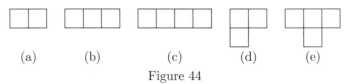

(a)   (b)   (c)   (d)   (e)

Figure 44

(i) Decide whether it is possible to tile a 20×20 chessboard with four 1×1 pieces, eight 2×2 pieces, twelve 3×3 pieces, and sixteen 4×4 pieces.

SOLUTION.   We show by induction that for any $k \geq 1$ a $k(k+1) \times k(k+1)$ chessboard can be tiled with four 1×1, eight 2×2, twelve 3×3, ..., $4k$ $k \times k$ pieces. For $k = 1$ the assertion is clear, since a 2×2 chessboard can be covered with four 1×1 pieces. Let us assume that the assertion holds for a $k(k + 1) \times k(k + 1)$ chessboard; we show that it is true for a $(k+1)(k+2) \times (k+1)(k+2)$ chessboard. This last board can be divided into two squares A, D and rectangles B, C (Figure 45). By the induction hypothesis the square A can be tiled with four pieces of size $1 \times 1, \ldots, 4k$ pieces of size $k \times k$; therefore, it suffices to show that the part $B \cup C \cup D$ can be tiled with

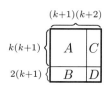

Figure 45

pieces of size $(k + 1) \times (k + 1)$. Indeed, B and C are rectangles of size $2(k+1) \times k(k+1)$ that can be tiled with $2k$ squares of size $(k+1) \times (k+1)$, and D is a square of side length $2(k + 1)$ that can be tiled with four $(k + 1) \times (k + 1)$ squares. Altogether, the area $B \cup C \cup D$ can therefore be tiled with $2k + 2k + 4 = 4(k + 1)$ pieces of size $(k + 1) \times (k + 1)$, which completes the proof.   □

(ii) An arbitrary field is cut from a 1993×1993 chessboard. Show that the remaining board can be tiled with pieces of type Γ (see Figure 44d).

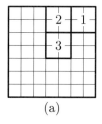   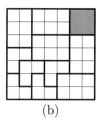   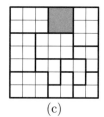   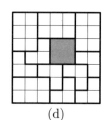

(a)   (b)   (c)   (d)

Figure 46

SOLUTION.   We prove by induction that this assertion holds in fact for any $(6n + 1) \times (6n + 1)$ chessboard; note that $1993 = 6 \cdot 332 + 1$.

(a) For $n = 1$ we have a 7×7 chessboard. This board must be tiled in such a way that one predetermined field remains uncovered. In view of symmetry it suffices to analyze the cases where the chosen field is in one of the squares 1, 2, or 3 (Figure 46a). Examples of the desired tiling are then given in Figures 46b–d; three fields in the shaded 2×2 squares are covered by a piece of type Γ.

(b) We assume that a $(6n + 1) \times (6n + 1)$ chessboard can be tiled in the desired fashion. It remains to show that this also holds for a $(6n + 7) \times (6n + 7)$ chessboard. Indeed, we place a $(6n + 1) \times (6n + 1)$ square into some corner of the $(6n + 7) \times (6n + 7)$ board such that the cut-out field lies in this square, which can then be tiled in the required way (Figure 47) by the induction hypothesis. It

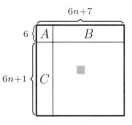

Figure 47

remains to show that the 6×6 square A, as well as each of the rectangles B and C of size $(6n + 1) \times 6$, can be tiled with pieces of type Γ; but this is easy to do.   □

3.10   Exercises

(i) Decide whether it is possible to tile an 8×8 chessboard with pieces of type T (see Figure 44e).

(ii) One field is removed from a $2^n \times 2^n$ chessboard. Show that the remaining part of the board can be tiled with pieces of type Γ (Figure 44d).

(iii) Remove one field from an 8×8 chessboard so that the remaining part of the board can be tiled with trominoes (Figure 44b).

(iv) Decide whether it is possible to tile a 6×6 chessboard with dominoes (Figure 44a) in such a way that the straight line that separates any two neighboring rows or columns of chessboard fields cuts through one or more of the dominoes.

*(v) Suppose that the integer n is not divisible by 3. Show that an $n \times n$ chessboard from which a corner field is removed can be tiled with pieces of type Γ (Figure 44d).

3.11   Auxiliary Tilings

In the final problems of this section we will show how we can take advantage of appropriately chosen auxiliary tilings of the objects under consideration. In particular, we will be dealing with assertions of the following type: If in a given bounded object "many" points are arbitrarily chosen, then some two of the chosen points have a sufficiently "small" distance.

(i) Five points are given inside a square of side length 2. Show that there are two of these points whose distance is not greater than $\sqrt{2}$.

SOLUTION.   We divide the given square by perpendicular bisectors into four squares of side length 1 that tile the given square. By the pigeonhole principle, one of these squares contains at least two of the given points, and therefore their distance is not greater than the length of the diagonals of the square, namely $\sqrt{2}$.   □

(ii) Six points are chosen in a rectangle $\mathcal{R}$ of size 3×4. Show that two of these points are separated by a distance of at most $\sqrt{5}$.

SOLUTION.   A rectangle of size 1×2 has diagonals of length $\sqrt{5}$, and therefore it suggests itself to tile $\mathcal{R}$ with six 1×2 rectangles. However, "too few" points are given for such a tiling, and a simple modification of the approach in (i) will there-fore not work. Instead, we consider a tiling of $\mathcal{R}$ with five pieces, as suggested in Figure 48. One of these pieces contains two of the chosen points, and since the distance between any two points in each of the five pieces is at most $\sqrt{5}$, this proves the desired assertion.   □

Figure 48

(iii) Six points are given in the plane such that the distance between any two of them is greater than 1. Show that these points cannot be covered by any disk of radius 1.

SOLUTION.   Let $A_1, A_2, \ldots, A_6$ be points with the property that $|A_i A_j| > 1$ for all $i \neq j$. We assume that the assertion does not hold, that is, that a disk $\mathcal{D}(S, 1)$ exists with the property $A_i \in \mathcal{D}$ for all $i = 1, 2, \ldots, 6$. Then clearly, $S \notin \{A_1, \ldots, A_6\}$, since $|SX| \leq 1$ for all points $X \in \mathcal{D}$. We now divide the disk $\mathcal{D}$ by the half-lines SA_i into six sectors that tile the disk. One of these sectors has an angle of size $60°$ at most; but then the points A_i, A_j $(i \neq j)$ lying on the boundary of this sector satisfy $|A_i A_j| \leq 1$, which is a contradiction.   □

3.12   Exercises

(i) In a rectangle of size 35×36 there are 55 arbitrarily chosen points. Show that some two of them are separated by a distance of at most $\sqrt{50}$.

(ii) Five arbitrarily chosen points lie inside an equilateral triangle of side length 1. Show that some two of them are separated by a distance of at most $\frac{1}{2}$.

(iii) A square of side length 1 contains 51 points. Show that some three of them lie in a disk of radius $\frac{1}{7}$.

(iv) We are given seven points inside a square of side length 1. Show that there exists a triangle of area at most $\frac{1}{16}$, each of whose vertices is one of the chosen points or one of the vertices of the square.

(v) Find the smallest number of pieces of type Γ (Figure 44d) that can be placed on an 8×8 chessboard such that no other congruent piece can be placed on the uncovered part of the board.

4   Colorings

In this section we will study situations where some set $\mathcal{M}$ of points in the plane is divided into several classes $\mathcal{T}_1, \mathcal{T}_2, \ldots, \mathcal{T}_n$. If we have n different colors at our disposal, we can visualize this situation by coloring all the points of the class $\mathcal{T}_i$ with the ith color; we then obtain a *coloring* of the set $\mathcal{M}$. In this connection the following two questions arise most frequently:

- For an arbitrary coloring of $\mathcal{M}$ with a given number of colors, does some class $\mathcal{T}_i$ have a certain prescribed geometrical property?
- For an appropriate coloring of $\mathcal{M}$ with a given number of colors, does every class $\mathcal{T}_i$ have a certain prescribed property? How can such a coloring be constructed?

In the area of colorings the best-known problem is without doubt the famous *four-color problem*, which will be mentioned in some detail in Section 4.3. As we will show in Section 4.7, the method of colorings can be successfully used even in the solution of problems whose statements make no reference to colorings.

4.1   Colorings of Points

We begin with problems in which there are, initially, no conditions on the sets of equally colored points (on the other hand, in all the later parts of this section these sets will normally form certain regions in the plane).

(i) Suppose that each point in the plane is colored with one of three colors. Show that there are two points of the same color whose distance from each other is equal to 1.

SOLUTION.   To begin with, we remark that the assertion is trivial if no more than two colors are used for coloring the points (it suffices to note that then at least two of the vertices of any equilateral triangle with side length 1 have the same color).

We assume that there is a coloring of the points of the plane such that the endpoints of any line segment of length 1 have different colors. We first show that any two points A, B with distance $\sqrt{3}$ have the same color. Indeed, if $|AB| = \sqrt{3}$, then the circles $c_1(A, 1)$ and $c_2(B, 1)$ intersect in the points $C \neq D$, and $\triangle ACD$ and $\triangle BCD$ are

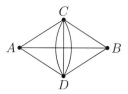

Figure 49

equilateral triangles of side length 1 (Figure 49). Since by our assumption the colors of the points in each of the triples $\{A, C, D\}$ and $\{B, C, D\}$ are distinct, the points A and B have the same color. Now it suffices to consider any triangle XYZ, with $|XY| = |XZ| = \sqrt{3}$ and $|YZ| = 1$. By what we just showed, Y has the color of X, and X has the color of Z. Hence the points Y and Z have the same color, while at the same time $|YZ| = 1$. This is a contradiction, which proves the assertion.   □

(ii) Suppose that each point of the plane is colored with one of two colors. Show that some three points of the same color form the vertices of an equilateral triangle.

SOLUTION.   We assume that for a certain coloring the assertion is false, that is, every equilateral triangle in the plane has two vertices of different colors. Let us consider a regular hexagon $ABCDEF$ with center S, whose color we declare to be black. One of the vertices, say B, of the equilateral triangle BDF must be black. From considering $\triangle ASB$ and $\triangle BSC$ it then follows that the points A, C have the other color (white). But then the point E is black ($\triangle ACE$) and F is white ($\triangle ESF$). Let us also consider the point G, namely the intersection of the lines AB and EF (see Figure 50). No matter which of the two colors it has, one of the equilateral triangles AFG and BEG will have all three vertices of the same color, which is a contradiction.   □

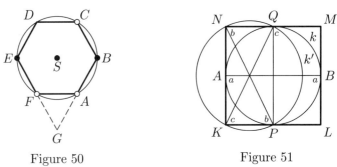

Figure 50   Figure 51

(iii) Suppose that each point of the plane is colored with one of three colors. Show that some three points of the same color form the vertices of a right triangle.

SOLUTION.   We choose two points A, B with the same color (say, a) in the plane. Let the circle k on the diameter AB be inscribed in the square $KLMN$, as shown in Figure 51 (the points of contact are the vertices of the square $APBQ$). If one of the points P and Q has the color a, then we are done (consider $\triangle ABP$ and $\triangle ABQ$). The same is true in the case where P and Q have the same color; then it suffices to consider the coloring of the vertices of some right triangle XYZ inscribed in the circle k such that

$$\{X, Y, Z\} \cap \{A, B, P, Q\} = \emptyset.$$

It remains to analyze the case where the points P and Q have different colors b, c, and where $X = A$ is the only point X on the segment KN with color a (otherwise, $\triangle BAX$ would be the desired triangle). For the color of the point K we distinguish between two possibilities. If K has color b, then it suffices to restrict ourselves to the case where each point $X \in KN$, $X \notin \{A, K\}$, has the color c (otherwise, it is enough to take $\triangle PKX$); choosing then such a point $X \neq N$ arbitrarily, we obtain a right triangle QNX with vertices of color c. If K has color c and if $\triangle KNQ$ does not have the desired property, then N has color b (see Figure 51), and therefore it suffices to consider the coloring of the vertices of some right triangle XYZ inscribed in the circle k' circumscribing the rectangle $KPQN$ such that

$$\{X, Y, Z\} \cap \{K, P, Q, N\} = \emptyset. \qquad \square$$

4.2   Exercises

  (i) Suppose that every point of a circle c is colored with one of two colors. Decide whether some three of its points that have the same color form the vertices of a triangle that is
  (a) right.   (b) equilateral.   (c) isosceles.

  (ii) Suppose that every point in the plane is colored with one of two colors. Show that the plane contains a parallelogram whose vertices have the same color.

 *__(iii)__ Suppose that every point in the plane is colored with one of seven colors. Decide whether there necessarily exist two points in the plane that have the same color and whose distance from each other is equal to 1.

4.3   Colorings of Regions

We now return to the situations discussed in Section 2, where a given region in the plane is to be divided into smaller regions by a system of curves. If we want to clearly distinguish the different regions from each other graphically (for example, to distinguish different countries on a map), we color them using several colors. However, in doing so we have to take care that no two "neighboring" regions have the same color. What is the smallest number of colors required for this?

From the middle of the nineteenth century a large number of outstanding mathematicians worked on this problem. At the end of the nineteenth century P.J. Heawood proved that five colors suffice for coloring any plane map. Subsequently, all those cases were described where two or three colors suffice. Finally, in 1976, W. Haken and K. Appel announced that they had proven the conjecture that every plane map can be colored with four colors. For the analysis of a large number of cases they had used computer methods.

For solving the map coloring problem it was convenient to abstract the geometrical properties of the individual regions and to represent the "neighborhood" relation by an appropriate *planar graph*. Many details, including a proof of Heawood's theorem, can be found in the book [3]. Here we will not be concerned with this approach; we will instead restrict ourselves to some problems that can be solved "geometrically," that is, without knowledge of graph theory.

(i) Show that if the plane is divided by n straight lines into several regions, then each of them can be colored with one of two colors such that any two areas of the same color have no more than one boundary point in common.

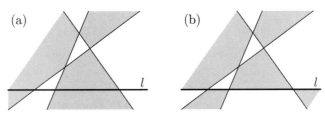

Figure 52

SOLUTION. We prove the assertion by induction on n. For $n = 1$ it is clear; we just color the two half-planes differently. Let us assume that the assertion is true for some $n = k \geq 1$, and consider the system $\mathcal{P}$ of $k + 1$ arbitrary lines in the plane. We choose an arbitrary line $l \in \mathcal{P}$ and then consider an appropriate coloring of the regions into which the plane is divided by the system of k lines $\mathcal{P}' = \mathcal{P} \setminus \{l\}$. Some of these regions lie completely inside one or the other of the two half-planes determined by the line l, while each of the remaining ones is cut into two new regions by l. These pairs of new areas are the reason why an allowable coloring for the system $\mathcal{P}'$ must be transformed into a coloring for the system $\mathcal{P}$. We proceed as follows: In one of the half-planes bounded by l (the upper one in Figure 52a) we keep the coloring, while we change the color of each region of the other half-plane (Figure 52b). We thus obtain an allowable coloring for the system $\mathcal{P}$, since pairs of "neighboring" regions have different colors also in the case where their common boundary (a line segment of positive length, a line, or a half-line) is part of the line l. This completes the proof. □

(ii) The plane is divided into a number of regions by n given straight lines ($n \geq 3$). Suppose that some of the regions are colored (with one color) such that any two colored regions have no more than one boundary point in common. Show that the number of colored regions does not exceed $\frac{n^2+n}{3}$.

SOLUTION. We write the number p of colored regions in the form $p = |\mathcal{S}_1| + |\mathcal{S}_2|$, where $\mathcal{S}_1$, respectively $\mathcal{S}_2$, denotes the class of all bounded, respectively unbounded, colored regions. The inequality $p \leq \frac{n^2+n}{3}$ clearly holds in the case where the system of lines consists of n parallels; in this case we have $\mathcal{S}_1 = \emptyset$ and $|\mathcal{S}_2| \leq n + 1$, and furthermore, the inequality

$n + 1 \leq \frac{n^2+n}{3}$ holds for all $n \geq 3$. In the remaining cases we obtain the desired bound by adding the inequalities

$$|\mathcal{S}_1| \leq \frac{n(n-2)}{3} \quad \text{and} \quad |\mathcal{S}_2| \leq n, \tag{23}$$

which we will now prove. We note that each of the n lines of the system is cut into exactly two half-lines and at most $(n-2)$ line segments (of positive length) by intersections with the other $(n-1)$ lines. In all we thus obtain a system of $2n$ half-lines and at most $n(n-2)$ segments, which form the boundaries of all the regions that occur, where each half-line and segment lies at the boundary of at most one colored region. Since each region $\mathcal{M} \in \mathcal{S}_1$ is a k-gon for some $k \geq 3$, we have $3|\mathcal{S}_1| \leq n(n-2)$, which leads to the first inequality in (23). The boundary of each region $\mathcal{M} \in \mathcal{S}_2$ contains two half-lines, and therefore we have $2|\mathcal{S}_2| \leq 2n$, which gives the second inequality in (23). This completes the proof. $\square$

4.4 Exercises

(i) Suppose that the plane is divided by a finite system of straight lines and circles into a number of regions. Show that they can be colored by two colors such that any two regions that have a common boundary of positive length will have different colors.

(ii) Suppose that n circles are placed in the plane, and in each of them we draw a chord such that no two chords lie on one line. This divides the plane into a number of regions; show that it can be colored with three colors such that any two regions that have a common boundary of positive length will have different colors.

(iii) An arbitrary (convex or nonconvex) polygon is divided into n nonoverlapping triangles by a finite system of line segments in such a way that no vertex of any triangle is an interior point of a side of another triangle. Show that these triangles can be colored with three colors such that any two triangles with a common side will have different colors.

4.5 Colorings of Chessboards

For the games of chess and checkers the 8×8 chessboard is colored in the *classical way*: In each row and column, black and white fields alternate in a regular fashion. In combinatorial geometry it is of interest also to study the more general situation where the fields of some $m \times n$ chessboard are arbitrarily colored with a given number of colors. We now present the following four problems.

(i) On a classically colored 8×8 chessboard we introduce the following operation: Choose any row or column and simultaneously change the color of all its fields to the opposite. Decide whether after a number of such operations one can obtain a chessboard with exactly one black field.

SOLUTION.   We assume that in some row (respectively column) there are exactly k black and $(8 - k)$ white fields, where $k \in \{1, 2, \ldots, 8\}$. If in this row (respectively column) we carry out the allowable operation, we obtain $(8-k)$ black and k white fields. The number of black fields on the chessboard thus has changed by $(8 - k) - k = 8 - 2k$, an even number. The operation in question therefore does not change the parity of the number of black fields, and since originally there were 32 black fields on the chessboard, the number of black fields remaining after any number of operations can never be odd.   □

(ii) Decide whether it is possible to color the fields of a 1990×1990 chessboard with two colors such that the fields symmetric with respect to the center of the chessboard have opposite colors and each row and each column has the same number of fields of both colors.

SOLUTION.   We assume that the colors used are black and white, and we divide the chessboard into four congruent squares A, B, C, D of size 995×995 (see Figure 53). For $X \in \{A, B, C, D\}$ we denote by w_X (respectively b_X) the number of white (respectively black) fields in the square X. In view of the required antisymmetry we have $w_A = b_D$. Since each row and column has the same number of fields of both colors, we have $w_A + w_B = \frac{1}{4} \cdot 1990^2 = 995^2$ and $w_B + w_D = 995^2$, that is, $w_A = w_D$, so together $995^2 = b_D + w_D = 2w_A$. But this is a contradiction, since 995^2 is odd. Hence the desired coloring does not exist.

Figure 53

□

(iii) Suppose that each field of a 12×12 chessboard is colored with one of three colors. Show that there are four fields of the same color that are corner fields of some rectangle.

FIRST SOLUTION.   We choose a column and denote by a, b, c the numbers of fields in this column with the individual colors; then $a + b + c = 12$, where in this column there are $\binom{a}{2} + \binom{b}{2} + \binom{c}{2}$ pairs of fields of the same color. From the Cauchy–Schwarz inequality for $n = 3$ (see, e.g., [5], inequality (48) on page 131) it follows that

$$a^2 + b^2 + c^2 \geq \frac{1}{3}(a + b + c)^2,$$

and therefore

$$\binom{a}{2} + \binom{b}{2} + \binom{c}{2} = \frac{1}{2}\left(a^2 + b^2 + c^2\right) - \frac{1}{2}(a + b + c)$$

$$\geq \frac{1}{6}(a + b + c)^2 - \frac{1}{2}(a + b + c) = \frac{1}{6} \cdot 12^2 - 6 = 18.$$

Hence in each column of the colored chessboard there are at least 18 pairs of fields with the same color. Therefore, in the whole chessboard there are at least $12 \cdot 18 = 216$ pairs of equally colored fields that lie in the same column, and of these at least $216/3 = 72$ pairs of fields must be of the same color. Since $\binom{12}{2} = 66 < 72$, this means that some two pairs of fields of the same color (lying in two different columns) occupy the same two rows, that is, they are corner fields of some rectangle. □

SECOND SOLUTION. From the pigeonhole principle it follows that at least 48 fields of the chessboard have the same color (say, black). We consider all possible pairs of black fields in the individual columns; it clearly suffices to show that there are always more than $\binom{12}{2}$ of them. If some column contains at most three black fields, we successively "move" to it some of the black fields from those columns that contain at least five black fields (such columns exist, since $48 = 4 \cdot 12$). This operation does not increase the total number of pairs of black fields (in the individual columns), since for $a < 4$ and $b > 4$ we have the inequality

$$\binom{a}{2} + \binom{b}{2} > \binom{a+1}{2} + \binom{b-1}{2}.$$

After a finite number of steps we thus obtain a chessboard that has in each column at least four black fields, and so there are at least $12 \cdot \binom{4}{2} = 72$ pairs of black fields, while in the original chessboard this number was not smaller. □

(iv) Given a rectangular $2 \times n$ chessboard, let $P(n)$ denote the number of all colorings of its fields, where in no 2×2 square are all four fields colored. Show that the number $P(1991)$ is divisible by 3, and find the highest power of 3 that divides $P(1991)$.

SOLUTION. We easily find $P(1) = 2^2 = 4$ and $P(2) = 2^4 - 1 = 15$; we now derive a recurrence relation for $P(n)$. We divide all $P(n)$ colorings of the $2 \times n$ chessboard into two groups according to whether or not both fields of the last (the nth) column are colored. In the first case at least one field in the $(n-1)$th column is not colored, and therefore we have $2^2 - 1 = 3$ possibilities for the coloring of this column; the "remainder" of the chessboard can be colored in $P(n-2)$ ways. With a similar argument we find that in the second case we have $3 \cdot P(n-1)$ colorings; therefore,

$$P(n) = 3 \cdot (P(n-1) + P(n-2)), \tag{24}$$

which means that $3 \mid P(n)$ for all $n \geq 2$, and thus also $3 \mid P(1991)$. Next we show by induction that for each $k \geq 1$ we have $3^k \mid P(2k)$, $3^k \mid P(2k+1)$, but $3^{k+1} \nmid P(2k+1)$. For $k = 1$ this is clear, since $P(2) = 15$ and $P(3) = 57$. We assume now that the assertion holds for some $k = m$. Then for $n =$

$2m+1$ it follows immediately from (24) and from the induction hypothesis that $3^{m+1} \mid P(2m+2)$, from which it follows again by (24) for $n = 2m+2$ that $3^{m+1} \mid P(2m+3)$. If we assume that we even have $3^{m+2} \mid P(2m+3)$, then from the identity $P(2m+1) = \frac{1}{3}P(2m+3) - P(2m+2)$ we would obtain $3^{m+1} \mid P(2m+1)$, which contradicts the induction hypothesis.

For $k = 995$ this assertion means that $3^{995} \mid P(1991)$ and $3^{996} \nmid P(1991)$. The desired largest power of 3 is therefore 3^{995}. $\qquad\qquad\square$

4.6 Exercises

(i) On a classically colored 8×8 chessboard one can carry out the following operations: Choose an arbitrary 2×2 square and simultaneously change the colors of all its fields to their opposites. Decide whether after a number of such operations one can obtain a chessboard with only one black field.

(ii) Find the largest possible number of fields of an $n \times n$ chessboard that can be colored such that no two colored fields have a point in common.

(iii) Decide whether it is possible to color 25 arbitrary fields of a 20×20 chessboard such that each colored field abuts (that is, has a common side with) an odd number of colored fields.

(iv) We are given a $2(2k - 1) \times 2(2k - 1)$ chessboard, where $k \geq 1$. Decide whether it is possible to color its fields with two colors in such a way that the fields symmetric to each other with respect to the center of the chessboard have opposite colors, and in any row and any column there is the same number of fields of both colors.

(v) Do Exercise (iv) with a $4k \times 4k$ chessboard.

(vi) Each field of a rectangular chessboard of size 3×9 has one of two colors. Show that on this chessboard there are four fields of the same color that form the corner fields of a rectangle.

(vii) Suppose that each field of a 5×5 chessboard has one of two colors. Show that on this chessboard there are four fields of the same color that form the corner fields of a rectangle. Show furthermore that in the case of a 4×4 chessboard this need not occur, that is, there is a coloring for which no four fields of the same color are the corner fields of a rectangle.

(viii) Suppose that we can color the fields of an 8×8 chessboard. However, in any 2×2 square we are not allowed to color all four fields. Let P be the number of all such colorings (including the case where no field is colored). Prove the estimate $P > 2^{48}$.

(ix) On a $1 \times n$ chessboard we color some sides of the square fields. We call a coloring admissible if at least one side of each field is colored.

Show that for each $n \geq 1$ the number $Q(n)$ of admissible colorings is odd.

4.7 Auxiliary Colorings

The solutions of a number of problems can be considerably simpler if we use the *method of auxiliary coloring*, where some points of the plane or parts of a geometrical object are appropriately colored, and thus divided into several classes (this is the same as the coloring of elements). With this method we can easily solve, for instance, a number of problems concerning the tiling of a chessboard with pieces of different types. In particular, one can show that tilings with certain properties cannot exist.

(i) Decide whether it is possible to tile an 8×8 chessboard, from which two opposite corner fields are removed, with 31 dominoes.

SOLUTION.   Such a tiling is not possible. We suppose that the chessboard is colored in the classical way; then the removed corner fields have the same color, say black. The "rest" of the chessboard then has 32 white and 30 black fields, while each domino covers exactly one black and one white field. Hence this modified chessboard of 62 fields cannot be tiled by 31 dominoes.   □

(ii) An $m \times n$ chessboard is tiled with some 2×2 pieces and some tetrominoes (Figure 44c). Show that if we exchange one 2×2 piece with a new tetromino, then the chessboard can no longer be tiled with this new collection of pieces.

SOLUTION.   We consider a coloring of the chessboard as shown in Figure 54. Then each tetromino covers either two or no colored fields, while a 2×2 piece always covers exactly one colored field. This means that the number of colored fields of the tiled chessboard and the number of 2×2 pieces have the same parity. If we exchange one such piece with a tetromino, we change the parity of the number of 2×2 pieces, and therefore it is impossible to tile the chessboard with this new collection of pieces.   □

(iii) We are given a convex n-gon $\mathcal{M}$ in a plane with a coordinate system. Show that if its vertices are lattice points (that is, points with integer coordinates), and if neither inside the polygon nor on its sides there is another lattice point, then $n \leq 4$.

SOLUTION.   We color all lattice points with four colors, as shown in Figure 55. If $n \geq 5$, then some two vertices of any convex n-gon $\mathcal{M}$ have the same color. The midpoint of a line segment with equally colored endpoints is clearly again a lattice point, and since $\mathcal{M}$ is a convex polygon, the midpoint of a line segment connecting two of its vertices is either an interior point of $\mathcal{M}$ or lies on some side of $\mathcal{M}$. But this contradicts the assumptions of the problem.   □

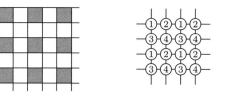

Figure 54   Figure 55   Figure 56

(iv) A convex n-gon $\mathcal{M}$ is divided into triangles by several nonintersecting diagonals, and moreover, each of the vertices of $\mathcal{M}$ belongs to an odd number of such triangles. Show that the number n is divisible by 3.

SOLUTION.   If the polygon $\mathcal{M}$ is divided into triangular parts by its diagonals, then these parts can be colored with two colors such that any two parts that have a side in common will have different colors (Figure 56). This can be done as follows: We construct the diagonals consecutively; each one of them divides $\mathcal{M}$ into two parts. In one of these parts we leave the original coloring, while in the other part we change the colors to their opposites (compare with 4.3.(i)). Since each vertex of $\mathcal{M}$ is a vertex of an odd number of triangles, with such a coloring the sides of $\mathcal{M}$ will belong to triangles with one color, say black. Let m denote the number of sides of the white triangles; it is clear that $3 \mid m$. Since every side of a white triangle and every side of the polygon $\mathcal{M}$ is also a side of some black triangle, the number of sides of black triangles is equal to $m + n$, and we have $3 \mid m + n$. Finally, with $3 \mid m$ this implies that $3 \mid n$.   □

4.8 Exercises

(i) In each field of a 5×5 chessboard there is a spider. Suppose that all the spiders crawl simultaneously to a neighboring field. Decide whether afterward at least one field will always be empty.

(ii) An $m \times n$ chessboard is tiled with dominoes, some of which are placed horizontally, while the remaining ones are placed vertically. Show that if both numbers m and n are even, then the number of "horizontal" dominoes and the number of "vertical" ones are both even.

(iii) Show that a 10×10 chessboard cannot be tiled with pieces of type T (Figure 44e).

(iv) Decide whether it is possible to tile a 10×10 chessboard with tetrominoes (Figure 44c).

(v) Suppose that 99 pieces of size 2×2 are placed on a 29×29 chessboard. Show that another such piece can always be added.

(vi) On a piece of checkered paper $4n$ arbitrary squares are marked off. Show that from these one can choose at least n squares such that no two of them have a point in common.

(vii) An equilateral triangle is divided into n^2 congruent equilateral triangles (see Figure 57 for $n = 4$). Suppose that some of them are numbered $1, 2, \ldots, m$ such that any two triangles with consecutive numbers have a side in common. Show that $m \leq n^2 - n + 1$.

Figure 57

4.9 Miscellaneous Problems

We close this section with a few more problems that can be solved with the help of colorings of geometrical objects.

(i) Suppose that on a line segment AB of length 2 a number of disjoint segments are colored in such a way that the distance between any two colored points is not equal to 1. Show that the sum of the lengths of the colored segments does not exceed 1.

SOLUTION.   On the segment AB we mark its midpoint S, and we shift all the colored line segments (or their parts) lying on AS by the vector $\overrightarrow{AS}$ (the segments or their parts lying on SB remain unshifted). Since $|AS| = 1$, we obtain a new system of disjoint colored segments lying on the segment SB, where the sum of the lengths of the segments in the new system is equal to the corresponding sum in the original system. But $|SB| = 1$, so this sum does not exceed 1.   □

(ii) We are given two circles with radius 1. On the first (respectively second) circle the arcs $\varphi_1, \ldots, \varphi_n$ (respectively $\psi_1, \ldots, \psi_m$) are marked off, and the sum of their lengths is α (respectively β), where $m\alpha + n\beta < 2\pi$. Show that the two circles can be superimposed in such a way that no two points of the marked arcs lie on top of each other.

SOLUTION.   We put the given circles on top of each other and place a painter on a fixed point of one of them. We rotate the circle with the painter, and each time any two marked points on the two circles coincide, the painter colors on the other circle the point he passes at that moment. We wish to show that at least one point remains uncolored. Note that the points that the painter colors because the arcs φ_i and ψ_j overlap ($1 \leq i \leq n, 1 \leq j \leq m$) themselves form an arc whose length is the sum of the lengths of φ_i and ψ_j. The set of all colored points is therefore the union of the mn such arcs. The sum of their lengths

$$\sum_{i=1}^{n}\sum_{j=1}^{m}(|\varphi_i| + |\psi_j|) = m \cdot \sum_{i=1}^{n}|\varphi_i| + n \cdot \sum_{j=1}^{m}|\psi_j| = m\alpha + n\beta$$

is, by the assumption of the problem, a number less than the arc length 2π of the whole circle. Therefore, the circle contains an uncolored point. □

(iii) On a circle with circumference $6n$ there are $3n$ colored points that divide it into $3n$ arcs, of which n have length 1, n have length 2, and the remaining n have length 3. Show that there are two colored points that lie on a diameter of the circle.

SOLUTION.   Note that the colored points are $3n$ vertices of some regular $6n$-gon that is inscribed in the given circle. We assume to the contrary that no two colored points lie on any diameter of the circle. Then "opposite" the colored endpoints of any arc of length 1 there are points that are not colored; that is, they are interior points of some arc of length 3. In this way we get a one-to-one correspondence between the sets of arcs of lengths 1 and those of length 3. Now fix one such arc of length 1 and its "opposite" arc of length 3, and remove them. This decomposes the circle into two arcs A_1 and A_2 of identical lengths $\frac{1}{2}(6n - 1 - 3) = 3n - 2$. If on A_1 there are exactly p arcs of length 1, q arcs of length 2, and r arcs of length 3, then A_2 contains exactly p arcs of length 3 (the ones "opposite" the arcs of length 1 in A_1). Hence $p + r = n - 1$, and the length of A_1 therefore satisfies $3n - 2 = p + 2q + 3r = n - 1 + 2(q + r)$, which implies $2(q + r) = 2n - 1$. This, however, is a contradiction, since there is an even integer on the left and an odd one on the right. □

4.10 Exercises

(i) Suppose that on a line segment of length 10 several disjoint segments are colored so that the distance between any two colored points is never equal to 1. Show that the sum of the lengths of the colored segments is not greater than 5.

*(ii) We are given two circles of radius 1, on each of which k disjoint arcs are colored such that the length of any colored arc is less than $\frac{\pi}{k^2 - k + 1}$. Furthermore, suppose it is possible to superimpose the circles in such a way that the colored arcs of both circles lie on top of each other. Show that the circles can also be superimposed in such a way that all colored arcs of one circle lie in the uncolored parts of the other circle.

*(iii) Suppose that the points $C_1, C_2, \ldots, C_n$ $(n \geq 2)$ divide the line segment AB of length 1 into a sequence of segments $AC_1, C_1C_2, \ldots, C_nB$ that are alternately colored with two colors ($C_1, C_2, \ldots, C_n$ are the only points having two colors). Show that one of the two colors, say color c, has the following property: For any d $(0 < d < \frac{1}{3})$, there are two points of color c on AB whose distance equals d. Is a stronger assertion valid for some longer interval $0 < d < \alpha$, where $\alpha > \frac{1}{3}$?

*(iv) Find all those α $(0 < \alpha \leq 180°)$ for which the following asser-
tion holds for any n: If a circle c is divided by arbitrary points
$A_1, A_2, \ldots, A_n$ into arcs $\widehat{A_1 A_2}, \widehat{A_2 A_3}, \ldots, \widehat{A_n A_1}$, and if each of
these arcs is colored with one of three colors, then on the circle
c there is an arc a of length α with endpoints of the same color (we
assume that a common endpoint of two arcs with different colors
has both of these colors).

4
Hints and Answers

1 Hints and Answers to Chapter 1

1.3 (i) $n^2 + (n-1)^2 + \cdots + 1^2 = \frac{1}{6}n(n+1)(2n+1)$ [see [5], page 11, equation (13)].

(ii) $8 + 24 + 16(n-3) + 24(n-4) + 8(n-4)^2 = 8(n-1)(n-2)$. [Proceed as in 1.2.(ii); treat $n = 3$ separately.]

(iii) From each field the queen can move horizontally in $(n-1)$ ways, and also vertically in $(n-1)$ ways; the total number of moves is therefore $2n^2(n-1)$. On a diagonal of length k the queen can make $k(k-1)$ different moves. The chessboard has four diagonals of each length $1, 2, \ldots, n-1$ and two of length n; hence the number of possible diagonal moves for the queen is $2(2\sum_{k=1}^{n-1} k(k-1) + n(n-1)) = 2(\frac{1}{3}(n-1)n(2n-1) - n(n-1) + n(n-1)) = \frac{2}{3}n(n-1)(2n-1)$, by equations (8) and (13) of [5], Chapter 1. For the total number of moves of the queen you then obtain $2n^2(n-1) + \frac{2}{3}n(n-1)(2n-1) = \frac{2}{3}n(n-1)(5n-1)$, and for $n = 8$ this gives 1456.

1.6 (i) $12 \cdot 9 \cdot 10 = 1080$.   (ii) $5 \cdot 5 = 25$; $5 \cdot 4 = 20$.

(iii) When Jirka chooses an apple, since $11 \cdot 10 > 12 \cdot 9$.

2.3 (i) $2 \cdot P(4) = 2 \cdot 4! = 48$.   (ii) $P(6) - P(5) = 5 \cdot 5! = 600$.   (iii) $32!$.

(iv) $2 \cdot (5!)^2 = 28\,800$. [The places for the men and the women can be chosen in 2 ways; on the chosen places the men can sit in 5! ways, and the same is true for the women.]

(v) $3!4!5!3! = 103,680$. [The groups of citizens of the same nationality can be placed in 3! ways. The number of placements of Canadians within their group is 4!, that of the Czechs is 5!, and the number of placements of the Germans is 3!. The result then follows from the multiplication rule.]

(vi) $n!$ [Each row and each column contains exactly one rook.]

2.6 (i) $3 \cdot 1 + 3 \cdot 4 + 3 \cdot 4 \cdot 3 + 3 \cdot 4 \cdot 3 \cdot 2 + 3 \cdot 4 \cdot 3 \cdot 2 \cdot 1 = 195$.

(ii) $V(2,5) = 5 \cdot 4 = 20$.

(iii) All $(k+1)$-element variations from n elements, of which there are $V(k+1, n)$, will be divided into groups such that each group contains all those variations that begin with the same k-tuple of elements; then each group contains exactly $(n-k)$ variations, and the number of groups is $V(k, n)$. By the addition rule you then have $V(k+1, n) = (n-k) \cdot V(k, n)$.

(iv) Subdivide all $(k+1)$-element variations from $(n+1)$ elements, of which there are $V(k+1, n+1)$, such that each group contains all those variations that begin with the same element; then each group has $V(k, n)$ variations, and the number of groups is $(n+1)$. Hence $V(k+1, n+1) = (n+1) \cdot V(k, n)$. Let us mention that the formula just obtained provides another proof of (4).

2.9 (i) $\binom{5}{2} = 10$.   (ii) $\binom{50}{5} = 2\,118\,760$.

(iii) $3 \cdot \binom{6}{2} \cdot \binom{60}{20}$, respectively $3 \cdot 6 \cdot 5 \cdot \binom{60}{20}$.

(iv) $\binom{3}{1} \cdot \binom{5}{3} = 30$.   (v) $\binom{10}{6} = 210$.   (vi) $\binom{7}{5} = 21$.

(vii) $\binom{15}{3} + \binom{15}{2} \cdot 15 = 2030$. [Either all three numbers are even, or two are odd and the third one is even.]

(viii) $2 \cdot \binom{50}{2} = 2450$. [If you choose two integers of the same parity, then their arithmetic mean is an integer, and is uniquely determined.]

(ix) $\binom{n}{k} \cdot \frac{n!}{(n-k)!} = \frac{(n!)^2}{k!((n-k)!)^2}$. [On k fixed rows the rooks can be placed in $n(n-1)\cdots(n-k+1)$ ways.]

2.12 (i) $3^6 = 729$.   (ii) $7 \cdot 8^3 = 3584$.   (iii) $2 + 2^2 + 2^3 + 2^4 = 30$.

(iv) $2 + 2^2 + 2^3 + 2^4 + 2^5 = 62$.   (v) $11 \cdot 12^2 = 1584$.   (vi) $6^2 \cdot 9 = 324$.

(vii) $5 \cdot 6 \cdot 9 = 270$.   (viii) $2 \cdot 6^3 = 432$.

(ix) n^n. [Each member can vote for any of the n candidates.]

(x) 3^n. [For each element $x \in M$ there is exactly one of the following three possibilities: either $x \notin A$, $x \notin B$, or $x \notin A$, $x \in B$, or $x \in A$, $x \in B$.]

2.16 (i) $P_0(3,2,2) = \frac{8!}{3!3!2!} = 560 = \binom{8}{3} \cdot \binom{5}{3}$.

(ii) $P_0(2,1,4,3) = \frac{10!}{2!1!4!3!} = 12\,600$.

(iii) Enumerate each m-element subset C of $A = \{a_1, \ldots, a_{m+n}\}$ by way of the ordered $(m+n)$-tuple $(\varepsilon_1, \varepsilon_2, \ldots, \varepsilon_{m+n})$ consisting of n zeros and m ones: $\varepsilon_i = 1$ exactly for those i for which $a_i \in C$. It is easy to verify that this "encoding" is a bijection. By 1.8, then, $C(m, m+n) = P_0(m, n)$.

(iv) $P_0(2,3,4) = \frac{9!}{2!3!4!} = 1260$.

(v) $P_0(2,1,2) = \frac{5!}{2!1!2!} = 30$. [The desired number must begin with a 7.]

(vi) $P_0(2,1,2) + P_0(3,2) = 30 + 10 = 40$. [To the result of (v) add the number of integers that begin with a 4.]

(vii) $P_0(2,2,2,1,1,1,1) = \frac{10!}{(2!)^3} = 453\,600$,

$2[P_0(2,2,1,1,1,1) + P_0(2,1,1,1,1,1,1)] = 2 \cdot 30\,240 = 60\,480.$

(viii) $P_0(4,4,2,1) - P_0(4,2,1,1) = \frac{11!}{(4!)^2 \cdot 2!} - \frac{8!}{4!2!} = 33\,810.$

(ix) 31. [16 points can be achieved in 1 way, 15 points in $P_0(3,1) = 4$ ways $(15 = 4 + 4 + 4 + 3)$, 14 points in $P_0(3,1) + P_0(2,2) = 10$ ways $(14 = 4 + 4 + 4 + 2 = 4 + 4 + 3 + 3)$, 13 points in $P_0(2,1,1) + P_0(3,1) = 16$ ways $(13 = 4 + 4 + 3 + 2 = 4 + 3 + 3 + 3)$.]

2.19 (i) $C_0(12,5) = P_0(12,4) = \binom{16}{4} = 1820;$ $C_0(2,5) = P_0(2,4) = \binom{6}{4} = 15.$ (ii) $C_0(5,7) = P_0(5,6) = \binom{11}{6} = 462.$

(iii) $C_0(30,5) = P_0(30,4) = \binom{34}{4} = 46\,376.$

(iv) (a) $\binom{9}{7} = \binom{9}{2} = 36.$ [It suffices to choose the two drawers in which the red balls are placed.]   (b) $C_0(7,9) \cdot C_0(2,9) = P_0(7,8) \cdot P_0(2,8) = \binom{15}{8} \cdot \binom{10}{8}.$ [First distribute the white and the red balls separately (this can be done in $C_0(7,9)$, respectively $C_0(2,9)$ ways), and then apply the multiplication rule.]

(v) $C_0(20,3) = P_0(20,2) = \binom{22}{2} = 231.$

(vi) (a) $C_0(9,4) = P_0(9,3) = \binom{12}{3} = 220;$ (b) $C_0(5,4) = P_0(5,3) = \binom{8}{3} = 56.$

(vii) (a) There are as many as there are numbers of solutions of the equation $x_1 + x_2 + x_3 + x_4 + x_5 = 9$ in $\mathbb{N}_0$; that is, $C_0(9,5) = P_0(9,4) = \binom{13}{4} = 715.$ (b) $\binom{9}{4} = 126.$ [Add to the number of solutions of the equality $x_1 + x_2 + x_3 + x_4 = 9$ in $\mathbb{N}$ the number of solutions of the strict inequality $x_1 + x_2 + x_3 + x_4 < 9$ in $\mathbb{N}$; this number is the same as the number of solutions of $x_1 + x_2 + x_3 + x_4 + x_5 = 9$ in $\mathbb{N}$. Alternative solution: The given inequality has as many solutions in $\mathbb{N}$ as the inequality $x_1 + x_2 + x_3 + x_4 \le 5$ does in $\mathbb{N}_0$, namely $\binom{9}{4}$.]

(viii) (a) $\binom{13}{4} - \binom{12}{3} = \binom{12}{4} = 495;$ (b) $\binom{9}{4} - \binom{8}{3} = \binom{8}{4} = 70.$

(ix) (a) $C_0(n, k+1) = P_0(n,k) = \binom{n+k}{k}.$ [Proceed as in (vii)(a).] (b) $C_0(n-k,k) + C_0(n-k-1, k+1) = \binom{n-1}{k-1} + \binom{n-1}{k} = \binom{n}{k}$ for $k \le n$, otherwise 0. [Proceed as in (vii)(b).]

(x) (a) $\binom{n+k}{k} - \binom{n+k-1}{k-1} = \binom{n+k-1}{k};$ (b) $\binom{n}{k} - \binom{n-1}{k-1} = \binom{n-1}{k}$ for $k \le n - 1$, otherwise 0.

(xi) $\binom{n+k}{k}.$ [Proceed as in 2.18.(v).]

3.2 (i) (a) The number does not change;   (b) The number decreases to one-sixth.

(ii) $\binom{7}{5} \cdot P_0(2,2,1,1) = 3780.$

(iii) (a) $2 \cdot P_0(2,1,1) \cdot P_0(2,1,1) = 288;$ (b) $\binom{5}{4} \cdot P_0(2,1,1) \cdot P_0(2,1,1) = 720.$

(iv) $6 \cdot 7 \cdot \binom{8}{3} = 2352.$ [First place all consonants in the unique alphabetical order, then choose places for the vowels I and E, and finally "fill in" the three letters O. Alternative solution: Divide all 282 240 anagrams from 3.1.(iii) into groups such that the same group contains all those anagrams that differ only in the distribution of consonants over a fixed set of places. Each group then contains exactly 5! anagrams, one of which has the consonants in alphabetical order. Hence the desired number is $\frac{282\,240}{5!} = 2352$.]

(v) $5! \cdot 4! \cdot \binom{6}{4} = 43\,200$.

(vi) $\binom{9}{6} = 84$. [Associate the digit 1 to the chosen books and 0 to those not chosen, and use the method of 3.1.(ii).]

(vii) $\binom{6}{4} + \binom{7}{5} = 36$. [Choose one of the knights, and divide all allowable quintuples into two groups according to whether the chosen knight is, or is not, among the "liberators." Then proceed as in 3.1.(iv).]

(viii) $\binom{2n-r-1}{r-1} + \binom{2n-r}{r} = \frac{2n}{2n-r}\binom{2n-r}{r}$. [Proceed as in 3.1.(iv).]

(ix) (a) $\binom{28}{8} = 3\,108\,105$;   (b) $\binom{31}{7} = 2\,629\,575$;   (c) $\binom{32}{8} - \binom{28}{8} = 7\,410\,195$;   (d) $\binom{4}{2} \cdot \binom{28}{6} + \binom{4}{3} \cdot \binom{28}{5} + \binom{4}{4} \cdot \binom{28}{4} = 2\,674\,035$;   (e) $\binom{8}{2}^4 = 614\,656$;   (f) $\binom{8}{2}^4 + \binom{4}{1} \cdot \binom{3}{2} \cdot \binom{8}{3} \cdot \binom{8}{2}^2 \cdot 8 + \binom{4}{2} \cdot \binom{8}{3}^2 \cdot 8^2 + \binom{4}{1} \cdot \binom{3}{1} \cdot \binom{8}{4} \cdot \binom{8}{2} \cdot 8^2 + \binom{4}{1} \cdot \binom{8}{5} \cdot 8^3 = 7\,653\,632$;   (g) $4[\binom{3}{1} \cdot \binom{8}{6} \cdot 8^2 + \binom{3}{1} \cdot \binom{2}{1} \cdot \binom{8}{5} \cdot \binom{8}{2} \cdot 8 + \binom{3}{1} \cdot \binom{8}{4} \cdot \binom{8}{2}^2 + \binom{3}{2} \cdot \binom{8}{3}^2 \cdot \binom{8}{2}] = 2\,787\,456$;   (h) $7\,653\,632 + 2\,787\,456 = 10\,441\,088$.

(x) $\frac{1}{2}\binom{12}{6} - \binom{8}{4} = \binom{4}{1} \cdot \binom{8}{5} + \frac{1}{2}\binom{4}{2} \cdot \binom{8}{4} = 434$.

(xi) (a) $\binom{3}{2} + 3 = 6$;   (b) $3 \cdot P_0(2,2) + 3 \cdot P_0(2,1,1) = 54$.

(xii) $\frac{P_0(2,2,1,1,1)}{3!} = \binom{7}{3} \cdot \binom{4}{2} = 210$.

(xiii) $P_0(3,3) = \binom{6}{3} = 20$. [It is enough to find all possible positions for the vowels and the consonants; the individual vowels and consonants can then be placed in a unique way.]

(xiv) $3 \cdot 2! \cdot 4! = 144$.

(xv) $1080 \cdot 5! \cdot 7! \cdot 10!$. [Proceed as in 3.1.(viii).]

(xvi) $\binom{n-k+1}{k}$ if $2k \le n+1$, 0 for $2k > n+1$. [Encode a chosen k-tuple of integers with a sequence of k ones and $n-k$ zeros such that for each $i = 1, \ldots, n$ a one is written in the ith position exactly when the integer i is chosen. Then use 3.1.(ii).]

3.4 (i) $\binom{5}{3}5^5 - \binom{4}{2}5^4 = 44 \cdot 5^4 = 27\,500$.

(ii) $\binom{2n}{n} \cdot 5^{2n} - \binom{2n-1}{n-1} \cdot 5^{2n-1} = \binom{2n-1}{n-1} \cdot 9 \cdot 5^{2n-1}$. [Generalize the arguments in 3.3.(i).]   (iii) $2 \left(1 + \frac{3!}{2!} + \frac{3!}{1!} + 3!\right) = 32$.

(iv) $9 \cdot 10^4 \cdot 5 = 450\,000$. [Choose all digits of the given integer, except for the last one; then this last digit can be chosen either from among the even or from among the odd digits, which would influence the parity of the digit sum. Alternative solution: Convince yourself that there are as many of these integers as there are integers with odd digit sum.]

(v) $9 \cdot 10^4$. [Choose all digits except for the last one; then this last digit is uniquely determined in view of the fact that the digit sum is divisible by 10.]

(vi) (a) $\binom{8}{4} \cdot \binom{4}{2} \cdot 2^4 = 6720$. [Proceed as in 3.3.(ii).]
(b) $\binom{6}{4} \cdot \binom{4}{2} \cdot 2^2 + 2 \cdot \binom{6}{3} \cdot \binom{3}{2} \cdot 2^3 + \binom{6}{2} \cdot 2^4 = 1560$. [Each integer ends in one of the digit pairs 12, 24, 32, 44.]

(c) $P_0(2,2,4)+P_0(2,2,3,1) = 2100$. [Derive the inequality $18 \le S(n) \le 22$ for the digit sum $S(n)$ of the integers in question; then $S(n) \in \{18, 21\}$.]

(vii) $3[P_0(2,1,1) + P_0(2,2)] = 54$.

(viii) $3P_0(4,1)+P_0(3,2)+3 \cdot P_0(3,1,1)+2 \cdot P_0(2,2,1)+2 \cdot P_0(2,1,1,1) = 265$. [Proceed as in 3.3.(iii).]

(ix) $(3 \cdot P_0(3,1) - 1) + P_0(2,2) + 2(3 \cdot P_0(2,1,1) - 2 \cdot P_0(2,1)) + (P(4) - P(3)) = 95$. [Proceed as in 3.3.(iii) and always subtract the number of integers that begin with the digit 0.]

(x) The number of integers that can be formed without the additional condition is $2 \cdot P_0(3,2)+\binom{3}{2} \cdot P_0(3,1,1)+\binom{3}{2} \cdot 2 \cdot P_0(2,2,1)+\binom{3}{1} \cdot P_0(2,1,1,1) = 440$, and in $24 = 2 \cdot 3 + 3 \cdot P(3)$ of them three digits 3 stand together. Hence the desired number is $440 - 24 = 416$.

(xi) $1 + 6 + 6 + (3 + 6) + (3 + 6) + 4 = 35$. [Proceed as in 3.3.(iv).]

(xii) Derive the inequality $1 \le S(n) \le 16$ for the digit sum $S(n)$ of the integers in question; then $S(n) = 9 = 4 + 4 + 1 + 0 = 4 + 3 + 2 + 0 = 4+3+1+1 = 4+2+2+1 = 3+3+3+0 = 3+3+2+1 = 3+2+2+2$. Hence the desired number of integers is $(P_0(2,1) + P(3)) + 3 \cdot P(3) + P_0(2,1,1) + P_0(2,1,1) + P_0(2,1) + P_0(2,1,1) + P_0(3,1) = 70$.

(xiii) (a) $(1 + 2 + \cdots + 6) \cdot 5 \cdot 4 \cdot (1 + 10 + 100) = 46\,620$;

(b) $(1 + 2 + \cdots + 6) \cdot 6^2 \cdot (1 + 10 + 100) = 83\,916$.

(xiv) (a) $(P_0(2,1) \cdot 1 + P(3) \cdot 2 + P_0(2,1) \cdot 5) \cdot (1+10+100+1000) = 33\,330$;

(b) $(1 \cdot 1 + P_0(2,1) \cdot 3) \cdot (1 + 10 + 100 + 1000) = 11\,110$;

(c) $(P_0(2,1) \cdot 1 + P_0(2,1) \cdot 4)(1 + 10 + 100 + 1000) = 16\,665$.

(xv) (a) If you take into account the contribution of the integers that begin with a 0, the sum is $(0+1+\cdots+5) \cdot 5 \cdot 4 \cdot 3 \cdot (1+10+100+1000) = 999\,900$, and the contribution mentioned is $(1+2+\cdots+5) \cdot 4 \cdot 3 \cdot (1+10+100) = 19\,980$. Hence the sum is $979\,920$.

(b) $(0+1+\cdots+5) \cdot 6^3 \cdot (1+10+100+1000) - (0+1+\cdots+5) \cdot 6^2 (1+10+100) = 3\,539\,700$.

(xvi) $(0 + 2 + 4) \cdot 5 \cdot 6^2 + (1 + 2 + \cdots + 5) \cdot 3 \cdot 6 \cdot 5 \cdot (10 + 100) + (1 + 2 + \cdots + 5) \cdot 3 \cdot 6^2 \cdot 1000 = 1\,769\,580$.

(xvii) $(2^5 - 2) \cdot \binom{10}{2} - (2^4 - 1) \cdot 9 = 1215$. [Proceed as in 3.3.(vii).]

(xviii) $(2^4 - 2) \cdot \binom{5}{2} + (2^3 - 1) \cdot 5 \cdot 5 - (2^3 - 1) \cdot 4 + (2^2 - 1) \cdot 5 = 272$. [First count the integers that begin with 0 and distinguish the cases where the integer in question has, or does not have, only even digits.]

(xix) Each such 6-digit integer occurs in one of the following three collections of digits: $[A, A, A, A, B, C]$, $[A, A, A, B, B, C]$, $[A, A, B, B, C, C]$. First, we also include integers that begin with the digit 0. Then there are $30 \cdot 10 \cdot \binom{9}{2}$ integers of the first type, $60 \cdot 10 \cdot 9 \cdot 8$ of the second type, and $90 \cdot \binom{10}{3}$ of the third type, and together $64\,800$. Similarly, the number of integers beginning with 0 is $5 \cdot 4 \cdot \binom{9}{2} + 9 \cdot 8 \cdot 5 \cdot 10 \cdot 8 \cdot (30 + 20 + 10) + 30 \cdot \binom{9}{2} = 6480$. By subtracting we get the result $58\,320$ (see 3.3.(vii)).

(xx) $\binom{k+8}{8}$. [Proceed as in 3.3.(viii).]

(xxi) Each k-digit strictly ascending integer is uniquely determined by the set of its digits, that is, by a k-element subset of the set $\{1,\ldots,9\}$. On the other hand, each k-element subset of $\{1,\ldots,9\}$ can be uniquely arranged into a k-digit strictly ascending integer. Hence there are $\binom{9}{k}$ such numbers for $k \le 9$, and none when $k > 9$.

3.6 (i) $\binom{33}{2} + 33 \cdot 34 = 1650$.

(ii) For $n = 3m$ you obtain $\binom{m}{2} + m^2 = \frac{1}{2}m(3m-1)$, for $n = 3m+1$ it is $\binom{m}{2} + m(m+1) = \frac{1}{2}m(3m+1)$, and for $n = 3m+2$ the number of choices is $\binom{m}{2} + (m+1)^2 = \frac{1}{2}\left(3m^2 + 3m + 2\right)$.

(iii) $3 \cdot \binom{m}{3} + m^3 = \frac{1}{2}m(3m^2 - 3m + 2)$.

(iv) For $n = 3m$ you obtain, by (iii), $\frac{1}{2}m\left(3m^2 - 3m + 2\right)$, for $n = 3m+1$ it is then $2\binom{m}{3} + \binom{m+1}{3} + m^2(m+1) = \frac{1}{2}m\left(3m^2 + 1\right)$, and finally for $n = 3m+2$ the number of possibilities is $\binom{m}{3} + 2 \cdot \binom{m+1}{3} + m(m+1)^2 = \frac{1}{2}m\left(3m^2 + 3m + 2\right)$.

(v) With four integers chosen, there are the following possibilities for the remainders upon division by 3: Either all are 0, or three are 1 (respectively 2) and the fourth one is 0; or two are 0, the third is 1, and the fourth is 2, or two are 1 and two are 2. The total number of choices is therefore $\binom{33}{4} + 33 \cdot \binom{34}{3} + 33 \cdot \binom{33}{3} + 34 \cdot 33 \cdot \binom{33}{2} + \binom{34}{2} \cdot \binom{33}{2} = 1\,307\,064$.

(vi) With the same argument as in (v) one shows that the number N of choices is as follows: When $n = 3m$, then $N = \binom{m}{4} + 2m\binom{m}{3} + m^2\binom{m}{2} + \binom{m}{2}^2 = \frac{1}{8}(3m-2)(3m-1)(m-1)m$; when $n = 3m+1$, then $N = \binom{m}{4} + m\binom{m+1}{3} + m\binom{m}{3} + m(m+1)\binom{m}{2} + \binom{m+1}{2}\binom{m}{2} = \frac{1}{8}\left(9m^2 + 3m + 2\right)(m-1)m$; when $n = 3m+2$, then $N = \binom{m}{4} + 2m\binom{m+1}{3} + (m+1)^2\binom{m}{2} + \binom{m+1}{2}^2 = \frac{1}{8}\left(9m^3 + 6m^2 - m - 6\right)m$.

(vii) With four integers chosen, there are the following possibilities for the remainders upon division by 4: Either all are the same; or two are 0 and two are 2; or two are 1 and two are 3; or two are 0, the third is 1, and the fourth is 3; or two are 1, the third is 0, and the fourth is 2; or two are 2, and the remaining ones are 1 and 3; or two are 3, with the remaining ones 0 and 2. The total number of choices is then $4 \cdot \binom{25}{4} + 2 \cdot \binom{25}{2}^2 + 4 \cdot \binom{25}{2} \cdot 25^2 = 980\,600$.

(viii) With the same argument as in (vii) one shows that the number N of choices is as follows: When $n = 4m$, then $N = 4 \cdot \binom{m}{4} + 2 \cdot \binom{m}{2}^2 + 4 \cdot \binom{m}{2}m^2 = \frac{1}{3}m(m-1)\left(8m^2 - 4m + 3\right)$; if $n = 4m+1$, then $N = 3 \cdot \binom{m}{4} + \binom{m+1}{4} + \binom{m}{2}^2 + \binom{m+1}{2}\binom{m}{2} + 2m(m+1)\binom{m}{2} + \binom{m+1}{2}m^2 + \binom{m}{2}m^2 = \frac{1}{3}m\left(8m^3 - 4m^2 + m - 2\right)$; when $n = 4m+2$, then $N = 2 \cdot \binom{m}{4} + 2 \cdot \binom{m+1}{4} + 2\binom{m}{2}\binom{m+1}{2} + 2\binom{m}{2}m(m+1) + 2\binom{m+1}{2}m(m+1) = \frac{1}{3}m\left(8m^3 + 4m^2 + m - 1\right)$; when $n = 4m+3$, then $N = \binom{m}{4} + 3 \cdot \binom{m+1}{4} + \binom{m}{2}\binom{m+1}{2} + \binom{m+1}{2}^2 + 2\binom{m+1}{2}m(m+1) + \binom{m+1}{2}(m+1)^2 + \binom{m}{2}(m+1)^2 = \frac{1}{3}m^2\left(8m^2 + 12m + 7\right)$.

(ix) If both variables are nonzero, there are $4(1 + 2 + \cdots + 98) = 19\,404$ solutions; if one is equal to zero, then the number is $4 \cdot 99 = 396$, and so together with the solution $(x,y) = (0,0)$ there is a total of $19\,801$ solutions.

(x) $4(1+2+\cdots+n-1)+4n+1 = 2n^2+2n+1$. [Generalize the argument of (ix).]

(xi) By 2.19.(ix), the inequality $a + b + c \leq 100$ has $\binom{103}{3} = 176\,851$ solutions in $\mathbb{N}_0$, where in one solution all the unknowns are 0, exactly two unknowns are 0 in $3 \cdot 100 = 300$ cases, a single unknown is 0 in $3 \cdot \binom{100}{2} = 14\,850$ cases, and all are nonzero in $\binom{100}{3} = 161\,700$ cases. If in the inequality $|x| + |y| + |z| \leq 100$ you consider all possible sign distributions, you get $1 + 2 \cdot 300 + 4 \cdot 14\,850 + 8 \cdot 161\,700 = 1\,353\,601$ solutions.

(xii) As in (xi) (where $n = 100$) the number of solutions is equal to $\binom{n}{3} \cdot 2^3 + 3 \cdot \binom{n}{2}2^2 + 3\binom{n}{1}2 + 1 = \frac{1}{3}\left(4n^3 + 6n^2 + 8n + 3\right) = \frac{1}{3}(2n + 1)\left(2n^2 + 2n + 3\right)$.

(xiii) Assume that among the values of the unknowns there are exactly s zeros $(s = 0, 1, \ldots, k)$; they can be chosen in $\binom{k}{s}$ ways. The remaining $(k - s)$ unknowns have nonzero values, and their signs can be chosen in 2^{k-s} ways. The inequality $y_1 + y_2 + \cdots + y_{k-s} \leq n$ has exactly $\binom{n}{k-s}$ solutions in $\mathbb{N}$; hence the number of all solutions of the given inequality is $\sum_{s=0}^{k} \binom{k}{s}\binom{n}{k-s} \cdot 2^{k-s} = \sum_{s=0}^{k} \binom{k}{s}\binom{n}{s} \cdot 2^s$.

(xiv) The substitutions $r = x$, $s = x + y$, $t = x + y + z$ transform the given equation into $r+s+t = n$, where $r \leq s \leq t$. The number of solutions of this equation is given in 3.5.(iv).

(xv) The given inequality has exactly as many solutions as the equation $3x + 2y + z = n$ in $\mathbb{N}_0$.

(xvi) The unknown x can assume values from the set $\{0, 1, \ldots, [\frac{n}{2}]\}$, and for a fixed x the equation has $(n - 2x + 1)$ solutions. The total number of solutions is therefore

$$\sum_{x=0}^{[n/2]}(n - 2x + 1) = \left(\left[\tfrac{n}{2}\right] + 1\right)(n + 1) - \left[\tfrac{n}{2}\right]\left(\left[\tfrac{n}{2}\right] + 1\right)$$
$$= \left(\left[\tfrac{n}{2}\right] + 1\right)\left(\left[\tfrac{n+1}{2}\right] + 1\right).$$

(xvii) Similar to (xvi) the total number of solutions is

$$\sum_{x=0}^{[n/2]}\left(\tfrac{n-2x+2}{2}\right) = \tfrac{1}{2}\left(\left[\tfrac{n}{2}\right] + 1\right)\left(\left[\tfrac{n}{2}\right] \cdot \left(\tfrac{4}{3}\left[\tfrac{n}{2}\right] - 2n - \tfrac{7}{3}\right) + (n + 2)(n + 1)\right).$$

(xviii) Analogously to (xvi), the total number of solutions is

$$\sum_{x=0}^{[n/3]}(n - 3x + 1) = \left(\left[\tfrac{n}{3}\right] + 1\right)\left(n + 1 - \tfrac{3}{2}\left[\tfrac{n}{3}\right]\right).$$

(xix) When x is fixed then y, with $y \in \mathbb{Z}$, is uniquely determined if and only if $2 \mid n - 3x$. Hence the desired number of solutions is equal to the number of elements in the set $\{x \in \mathbb{Z};\ 0 \leq x \leq \frac{n}{3},\ 2 \mid x - n\}$, namely $\left[\frac{n}{6}\right] + 1$ for even n, and $\left[\frac{n+3}{6}\right]$ for odd n. The final result can be written in the form $\left[\frac{n+4}{6}\right] + \left[\frac{n}{6}\right] - \left[\frac{n-1}{6}\right]$ (consider the remainders upon dividing n by 6).

(xx) (a) $\tau(n) = \prod_{i=1}^{k}(1+w_i) \neq w$.   (b) When n is not a perfect square, the number of decompositions is $\frac{1}{2}\tau(n)$; otherwise (that is, when $2 \mid w_i$ for all $i = 1, 2, \ldots, k$), the number is $\frac{1}{2}(\tau(n) + 1)$.

(xxi) Set $x_1 = p_1^{\alpha_1} \cdot p_2^{\alpha_2} \cdots p_k^{\alpha_k}$, $x_2 = p_1^{\beta_1} \cdot p_2^{\beta_2} \cdots p_k^{\beta_k}$, where $\alpha_i \in \mathbb{N}_0$ and $\beta_i \in \mathbb{N}_0$ for all $i = 1, 2, \ldots, k$. Then $\omega_i = \alpha_i + \beta_i$ and $\alpha_i \beta_i = 0$ has to hold for all $i = 1, 2, \ldots, k$, that is, $(\alpha_i, \beta_i) \in \{(\omega_i, 0), (0, \omega_i)\}$ holds in case (a), and so the number of all decompositions is 2^k. If the order of the factors x_1, x_2 is irrelevant, then this number is 2^{k-1}.

(xxii) Use the notation from (xxi) and set, in addition, $x_3 = p_1^{\gamma_1} \cdot p_2^{\gamma_2} \cdots p_k^{\gamma_k}$. Then $\omega_i = \alpha_i + \beta_i + \gamma_i$ holds for each $i = 1, 2, \ldots, k$, and furthermore, $\alpha_i \beta_i = \alpha_i \gamma_i = \beta_i \gamma_i = 0$. Hence two of the numbers $\alpha_i, \beta_i, \gamma_i$ are equal to 0, and the third one is ω_i. Thus, in case (a) there are 3^k possibilities. Of these, three decompositions have the factors $1, 1, n$, while all other decompositions have three distinct factors. Therefore, in case (b) the total number of decompositions is $1 + \frac{1}{3!} (3^k - 3) = 1 + \frac{1}{2} (3^{k-1} - 1)$.

(xxiii) Use the notation of (xxii); then $\omega_i = \alpha_i + \beta_i + \gamma_i$ and $\alpha_i \beta_i \gamma_i = 0$ must hold for $i = 1, 2, \ldots, k$, and so one of the integers $\alpha_i, \beta_i, \gamma_i$ has to be 0, and the sum of the remaining two ω_i. Exactly one 0 is obtained in $3(\omega_i - 1)$ cases, and in three cases two of the integers $\alpha_i, \beta_i, \gamma_i$ are equal to zero. Hence by the multiplication rule there are $3^k \cdot \omega_1 \omega_2 \cdots \omega_k$ possibilities in the situation (a). For case (b) note that each "unordered" decomposition into three different factors gives six decompositions from case (a), while each decomposition into three factors of which two are identical (and because of $(x, y, z) = 1$, the third one must be different), gives three decompositions from case (a). Determine now the number of decompositions of the integer n into a product $n = x^2 y$, where $(x, y) = 1$: If $x = p_1^{\varphi_1} \cdot p_2^{\varphi_2} \cdots p_k^{\varphi_k}$, $y = p_1^{\psi_1} \cdot p_2^{\psi_2} \cdots p_k^{\psi_k}$, then both $2\varphi_i + \psi_i = \omega_i$ and $\varphi_i \cdot \psi_i = 0$ must hold for $i = 1, 2, \ldots, k$. This has one solution when $2 \nmid \omega_i$, and two solutions when $2 \mid \omega_i$. If among the integers $\omega_1, \omega_2, \ldots, \omega_k$ exactly t are even, then the system has 2^t solutions. If N denotes the desired number of unordered decompositions, then comparing with part (a) gives $3^k \omega_1 \omega_2 \cdots \omega_k = 6(N - 2^t) + 3 \cdot 2^t$, and thus $N = \frac{1}{6} (3 \cdot 2^t + 3^k \omega_1 \omega_2 \cdots \omega_k)$.

(xxiv) Set $x_1 = p_1^{\alpha_1} \cdot p_2^{\alpha_2} \cdots p_k^{\alpha_k}$, $x_2 = p_1^{\beta_1} \cdot p_2^{\beta_2} \cdots p_k^{\beta_k}$, $x_3 = p_1^{\gamma_1} \cdot p_2^{\gamma_2} \cdots p_k^{\gamma_k}$, where $\alpha_i, \beta_i, \gamma_i \in \mathbb{N}_0$ for all $i = 1, 2, \ldots, k$. Proceed as in 3.5.(vii); then the number U of "ordered" decompositions from (a) is equal to $\prod_{i=1}^{k} \binom{\omega_i + 2}{2}$. If you divide these decompositions into classes R_1, R_2, R_3 as in 3.5.(vii), then the numbers r_1, r_2, r_3 of "unordered" decompositions in these classes satisfy $U = r_1 + 3r_2 + 6r_3$, $N = r_1 + r_2 + r_3$, where N is the total number of "unordered" decompositions. The numbers r_1, r_2, and r_3 are determined successively: If n is the third power of a natural number, then $r_1 = 1$; otherwise, $r_1 = 0$. The number $r_1 + r_2$ counts all the "unordered" decompositions into three factors, where at least two factors agree. Hence $r_1 + r_2 = \tau \left(\prod_{i=1}^{k} p_i^{\left[\frac{\omega_i}{2} \right]} \right) = \prod_{i=1}^{k} \left(\left[\frac{\omega_i}{2} \right] + 1 \right)$, and finally, $r_3 = \frac{1}{6}(U - r_1 - 3r_2)$ and $N = \frac{1}{6}(U + 5r_1 + 3r_2)$.

(xxv) Set $x = p_1^{\alpha_1} \cdot p_2^{\alpha_2} \cdots p_k^{\alpha_k}$, $y = p_1^{\beta_1} \cdot p_2^{\beta_2} \cdots p_k^{\beta_k}$. Then $2\alpha_i + \beta_i = \omega_i$ must hold for $i = 1, 2, \ldots, k$. For a given i this equation has $\left[\frac{\omega_i}{2} \right] + 1$

solutions and therefore, by the multiplication rule, the given equation has exactly $\prod_{i=1}^{k}\left(\left[\frac{\omega_i}{2}\right]+1\right)$ solutions. A different approach: Substitute $u = x$ and $v = xy$ and use 3.5.(v) for the equation $n = uv$ under the condition $u|v$.

(xxvi) With the same notation as in (xxv), $3\alpha_i + \beta_i = \omega_i$ must hold for $i = 1, 2, \ldots, k$. For a given i this equation has exactly $\left(\left[\frac{\omega_i}{3}\right]+1\right)$ solutions, and therefore by the multiplication rule the given equation has exactly $\prod_{i=1}^{k}\left(\left[\frac{\omega_i}{3}\right]+1\right)$ solutions.

(xxvii) Use the substitutions $r = xyz$, $s = xy$, and $t = x$ to transform the given problem to that of determining the number of all decompositions $n = rst$, where $t \mid s$ and $s \mid r$; but this is 3.5.(vi).

3.9 (i) There are 5 possibilities in total: $(5, 5, 5, 1, 1)$, $(5, 5, 1, 5, 1)$, $(5, 5, 1, 1, 5)$, $(5, 1, 5, 5, 1)$, $(5, 1, 5, 1, 5)$.

(ii) The result is the same as in (i); if the first digit 5 is not written, then the possibilities are $(5, 5, 1, 1)$, $(5, 1, 5, 1)$, $(5, 1, 1, 5)$, $(1, 5, 5, 1)$, $(1, 5, 1, 5)$.

(iii) $f(10, 20) = \frac{11}{21}\binom{30}{20} = 15\,737\,865$.

(iv) $g(10, 20, 5) = \binom{30}{10} - \binom{30}{4} = 30\,017\,610$.

(v) $f(10, 10) = \frac{1}{11} \cdot \binom{20}{10} = 16\,796$. [Encode Alenka's daily decisions into a sequence of letters N, I, where I (respectively N) stands for a day when she buys (respectively does not buy) herself ice cream. The allowable sequences are exactly those where in front of each I there are more letters N than I.]

(vi) For $d \le r$ the result is $d!r!f(d, r) = \frac{r-d+1}{r+1} \cdot (d + r)!$; otherwise, it is 0.

(vii) (a) $\binom{m+k-1}{m} - \binom{m+k-1}{m-1} = \frac{k-m}{k}\binom{m+k-1}{m}$. [Proceed exactly as in 3.7; convince yourself that each "good" sequence begins with a 5.]
(b) $\frac{1}{k}\binom{2k-2}{k-1}$. [Note that each "good" sequence begins with a 5 and ends with a 1; then the problem can easily be reduced to 3.7, and the identity (13) can be used.]

3.11 The number of desired sequences is $f(k, k) = \frac{1}{k+1}\binom{2k}{k}$. [Encode an arbitrary sequence $(a_1, a_2, \ldots, a_k)$ by way of a queue with $2k$ persons, where k have a five-dollar bill, and k have a ten-dollar bill, and such that the rth person with a five-dollar bill stands in the $(a_r + r - 1)$th position. Note that all "good" arrangements of the queue are obtained in this way.]

4.3 (i) Let $a_1, \ldots, a_N$ be those N distinct objects from the situation corresponding to identity (15), and to each distribution of these objects over the bins $T_1, \ldots, T_k$ associate an ordered N-tuple as follows: In the first place write the number of the bin that contains the object a_1, in the second place that of the bin containing a_2, etc. It is clear that this defines a bijection between the set of all distributions under consideration and the set of all arrangements with repetition of n_1 elements of the first type T_1, n_2 elements of the second type T_2, etc.

(ii) $\frac{32!}{(10!)^3 \cdot 2!}$.   (iii) (a) $\frac{32!}{(10!)^2 12!}$;   (b) $\frac{32!}{(10!)^3 \cdot 2!}$.   (iv) $\frac{30!}{3!(10!)^3}$, $\frac{30!}{10!(3!)^{10}}$.

(v) $\frac{(3n)!}{(n!)^3}$.   (vi) $\frac{16!}{8!}$.   (vii) (a) $\frac{(\frac{n(n+1)}{2})!}{1!2!\cdots(n-1)!}$;   (b) $\frac{(\frac{n(n+1)}{2})!}{1!2!\cdots(n-1)!n!}$.

(viii) $\frac{10!}{5!(2!)^5} = 945$.   (ix) $\frac{9!}{(3!)^4} = 280$.   (x) $\frac{9!}{4!(2!)^4} = 945$.

(xi) The deck of cards without any aces can be divided into two parts of equal size in $\frac{28!}{(14!)^2 \cdot 2}$ ways. Two aces can be added to the first part in $\binom{4}{2}$ ways, and the two remaining aces are then added to the second part. By the multiplication rule the deck of cards can therefore be divided in $\frac{3 \cdot 28!}{(14!)^2}$ ways.

(xii) $\frac{9!}{4!(2!)^4} = 945$.   (xiii) $\frac{9!}{4!} = 15\,120$.

(xiv) 16!. [For the first round there are $\frac{16!}{8!}$ possibilities, for the second round $\frac{8!}{4!}, \dots$, for the fourth round 2!; then use the multiplication rule.]

(xv) $17 \cdot \frac{16!}{8!} \cdot 9 \cdot \frac{8!}{4!} \cdot 5 \cdot \frac{4!}{2!} \cdot 3 \cdot 2! \cdot 2! = 270 \cdot 17!$.

(xvi) $17 \cdot (16!)^2$.   (xvii) 12!.

(xviii) (a) $\frac{52!}{(4!)^{13}}$;   (b) $(13!)^4$;   (c) $13 \cdot 13^4 \cdot 12! \left(\frac{12!}{3!(4!)^3}\right)^4 = \frac{(13!)^5}{(3!)^4 \cdot (4!)^{12}}$.

4.6 (i) $\binom{6}{3} \cdot \binom{13}{3} = 5720$.   (ii) $(n_1 + 1) \cdot (n_2 + 1) \cdots (n_k + 1)$.

(iii) If $n = p_1^{\alpha_1} \cdot p_2^{\alpha_2} \cdots p_k^{\alpha_k}$ and $a \mid n$, then $n = a \cdot b$, where $a = p_1^{\beta_1} \cdot p_2^{\beta_2} \cdots p_k^{\beta_k}$, $b = p_1^{\gamma_1} \cdot p_2^{\gamma_2} \cdots p_k^{\gamma_k}$, and the exponents β_i, γ_i satisfy $0 \leq \beta_i$, $0 \leq \gamma_i$, $\beta_i + \gamma_i = \alpha_i$ for all $i = 1, 2, \dots, k$. Therefore, the number $p(n)$ of all desired divisors of n is equal to the number of possible distributions of α_1 elements of the first type, α_2 elements of the second type, $\dots, \alpha_k$ elements of the kth type over two different bins. Therefore, by (ii), $p(n) = (\alpha_1 + 1)(\alpha_2 + 1) \cdots (\alpha_k + 1)$.

(iv) (a) $\binom{13}{4} \cdot \binom{15}{6} = 3\,578\,575$;   (b) $\binom{10}{4} = 210$.

(c) Two empty bins can be chosen in $\binom{10}{2} = 45$ ways; then 4 blue and 6 white balls have to be distributed over 8 bins such that none of these bins remains empty. All the distributions can be divided into 4 groups S_1, S_2, S_3, S_4 according to the placements of the blue balls. Not all of them can be in the same bin. The group S_1 contains those placements where the blue balls are distributed according to the scheme $3 + 1$; the bins for them can be chosen in $8 \cdot 7 = 56$ ways, and the remaining bins will then contain one white ball each. S_2 will have those distributions where two bins contain two blue balls each, for which there are $\binom{8}{2} = 28$ possibilities. In S_3 the blue balls are distributed according to the scheme $2 + 1 + 1$, the 5 remaining bins will contain one white ball each, and for the sixth white ball there are 8 possibilities; hence $|S_3| = 8 \cdot \binom{7}{2} \cdot 8 = 1344$. Similarly, the group S_4, where the blue balls are distributed according to the scheme $1 + 1 + 1 + 1$, contains $\binom{8}{4} \cdot \binom{9}{2} = 2520$ distributions. Therefore, the number of all possible distributions is $45 \cdot (56 + 28 + 1344 + 2520) = 177\,660$.

(v) (a) $\binom{15}{6} \cdot \binom{11}{2}^2 = 15\,140\,125$;   (b) $P_0(6, 2, 2) = \binom{10}{6} \cdot \binom{4}{2} = 1260$.

(vi) (a) $\binom{10}{3}^3 = 1\,728\,000$.   (b) One bin must contain two balls, and the remaining ones one ball each. If both balls have the same color, then the number of possibilities is $3 \cdot 8 \cdot P_0(3, 3, 1)$; if the two balls are of different colors, then there are $3 \cdot 8 \cdot P_0(3, 2, 2)$ possibilities. The total number of placements is therefore 8400.

(vii) $3!(p_2(6,3)+p_2(6,2,1)+p_2(5,4)+p_2(5,3,1)+p_2(5,2,2)+p_2(4,4,1)+p_2(4,3,2)+p_2(3,3,3)) = 19\,194$. [Proceed as in 4.5.(iii).]

(viii) $2 \cdot P_0(2,1) = 6$. [Only distributions according to the schemes $4+4+2$ or $4+3+3$ are possible.]

(ix) (a) 264. [There are two possible distributions, namely, $3+3+3+1$, $3+3+2+2$. In the first case one can choose a "disadvantaged" child in 4 ways and hand him or her either an apple or a pear, but always the four remaining pieces of fruit can be distributed in 12 ways. Similarly, for the scheme $3+3+2+2$ two "disadvantaged" children can be chosen in 6 ways, and for each choice there are 28 ways of distributing the fruit.
(b) If p children get apples and q get pears ($1 \le p \le 4, 1 \le q \le 4, p+q \le 4$) then in analogy with 4.5.(v) the number n of all distributions is

$$n = \sum_{(p,q)} P_0(p,q,4-(p+q)) \cdot \binom{4}{p-1} \cdot \binom{4}{q-1},$$

where $(p,q) \in \{(1,1);(1,2);(2,1);(1,3);(3,1);(2,2)\}$. The final result is then $n = 252$.

(x) $P_0(6,4,5,4) = \frac{19!}{6!5!(4!)^2}$.   (xi) $\binom{10}{2} \cdot P_0(5,2,2,1,1) = \frac{11!10!}{8!5!2^3}$.

(xii) $(p+1)\binom{n+1}{p}$. [In $\binom{n+1}{p}$ cases the black ball is not used, and in $p \cdot \binom{n+1}{p}$ cases it is used.]

(xiii) Let x_i ($i = 1,2,3$) denote the number of elements of the ith kind in the first bin; then the number distributions of the object is equal to the number of all nonnegative integer solutions of the equation $x_1+x_2+x_3 = 3n$ under the conditions $0 \le x_i \le 2n$. Without these conditions, the equation has $\binom{3n+2}{2}$ solutions. Now determine the number of solutions with $x_1 > 2n$; this number is the same as the number of solutions of the inequality $x_2 + x_3 < n$ in $\mathbb{N}_0$, namely, according to 2.19.(x), $\binom{n+1}{2}$. The number of solutions is the same for $x_2 > 2n$, and for $x_3 > 2n$. Hence the given equation has $\binom{3n+2}{2} - 3 \cdot \binom{n+1}{2} = 3n^2 + 3n + 1$ solutions, which is the same as the desired number of distributions.

(xiv) Similar to (xiii), the number of distributions is

$$\binom{4n+3}{3} - 4\binom{2n+2}{3} = \frac{(2n+1)(8n^2+8n+3)}{3}.$$

(xv) For $i = 1,2,3$ let x_i (respectively y_i) denote the number of white (respectively black) balls in the ith bin. Then $x_1+x_2+x_3 = 5$, $y_1+y_2+y_3 = 5$ under the conditions $0 \le x_i+y_i \le 5$. Without these conditions, each of the two equations has $\binom{7}{2} = 21$ solutions, and thus the system has $21^2 = 441$ solutions. The number of solutions for which $x_1 + y_1 \le 5$ does not hold is equal to the number of nonnegative integer solutions of the system of equations $x_2 + x_3 = r$, $y_2 + y_3 = s$, where $0 \le r, s < 5$ and $r + s < 5$. Since for fixed r, s this number is equal to $(r + 1)(s + 1)$, there are 70 solutions in total, which is the same number of solutions for which $x_2 + y_2 > 5$, respectively $x_3 + y_3 > 5$, holds. Therefore, the number of all distributions of the balls is $441 - 3 \cdot 70 = 231$.

4.9 (i) $r_1(8,4) = 4^8 = 65\,536$; $P_1(2,4) = P_0(2,2,2,2) = \frac{8!}{(2!)^4} = 2520$.

(ii) $\frac{r_5(24,6)}{6!} = \frac{24!}{6!}\binom{23}{5}$.

(iii) $3 \cdot \binom{33}{11} \cdot 2^{22} - 2 \cdot \frac{33!}{(11!)^3}$. [Proceed as in 4.8.(iii).]

(iv) (a) $r_4(7,3) = \frac{9!}{2!} = 181\,440$;   (b) $r_5(7,3) = 7! \cdot \binom{6}{2} = 75\,600$;

(c) $\binom{7}{1} \cdot r_4(1,3) + \binom{7}{2} \cdot r_4(2,3) + \cdots + \binom{7}{7} \cdot r_4(7,3) = \binom{7}{1} \cdot \frac{3!}{2!} + \binom{7}{2} \cdot \frac{4!}{2!} + \cdots + \binom{7}{7} \cdot \frac{9!}{2!} = 390\,453$;

(d) $\binom{7}{3} \cdot \binom{2}{2} \cdot 3! + \binom{7}{4} \cdot \binom{3}{2} \cdot 4! + \binom{7}{5} \cdot \binom{4}{2} \cdot 5! + \binom{7}{6} \cdot \binom{5}{2} \cdot 6! + \binom{7}{7} \cdot \binom{6}{2} \cdot 7! = 143\,850$.

(v) (a) $r_4(n,k) = \frac{(n+k-1)!}{(k-1)!}$; (b) $r_5(n,k) = n!\binom{n-1}{k-1}$ for $n \geq k$, otherwise 0;

(c) $\binom{n}{1} \cdot \frac{k!}{(k-1)!} + \binom{n}{2} \cdot \frac{(k+1)!}{(k-1)!} + \cdots + \binom{n}{n} \cdot \frac{(n+k-1)!}{(k-1)!}$;

(d) $\binom{n}{k}\binom{k-1}{k-1} \cdot k! + \binom{n}{k+1}\binom{k}{k-1} \cdot (k+1)! + \binom{n}{k+2}\binom{k+1}{k-1} \cdot (k+2)! + \cdots + \binom{n}{n}\binom{n-1}{k-1} \cdot n!$ for $n \geq k$, otherwise 0.

5.3 (i) For a given n-element set A compare the list of all pairs (a, B), where $B \subseteq A$, $|B| = k + 1$, $a \in B$, with the list of all pairs (a, C), where $C \subseteq A$, $|C| = k$, $a \in A \setminus C$. The mapping $(a, B) \mapsto (a, B \setminus \{a\})$ gives a bijection.

(ii) There are a total of $(2^n)^2 = 4^n$ ordered pairs (A, B) of subsets of X. Associate to each such pair three further pairs $(X \setminus A, B)$, $(A, X \setminus B)$, $(X \setminus A, X \setminus B)$; this divides the system of all pairs into quadruples. Each element $x \in X$ lies in exactly one of the sets $A \cap B$, $(X \setminus A) \cap B$, $A \cap (X \setminus B)$, $(X \setminus A) \cap (X \setminus B)$, which means that each of the quadruples counts exactly once. Hence the number of these quadruples is $\frac{1}{4} \cdot 4^n = 4^{n-1}$, and the sum of the number of elements in all sets of the form $A \cap B$ is equal to $n \cdot 4^{n-1}$.

5.5 (i) As in 5.4.(i), consider all m-element subsets of the set $A = B \cup C$, where $B = \{b_1, b_2, \ldots, b_p\}$, $C = \{c_1, c_2, \ldots, c_m\}$, and divide them into classes $T_0, T_1, \ldots, T_m$ such that the class T_i contains exactly i elements of the set B. Then $|T_i| = \binom{p}{i}\binom{m}{m-i} = \binom{p}{i}\binom{m}{i}$.

(ii) Proceed exactly as in 5.4.(iii) and note that for $m \leq n$ the number of nonempty classes under consideration is exactly m.

5.7 (i) Consider all words of length n in which k copies of the letter A and $(n - k)$ copies of B are used.

(ii) Consider all words of length n in which k copies of the letter A, $(n - k - 1)$ copies of B, and one copy of C are used.

(iii) Consider all words of length n in which k copies of the letter A, $(m - k)$ copies of B, and $(n - m)$ copies of C are used.

(iv) Consider all words of length n formed with the letters A, B, C and then divide them into groups according to how often the letter A is used.

(v) Consider all words formed with n copies of A and n copies of B and divide them into groups according to how often the letter A occurs in the first n positions of the word.

(vi) Proceed as in 5.6.(v) for the proof of (39).

(vii) Consider all words of length n in which the letter C is used once and the remaining letters are copies of of the letters A and B. Show, as in

5.6.(iii), that the number of words in which the number of copies of A is even is the same as the number of words in which the number of copies of A is odd.

(viii) Consider all words of length n in which the letter C is used $(n-m)$ times and the remaining letters are copies of A and B; divide the words into groups according to how many copies of A they contain.

(ix) Proceed as in (viii) and show that the number of words in which the letter A occurs an even number of times is equal to the number of those in which A occurs an odd number of times.

(x) Consider all words of length n that contain exactly one letter C and at most one letter D, and the remaining letters are copies of A and B. Subdivide these words in two different ways: To determine the right-hand side, divide them according to whether or not a word contains the letter D; and for the sum on the left-hand side, divide them into n groups $T_1, \ldots, T_n$ such that T_i contains those words in which A occurs in exactly $(n-i)$ copies. Each group T_i can be further subdivided according to whether or not a given word contains the letter D.

(xi) Consider all words of length $(n+m-1)$ in which $(n-1)$ copies of the letter C are used, with the remaining positions occupied by copies of A and B, such that A can occur only in the first n positions. For $k = 0, 1, \ldots, \alpha$, where $\alpha = \min(m, n)$, let M_k denote the set of all words that contain exactly k copies of A, and thus $(m-k)$ copies of B. The group M_k contains exactly $\binom{n}{k} \cdot \binom{n+m-1-k}{m-k}$ words, where it is clear that there are as many words with an even number of copies of A as there are those with an odd number of copies of A.

(xii) Let U denote the set of all words of length $2n$ formed with n copies of the letter X and n copies of Y; clearly, $|U| = \binom{2n}{n}$. Each word $u \in U$ can be written as n pairs of neighboring letters, where the pairs are of the form XX (respectively YY, respectively XY, respectively YX). Replace them with the letters A (respectively B, respectively C, respectively D); this gives a word of length n consisting of the letters A, B, C, D, where the number of copies of A and B are the same. Denote the set of these words by V and justify that $|U| = |V|$. To determine $|V|$, subdivide all words in V into groups S_k ($k = 1, 2, \ldots, \left[\frac{n}{2}\right]$) such that S_k contains all those words in V that contain exactly k copies of the letter A (and thus also of B). Positions for these two letters can then be chosen in $\binom{n}{2k}$ ways from the n places, and for each of these the letters A can be written there in any of the $\binom{2k}{k}$ ways; the remaining $(n-2k)$ positions are then occupied by the letters C and D. Therefore, $|S_i| = \binom{n}{2k}\binom{2k}{k}2^{n-2k}$, and upon summing over all $k \in \left\{0, 1, \ldots, \left[\frac{n}{2}\right]\right\}$ you obtain the desired result.

5.9 (i) As in 5.8.(iii) divide all walks from the point $P[0,0]$ to the point $E[m,p]$ into classes according to which one of the points $D_i[m-i,i]$ it traverses.

(ii) As in 5.8.(iv) divide all walks from the point $P[0,0]$ to the point $Q[m,n]$ into classes according to the point $F_i[k,i]$ at which it leaves the "barrier" of points $F_0, F_1, \ldots, F_n$.

(iii) Visualize an arbitrary "bad" arrangement of the queue by a walk from $U[-q-1, q+1]$ to $T[k,m]$.

(iv) Visualize all "bad" arrangements of the queue by walks from $W[0,1]$ to $T[k,m]$.

(v) Divide all walks from $A[-1,1]$ to $T[k,m]$ into classes M_s according to which point $R_s[s, s+1]$ $(s = 0, 1, \ldots, m-1)$ is the *first* point the walk has in common with the straight line r of 5.8.(v). Determine the number of walks in each of these classes M_s: All walks from A to R_s end in the segment XR_s, where $X[s-1, s+1]$, so there are as many "good" walks from A to R_s as there are walks from $P[0,0]$ to $Y[s,s]$ that have no points in common with the straight line r; this number is $P_0(s,s) - P_0(s+1, s-1) = \frac{1}{s+1}\binom{2s}{s}$. Each walk from A to R_s can then be "extended" by a walk from R_s to T, where $c(R_s, T) = P_0(k-s, m-s-1) = \binom{k+m-2s-1}{m-s-1}$. Therefore, $|M_s| = \frac{1}{s+1}\binom{2s}{s}\binom{k+m-2s-1}{m-s-1}$, and upon summing over all $s = 0, 1, \ldots, m-1$ you obtain the assertion.

(vi) If you divide the desired identity by $k+1$, the expression on the left will, by (13) of 3.7, give the number of walks from $P[0,0]$ to $T[k,k]$ that have no point in common with the straight line r from 5.8.(v). Divide these walks into groups M_s $(s = 1, 2, \ldots, k)$ according to which one of the points $R_s[s,s]$ is the *first* point (different from P) the walk has in common with the straight line p of 5.8.(v). Then $|M_s| = n_1(s) \cdot n_2(s)$, where $n_1(s)$ is the number of walks from P to R_s that have no "inner points" in common with the straight line p, and $n_2(s)$ the number of walks from R_s to T that do not touch the straight line r. To determine $n_1(s)$, note that each "good" walk goes through the points $X[1,0]$ and $Y[s, s-1]$; hence $n_1(s) = c(X, Y) - c(Z, Y)$, where $Z[0,1]$, and $c(X, Y) = \binom{2s-2}{s-1}$, $c(Z, Y) = \binom{2s-2}{s-2}$, which means that $n_1(s) = \frac{1}{s}\binom{2s-2}{s-1}$. Furthermore, $n_2(s) = c(R_s, T) - c(U, T)$, where $U[s-1, s+1]$, and $c(R_s, T) = \binom{2k-2s}{k-s}$, $c(U,T) = \binom{2k-2s}{k-s+1}$, and therefore $n_2(s) = \frac{1}{k-s+1}\binom{2k-2s}{k-s}$. Together, $|M_s| = \frac{1}{s}\binom{2s-2}{s-1} \cdot \frac{1}{k-s+1}\binom{2k-2s}{k-s} = \frac{1}{s(k-s+1)} \cdot \binom{2s-2}{s-1} \cdot \binom{2k-2s}{k-s}$, and the result is obtained by summing over all $s = 1, 2, \ldots, k$.

6.4 (i) With the notation from 6.3.(i):

(a) $|E \cup G \cup F| = 11$;

(b) $|E| - |E \cap G| - |E \cap F| + |E \cap F \cap G| = 1$;

(c) $|F| - |F \cap E| - |F \cap G| + |E \cap F \cap G| = 3$.

(ii) The teacher is not telling the truth. Let T_i denote the set of students in the ith club; then in this case the identity (47) has the form

$$|T_1 \cup T_2 \cup T_3 \cup T_4| = |T_1| + \cdots + |T_4| - |T_1 \cap T_2| - \cdots - |T_3 \cap T_4|.$$

The inequalities $|T_1 \cup T_2 \cup T_3 \cup T_4| \leq 30$ and $|T_i \cap T_j| \leq 5$ then lead to a contradiction.

(iii) $\frac{6!}{2^3} - 3 \cdot \frac{5!}{2^2} + 3 \cdot \frac{4!}{2} - 3! = 30$. [Proceed as in 6.3.(iii).]

(iv) $2n^n - n!$. [Let A, respectively B, denote the set of all placements that have exactly one rook in each row, respectively column; use (45) to determine $|A \cup B|$.]

(v) For $n = 1$ the assertion is trivial, and for $n = 2$ it also holds. Assume that (47) holds for $n > 1$; you then have to prove it for the $(n + 1)$ sets $M_1, \ldots, M_{n+1}$. By the induction hypothesis, $|M_1 \cup M_2 \cup \cdots \cup M_{n+1}| = |(M_1 \cup M_2 \cup \cdots \cup M_n) \cup M_{n+1}| = |M_1 \cup M_2 \cup \cdots \cup M_n| + |M_{n+1}| - |(M_1 \cup M_2 \cup \cdots \cup M_n) \cap M_{n+1}| = |M_1 \cup M_2 \cup \cdots \cup M_n| + |M_{n+1}| - |(M_1 \cap M_{n+1}) \cup (M_2 \cap M_{n+1}) \cup \cdots \cup (M_n \cap M_{n+1})| = \sum (-1)^{r+1} |M_{j_1} \cap M_{j_2} \cap \cdots \cap M_{j_r}| + |M_{n+1}| - \sum (-1)^{r+1} |(M_{j_1} \cap M_{n+1}) \cap (M_{j_2} \cap M_{n+1}) \cap \cdots \cap (M_{j_r} \cap M_{n+1})|$, where the sum is taken over all nonempty subsets $\{j_1, j_2, \ldots, j_r\}$ of the set $\{1, 2, \ldots, n\}$. Since $(M_{j_1} \cap M_{n+1}) \cap (M_{j_2} \cap M_{n+1}) \cap \cdots \cap (M_{j_r} \cap M_{n+1}) = M_{j_1} \cap M_{j_2} \cap \cdots \cap M_{j_r} \cap M_{n+1}$, you obtain

$$|M_1 \cup M_2 \cup \cdots \cup M_{n+1}| = \sum (-1)^{r+1} |M_{j_1} \cap M_{j_2} \cap \cdots \cap M_{j_r}|,$$

where the sum is now taken over all nonempty subsets $\{j_1, j_2, \ldots, j_r\}$ of the set $\{1, 2, \ldots, n+1\}$; that is, (47) holds also for $n + 1$.

(vi) Proceed as in the proof of (47) in 6.1; note that any element $m \in M_1 \cup M_2 \cup \cdots \cup M_n$ contributes to both sides of (48) with the contribution $f(m)$.

(vii) $4\left(\binom{24}{8} - 3\binom{16}{8} + 3\right) = 2\,787\,456$.

6.6 (i) Let M_i denote the set of students in the ith club. Then $|M_i| \geq 15$, $|M_i \cap M_j \cap M_k| \geq 1$, and the summands in (47) that correspond to the intersection of four or more sets are 0. Hence $54 \geq \binom{11}{1} \cdot 15 - \sum |M_i \cap M_j| + \binom{11}{3} \cdot 1$, which means that $\sum |M_i \cap M_j| \geq 11 \cdot 15 + \binom{11}{3} - 54 = 276$. Because the sum $\sum |M_i \cap M_j|$ has $\binom{11}{2} = 55$ summands, at least one of them is 6 or larger, since $55 \cdot 5 = 275 < 276$.

(ii) $\frac{8!}{2^2} - 2 \cdot \frac{7!}{2} + 6! = 5760$.   (iii) $\sum_{r=0}^{6} (-1)^r \cdot \binom{6}{r} \cdot \frac{(13-r)!}{2^{6-r}} = 37\,079\,280$.

(iv) $9! - \binom{3}{1} \cdot 3! \cdot 7! + \binom{3}{2} \cdot (3!)^2 \cdot 5! - (3!)^4 = 283\,824$.

(v) The number of placements of the couples is 2^n-times larger than the value a_n determined in 6.5.(i) with the identity (50). [As a first step in the solution consider a couple as a pair of elements of the same type.]

In the solutions (vi)–(x), $v_0(k, n)$ is the symbol introduced in 6.5.(ii).

(vi) $v_0(6, 4) = \sum_{r=0}^{4} (-1)^r \binom{4}{r} (4 - r)^6 = 1560$.

(vii) $v_0(12, 3) = \sum_{r=0}^{3} (-1)^r \binom{3}{r} (3 - r)^{12} = 519\,156$.

(viii) $v_0(8, 5) = \sum_{r=0}^{5} (-1)^r \binom{5}{r} (5 - r)^8 = 126\,000$.

(ix) Set $|P| = p$, $|Q| = q$, and let $n(p, q)$ denote the desired number. If $p < q$, then $n(p, q) = 0$. If $p \geq q$, then $n(p, q) = v_0(p, q)$, since if you enumerate the elements of the set P, then each mapping from P onto Q

is uniquely determined by an ordered p-tuple of elements of Q, where each elements from Q occurs at least once.

(x) Denote the desired number by $m(p,q)$, where $p = |P|$ and $q = |Q|$. Let $p \geq q$ (otherwise, $m(p,q) = 0$).

First solution. Divide all mappings into two disjoint groups S_1, S_2, where S_1 contains all mappings from P onto Q, the number of which is $v_0(p,q)$, by (ix). S_2 will contain all those mappings f from P onto Q for which there exists an $x \in P$ that is not the preimage of any $y \in Q$. Then "extend" the set Q by an element $y_0 \notin Q$ and define the "extended" mapping f' from P onto $Q' = Q \cup \{y_0\}$ as follows: If $x \in D(f)$, then $f'(x) = f(x)$; if $x \notin D(f)$, then set $f'(x) = y_0$. Each mapping f can thus be "extended" in a unique way, and therefore S_2 has as many mappings as there are mappings from P onto Q'; but this number is, by (ix), $v_0(p, q+1)$. Hence $m(p,q) = v_0(p,q) + v_0(p,q+1)$.

Second solution. Divide all mappings f into groups $S_q, S_{q+1}, \ldots, S_p$ according to the number of elements of the set $D(f)$. The group S_i contains exactly $\binom{p}{i} v_0(i,q)$ mappings for $i \in \{q, q+1, \ldots, p\}$, and together this gives $m(p,q) = \binom{p}{q} v_0(q,q) + \binom{p}{q+1} v_0(q+1,q) + \cdots + \binom{p}{p} \cdot v_0(p,q)$.

In both solutions you have, of course, the same result in different forms, connected by the identity

$$v_0(p, q+1) = \binom{p}{q} v_0(q,q) + \binom{p}{q+1} v_0(q+1,q) + \cdots + \binom{p}{p-1} \cdot v_0(p-1,q),$$

which can be verified as follows: For $p > q$ divide all $v_0(p, q+1)$ mappings from P onto Q, where $|P| = p$ and $|Q| = q+1$, into groups T_i ($i = 0, 1, \ldots, p-q-1$). The group T_i will contain all those mappings for which the preimage of a fixed element $y \in Q$ is a $(p-q-i)$-element set.

In the solutions of (xi) and (xii), c_n is the symbol defined in 6.5.(iii).

(xi) $c_9 = 8! \sum_{r=0}^{8} \frac{(-1)^r}{r!} (9-r)$.   (xii) $c_{100} = 99! \sum_{r=0}^{99} \frac{(-1)^r}{r!} (100-r)$.

(xiii) $\sum_{k=0}^{r} (-1)^k \cdot \binom{r}{k} (n-k)!$. [Proceed as in 6.5.(iii).]

(xiv) $\sum_{r=0}^{n-1} (-1)^r \binom{n}{r} (n-r)!$. [Proceed as in 6.5.(iii). Let M_n denote the set of all arrangements in which the element a_1 follows directly after a_n. Then $M_1 \cap M_2 \cap \cdots \cap M_n = \emptyset$, since no arrangement can contain all n "forbidden" pairs.]

(xv) $\sum_{r=0}^{n-1} (-1)^r \binom{n}{r} (n-r-1)! + (-1)^n$. [Choose a certain child and denote it by d_1, and denote the remaining children by $d_2, d_3, \ldots, d_n$ (for instance, cyclically in the positive direction). Let M_i ($i = 1, 2, \ldots, n$) be the set of all placements in which the child d_i sits directly in front of d_{i+1} (where $d_{n+1} = d_1$). There are $\frac{n!}{n} = (n-1)!$ "essentially different" placements of the children, since two placements are considered different if at least one child has a different neighbor on at least one side. For these reasons,

$$|M_{j_1} \cap M_{j_2} \cap \cdots \cap M_{j_r}| = \frac{(n-r)!}{n-r} = (n-r-1)!$$

for each r-element subset $\{j_1, \ldots, j_r\}$ of $\{1, \ldots, n\}$, where $0 < r < n$. Finally, note that (compare with (xiv)) $|M_1 \cap M_2 \cap \cdots \cap M_n| = 1$. (For example, the arrangement $(d_1, d_2, \ldots, d_n)$ is acceptable.)]

(xvi) Use (56) with $k = 10$, $n = 4$, $\alpha_1 = 3$, $\alpha_2 = 2$, $\alpha_3 = 10$, $\alpha_4 = 15$; then $\binom{13}{3} - \binom{9}{3} - \binom{10}{3} + \binom{6}{3} = 102$.

(xvii) For $n > 9k$ there is no solution. If $n \leq 9k$, the equation has exactly as many solutions as there are n-element combinations with repetitions from k elements in which no element is repeated more than nine times. By (56) this number is equal to (with $\beta = [\frac{n}{10}]$) $\sum_{r=0}^{\beta}(-1)^r \binom{k}{r}\binom{n-10r+k-1}{k-1}$. For (a), set $k = 5$, $n = 12$, $\beta = 1$; then the number is 1745.

(xviii) Let $A(n)$ denote the desired number, and $A_0(n)$ the number of placements of n distinguishable rooks, at least one of which is not threatened; then clearly, $A(n) = \frac{1}{n!}A_0(n)$. Let M_i be the set of all placements in which the ith rook is not threatened ($i = 1, 2, \ldots, n$). Then $A_0(n) = |M_1 \cup \cdots \cup M_n|$, and this number can be determined by using (49) as follows. The identity $|M_{j_1} \cap M_{j_2} \cap \cdots \cap M_{j_r}| = \binom{n}{r}^2 (r!)^2 \binom{(n-r)^2}{n-r}(n-r)!$ holds for all nonempty subsets $\{j_1, j_2, \ldots, j_r\}$ of $\{1, 2, \ldots, n\}$: First, choose r columns and r rows of the chessboard on which to place the unthreatened rooks $V_{j_1}, V_{j_2}, \ldots, V_{j_r}$; each column $s_{j_1}, s_{j_2}, \ldots, s_{j_r}$ and each row $t_{j_1}, t_{j_2}, \ldots, t_{j_r}$ must contain exactly one of the rooks, and this placement can be done in $\binom{n}{r}^2 (r!)^2$ ways. The remaining $(n - r)$ rooks can be placed on any field of the chessboard with the r columns $s_{j_1}, s_{j_2}, \ldots, s_{j_r}$ and the r rows $t_{j_1}, t_{j_2}, \ldots, t_{j_r}$ removed; so there are $\binom{(n-r)^2}{n-r}(n - r)!$ possibilities for their placement. (49) implies $A_0(n) = \sum_{r=1}^{n}(-1)^{r+1}\binom{n}{r}^3 (r!)^2 \binom{(n-r)^2}{n-r}(n - r)! = \sum_{r=1}^{n}(-1)^{r+1}\binom{n}{r}\frac{(n!)^2}{(n-r)!}\binom{(n-r)^2}{n-r}$, and therefore we have $A(n) = \frac{1}{n!}A_0(n) = \sum_{r=1}^{n}(-1)^{r+1}\binom{n}{r}\frac{n!}{(n-r)!}\binom{(n-r)^2}{n-r}$.

(xix) Let $k \leq n$ (otherwise, there is no such placement). The columns $s_1, \ldots, s_k$ on which the rooks are placed can be chosen in $\binom{n}{k}$ ways. It remains to determine the number of ways in which the k rooks can be placed on n rows and the k columns $s_1, \ldots, s_k$ such that each column contains exactly one rook, and at the same time at least one row contains exactly one rook. For $i \in \{1, \ldots, n\}$ let T_i denote the set of all those placements where the ith row has exactly one rook. Show that for any nonempty subset $\{j_1, \ldots, j_r\}$ of $\{1, \ldots, n\}$ the identity $|T_{j_1} \cap \cdots \cap T_{j_r}| = \frac{k!}{(k-r)!}(n - r)^{k-r}$ holds if $r \leq k$ (in the case $r = k = n$ set $0^0 = 1$), and $T_{j_1} \cap \cdots \cap T_{j_r} = \emptyset$ if $r > k$. By (49) the desired number of all placements is

$$\binom{n}{k}\sum_{r=1}^{k}(-1)^{r+1}\binom{n}{r}\frac{k!}{(k - r)!}(n - r)^{k-r}.$$

(xx) Show that the set of all the desired placements is $A \cup B$, where A (respectively B) is the set of all placements of the n rooks on the chessboard, when each column (respectively row) contains exactly one rook and each field of the chessboard is threatened. Show that $A \cap B = \emptyset$ and that $|A| = |B|$. Now n rooks can be placed in n^n ways on the chessboard such that each column contains exactly one rook. Hence $|A| = n^n - |C|$, where C is the set of all placements of n rooks such that each column has a single rook and at least one rook is placed on an unthreatened field. By (xix), again with $0^0 = 1$,

$$|A \cup B| = 2 \cdot |A| = 2 \cdot \left(n^n - \sum_{r=1}^{n} (-1)^{r+1} \binom{n}{r} \frac{n!}{(n-r)!} (n-r)^{n-r} \right)$$

$$= 2 \cdot \sum_{r=0}^{n} (-1)^r \binom{n}{r}^2 (n-r)^{n-r} r!.$$

6.8 In what follows the symbols q_4, S_0, and S from 6.7 will be used.

(i) $q_4(6,4,2) = \sum_{r=0}^{2} (-1)^r \binom{4}{r} \binom{9-3r}{3} = 10$,
$q_4(10,3,4) = \sum_{r=0}^{2} (-1)^r \binom{3}{r} \binom{12-5r}{2} = 6$.

(ii) Consider an arbitrary distribution of n identical balls over k numbered bins, where each bin can contain at most s balls. Let x_i denote the number of balls in the ith bin. Then clearly, the number $q_4(n,k,s)$ of all distributions is equal to the number of all solutions of the equation $x_1 + x_2 + \cdots + x_k = n$ in $\mathbb{N}_0$ under the condition $x_i \leq s$ for all $i = 1,2,\ldots,n$. Therefore, the given equation has exactly $q_4(n,k,s) = \sum_{r=0}^{\alpha} (-1)^r \binom{k}{r} \binom{n+k-rs-r-1}{k-1}$ solutions (where $\alpha = \left\lfloor \frac{n}{s+1} \right\rfloor$ and $n \leq ks$).
In Exercise 6.6.(xvii)(b) then $s = 9$, and for the number N of all solutions this means that $N = \sum_{r=0}^{\alpha} (-1)^r \binom{k}{r} \binom{n+k-10r-1}{k-1}$, where $\alpha = \left\lfloor \frac{n}{10} \right\rfloor$ in the case $n \leq 9k$; in the case $n > 9k$ the result is $N = 0$.

(iii)(a) $S(6,4) = \frac{1}{4!} \sum_{r=0}^{4} (-1)^r \binom{4}{r} (4-r)^6 = 65$;

(b) $S_0(6,4) = \sum_{r=1}^{4} S(6,r) = 1 + 31 + 90 + 65 = 187$.

(iv) If the bins are numbered $1, 2, \ldots, k$, then each distribution of n different objects over k distinct bins, none of which may remain empty, can be described with the help of an ordered n-tuple of elements of the set $A = \{1, 2, \ldots, k\}$, where each element $a \in A$ occurs at least once. But the number of these is exactly $v_0(n,k)$.

(v) For $k \geq n$ the result is $\sum_{r=0}^{n} (-1)^r \binom{n}{r} (n+m-r)^k$, and for $k < n$ it is 0.

6.10 (i) $\sum_{r=0}^{4} (-1)^r \binom{4}{r} (10-r)^5 = 960$.

(ii)(a) $\sum_{r=0}^{4} (-1)^r \binom{4}{r} \frac{(8-r)!}{2^{4-r}} = 864$; (b) $\sum_{r=0}^{3} (-1)^r \binom{3}{r} \frac{(8-r)!}{2^{3-r}} = 2220$;

(c) $\sum_{r=0}^{2} (-1)^r \binom{2}{r} \frac{(8-r)!}{2^{2-r}} - \sum_{r=0}^{2} (-1)^r \binom{2}{r} \frac{(7-r)!}{2^{2-r}} = 5100$.

(iii) Divide all integers into groups S_0, S_1, S_2 such that S_i contains the integers with exactly i pairs of identical digits. Then S_0 has $5! = 120$ integers, S_1 has $5 \cdot \binom{4}{3} [P_0(2,1,1,1) - P(4)] = 720$, and S_2 has $\binom{5}{2} \cdot \binom{3}{1} \cdot [P_0(2,2,1) - 2 \cdot P_0(2,1,1) + P(3)] = 360$ integers, and the total is 1200.

(iv) Generalize the argument in 6.9.(iv); then the number N of the integers is

$$N = \binom{9}{k} \sum_{r=0}^{k-1} (-1)^r \binom{k}{r} (k-r)^m$$

$$+ (k-1) \binom{9}{k-1} \sum_{r=0}^{k-1} (-1)^r \binom{k-1}{r} (k-r)^{m-1},$$

which can be written in the form $N = \frac{9}{10-k} \binom{9}{k} \sum_{r=0}^{k-1} (-1)^r \binom{k}{r} (k-r)^m$. For $m = 4$ and $k = 2$ one obtains 567.

6.12 The numbers d_i are as determined by the identity (61).
(i) $d_1 = 0$, $d_2 = 1$, $d_3 = 2$, $d_4 = 9$, $d_5 = 44$.
(ii)(a) $d_5 = 44$;   (b) $\binom{5}{1} d_4 = 45$, respectively $\binom{5}{2} d_3 = 20$, respectively $\binom{5}{3} d_2 = 10$.
(iii) $12! d_{12}$.   (iv) d_n.   (v) $2n! d_n$.
(vi) $\sum_{r=0}^{n} (-1)^r \binom{n}{r} (2n-r)!$.   (vii) $\sum_{r=0}^{s} (-1)^r \binom{s}{r} (n-r)!$.

6.14 (i) Apply 6.13.(iv) to $f : \mathbb{N} \to \mathbb{R}$ defined by $f(1) = 1$ and $f(m) = 0$ for $m > 1$.
(ii) Use 6.13.(iv) and the facts that $\tau(m) = \sum_{d|m} 1$, $\sigma(m) = \sum_{d|m} d$, and $\log \nu(m) = \sum_{d|m} \log d$.
(iii) Set $G(k) = \sum_{d|k} f(d)$ and use induction on k to show with 6.13.(iv) that $G(k) = g(k)$: If $G(k_0) = g(k_0)$ for all $k_0 \in \mathbb{N}$, $k_0 < k$, then the identity

$$\sum_{d|k} \mu(d) g\left(\frac{k}{d}\right) = f(k) = \sum_{d|k} \mu(d) G\left(\frac{k}{d}\right)$$

with $\mu(1) = 1$ implies $G(k) = g(k)$.
(iv)(a) Show that for any $a \in \{1, \ldots, m\}$, $(a, m) = d$ holds if and only if $a = d \cdot b$, where $b \in \{1, \ldots, \frac{m}{d}\}$ and $(b, \frac{m}{d}) = 1$. The number of such b is exactly $\varphi(\frac{m}{d})$.
(b) Show that (62) implies $\varphi(m) = \sum_{d|m} \mu(d) \frac{m}{d}$ and use (iii).
(v)(a) Show that for a fixed m the sets $H(m, d)$, $d \mid m$, form a decomposition of the Cartesian product $\{1, \ldots, m\} \times \{1, \ldots, m\}$.
(b) Show that the mapping that associates to the pair $(a, b) \in H(\frac{m}{d}, 1)$ the pair (da, db), is a bijection $H(\frac{m}{d}, 1) \to H(m, d)$.
(c) By (a) and (b), $\sum_{d|m} h(d, 1) = m^2$ holds, since if d runs through all positive divisors of m, then so does $\frac{m}{d}$. Apply 6.13.(iv) to $f(d) = h(d, 1)$ and $g(k) = k^2$. Derive from the definition of the function μ that

$$\sum_{d|m} \mu(d) \frac{m^2}{d^2} = m^2 \left(1 - \frac{1}{p_1^2}\right) \cdots \left(1 - \frac{1}{p_n^2}\right),$$

where $p_1, \ldots, p_n$ are all the primes dividing m.

(d) Show that $\sum_{d|m} h_r(m, d) = m^r$, $h_r(m, d) = h_r\left(\frac{m}{d}, 1\right)$, and further proceed as in (c), with the fact that

$$\sum_{d|m} \mu(d)\frac{m^r}{d^r} = m^r\left(1 - \frac{1}{p_1^r}\right)\cdots\left(1 - \frac{1}{p_n^r}\right).$$

6.17 (i) Proceed as in 6.16.(i).

(ii) Decompose an arbitrary $m > 1$ into prime divisors. Thus, let $m = \prod_{i=1}^{n} p_i^{\alpha_i}$ and set $F(x_1, \ldots, x_n) = f(\prod_{i=1}^{n} p_i^{x_i})$, $G(x_1, \ldots, x_n) = g(\prod_{i=1}^{n} p_i^{x_i})$, where $x_i \in \{0, 1, \ldots, \alpha_i\}$ for all $i = 1, \ldots, n$. Then (65) is satisfied for $\nu_i(x_i, y_i) = 1$ for all $0 \le y_i \le x_i \le \alpha_i$. If you set $\mu_i(\alpha_i) = 1$, $\mu_i(\alpha_i - 1) = -1$, and $\mu_i(t) = 0$ for all nonnegative integers $t < \alpha_i - 1$, then (67) assumes the form (64), since $\prod_{i=1}^{n} \mu_i(x_i) = \mu\left(\prod_{i=1}^{n} p_i^{\alpha_i - x_i}\right)$.

(iii) First, make no distinction between the members of the individual nationalities and consider, as in 6.16.(ii), the values $F(x_1, x_2, x_3)$, $G(x_1, x_2, x_3)$. Compute the values $G(x_1, x_2, x_3)$ and use the theorem in 6.15. The identity (67) will be in the form (70). The result 174 finally has to be multiplied by $(3!)^3$ in order to distinguish the individual persons. The final result is $37\,584$.

(iv) Proceed in a similar fashion as in (iii) and use the fact that $G(x_1, \ldots, x_n)$ is the number of arrangements of $\sum_{i=1}^{n}(1 + x_i)$ objects of which exactly as many triples are indistinguishable as there are $i \in \{1, \ldots, n\}$ with $x_i = 2$. Set $h(0) = h(1) = 1$, $h(2) = 6$; then the result can be written in the form

$$F(2, \ldots, 2) = 6^n \cdot \sum_{x_1=0}^{2} \cdots \sum_{x_n=0}^{2} (-1)^{\left(\sum_{i=1}^{n} x_i\right)} \frac{(n + \sum_{i=1}^{n} x_i)!}{\prod_{i=1}^{n} h(x_i)}.$$

(v) Suppose that $k \ge 8$ (for $k < 8$ there is no such placement). There are exactly as many of the desired placements as there are integers that can be expressed in 4 copies of the digit 1, 4 copies of the digit 2, and $(k - 8)$ copies of the digit 3 such that neither two digits 1 nor two digits 2 are placed next to each other. As in 6.16.(ii), define $F(x_1, x_2)$, $G(x_1, x_2)$ for $x_i \in \{0, 1, \ldots, 4\}$, where $F(x_1, x_2)$ denotes the number of integers formed from the given digits such that for $i = 1, 2$

• all four digits i are next to each other when $x_i = 0$;

• three digits i are next to each other and the fourth one is elsewhere when $x_i = 1$;

• two pairs of adjacent digits i are isolated from each other when $x_i = 2$;

• there is exactly one pair of adjacent digits i when $x_i = 3$;

• no two of the digits i are adjacent when $x_i = 4$.

Similarly, $G(x_1, x_2)$ denotes the number of those integers such that for $i = 1, 2$

- a quadruple of digits i next to each other is underlined when $x_i = 0$;
- a triple of digits i next to each other is underlined when $x_i = 1$;
- two pairs of adjacent digits i are underlined when $x_i = 2$;
- one pair of adjacent digits i is underlined when $x_i = 3$;
- no digits i are underlined when $x_i = 4$,

where two integers of the same value are considered different if they are underlined differently. Show that for any $x_1, x_2 \in \{0, 1, \ldots, 4\}$,

$$G(x_1, x_2) = \sum_{y_1=0}^{x_1} \sum_{y_2=0}^{x_2} \nu(x_1, y_1)\nu(x_2, y_2)F(y_1, y_2),$$

if you define

$$
\nu(r, s) = \begin{cases}
0 & \text{if } r = 2, s = 1; \\
1 & \text{if } r = s \text{ or } r = 4 \text{ or } r = 2, s = 0; \\
2 & \text{if } r = 1, s = 0 \text{ or } r = 3, s \in \{1, 2\}; \\
3 & \text{if } r = 3, s = 0.
\end{cases}
$$

For the condition (66) in 6.15 to be satisfied, set

$$
\mu(r) = \begin{cases}
1 & \text{if } r \in \{1, 2, 4\}; \\
-1 & \text{if } r \in \{0, 3\}.
\end{cases}
$$

The theorem in 6.15 now gives

$$F(4, 4) = \sum_{x_1=0}^{4} \sum_{x_2=0}^{4} \mu(x_1)\mu(x_2)G(x_1, x_2)$$

with $G(x_1, x_2) = \frac{(k-8+h(x_1)+h(x_2))!}{(k-8)!d(x_1)d(x_2)}$ for any $x_1, x_2 \in \{0, \ldots, 4\}$, where

$$
h(r) = \begin{cases}
r+1 & \text{if } r \in \{0, 1\}, \\
r & \text{if } r \in \{2, 3, 4\},
\end{cases}
\qquad
d(r) = \begin{cases}
1 & \text{if } r \in \{0, 1\}, \\
2 & \text{if } r \in \{2, 3\}, \\
24 & \text{if } r = 4.
\end{cases}
$$

(vi) Proceed in a similar fashion as in (iv) and (v); the result is

$$24^n \cdot \sum_{x_1=0}^{4} \cdots \sum_{x_n=0}^{4} \left(\prod_{i=1}^{n} \mu(x_i)\right) \frac{(\sum_{i=1}^{n} h(x_i))!}{\prod_{i=1}^{n} d(x_i)},$$

where the functions μ, h, d are defined in the solution of (v).

(vii) $\sum_{x_1=0}^{4} \sum_{x_2=0}^{4} \mu(x_1)\mu(x_2)\frac{(3+h(x_1)+h(x_2))!-2\cdot(2+h(x_1)+h(x_2))!}{2d(x_1)d(x_2)} = 2016.$
Use (v) with $k - 11$ and $k - 10$ (in the second case write the letter X instead of the pair of letters PP).

7.4 (i) $\frac{1}{24}\left(3^6 + 8 \cdot 3^2 + 6 \cdot 3^3 + 6 \cdot 3^3 + 3 \cdot 3^4\right) = 57.$

(ii) $\frac{1}{24}\left(2^6 + 8 \cdot 2^2 + 6 \cdot 2^3 + 6 \cdot 0 + 3 \cdot 2^4\right) = 8.$

(iii) $\frac{1}{24}\left(4^6 + 8 \cdot 4^2 + 6 \cdot 4^3 + 6 \cdot 0 + 3 \cdot 0\right) = 192$.

(iv) $\frac{1}{24}\left(5^6 + 8 \cdot 5^2 + 6 \cdot 5^3 + 6 \cdot 5 + 3 \cdot 5^2\right) = 695$.

(v) $\frac{1}{24}\left(\frac{6!}{2! \cdot 2! \cdot 2!} + 8 \cdot 0 + 6 \cdot 3! + 6 \cdot 0 + 3 \cdot 3!\right) = 6$.

(vi) $\frac{1}{24}\left(2^{12} + 8 \cdot 2^4 + 6 \cdot 2^7 + 6 \cdot 2^3 + 3 \cdot 2^6\right) = 218$.

(vii) $\frac{1}{24}\left(3^8 + 8 \cdot 3^4 + 6 \cdot 3^4 + 6 \cdot 3^2 + 3 \cdot 3^4\right) = 333$.

(viii) One can rest a dodecahedron on any of its twelve pentagonal faces, and this face can be rotated in five ways such that one of its vertices points forward. Hence the set of bijections Φ has $12 \cdot 5 = 60$ elements. One of them is the identity, which contributes a summand of $v(\varphi) = 2^{12} = 4096$. Each nonidentical rotation φ about an axis that goes through the centers of two opposite faces (of which there are $4 \cdot 6 = 24$) contributes a summand of $v(\varphi) = 2^4 = 16$. Each nonidentical rotation φ about an axis through the centers of two opposite edges (of which there are $1 \cdot 15 = 15$) contributes a summand of $v(\varphi) = 2^6 = 64$. Finally, each nonidentical rotation φ about an axis through opposite vertices (of which there are $2 \cdot 10 = 20$) contributes a summand of $v(\varphi) = 2^4 = 16$. These are all of the 60 rotations, and by Burnside's lemma the desired number of different colorings of a dodecahedron is $\frac{1}{60}(4096 + 24 \cdot 16 + 15 \cdot 64 + 20 \cdot 16) = 96$.

(ix) Proceed as in (viii); the desired number is, by Burnside's lemma, $\frac{1}{60}\left(3^{30} + 24 \cdot 3^6 + 15 \cdot 3^{16} + 20 \cdot 3^{10}\right)$.

(x) Proceed as in (viii); you have to restrict yourself to coloring ten vertices white and ten vertices black. The desired number is then $\frac{1}{60}\left(\binom{20}{10} + 24\binom{4}{2} + 15 \cdot \binom{10}{5} + 20 \cdot 2 \cdot \binom{6}{3}\right)$.

(xi) This problem can be solved as in 7.3.(ii). By 6.16.(iii) there are exactly 132 different colorings of the vertices of a 9-gon, with three vertices colored white, three red, and three black, and where no two neighboring vertices have the same color. In this problem, however, you do not distinguish between colorings when one can be obtained from the other by rotation or by reflection about an axis. Apart from the identity there are 8 rotations and 9 reflections. Colorings that do not change upon nonidentical mappings exist only for rotations by $120°$ and $240°$; in both cases there are six different colorings (three adjacent vertices have different colors). By Burnside's lemma the desired number of bracelets is $\frac{1}{18}(132 + 2 \cdot 6) = 8$.

7.6 (i) With an approach similar to 7.5.(ii) (use the result of 3.5.(iv) on the number of solutions of $2u + v = n$) you obtain the expression

$$\frac{1}{6}\left(\prod_{j=1}^{k}\binom{\omega_j+2}{2} + 3 \cdot \prod_{j=1}^{k}\left(\left[\frac{\omega_j}{2}\right] + 1\right) + 2 \cdot \prod_{j=1}^{k}\left(\left[\frac{\omega_j}{3}\right] - \left[\frac{\omega_j-1}{3}\right]\right)\right),$$

where $n = \prod_{j=1}^{k} p_j^{\omega_j}$, $p_1, \ldots, p_k$ are different primes, and $\omega_1, \ldots, \omega_k \in \mathbb{N}$.

(ii) As in 7.5.(ii), write the integers n, x_1, $\ldots$, x_4 in their prime decompositions: $n = \prod_{j=1}^{k} p_j^{\omega_j}$, $x_i = \prod_{j=1}^{k} p_j^{\alpha_{ij}}$, $i = 1, 2, 3, 4$. You have to determine the number of solutions of the following system of k equations in the $4k$

unknowns $\alpha_{ik} \in \mathbb{N}_0$:

$$\alpha_{11} + \alpha_{21} + \alpha_{31} + \alpha_{41} = \omega_1, \quad \dots, \quad \alpha_{1k} + \alpha_{2k} + \alpha_{3k} + \alpha_{4k} = \omega_k$$

under the conditions $\alpha_{1j} \le \alpha_{2j} \le \alpha_{3j} \le \alpha_{4j}$ for each $j = 1, \dots, k$. By the multiplication rule and 7.5.(i) the desired number of solutions is

$$\prod_{j=1}^{k} \left(\frac{1}{24} \left(\binom{\omega_j + 3}{3} + 6 \cdot \left(\left[\frac{\omega_j}{2} \right] + 1 \right) \left(\left[\frac{\omega_j + 1}{2} \right] + 1 \right) \right. \right.$$
$$+ 3 \cdot \left(\left[\frac{\omega_j}{2} \right] - \left[\frac{\omega_j - 1}{2} \right] \right) \left(\left[\frac{\omega_j}{2} \right] + 1 \right)$$
$$\left. \left. + 8 \cdot \left(\left[\frac{\omega_j}{3} \right] + 1 \right) + 6 \cdot \left(\left[\frac{\omega_j}{4} \right] - \left[\frac{\omega_j - 1}{4} \right] \right) \right) \right).$$

(iii) This problem can be solved in analogy to 7.5.(i). Set $X = \{(x_1, x_2, \dots, x_5) : x_1, x_2, \dots, x_5 \in \mathbb{N}_0, x_1 + x_2 + \dots + x_5 = n\}$ and let Φ be the set of bijections on X induced by the various permutations of the index set $\{1, 2, 3, 4, 5\}$. Denote the values $v(\varphi)$ for the different types of bijections by $V_1(n), \dots, V_7(n)$: If $\varphi \in \Phi$ is induced by the identity, then $V_1(n) = v(\varphi) = |X| = \binom{n+4}{4}$. If $\varphi \in \Phi$ is induced by the interchange of one pair of indices (there are $\binom{5}{2} = 10$ such φ), then $v(\varphi)$ is equal to the number of solutions of the equation $2x + y + z + w = n$ in $\mathbb{N}_0$, and by 3.6.(xvii) you get $V_2(n) = v(\varphi) = \frac{1}{2}([\frac{n}{2}]+1)([\frac{n}{2}] \cdot (\frac{4}{3}[\frac{n}{2}] - 2n - \frac{7}{3}) + (n+2)(n+1))$. If $\varphi \in \Phi$ is induced by the interchange of two pairs of indices (there are $\frac{1}{2}\binom{5}{2}\binom{3}{2} = 15$ such φ), then $v(\varphi)$ is equal to the number of solutions of $2x + 2y + z = n$ in $\mathbb{N}_0$, that is, the number of solutions of the inequality $x + y \le [\frac{n}{2}]$, and so $V_3(n) = v(\varphi) = \binom{2+[\frac{n}{2}]}{2}$. If $\varphi \in \Phi$ is induced by a permutation that leaves exactly two indices fixed (there are $2 \cdot \binom{5}{2} = 20$ such φ), then $v(\varphi)$ is equal to the number of solutions of $3x + y + z = n$ in $\mathbb{N}_0$, and by 3.6.(xviii) you obtain $V_4(n) = v(\varphi) = ([\frac{n}{3}] + 1)(n + 1 - \frac{3}{2}[\frac{n}{3}])$. If $\varphi \in \Phi$ is induced by any cyclic permutation of four indices (there are $5 \cdot \frac{4!}{4} = 30$ of them), then $v(\varphi)$ is equal to the number of solutions of $4x + y = n$ in $\mathbb{N}_0$; that is, $V_5(n) = v(\varphi) = [\frac{n}{4}] + 1$. If $\varphi \in \Phi$ is induced by a cyclic permutation of three indices and an interchange of the remaining two (there are $2 \cdot \binom{5}{3} = 20$ such permutations), then $v(\varphi)$ is equal to the number of solutions of $3x + 2y = n$ in $\mathbb{N}_0$. By 3.6.(xix), $V_6(n) = v(\varphi) = [\frac{n+4}{6}] + [\frac{n}{6}] - [\frac{n-1}{6}]$. Finally, if $\varphi \in \Phi$ is induced by a cyclic permutation of all five indices (there are $\frac{5!}{5} = 24$ of them), then $v(\varphi)$ is equal to the number of solutions of $5x = n$ in $\mathbb{N}_0$. There is one such solution if $5 \mid n$, and none if $5 \nmid n$; that is, $V_7(n) = v(\varphi) = [\frac{n}{5}] - [\frac{n-1}{5}]$. This exhausts all 120 permutations of a 5-element set, and by Burnside's lemma the desired number of solutions is

$$\frac{1}{120} \big(V_1(n) + 10 \cdot V_2(n) + 15 \cdot V_3(n) + 20 \cdot V_4(n)$$
$$+ 30 \cdot V_5(n) + 20 \cdot V_6(n) + 24 \cdot V_7(n) \big).$$

(iv) Use a similar approach as in (ii) for $n = \prod_{j=1}^{k} p_j^{\omega_j}$, where $p_1, \ldots, p_k$ are distinct primes and $\omega_1, \ldots, \omega_k \in \mathbb{N}$. Then it follows from (iii) with the multiplication rule that the desired number of solutions is

$$\prod_{j=1}^{k} \Big(\tfrac{1}{120} \big(V_1(\omega_j) + 10V_2(\omega_j) + 15V_3(\omega_j) + 20V_4(\omega_j)$$

$$+ 30V_5(\omega_j) + 20V_6(\omega_j) + 24V_7(\omega_j) \big) \Big).$$

(v) If you use the multiplication rule and proceed as in (ii) with $n = \prod_{j=1}^{k} p_j^{\omega_j}$, where $p_1, \ldots, p_k$ are primes and $\omega_1, \ldots, \omega_k \in \mathbb{N}$, you obtain the desired number of solutions in the form

$$\frac{1}{120} \Big(\prod_{j=1}^{k} V_1(\omega_j) + 10 \prod_{j=1}^{k} V_2(\omega_j) + 15 \prod_{j=1}^{k} V_3(\omega_j) + 20 \prod_{j=1}^{k} V_4(\omega_j)$$

$$+ 30 \prod_{j=1}^{k} V_5(\omega_j) + 20 \prod_{j=1}^{k} V_6(\omega_j) + 24 \prod_{j=1}^{k} V_7(\omega_j) \Big).$$

8.2 (i)(a) Substitute (61) and simplify.

(b) Divide all d_{n+1} relevant permutations of the set $A = \{a_1, a_2, \ldots, a_{n+1}\}$ into n classes T_i, $i = 2, 3, \ldots, n+1$, such that T_i contains the permutation $\varphi : A \to A$ with the property $\varphi(a_1) = a_i$ (recall why $\varphi(a_1) \neq a_1$). In each class T_i consider the subset $R_i = \{\varphi \in T_i : \varphi(a_i) = a_1\}$. Show that $|R_i| = d_{n-1}$ (to do this, remove the elements a_1 and a_i from A) and that $|T_i \setminus R_i| = d_n$ (for each $\varphi \in T_i \setminus R_i$ consider $\psi : A \setminus \{a_i\} \to A \setminus \{a_i\}$ defined as follows: $\psi(a_j) = \varphi(a_j)$ for $j \neq 1$ and $\psi(a_1) = \varphi(a_i)$).

(ii) Let $k \geq n \geq 1$. Divide the set of all $v_0(k+1, n+1)$ relevant $(k+1)$-element variations from elements of $(n+1)$ types into $(n+1)$ classes according to which element is in first place. Then divide each of these classes into two subclasses according to whether or not the element in first place occurs elsewhere. The first subclass contains exactly $v_0(k, n+1)$ variations, and the second one contains $v_0(k, n)$.

(iii) Let $n > 2$. Divide the set C_n of all c_n relevant arrangements into two classes C_n', respectively C_n'', according to whether upon removal of the element a_n from the given arrangement you obtain an arrangement in which for no index, respectively for some index $i \in \{1, \ldots, n-2\}$, the element a_{i+1} follows a_i immediately. From each arrangement in the class C_n' you obtain in this way some arrangement $p \in C_{n-1}$, where this same p always originates from $(n-1)$ different elements of C_n' (the element a_n can be at the beginning, at the end, or in any space in p, but not after a_{n-1}); that is, $|C_n'| = (n-1)c_{n-1}$. Each arrangement from C_n'' contains a triple of consecutive elements a_j, a_n, a_{j+1} for some $j \in \{1, \ldots, n-2\}$. Consider now the following process: Remove the elements a_n and a_{j+1}, and for each $i \in \{j+2, \ldots, n-1\}$ replace the element a_i by a_{i-1}. In this way you obtain some arrangement $p \in C_{n-2}$, where this same p always originates from $(n-2)$ different elements C_n'''; that is, $|C_n''| = (n-2)c_{n-2}$.

(iv) Divide the set M_k of all f_k relevant sequences into k classes T_i $(i = 1, \ldots, k)$ as follows: The sequence $a_1, \ldots, a_{2k}$ belongs to T_i when i is the smallest positive integer such that the sequence $a_1, \ldots, a_{2i}$ contains exactly i ones and i fives. Then, of course, $a_1 = 5$ and $a_{2i} = 1$. Note that both sequences $a_2, \ldots, a_{2i-1}$ and $a_{2i+1}, \ldots, a_{2k}$ satisfy the conditions of the problem concerning the number of fives and of ones, and derive the relationship $|T_i| = f_{i-1} \cdot f_{k-i}$.

(v) The equations are $p_k = (d+1)^k$ $(1 \le k \le s)$ and $p_{k+s} = d(p_k + p_{k+1} + \cdots + p_{k+s-1})$ for $k \ge 1$. Divide all allowable sequences of length $k + s$ into groups according to the number j of equal terms in which the individual sequences end, $1 \le j \le s$. For each such j the corresponding group has exactly $d \cdot p_{k+s-j}$ sequences, since each one of them arises such that to an allowable sequence of length $k + s - j$ (its last term denoted by x) you add j identical integers y, where y is an arbitrary integer from the set $\{0, 1, \ldots, d\}$, and is different from x.

2   Hints and Answers to Chapter 2

1.2 (i) Show that for some odd k the number $a_k + k$ is even, that is, among the $\frac{n+1}{2}$ numbers $a_1, a_3, \ldots, a_n$, at least one is odd. This holds because only $\frac{n-1}{2}$ of the numbers $1, 2, \ldots, n$ are even.

(ii) An arrangement $a_1, a_2, \ldots, a_{n-1}$ of the numbers 1, 2, $\ldots$, $n - 1$ has the desired property if and only if the sums $s_k = a_1 + a_2 + \cdots + a_k$ $(1 \le k \le n - 1)$ have distinct nonzero remainders upon division by n. Convince yourself that for the arrangement $n - 1$, 2, $n - 3$, 4, $n - 5$, 6, $\ldots$, 3, $n - 2$, 1 (that is, $a_i = n - i$ for odd i and $a_i = i$ for even i) the numbers s_k are determined by the identities $s_{2i-1} = in - i$ $(1 \le i \le \frac{n}{2})$ and $s_{2i} = in + i$ $(1 \le i \le \frac{n}{2} - 1)$; their remainders upon division by n are therefore different and nonzero.

(iii) $p_n = 3k$ for $n = 4k$, $p_n = 3k + 1$ for $n = 4k + 1$, and $p_n = 3k + 2$ for $n \in \{4k+2, 4k+3\}$. Set $s_0 = 0$ and note that the number of pairs (s_{i-1}, s_i) of neighboring elements with the property $s_{i-1} \not\equiv s_i \pmod 2$ in the sequence $s_0, s_1, \ldots, s_n$ is equal to the number of odd integers in $\{1, 2, \ldots, n\}$. Each even integer s_i $(i \ne 0, n)$ is an element of at most two such pairs, and each of the numbers s_0, s_n belongs to at most one pair. If you take into account that $s_n = \frac{n(n+1)}{2}$ is even if and only if $n = 4k$ or $n = 4k + 3$, you obtain for the number q of even integers in the sequence $s_0, s_1, \ldots, s_n$ the lower bounds $2q - 2 \ge 2k$ $(n = 4k)$, $2q - 1 \ge 2k + 1$ $(n = 4k + 1, n = 4k + 2)$, and $2q - 2 \ge 2k + 2$ $(n = 4k + 3)$. These imply the required upper bounds for the number of odd integers s_i (which is equal to $n + 1 - q$). The sharpness of the bounds is shown by the examples of the arrangements

$$1, 2, 4, 6, \ldots, n, 3, 5, 7, \ldots, n - 1 \quad (n \text{ even}),$$
$$1, 2, 4, 6, \ldots, n - 1, 3, 5, 7, \ldots, n \quad (n \text{ odd}).$$

(iv) For $n = 5$ the quintuple $(2, 3, 7, 43, 1807)$ was already found; verify that $(2, 3, 7, 47, 395)$ also has the desired property. Further, it suffices to show that for $n = 4$ the quadruple $2 \leq a_1 < a_2 < a_3 < a_4$ is unique; thanks to the recursive construction in the solution of 1.1.(ii) you then obtain the desired result also for $n < 4$, as well as for $n > 5$. So, set $s = a_1 a_2 a_3 a_4$. Multiplying the four identities $k_i a_i = 1 + \frac{s}{a_i}$, where $k_i \in \mathbb{N}$, $1 \leq i \leq 4$, you obtain after simplification,

$$\frac{1}{a_1} + \frac{1}{a_2} + \frac{1}{a_3} + \frac{1}{a_4} + \frac{1}{a_1 a_2 a_3 a_4} = A, \quad A \in \mathbb{N}.$$

Since $a_i \geq i + 1$ $(1 \leq i \leq 4)$, the number A satisfies the bound

$$A \leq \frac{1}{2} + \frac{1}{3} + \frac{1}{4} + \frac{1}{5} + \frac{1}{2 \cdot 3 \cdot 4 \cdot 5} = \frac{31}{24},$$

and therefore $A = 1$. Similarly, one can show that neither $a_1 > 2$ nor $a_2 > 3$ can hold. By substituting $a_1 = 2$ and $a_2 = 3$ you obtain $\frac{1}{a_3} + \frac{1}{a_4} + \frac{1}{6 a_3 a_4} = \frac{1}{6}$, which implies $a_3 = 6 + 37/(a_4 - 6)$. Therefore, $|a_4 - 6| \in \{1, 37\}$, and with an easy argument you find that $(a_3, a_4) = (7, 43)$.

(v) The integer $n^2 + n$ is a multiple of 4 if and only if $n = 4m$ or $n = 4m - 1$. For $n = 3$ and $n = 4$ consider the arrangement $(2, 3, 1, 2, 1, 3)$, respectively $(2, 3, 4, 2, 1, 3, 1, 4)$. To write down appropriate arrangements for larger values of n in a suggestive way, use the following notation: If $a - b = 2j$, where $j \in \mathbb{Z}$, then in the case $j > 0$ the notation $a \searrow b$ stands for the sequence $(a, a - 2, a - 4, \ldots, b)$, and in the case $j < 0$ set $a \nearrow b = (a, a + 2, a + 4, \ldots, b)$. For $n = 4m$, $m \geq 2$, an appropriate arrangement has the form

$$4m - 4 \searrow 2m, 4m - 2, 2m - 3 \searrow 1, 4m - 1, 1 \nearrow 2m - 3,$$
$$2m \nearrow 4m - 4, 4m, 4m - 3 \searrow 2m + 1, 4m - 2, 2m - 2 \searrow 2, 2m - 1,$$
$$4m - 1, 2 \nearrow 2m - 2, 2m + 1 \nearrow 4m - 3, 2m - 1, 4m;$$

for $n = 4m - 1$, $m \geq 2$, an appropriate arrangement is of the form

$$4m - 4 \searrow 2m, 4m - 2, 2m - 3 \searrow 1, 4m - 1, 1 \nearrow 2m - 3,$$
$$2m \nearrow 4m - 4, 2m - 1, 4m - 3 \searrow 2m + 1, 4m - 2, 2m - 2 \searrow 2,$$
$$2m - 1, 4m - 1, 2 \nearrow 2m - 2, 2m + 1 \nearrow 4m - 3.$$

It is worth mentioning a nice trick that not only simplifies the checking of the properties of both the above arrangements, but also indicates how such arrangements can be discovered: If in the arrangement

$$\ldots, a \searrow b, \underbrace{\ldots}_{b}, b \nearrow a, \ldots$$

the dots between the numbers b stand for exactly b terms, then for all k from the sequence $a \searrow b$ there are exactly k terms between the k from

$a \searrow b$ and the k from $b \nearrow a$. Thus, for instance, for the above arrangement with $n = 4m$ it suffices to verify this property for $k \in \{2m, 4m - 2, 1, 4m - 1, 4m, 2m + 1, 2, 2m - 1\}$.

1.4 (i) Suppose that $a_j > \max\{b_1, b_2, \ldots, b_j\}$ for some j; that is, $a_j > b_i$ for all $i = 1, 2, \ldots, j$. But then all numbers $b_1, b_2, \ldots, b_j$ are in the arrangement $a_1, a_2, \ldots, a_n$ in front of the number a_j, which is a contradiction.

(ii) The assertion remains unchanged if you apply the same permutation simultaneously to both n-tuples $x_1, x_2, \ldots, x_n$ and $y_1, y_2, \ldots, y_n$. Therefore, you may as well assume that $y_1 \geq y_2 \geq \cdots \geq y_n$, that is, $v_k = y_k$ $(1 \leq k \leq n)$. If there exist indices $i < j$ such that $x_i > x_j$, then in view of the inequality $y_i \geq y_j$ you obtain $x_i + y_i \geq \max\{x_i + y_j, x_j + y_i\}$, which means that an interchange of the elements x_i, x_j in the n-tuple $x_1, x_2, \ldots, x_n$ does not increase the value of the largest of the numbers $x_k + y_k$ $(1 \leq k \leq n)$. Since there are at most $\binom{n}{2}$ pairs of indices $i < j$ with the property $x_i > x_j$ in the original n-tuple $x_1, x_2, \ldots, x_n$, a finite number of the changes described will transform the original n-tuple into the nondecreasing n-tuple $u_1 \leq u_2 \leq \cdots \leq u_n$. So $\max\{x_k + y_k\} \geq \max\{u_k + y_k\} = \max\{u_k + v_k\}$.

(iii) It can be shown that such an arrangement exists for each n and is determined by the conditions $a_{n+1} = 2n + 1$ and $a_{n+2} > a_n$. For then you will have (choose $k = n + 1$ and $k = 1$ in the conditions) $a_{n+2} > a_n$, $a_{n+3} > a_{n-1}$, $\ldots$, $a_{2n+1} > a_1$, and (in view of $a_{n+2} < a_{n+1}$) $a_{n+3} < a_n$, $a_{n+4} < a_{n-1}$, $\ldots$, $a_{2n+1} < a_2$, so altogether,

$$a_{n+1} > a_{n+2} > a_n > a_{n+3} > a_{n-1} > a_{n+4} > a_{n-2} > \cdots > a_{2n+1} > a_1.$$

You then have the arrangement $a_k = 2k - 1$ $(1 \leq k \leq n+1)$, $a_k = 4n + 4 - 2k$ $(n + 2 \leq k \leq 2n + 1)$, which does indeed possess the required property: $a_{k-j} < a_{k+j}$ $(1 \leq j \leq n, 1 \leq k \leq n + 1)$, $a_{k-j} > a_{k+j}$ $(1 \leq j \leq n, n + 2 \leq k \leq 2n + 1)$.

(iv) Let $\mathcal{P}$ denote the set of all arrangements $p = (x_1, x_2, \ldots, x_n)$ of the numbers $a_1, a_2, \ldots, a_n$. If you set $S_p = x_1 x_2 + \cdots + x_n x_1$ for all $p = (x_1, x_2, \ldots, x_n)$, then clearly

$$\sum_{p \in \mathcal{P}} S_p = 2n \cdot (n - 2)! \cdot S, \quad \text{where} \quad S = \sum_{1 \leq i < j \leq n} a_i a_j.$$

From the identity $(a_1 + \cdots + a_n)^2 = a_1^2 + \cdots + a_n^2 + 2S$ and Cauchy's inequality $n \left(a_1^2 + \cdots + a_n^2\right) \geq (a_1 + \cdots + a_n)^2$ (see, e.g., [5], Chapter 2, Section 5) it follows that $2S \leq \frac{n-1}{n}(a_1 + \cdots + a_n)^2 = \frac{n-1}{n}$, and thus

$$\sum_{p \in \mathcal{P}} S_p \leq n \cdot (n - 2)! \cdot \frac{n - 1}{n} = (n - 1)!.$$

The arithmetic mean of all $n!$ numbers S_p is therefore at most $\frac{1}{n}$, which implies $S_p \leq \frac{1}{n}$ for an appropriate $p \in \mathcal{P}$.

(v) Proceed as in 1.3.(v). For a fixed $k \in \{1, 2, \ldots, n\}$ let, for instance, the inequality $a_j \geq b_j$ hold for $\alpha \geq \frac{n-k+1}{2}$ values of the index $j \in \{k, k+1, \ldots, n\}$, and let b_s $(k \leq s \leq n)$ be the *largest* of these α numbers b_j. Then $b_s \geq \frac{1}{n+1-\alpha} \geq \frac{2}{n+k+1}$, which, together with $a_k + b_k \geq a_s + b_s \geq 2b_s$, gives the desired bound.

(vi) With the help of the recurrence relations

$$Q(n) = Q(2^{k-1} - 1) \, Q(n - 2^{k-1}) \cdot \binom{n-1}{n-2^{k-1}} \quad (2^k \leq n < 2^k + 2^{k-1}),$$

$$Q(n) = Q(2^k - 1) \, Q(n - 2^k) \cdot \binom{n-1}{n-2^k} \quad (2^k + 2^{k-1} \leq n < 2^{k+1})$$

you obtain $Q(100) = Q(63)Q(36) \cdot \binom{99}{36}$, $Q(63) = Q(31)^2 \cdot \binom{62}{31}$, $Q(36) = Q(15)Q(20) \cdot \binom{35}{20}$, $Q(31) = Q(15)^2 \cdot \binom{30}{15}$, $Q(20) = Q(7)Q(12) \cdot \binom{19}{12}$, $Q(15) = Q(7)^2 \cdot \binom{14}{7}$, $Q(12) = Q(7)Q(4)\binom{11}{4}$, $Q(7) = Q(3)^2 \cdot \binom{6}{3}$, $Q(4) = Q(1)Q(2) \cdot \binom{3}{2}$. Since $Q(1) = Q(2) = 1$ and $Q(3) = 2$, you obtain, upon successive substitutions, $Q(100) = 3 \cdot 4^{12} \cdot \binom{99}{36} \cdot \binom{62}{31} \cdot \binom{35}{20} \cdot \binom{30}{15}^2 \cdot \binom{19}{12} \cdot \binom{14}{7}^5 \cdot \binom{11}{4} \cdot \binom{6}{3}^{12}$.

(vii) (a) $\binom{6}{2} \cdot \binom{4}{2} = 90$ ways. The placement of the numbers 1 and 8 is unique, and from the remaining six numbers choose consecutively three pairs for the three "branches" connecting 1 and 8.

(b) $3! \cdot \binom{6}{2} \cdot \binom{4}{2} = 540$ ways. The numbers $8, 9, 10$ lie necessarily on the upper "branches" (3! ways), and the number 7 directly underneath. The number of placements of the integers $1, \ldots, 6$ in the lower part is then determined as in (a).

(c) $\binom{4}{2} \cdot \binom{6}{3} = 120$ ways. The placement of the numbers 5 and 12 is unique, and from among the numbers $6, 7, \ldots, 11$ one can choose three for one branch connecting 5 and 12 in exactly $\binom{6}{3}$ ways (the other branch is then formed with the remaining three numbers). The numbers 1, 2, 3, 4 can be distributed in the lower part in $\binom{4}{2}$ ways.

The borrowed "biological terms" in the solutions of (d)–(j) below will be used without quotation marks.

(d) $\binom{9}{2} \cdot \binom{7}{2} \cdot \binom{4}{2} = 4536$ ways. The head and neck are occupied by the numbers 11, respectively 10. For the two arms there are $\binom{9}{2} \cdot \binom{7}{2}$ choices in total. Of the remaining five numbers, the largest belongs to the belly, and for the left leg there are $\binom{4}{2}$ possibilities; the right leg is then formed with the remaining two numbers.

(e) $\binom{5}{2} \cdot \binom{3}{2} \cdot \binom{4}{2} = 180$ ways. Neck and belly are occupied by the numbers 6, respectively 5. Four of the numbers $7, 8, \ldots, 11$ are placed on the arms in $\binom{5}{2} \cdot \binom{3}{2}$ ways, while the remaining fifth number forms the head. For placing the numbers 1, 2, 3, 4 on the legs, there are $\binom{4}{2}$ possibilities.

(f) $7 \cdot 5 \cdot 4 \cdot 2 = 280$ ways. The head is formed by the number 8, while for the snout there are seven possibilities. The largest of the remaining six numbers goes to the lower end of the neck, and the front legs can be chosen in 5·4 ways. The belly then gets the largest of the three remaining numbers, and finally, the hind legs can obtain their numbers in two ways.

(g) $15 \cdot 14 \cdot \binom{12}{2}\binom{10}{2} \cdot \binom{7}{2} \cdot \binom{5}{2} = 130\,977\,000$ ways. The number 16 goes to the tip of the tree. Then consecutively choose the right and left top branches, the right and left middle branches, and the right and left lower branches.

(h) $\left(\binom{11}{2} - 3\right) \cdot \binom{7}{2} \cdot \binom{4}{2} = 6552$ ways. Elbow and hand of the right (bent) arm can be occupied in $\binom{11}{2} - 3$ ways (the pairs $(10, 11)$, $(9, 11)$, and $(9, 10)$ have to be excluded). Then neck and head are uniquely determined. Next you can choose the left hand in $\binom{7}{2}$ ways; the largest of the remaining numbers belongs to the belly, and finally the left leg can be chosen in $\binom{4}{2}$ ways. A different approach is based on distinguishing three cases:

$$11 = \text{hand} \qquad \cdots \qquad \binom{8}{2} \cdot 6 \cdot \binom{4}{2} = 1008 \quad \text{ways,}$$
$$11 = \text{head, } 10 = \text{hand} \quad \cdots \qquad \binom{8}{2} \cdot 6 \cdot \binom{4}{2} = 1008 \quad \text{ways,}$$
$$11 = \text{head, } 10 = \text{neck} \quad \cdots \qquad \binom{9}{2} \cdot \binom{7}{2} \cdot \binom{4}{2} = 4536 \quad \text{ways.}$$

(j) $10\,080$ ways. Distinguish the possible cases for placing the number k on the neck:

$$k = 10 \quad \cdots \qquad \binom{9}{2} \cdot \binom{7}{2} \cdot \binom{4}{2} = 4536 \quad \text{ways,}$$
$$k = 9 \quad \cdots \quad 2 \cdot 2 \cdot 8 \cdot \binom{7}{2} \cdot \binom{4}{2} = 4032 \quad \text{ways,}$$
$$k = 8 \quad \cdots \quad 3 \cdot 2 \cdot 7 \cdot 6 \cdot \binom{4}{2} = 1512 \quad \text{ways.}$$

(viii) 2^{n-1} arrangements. Show that if $a_i = 1$, then the sequence $a_1, \ldots, a_{i-1}, a_i$ is decreasing and the sequence $a_i, a_{i+1}, \ldots, a_n$ is increasing. Thus our arrangement is uniquely determined by the subset $\{a_1, \ldots, a_{i-1}, a_i\}$ (which is empty if $i = 1$) of the set $\{2, 3 \ldots, n\}$. Conversely, each of the 2^{n-1} subsets of the set $\{2, 3 \ldots, n\}$ determines the arrangement satisfying the given condition.

1.6 (i) $\max\{\min_k (x_k y_k)\} = n$, $\min\{\max_k (x_k y_k)\} = \left[\left(\frac{n+1}{2}\right)^2\right]$. The collection of numbers $x_k y_k$ does not change if you apply the same permutation to both n-tuples $x_1, \ldots, x_n$ and $y_1, \ldots, y_n$; you may therefore assume that $x_k = k$ ($1 \le k \le n$). The bound $\min_k (k y_k) \le n$ comes from the fact that $y_k = 1$ for an appropriate k, and for a bound for $\max_k (k y_k)$ note that $y_k \ge \frac{n}{2} + 1$ for some $k \ge \frac{n}{2}$ (if n is even), respectively $y_k \ge \frac{n+1}{2}$ for some $k \ge \frac{n+1}{2}$ (if n is odd). The sharpness of both bounds is illustrated by the arrangement $y_k = n + 1 - k$ ($1 \le k \le n$).

(ii) $\max S = \left[\frac{n^2}{2}\right]$. Upon removal of the absolute values you obtain the expression $S = \pm 1 \pm 1 \pm 2 \pm 2 \pm \cdots \pm n \pm n$ with appropriate signs (n-times "+", n-times "−"). This sum is maximal if the "−" sign is given to exactly the n first terms of the sequence $1, 1, 2, 2, \ldots, n, n$ (for determining this value, distinguish between n even and n odd). Such a choice of signs is possible: It corresponds, for instance, to the arrangement $x_k = n + 1 - k$ ($1 \leq k \leq n$).

(iii) For $n = 4k + 1$ the bound $d_* \leq 2k$ follows from the fact that the inequality $|2k + 1 - x| \leq 2k$ holds for all $x \in \{1, 2, \ldots, 4k + 1\}$; equality $d_* = 2k$ occurs if you choose the arrangement $x_1, x_2, \ldots, x_n$ such that $x_{j+1} \equiv 2kj \pmod{n}$, $0 \leq j \leq n-1$. But this is the arrangement $\underbrace{4k + 1, 2k,}$ $\underbrace{4k, 2k - 1,} \ldots, \underbrace{2k + 2, 1,} 2k + 1$.

For $n = 4k + 2$ the bound $d_* \leq 2k$ follows from the fact that the inequality $|2k + 1 - x| \geq 2k + 1$ holds for a unique $x \in \{1, 2, \ldots, 4k + 2\}$, namely, for $x = 4k$; equality $d_* = 2k$ occurs for the arrangement $\underbrace{1, 2k + 3,} \underbrace{3, 2k + 5,}$ $\underbrace{5, 2k + 7,} \ldots, \underbrace{2k - 1, 4k + 1,} 2k + 1, \underbrace{4k + 2, 2k,} \underbrace{4k, 2k - 2,} \ldots, \underbrace{2k + 4, 2,}$ $2k + 2$. (Writing it with the help of congruences is now not as easy as before, since the numbers $2k$ and $4k + 2$ are not relatively prime.)

For $n = 4k + 3$ the bound $d_* \leq 2k + 1$ follows from the fact that the inequality $|2k + 2 - x| \leq 2k + 1$ holds for all $x \in \{1, 2, \ldots, 4k + 3\}$; equality $d_* = 2k + 1$ occurs if you choose the arrangement $x_1, x_2, \ldots, x_n$ such that $x_{j+1} \equiv (2k + 1)j \pmod{n}$, $0 \leq j \leq n - 1$. But this is the arrangement $\underbrace{4k + 3, 2k + 1,} \underbrace{4k + 2, 2k,} \underbrace{4k + 1, 2k - 1,} \ldots, \underbrace{2k + 3, 1,} 2k + 2$.

(iv) $\min d^* = 2$. Since $|x - 1| \leq 1$ holds for a unique $x \in \{2, 3, \ldots, n\}$, you have $d^* \geq 2$. Equality $d^* = 2$ occurs, for instance, with the arrangement $1, 3, 5, \ldots, 2k - 1, 2k, 2k - 2, 2k - 4, \ldots, 2$ ($n = 2k$), respectively $1, 3, 5, \ldots, 2k + 1, 2k, 2k - 2, 2k - 4, \ldots, 2$ ($n = 2k + 1$).

(v) In the cyclic arrangement $x_1, x_2, \ldots, x_{4k}$ under consideration no two elements of the set $M = \{1, 2, \ldots, k, 3k + 1, 3k + 2, \ldots, 4k\}$ can be neighbors. Since $|M| = 2k$, this means that M is one of the sets $\{x_1, x_3, x_5, \ldots, x_{4k-1}\}$, $\{x_2, x_4, x_6, \ldots, x_{4k}\}$. From $k + 1 \notin M$ it follows that M contains both numbers, which in the arrangement are the neighbors of $k + 1$. According to the condition, however, only the numbers $1, 2k + 1, 2k + 2, \ldots, 3k$ can be neighboring on $k + 1$, but among these numbers there is only one element of M. Hence there is no arrangement with the desired property.

(vi) For $n = 4k$ and $n = 4k - 1$. The sum $S = |a_1 - a_2| + |a_2 - a_3| + \cdots + |a_n - a_1|$ must consist of the summands $1, 2, 3, \ldots, n$ in some order, that is, $S = \frac{n(n+1)}{2}$. On the other hand, from the expression $S = \pm a_1 \pm a_2 \pm \cdots \pm a_n \pm a_1$ with an appropriate choice of signs it follows that $2 \mid S$, since $\pm x \pm x$ is even for any $x \in \mathbb{Z}$. Hence the number $n(n+1)$ is divisible by 4, which holds only when $n = 4k$ or $n = 4k - 1$. For $n = 4k$ an appropriate arrangement is $4k + 1, 1, 4k, 2, \ldots, 3k + 2, k, 3k + 1, k + 2, 3k, k + 3, \ldots, 2k + 2, 2k + 1$ (the number $k + 1$ is not used), and for $n = 4k - 1$ the arrangement $4k, 1, 4k - 1, 2, \ldots, 3k + 1, k, 3k - 1, k + 1, 3k - 2, k + 2, \ldots, 2k + 1, 2k - 1, 2k$ (the number $3k$ is not used).

(vii) For the entire solution below, let $x_1, x_2, \ldots, x_n$ be an arbitrary arrangement of the given numbers $a_1, a_2, \ldots, a_n$, for which you can assume the ordering $a_1 \leq a_2 \leq \cdots \leq a_n$. The approach in 1.5.(i) can easily be adapted $(x_1 = a_1,\ x_k = a_n)$ to obtain the bound

$$|x_2 - x_1| + |x_3 - x_2| + \cdots + |x_n - x_{n-1}| + |x_1 - x_n| \geq 2(a_n - a_1),$$

where equality occurs for the arrangement $a_1, a_n, a_{n-1}, \ldots, a_2$. If you denote by $M_n(a_1, a_2, \ldots, a_n)$ the maximum of the expression $S_n(x_1, x_2, \ldots, x_n)$ in 1.5.(ii), then from the identity $S_n(a_n, x_2, \ldots, x_n) = S_{n-1}(x_2, \ldots, x_n) + a_n^2 - (a_n - x_2)(a_n - x_n)$ you obtain $M_n(a_1, a_2, \ldots, a_n) = M_{n-1}(a_1, a_2, \ldots, a_{n-1}) + a_n^2 - (a_n - a_{n-1})(a_n - a_{n-2})$. Finding the numbers d_* and d^* in 1.5.(iii) and 1.6.(iv) is difficult in general. In the case of *positive* numbers a_i you can prove the following generalization of 1.6.(i):

$$\min\left\{\max_k (x_k y_k)\right\} = \max_k (a_k a_{n+1-k}), \quad \max\left\{\min_k (x_k y_k)\right\} = \min_k (a_k a_{n+1-k}).$$

Once again, we can assume that $x_k = a_k$ $(1 \leq k \leq n)$. If $a_p a_{n+1-p} = \max_k(a_k a_{n+1-k})$, then for an arbitrary arrangement $y_1, y_2, \ldots, y_n$ the inequality $y_i < a_{n+1-p}$ holds for at most $n - p$ indices $i \in \{1, 2, \ldots, n\}$. Therefore, among the $n - p + 1$ indices $i \in \{p, p+1, \ldots, n\}$ there exist those for which $y_i \geq a_{n+1-p}$, and thus $a_i y_i \geq a_p y_i \geq a_p a_{n+1-p}$. This implies $\max_k(a_k y_k) \geq a_p a_{n+1-p}$. In a similar way one can also find an upper bound for $\min_k(a_k y_k)$. Finally, in Exercise 1.6.(ii) about the maximum of the sum $S = |x_1 - a_1| + \cdots + |x_n - a_n|$ the expression $S = \pm a_1 \pm a_1 \pm \cdots \pm a_n \pm a_n$ with an equal number of both signs leads to the bound $S \leq 2(a_1 + a_2 + \cdots + a_n - K)$, where K is the sum of the first n terms of the sequence $a_1, a_1, a_2, a_2, \ldots$. Equality occurs for the arrangement $x_k = a_{n+1-k}$ $(1 \leq k \leq n)$.

1.8 (i) $A = 15$. The example of the arrangement $0, 9, 5, 1, 8, 3, 4, 2, 7, 6$ on the circumference of a circle shows that $A \leq 15$. On the other hand, in any arrangement one can delete 0 and divide the remaining nine digits into three triples (of consecutive elements) with sums s_1, s_2, s_3. From the identity $s_1 + s_2 + s_3 = 45$ it follows that $s_i \geq 15$ for some index $i \in \{1, 2, 3\}$.

(ii) Verify that the sequence $Q = (n, P_{n-1}, P_{n-2}, \ldots, P_2, 1, n, 2)$ is n-universal, where $P_k = (1, 2, \ldots, k, n, k+1, k+2, \ldots, n-1)$ for all $k \leq n-1$. To do this, show for all $k = 1, 2, \ldots, n$ that to the left of the kth occurrence of the number n in Q one can obtain by deletion an arbitrary sequence of $k - 1$ numbers from the set $\{1, 2, \ldots, n-1\}$, and to the right an arbitrary sequence of $n - k$ numbers from $\{1, 2, \ldots, n-1\}$.

(iii) $A(k, n) \geq \frac{(n+1)k}{2}$ (if k is a divisor of n), $A(k, n) \leq \frac{(n+1)k}{2}$ (if k is even). In particular, $A(k, n) = \frac{(n+1)k}{2}$ if k is an even divisor of n.

(iv) $A(2, n) = n + 1$, $A(n-1, n) = \frac{n(n-1)}{2} + 1$, $A(n-2, n) = \frac{n(n-3)}{2} + 3$.
Hint: Let $x_1, \ldots, x_n$ be any arrangement of the numbers $1, 2, \ldots, n$. In the case $k = 2$ note that $x_i + x_{i+1} \geq n + 1$ if $n \in \{x_i, x_{i+1}\}$; then consider the arrangement

$$n, 1, n-1, 2, n-2, 3, \ldots, a_n, \tag{45}$$

where $a_n \in \{\frac{n}{2}, \frac{n+1}{2}\}$. For $k = n - 1$ note that $x_1 + x_2 + \cdots + x_{n-1} = \frac{n(n+1)}{2} - x_n$, $x_2 + x_3 + \cdots + x_n = \frac{n(n+1)}{2} - x_1$, and $\min\{x_1, x_n\} \le n - 1$; then take any arrangement with $\{x_1, x_n\} = \{n-1, n\}$. In the case $k = n-2$ it is enough to verify that the largest possible value of the smallest of the numbers $x_1 + x_n$, $x_1 + x_2$, $x_{n-1} + x_n$ is equal to $2n - 3$ (which is the value in the case $x_1 = n$, $x_2 = n - 3$, $x_{n-1} = n - 1$, $x_n = n - 2$).

(v) If k is odd, then the largest of the sums of k neighboring terms in the arrangement (45) is equal to $\frac{n(k+1)}{2}$. In the case of even n also consider the arrangement

$$n, 1, n - 2, 3, n - 4, 5, \ldots, n - 3, 2, n - 1 \tag{46}$$

and show that (for odd k) the example (46) leads to the sharper bound $A(k, n) \le \frac{n(k+1)}{2} - \frac{k-1}{2}$.

(vi) A formula for $A(k, 2k)$ with even k is included in the answer to (iii); furthermore, the bound $A(k, 2k) \ge \frac{(2k+1)k}{2} = k^2 + \frac{k}{2}$ is given there, and from this it follows that $A(k, 2k) \ge k^2 + \frac{k+1}{2}$ in the case of an odd k. In this last relation you have equality, as is shown by the example (46).

(vii) If $\frac{n}{2} < k < n$, then the largest of the sums of k neighbors in the arrangement $x_1, x_2, \ldots, x_n$ of the numbers $1, 2, \ldots, n$ is equal to $\frac{n(n+1)}{2} - s_1$, where s_1 is the smallest of the sums of $(n - k)$ neighbors in the sequence of the $2(n-k)$ numbers $x_{k+1}, x_{k+2}, \ldots, x_n, x_1, x_2, \ldots, x_{n-k}$. If you rewrite this last sequence as $n+1-y_1, n+1-y_2, n+1-y_3, \ldots, n+1-y_{2n-2k}$, then $1 \le y_i \le n$ and $y_i \ne y_j$ ($1 \le i < j \le 2n-2k$). Therefore, $s_1 = (n+1)(n-k)-s_2$, where s_2 is the largest of the sums of $(n - k)$ neighboring numbers in the sequence $y_1, y_2, \ldots, y_{2n-2k}$. Explain why the smallest possible value s_2 occurs in the case $\{y_1, y_2, \ldots, y_{2n-2k}\} = \{1, 2, \ldots, 2n - 2k\}$ and is equal to $A(n - k, 2n - 2k)$.

(viii) Such an arrangement exists for all $n \ge 3$. Call every sequence of numbers $a_1, a_2, \ldots, a_s$ (of arbitrary length) a *good* sequence if it has the property $a_j \ne \frac{a_i + a_k}{2}$ ($1 \le i < j < k \le s$). If the sequence $a_1, a_2, \ldots, a_s$ is good, then $2a_1, 2a_2, \ldots, 2a_s$ and $2a_1 - 1, 2a_2 - 1, \ldots, 2a_s - 1$ are also good. This leads to the following construction: If $x_1, x_2, \ldots, x_n$ is a good arrangement of the numbers $1, 2, \ldots, n$, then $2x_1, 2x_2, \ldots, 2x_n, 2x_1 - 1, 2x_2 - 1, \ldots, 2x_n - 1$ is a good arrangement of $1, 2, \ldots, 2n$ (to see this, you still have to note that $\frac{a+b}{2} \notin \mathbb{Z}$ if $a \not\equiv b \pmod{2}$). Therefore, starting with the good arrangement $1, 3, 2$ for $n = 3$ you can recursively construct good arrangements of $1, 2, \ldots, n$ for all $n = 3 \cdot 2^k$. If $3 < n' < n$, then by deleting the numbers $n' + 1, n' + 2, \ldots, n$ from the good arrangement of $1, 2, \ldots, n$ you obtain a good arrangement of the numbers $1, 2, \ldots, n'$.

2.2 (i) For fixed $x_2, \ldots, x_n$ the sum S is a linear function in the variable $x_1 \in [-1, 1]$; therefore, S is minimal if $x_1 = -1$ or $x_1 = 1$. Since a similar consideration is also true for the remaining variables $x_2, \ldots, x_n$, the answer $S_{\min} = -\left\lceil \frac{n}{2} \right\rceil$ follows from the solution of 2.1.(i).

(ii) Assume that such numbers exist. Then $a_4 + a_2 > a_3\sqrt{3} = (a_1 + a_3)\sqrt{3} > a_2 \left(\sqrt{3}\right)^2 = 3a_2$; that is, $a_4 > 2a_2$. Similarly, $a_4 > 2a_6$, and upon adding, $a_4 > a_2 + a_6$. On the other hand, $a_6 + 2a_4 + a_2 > (a_3 + a_5)\sqrt{3} > a_4 \left(\sqrt{3}\right)^2 = 3a_4$, that is, $a_6 + a_2 > a_4$, which is a contradiction.

(iii) If all the numbers $S_j = x_1 + x_2 + \cdots + x_j$ $(1 \leq j \leq n)$ are nonnegative, you can set $k = 1$. Assume now that some numbers S_j are negative, and choose the *smallest* of them: $S_{k-1} = \min\{S_i; 1 \leq i \leq n\} < 0, 1 < k \leq n$. Since $S_n = 0$, the inequality $S_{k-1} \leq x_1 + x_2 + \cdots + x_m$ holds not only for all $m \in \{1, 2, \ldots, n\}$, but also for all $m > n$. If you apply this inequality consecutively for $m = k, k+1, \ldots, k+n-1$, you obtain the desired result.

(iv) Set $S_k = \sum_{i=1}^{k} x_i - \sum_{i=k+1}^{n} x_i$ $(0 \leq k \leq n)$ and assume that none of the numbers $S_0, S_1, \ldots, S_n$ is equal to zero (otherwise, it suffices to take the index k with the property $S_k = 0$). Since $S_0 = -S_n \neq 0$, some neighboring numbers S_j, S_{j+1} in the sequence $S_0, S_1, \ldots, S_n$ have opposite signs. But then $|S_j| + |S_{j+1}| = |S_{j+1} - S_j| = |2x_{j+1}| = 2|x_{j+1}|$, which implies $\min\{|S_j|, |S_{j+1}|\} \leq |x_{j+1}| \leq \max\{|x_i| : 1 \leq i \leq n\}$.

(v) Set $\varepsilon_1 = 1$. If for some index $n > 1$ the numbers $\varepsilon_1, \varepsilon_2, \ldots, \varepsilon_{n-1}$ are already chosen such that $|S_i| \leq 1$, where $S_i = \varepsilon_1 x_1 + \varepsilon_2 x_2 + \cdots + \varepsilon_i x_i$ for all $i < n$, then choose $\varepsilon_n = \pm 1$ such that $\varepsilon_n x_n S_{n-1} \leq 0$. Then the numbers $\varepsilon_n x_n$ and S_{n-1} are neither simultaneously positive nor simultaneously negative, and thus $|\varepsilon_1 x_1 + \varepsilon_2 x_2 + \cdots + \varepsilon_n x_n| = |S_{n-1} + \varepsilon_n x_n| \leq \max\{|S_{n-1}|, |\varepsilon_n x_n|\} \leq 1$.

(vi) Verify by induction on $k \geq 0$ that in the sum $S_k = a_{n-k} \pm a_{n-k+1} \pm \cdots \pm a_n$ one can choose the signs such that $|S_k| \leq n - k, k = 0, 1, \ldots, n-1$ (note that $S_0 = a_n$, and set $S_{k+1} = a_{n-k-1} + S_k$, respectively $S_{k+1} = a_{n-k-1} - S_k$, according to whether $S_k \leq 0$, respectively $S_k > 0, 0 \leq k \leq n - 2$). Since in the sequence $a_1, a_2, \ldots, a_n$ there is an even number of odd integers, the expression V is even for any choice of signs. If you choose $V = S_{n-1}$, then the inequality $|V| = |S_{n-1}| \leq 1$ implies $V = 0$.

(vii) $S_{\max} = 2^n$. If for fixed $x_1, x_2, \ldots, x_n$ all the sums under consideration are denoted by $s_1, s_2, \ldots, s_N$, where $N = 2^n$, then explain why the identity $s_1^2 + s_2^2 + \cdots + s_N^2 = 2^n \left(x_1^2 + x_2^2 + \cdots + x_n^2\right) = 2^n$ holds. Then the bound $S = |s_1| + |s_2| + \cdots + |s_N| \leq 2^n$ follows from Cauchy's inequality

$$(|s_1| + |s_2| + \cdots + |s_N|)^2 \leq N \cdot \left(s_1^2 + s_2^2 + \cdots + s_N^2\right) = N \cdot 2^n = 2^{2n}$$

(see, e.g., [5], Chapter 2, 5.5.(vii)). Equality $S = 2^n$ occurs, for instance, for $x_1 = 1, x_2 = \cdots = x_n = 0$.

(viii) The assertion is clear when $A = 0$. In the case $A > 0$ there is an index k $(1 \leq k < n)$ such that $x_k < 0 \leq x_{k+1}$. Then the sums $s^- = x_1 + \cdots + x_k$ and $s^+ = x_{k+1} + \cdots + x_n$ satisfy $s^- + s^+ = 0, s^+ = -s^- = A$, $s^- \geq kx_1$, and $s^+ \leq (n - k)x_n$, which implies $s^+ = -s^- = \frac{A}{2}$ and

$$x_n - x_1 \geq \frac{s^+}{n - k} - \frac{s^-}{k} = \frac{An}{2k(n - k)} \geq \frac{2A}{n},$$

since $k(n-k) = \frac{n^2}{4} - \left(\frac{n}{2} - k\right)^2 \le \frac{n^2}{4}$.

(ix) $N_{\max} = 2k + 2$, where k is determined by the inequalities $S_k \le n < S_{k+1}$, with $S_i = 1 + 2 + \cdots + i$. Equality $N = 2k + 2$ occurs, for instance, for the sequence $S_k, S_{k-1}, \ldots, S_1, 0, 0, S_1, S_2, \ldots, S_k$. For any sequence $a_1, a_2, \ldots, a_N$ set $d_i = a_{i+1} - a_i$ $(1 \le i < N)$. The condition of the problem can be written as $d_{i-1} < d_i$; then $d_1 < d_2 < \cdots < d_{N-1}$. Show that in this sequence of integers d_i at most k numbers are negative, at most one is 0, and at most k numbers are positive (from this it follows that $N - 1 \le k + 1 + k$, that is, $N \le 2k + 2$). For if $d_{k+1} < 0$ were to hold, then you would have $a_{k+2} - a_1 = d_1 + d_2 + \cdots + d_{k+1} \le -S_{k+1} < -n$, which is a contradiction to $|a_{k+2} - a_1| \le n$; in a similar way the inequality $d_{N-1-k} > 0$ is also excluded.

(x) The smallest possible value is $\frac{1}{2n-2}$. This value is attained when you set $x_1 = x_2 = \cdots = x_{n-2} = \frac{1}{2}$, $x_{n-1} = 1$, and $x_n = 0$; for then you have $y_1 = y_2 = \cdots = y_{n-2} = \frac{1}{2}$, $y_{n-1} = \frac{n}{2n-2}$, $y_n = \frac{1}{2}$, and $\frac{n}{2n-2} - \frac{1}{2} = \frac{1}{2n-2}$. As in the solution of 2.1.(iv) you may restrict yourself to those sequences $x_1, x_2, \ldots, x_n$, for which $\min_k x_k = x_p = 0$ and $\max_k x_k = x_q = 1$, and furthermore assume that $p > q$ (in the case $p < q$ change the sequence $x_1, x_2, \ldots, x_n$ to $1 - x_1, 1 - x_2, \ldots, 1 - x_n$). If $q = 1$, then $y_1 = 1$ and $y_n \le \frac{n-1}{n}$, which implies $y_1 - y_n \ge \frac{1}{n} \ge \frac{1}{2n-2}$ (for all $n \ge 2$). Since $\frac{1}{2n-2} \le \frac{1}{p+q-1}$, in the case $q > 1$ it suffices to show that at least one of the three numbers $d_1 = y_q - y_{q-1}$, $d_2 = y_{p-1} - y_p$, $d_3 = y_q - y_p$ is not less than $\frac{1}{p+q-1}$. Note that $(q-1)y_{q-1} = x_1 + \cdots + x_{q-1} = qy_q - 1$ and $(p-1)y_{p-1} = x_1 + \cdots + x_{p-1} = py_p$; hence $d_1 = \frac{1-y_q}{q-1}$ and $d_2 = \frac{y_p}{p-1}$. Therefore, if for some c you have $d_1 < c$ and $d_2 < c$, then $y_q > 1 - c(q-1)$ and $y_p < c(p-1)$, which implies $d_3 > 1 - (p+q-2)c$. This consideration is optimal if $c = 1 - (p + q - 2)c$, that is, if $c = \frac{1}{p+q-1}$.

(xi) In the solution of 2.1.(v) we have demonstrated the bound $\varepsilon \ge \frac{4}{n}$ for the largest of the numbers in (3) of Chapter 2. If you use the same result also for the n-tuple of numbers $\frac{x_2-x_1}{\varepsilon}, \frac{x_3-x_2}{\varepsilon}, \ldots, \frac{x_1-x_n}{\varepsilon}$, then in view of the identity $x_k - \frac{x_{k-1}+x_{k+1}}{2} = \frac{x_k-x_{k-1}}{2} - \frac{x_{k+1}-x_k}{2}$ you will easily obtain the desired bound.

2.4 (i) The statement does not hold. A counterexample is the sequence of the 100 numbers $9, 8, \ldots, 1, 0, 19, 18, \ldots, 11, 10, \ldots, 99, 98, \ldots, 91, 90$. The terms of each of its decreasing subsequences have the same digit in the tens position, while the terms of each of its increasing subsequences have distinct digits in the tens position. Therefore, each of the decreasing or increasing subsequences has length at most 10.

(ii) The desired numbers are exactly all the even integers $k > 8$. For odd $k > 8$ the sequence $a_1 = a_2 = \cdots = a_k = -2$ is a counterexample. Let therefore $k = 2n$, $n \ge 5$, and let the product of any eight terms of the sequence $a_1, a_2, \ldots, a_{2n}$ be greater than 1. Then all numbers a_i are different from zero and have the same sign: If you suppose, for instance, that

$a_1 a_2 \leq 0$, and choose from $a_3, a_4, \ldots, a_{2n}$ six terms $a_{i_1}, a_{i_2}, \ldots, a_{i_6}$ such that $a_{i_1} a_{i_2} \geq 0$, $a_{i_3} a_{i_4} \geq 0$, and $a_{i_5} a_{i_6} \geq 0$ (this is possible, since from any three numbers one can choose two whose product is not negative), then you obtain $a_1 a_2 a_{i_1} a_{i_2} a_{i_3} a_{i_4} a_{i_5} a_{i_6} \leq 0$, which is a contradiction. Let therefore all numbers $a_1, a_2, \ldots, a_{2n}$ be positive (if they are all negative, replace them with their opposites), and furthermore, let them be ordered such that $0 < a_1 \leq a_2 \leq \cdots \leq a_{2n}$. From the inequality $a_8^8 \geq a_1 a_2 \cdots a_8 > 1$ it follows that $a_8 > 1$, which implies $a_9 a_{10} \cdots a_{2n} > 1$, and thus $(a_1 a_2 \cdots a_8) \cdot (a_9 a_{10} \cdots a_{2n}) > 1$.

(iii) $n_{\max} = 50$. The sequence $x_1 = x_2 = \cdots = x_{50} = 2$ has the property in question; suppose now that some sequence of length $n \leq 49$ also has this property. Since its terms can be divided into seven groups such that each one contains at most seven terms, the sum of all its terms is at most $7 \cdot 14 = 98$, which is a contradiction.

(iv) Show that $S \leq 3$. To do this, assume that $S > 2$, and take the *smallest* index i $(1 < i < n)$ such that $S_i = x_1 + x_2 + \cdots + x_i > 1$ (such an index exists, since $S_1 \leq 1$ and $S_{n-1} = S - x_n > 2 - 1 = 1$). By hypothesis $S - S_i \leq 1$ must hold. Since $S_i - x_i = S_{i-1} \leq 1$ and $x_i \leq 1$, you will also have $S_i = (S_i - x_i) + x_i \leq 1 + 1 = 2$, so in total, $S = S_i + (S - S_i) \leq 2 + 1 = 3$. Equality $S = 3$ is possible for all $n \geq 3$: Set $x_1 = x_2 = x_3 = 1$ and $x_i = 0$ $(3 < i \leq n)$.

(v) You can certainly assume that $a_1 \geq a_2 \geq \cdots \geq a_n > 0$. Consider the sums $S = a_1 + a_2 + \cdots + a_n$ and $S_1 = a_1 + a_k + a_{2k-1} + a_{3k-2} + \cdots$. Clearly, $S \leq (k-1)S_1$. By the hypothesis, $S > (2k-2)a_1$; so in total $2(k-1)a_1 < S \leq (k-1)S_1$, and thus $2a_1 < S_1$. If for some $i \in \{1, 2, \ldots, n+1-k\}$ the inequality $a_{i+k-1} > \frac{a_i}{2}$ holds, then the choice of the k terms $a_i, a_{i+1}, \ldots, a_{i+k-1}$ has the desired property. In the opposite case it follows from the inequality $a_{i+k-1} \leq \frac{a_i}{2}$ $(1 \leq i \leq n+1-k)$ that $a_k \leq \frac{a_1}{2}$, $a_{2k-1} \leq \frac{a_k}{2} \leq \frac{a_1}{4}$, $a_{3k-2} \leq \frac{a_{2k-1}}{2} \leq \frac{a_1}{8}$, etc., which leads to the bound $S_1 = a_1 + a_k + a_{2k-1} + a_{3k-2} + \cdots \leq a_1 \left(1 + \frac{1}{2} + \frac{1}{4} + \frac{1}{8} + \ldots\right) \leq 2a_1$; but this is a contradiction to the inequality $2a_1 < S_1$ derived above.

(vi) The necessity of the condition $2 \mid S$ is clear. For the sufficiency it is enough to show that if $2 \mid S$, then in the sum $S_1 = \pm a_1 \pm a_2 \pm \cdots \pm a_n$ one can choose the signs such that $S_1 = 0$ (the terms with equal signs then form the two sets of the desired subdivision). By 2.1.(iii) you have $S_1 \in \{0, 1\}$ for an appropriate choice of signs, and thus $S_1 = 0$, since both S and $S - S_1$ are even.

(vii) Generalization: For a given p the selection can be done in such a way that the absolute value of the sum of the chosen numbers is not less than $\frac{p-2}{2p}(|a_1| + |a_2| + \cdots + |a_n|)$. Hint: Assume that $a_1 + a_2 + \cdots + a_n \geq 0$ and show that one of the $p(p-1)$ subsequences

$$Y_{ij} = X_i \cup \left(\bigcup_{\alpha \notin \{i,j\}} (X_\alpha \cap \mathbb{R}^+)\right) \quad (i, j \in \{1, 2, \ldots, p\}, i \neq j)$$

has the desired property, where $X_i = (a_m : m \equiv i \pmod{p})$, $i = 1, 2, \ldots, p$. Denote by s_i^+, respectively s_i^-, the sum of the positive, respectively negative, terms of the sequence X_i and show that the sum of all $p(p-1)$ sums of the elements of Y_{ij} is equal to

$$(p-1)^2 \sum_{i=1}^{p} s_i^+ + (p-1) \sum_{i=1}^{p} s_i^-,$$

which together with the assumption $\sum_{i=1}^{p}(s_i^+ + s_i^-) \geq 0$ gives the required bound.

2.6 (i) Proof by induction: For $n = 1$ the sequence a_1, a_2, a_3 contains two equal terms (they form a parity-balanced subsequence of length 2). For the step from n to $n + 1$ we distinguish, for an arbitrary sequence $a_1, a_2, \ldots, a_{2n+3}$, two cases: $a_1 = a_{2n+3}$ and $a_i = a_{i+1}$ for some $i = 1, 2, \ldots, 2n + 2$ (explain why one of these cases always occurs). In the first case apply the induction hypothesis to the sequence $a_2, a_3, \ldots, a_{2n+2}$, and in the second case to the sequence $a_1, a_2, \ldots, a_{i-1}, a_{i+2}, \ldots, a_{2n+3}$, and then return to their original places the elements a_1, a_{2n+3}, respectively a_i, a_{i+1}, in the subsequences thus obtained.

(ii) The same assertion holds also for odd n if one adds to the hypothesis of the problem, for instance, the condition that $a_i \neq a_j$ for some $i, j \in \{1, 2, \ldots, n\}$.

(iii) $d_k = \frac{k}{k+1}$. Order the numbers $1 \geq x_1 \geq x_2 \geq \cdots \geq x_n \geq 0$ and choose an index j such that $S = x_1 + x_2 + \cdots + x_j \leq k < x_1 + x_2 + \cdots + x_{j+1}$. Then $j \geq k$, $S \geq jx_{j+1}$, and thus $k < S + x_{j+1} \leq S + \frac{1}{j}S = \left(1 + \frac{1}{j}\right)S \leq \left(1 + \frac{1}{k}\right)S$; this implies $S > \left(1 + \frac{1}{k}\right)^{-1} k = k - \frac{k}{k+1}$, and therefore $d_k \leq \frac{k}{k+1}$. On the other hand, if you choose $n = k+1$ and $x_i = \frac{k+\varepsilon}{k+1}$ for some $0 < \varepsilon < 1$, then any sum $S \leq k$ of some terms x_i satisfies $S \leq k \cdot \frac{k+\varepsilon}{k+1} = k - \frac{k(1-\varepsilon)}{k+1}$; this implies $d_k \geq \frac{k(1-\varepsilon)}{k+1}$, which in view of the possible choices of ε means that $d_k \geq \frac{k}{k+1}$.

(iv) $d_k = \frac{k}{2k+1}$. Set $\varepsilon_k = \frac{k}{2k+1}$ and assume first of all that there exist n numbers $x_i \in [0, 1]$, $1 \leq i \leq n$, such that $x_1 + x_2 + \cdots + x_n > k$ and at the same time $|S - k| > \varepsilon_k$ for every sum S of the terms of an arbitrary subsequence $x_{i_1}, x_{i_2}, \ldots, x_{i_m}$. Among these sums there are numbers greater than k; choose the *smallest* of them. Let this be, for instance, the sum $S_m = x_1 + x_2 + \cdots + x_m$ with no zero terms x_i (if necessary, change the indexing of the numbers x_i). Then $S_m > k + \varepsilon_k$ and $S_m - x_i < k - \varepsilon_k$ $(1 \leq i \leq m)$, which implies that $m > k$ (since $x_i \leq 1$ for all i) and $x_i > 2\varepsilon_k$ $(1 \leq i \leq m)$. But this means that the sum $S_k = x_1 + x_2 + \cdots + x_k$ satisfies $2k\varepsilon_k < S_k < k - \varepsilon_k$, which is a contradiction, since $2k\varepsilon_k = k - \varepsilon_k$. Thus you have shown that the desired number d_k satisfies the bound $d_k \leq \varepsilon_k$. Furthermore, for $n \geq k + 1$ consider the sequence $x_i = \frac{2k}{2k+1}$ $(1 \leq i \leq n)$. Then $x_1 + x_2 + \cdots + x_n = \frac{2kn}{2k+1} > k$, and also the sum S_m of any m chosen terms x_i satisfies

$$|S_m - k| = \left| \frac{2km}{2k+1} - k \right| = \frac{k}{2k+1}|2m - 2k - 1| \geq \frac{k}{2k+1} = \varepsilon_k,$$

since the number $2m - 2k - 1$ is odd. Altogether, you then obtain $d_k = \varepsilon_k$.

(v) $d_n = \frac{1}{n+1}$. For any $x_1, x_2, \ldots, x_n \in \mathbb{R}$ set $y_i = \langle x_1 + x_2 + \cdots + x_i \rangle$, where $1 \leq i \leq n$, and $\langle x \rangle = x - [x]$ is the fractional part of the number x (see [5], Chapter 3, Section 7). If $0 \leq y_i \leq \frac{1}{n+1}$ or $\frac{n}{n+1} \leq y_i < 1$ for some i, then $|x_1 + x_2 + \cdots + x_i - k| \leq \frac{1}{n+1}$ holds for an appropriate $k \in \mathbb{Z}$; otherwise, the n numbers y_i lie in the interval $\left(\frac{1}{n+1}, \frac{n}{n+1} \right)$, and thus $|y_j - y_i| < \frac{1}{n+1}$ for appropriate $1 \leq i < j \leq n$. But the sum $S = x_{i+1} + x_{i+2} + \cdots + x_j$ is of the form $S = y_j - y_i + k$, and thus $|S - k| < \frac{1}{n+1}$, where $k \in \mathbb{Z}$. On the other hand, the sequence $x_i = \frac{1}{n+1}$ ($1 \leq i \leq n$) shows that $d_n \geq \frac{1}{n+1}$.

(vi) Set $S(c) = x_{i_1} + x_{i_2} + \cdots + x_{i_k}$ for each sequence $c = (x_{i_1}, x_{i_2}, \ldots, x_{i_k})$ of length k, with $1 \leq i_1 < i_2 < \cdots < i_k \leq n$. Verify that the arithmetic mean of all $\binom{n}{k}$ numbers $S(c)$ is equal to $\frac{kA}{n}$, and therefore there exist subsequences c and c' of length k such that $S(c) \leq \frac{kA}{n} \leq S(c')$. If both inequalities are strict (otherwise, there is nothing to show), connect c and c' with a sequence of k-tuples whose neighboring terms are adjacent in the sense of the solution of 2.5.(iii). The conclusion of the proof is analogous.

(vii) It suffices to show how the given numbers $x_1, x_2, \ldots, x_n \in [-1, 1]$ can be subdivided into three nonempty sets T_1, T_2, T_3 such that the difference between the largest and the smallest of the three sums s_i of the terms of T_i ($i = 1, 2, 3$) does not exceed 1. (If, for instance, $s_1 \leq s_2 \leq s_3$ and $s_3 - s_1 \leq 1$ hold, then from $3s_2 \leq (s_1 + 1) + s_2 + s_3 = A + 1$ and $3s_2 \geq s_1 + s_2 + (s_3 - 1) = A - 1$ you obtain $\left| s_2 - \frac{A}{3} \right| \leq \frac{1}{3}$, and thus the terms of T_2 were appropriately chosen.) Since $n \geq 5$, by the pigeonhole principle you may assume a numbering of the x_i that is such that all three numbers x_1, x_2, x_3 are either nonnegative or nonpositive. Then begin by setting $T_i = (x_i)$ and $s_i = x_i$ ($i = 1, 2, 3$), and then adjoin the terms x_j ($j \geq 4$) consecutively to the subsequences T_i as follows: If the terms $x_1, x_2, \ldots, x_{j-1}$ have already been used and the sets T_1, T_2, T_3 formed with them have sums s_1, s_2, s_3, then adjoin the term x_j to the end of the sequence T_i that has the smallest (if $x_i > 0$), respectively the largest (if $x_i \leq 0$), sum s_i; in the case where such a T_i is not unique, choose an arbitrary one. It is not difficult to verify that the sums s_1, s_2, s_3 have the required property at each step: If $s_1' \leq s_2' \leq s_3'$ is a nondecreasing arrangement of them, then $s_3' - s_1' \leq 1$. After adjoining the last term x_n we obtain the desired sets T_1, T_2, T_3.

(viii) For each $k \in M$ let v_k denote the number of occurrences of the integer k in the sequence A. By hypothesis, $v_1 + 2v_2 + \cdots + mv_m = 2S$. The number S is a common multiple of all integers from M, and thus also of the three integers $m - 2, m - 1, m$; therefore, you obtain the bound $S \geq \frac{(m-2)(m-1)m}{2} = (m-2)(1 + 2 + \cdots + m)$. You can therefore choose a $k \in M$ such that $v_k \geq m - 1$ (and even $v_k \geq 2m - 4$, if $m \geq 3$). Distinguish now two cases: $kv_k \geq S$ and $kv_k < S$. In the first case everything is clear: Since $k \mid S$, you can choose $\frac{S}{k}$ terms from A that are equal to k. In the second case the subsequence $B = (b_1, b_2, \ldots, b_{n-v_k})$ of all elements A different from k satisfies $b_1 + b_2 + \cdots + b_{n-v_k} = 2S - kv_k > S$. It therefore suffices to show that from B one can choose some terms such that their sum S' satisfies $k \mid S'$ and $S - mk < S' \leq S$ (since from $k \mid S$ it follows that $S' = S - kj$ for an appropriate $j \in \{0, 1, \ldots, m-1\}$; the chosen terms from B can therefore be supplemented with j copies of the number k to obtain the desired subsequence of A). The assertion about the choice of terms from the sequence B can be proven as follows: By the pigeonhole principle one can choose from any sequence of integers $c_1, c_2, \ldots, c_k$ a subsequence of length k_1 ($1 \leq k_1 \leq k$) such that the sum of its terms is divisible by k. Indeed, if none of the numbers $c_1, c_1 + c_2, \ldots, c_1 + c_2 + \cdots + c_k$ is a multiple of k, then the difference of two of those that have the same remainder upon division by k is a multiple of k. From the sequence B you can therefore repeatedly choose subsequences such that the sum of the numbers in each of them is equal to one of the numbers $k, 2k, \ldots, mk$; if at some time the sum S' of terms chosen up to then satisfies $S' \leq S - mk$, then in view of $b_1 + b_2 + \cdots + b_{n-v_k} > S$ at least k unchosen terms remain in B, and then it is possible to choose a further group.

2.8 (i) $k = 7$; an example is the arrangement 1, 6, 7, 4, 3, 2, 9. The graph of the possible neighboring pairs can be seen in Figure 1. It follows from this graph that if two digits x, y from the set $\{2, 5, 8\}$ are chosen, then the set of digits that can be put down is $\{x, y, 3, 9\}$. Otherwise at most one number is chosen from $\{2, 5, 8\}$, and therefore $k \leq 7$.

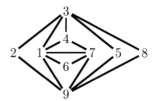

Figure 1

(ii) Denote the numbers consecutively by $x_1, x_2, \ldots, x_{n+1}$. The numbers $s_i = x_1 + x_2 + \cdots + x_i$ satisfy $1 \leq s_1 < s_2 < \cdots < s_{n+1} = 3n$. By the pigeonhole principle you have $s_i \equiv s_j \pmod{n}$ for appropriate $1 \leq i < j \leq n+1$. Then either $s_j - s_i = n$ or $s_j - s_i = 2n$. The desired collection is then either $(x_{i+1}, x_{i+2}, \ldots, x_j)$, or its "complement."

(iii) $n = 6$. Suppose the terms of a cyclic sequence $x_1, x_2, \ldots, x_n$ are used. Clearly, $x_i \notin \{0, 1\}$ for all i, and therefore it follows from $x_{i+1} = x_i x_{i+2}$ and $x_{i+2} = x_{i+1} x_{i+3}$ that $x_{i+3} = \frac{1}{x_i}$, which leads to $x_{i+6} = x_i$; this implies $n \mid 6$. The case $n = 3$ is easily excluded, and by the above, for $n = 6$ the sequence has the form $a, b, \frac{b}{a}, \frac{1}{a}, \frac{1}{b}, \frac{a}{b}$; these are distinct numbers, for instance, for $a = 2$ and $b = 3$.

(iv) There is a unique cyclic sequence $1, 1, 0, 1, 1, 0, \ldots, 1, 1, 0$. You can "turn" the desired sequence such that $x_1 \geq x_i$ for all i. Then in view of the fact that $x_i \geq 0$ for all i, the equality $x_1 = |x_2 - x_3|$ implies $x_2 = 0$ (and $x_1 = x_3$) or $x_3 = 0$ (and $x_1 = x_2$). Let therefore $x_2 = 0$ (otherwise, turn the sequence back by one place). If you set $x_1 = a$, then it is easy to determine consecutively $x_{30} = |x_2 - x_1| = a$, $x_{29} = |x_1 - x_{30}| = 0$, $x_{28} = |x_{30} - x_{29}| = a$, etc. The sequence has therefore the form $a, 0, a, a, 0, a, \ldots, a, 0, a$. The sum of its elements is $20a$, which by hypothesis is equal to 20, and so $a = 1$.

(v) 1000. The number $1000a + b$ on an unlucky ticket, with $a, b \in \{0, 1, \ldots, 999\}$ and $a \neq b$, lies between the numbers (on lucky tickets) $1000c + c$ and $1000(c + 1) + (c + 1)$, where $c = a$ (if $a < b$) or $c = a - 1$ (if $a > b$). Since $[1000(c + 1) + (c + 1)] - [1000c + c] = 1001$, there are no more than 1000 consecutive unlucky ticket numbers. On the other hand, $000001, 000002, \ldots, 001000$ is such a sequence of 1000 numbers.

(vi) Restrict yourself to those ways of cutting the strip where among the numbers obtained there are exactly 4 three-digit numbers and $(80 - 4 \cdot 3)/2 = 34$ two-digit numbers. The number of these ways is equal to $P_0(4, 34) = \frac{38!}{4! 34!} = 73\,815$. The sum of the 38 numbers for each such cutting does not exceed $4 \cdot 999 + 34 \cdot 99 = 7362$. Since $73\,815 > 7362 \cdot 10$, the pigeonhole principle shows that among the $73\,815$ sums at least 11 are identical.

(vii) Prove the assertion for a pair (n, k), $n \neq k$; by induction on $p = n + k \geq 3$. In the case $p = 3$, that is, $\{n, k\} = \{1, 2\}$, the bound $m < 2$ is clear. Suppose that for given relatively prime n, k $(n + k > 3)$ the sequence $a_1, a_2, \ldots, a_m$ has the required property. Now restrict yourself to the case $1 < n < k < m$ (if $n > k$, then consider instead the sequence $-a_1, -a_2, \ldots, -a_m$; for $n = 1$ the bound $m < k$ is clear, and if $m \leq k$, then the bound $m < n + k - 1$ follows from the assumption $n > 1$). If you set $S(i, j) = a_i + a_{i+1} + \cdots + a_j$, then for all $i = 1, 2, \ldots, m - k + 1$ you will have $S(i, i + k - n - 1) = S(i, i + k - 1) - S(i + k - n, i + k - 1) > 0$, since $S(i, i + k - 1) > 0$ and $S(i + k - n, i + k - 1) < 0$. This means that the sequence $a_1, a_2, \ldots, a_{m-n}$ has the property that the sum of any $k - n$ consecutive terms is positive, and the sum of any n consecutive terms is negative. The numbers n and k are relatively prime, so the same is true for $k - n$ and n; since $(k - n) + n = k < n + k$, by the induction hypothesis for the length $m - n$ of the shortened sequence the bound $m - n < (k - n) + n - 1$ holds, which implies $m < n + k - 1$. This completes the proof by induction. For $n = 13$ and $k = 7$ an appropriate sequence of length 18 is, for instance, as follows:

$$\underbrace{4, 4, \ldots, 4}_{5}, -23, \underbrace{4, 4, \ldots, 4}_{6}, -23, \underbrace{4, 4, \ldots, 4}_{5}.$$

Each sum of 13 consecutive terms is equal to $2 \cdot (-23) + 11 \cdot 4 = -2$, and each sum of 7 consecutive terms is equal to $(-23) + 6 \cdot 4 = 1$. In the case where n, k are not relatively prime, the bound is of the form $m < n + k - d$, where d is the greatest common divisor of n, k. To prove this, proceed as in the case $d = 1$, but don't forget that the case $n = k$ also needs to be considered.

2.10 (i) Yes; for instance, the sequence with first term $a_1 = 2^{1990}$ and quotient $q = \frac{1}{2}$.

(ii) Write the rational number $q = \frac{a_2}{a_1}$ as the fraction $q = \frac{m}{n}$ with relatively prime integers m, n. For every $k \geq 0$ the number $a_{k+1} = \frac{a_1 m^k}{n^k}$ is an integer, that is, $n^k \mid a_1 m^k$, which means that $n^k \mid a_1$. This is possible only when $n = \pm 1$, that is, when $q \in \mathbb{Z}$.

(iii) No, as the example of the sequence $a_n = \left(\sqrt{2}\right)^n$ shows.

(iv) No. Derive from the equations $3 = aq^k$, $5 = aq^m$, $7 = aq^n$ the new equation $7^{m-k} 3^{n-m} 5^{k-n} = 1$, which by the unique factorization theorem of integers into prime powers is possible only when $m - k = n - m = k - n$, that is, when $k = m = n$, which is a contradiction.

(v) For every $k \geq 3$. Consider the sequence of length k of the form $\frac{k}{k!}, \frac{k-1}{k!}, \ldots, \frac{2}{k!}, \frac{1}{k!}$.

(vi) It clearly suffices to show that the terms of the given sequence are at most four different numbers. First note that in each geometric sequence of four different positive numbers $a_1 = a$, $a_2 = aq$, $a_3 = aq^2$, $a_4 = aq^3$, the terms are ordered according to size: Either $a_1 < a_2 < a_3 < a_4$ (if $q > 1$), or $a_1 > a_2 > a_3 > a_4$ (if $0 < q < 1$). Furthermore, $a_1 a_4 = a_2 a_3$ $(= a^2 q^3)$ holds in both cases. Then assume that the given sequence of length $4n$ contains five different numbers; denote them by $a < b < c < d < e$. Since the quadruples $a < b < c < e$ and $a < b < d < e$ form geometric sequences, according to the above you have $ae = bc$ and $ae = bd$, which implies $c = d$, but this is a contradiction.

(vii) Show that the equation $2\binom{n}{k} = \binom{n}{k-1} + \binom{n}{k+1}$ holds if and only if $4k^2 - 4nk + n^2 - n - 2 = 0$. Solving this equation in k, you get

$$k_{1,2} = \frac{n \pm \sqrt{n+2}}{2} \tag{47}$$

(explain what the identity $k_2 = n - k_1$ means for the original equation with the binomial coefficients). For each $n = m^2 - 2$ $(m \geq 3)$ you obtain an integer root $k_1 = \frac{m^2 + m - 2}{2}$ satisfying $\frac{n}{2} < k_1 < n$. If you assumed that the quadruple $\binom{n}{k-1}$, $\binom{n}{k}$, $\binom{n}{k+1}$, $\binom{n}{k+2}$ formed an arithmetic sequence, then $\{k, k+1\} = \{k_1, k_2\}$ would have to hold, where k_1, k_2 are the numbers in (47). Then you would have $k + (k+1) = k_1 + k_2 = n$, which would imply $k_{1,2} = \frac{n \pm 1}{2}$. Equating this with (47) you would obtain the condition $\sqrt{n+2} = 1$, which, however, is never satisfied. Finally, no triple $\binom{n}{k-1}$, $\binom{n}{k}$, $\binom{n}{k+1}$ forms a geometric sequence, since the inequality $\binom{n}{k}^2 > \binom{n}{k-1}\binom{n}{k+1}$ holds. (The last relation is equivalent to the inequality $\frac{n-k+1}{k} > \frac{n-k}{k+1}$.)

(viii) An arbitrary term a_k of the given sequence can be the middle term of at most $\min\{k - 1, n - k\}$ three-term arithmetic subsequences. Hence the number of such subsequences is at most

$$2(0 + 1 + \cdots + (m - 1)) = m(m - 1) \quad \text{(if } n = 2m),$$
$$2(0 + 1 + \cdots + (m - 1)) + m = m^2 \quad \text{(if } n = 2m + 1).$$

Exactly this many three-element arithmetic subsequences can be chosen from $1, 2, \ldots, n$.

(ix) Explain why it suffices to prove the following assertion: If for some numbers $d > 0$ and $q > 1$ there are indices $1 = m_1 < m_2 < \cdots < m_k = n$ such that $1 + (m_j - 1)d = q^{j-1}$ $(1 \le j \le k)$, then $n \ge 2^{k-1}$. From the identity $m_j = 1 + \left(q^{j-1} - 1\right)/d$ for $j \in \{2, 3\}$ it follows that both numbers d, q are in $\mathbb{Q}$ ($\frac{q-1}{d}$ and $\frac{q^2-1}{d}$ are rational, and therefore $q + 1$ is as well, and so is q, and finally also d). Write them in the form of quotients of relatively prime positive integers, $q = \frac{u}{v}$, $d = \frac{a}{b}$. The two fractions

$$\frac{q-1}{d} = \frac{u-v}{a} \cdot \frac{b}{v} \quad \text{and} \quad \frac{q^{k-1}-1}{d} = \frac{u^{k-1}-v^{k-1}}{a} \cdot \frac{b}{v^{k-1}}$$

are integers, which implies that $a \mid (u - v)$ and $v^{k-1} \mid b$, that is, $u - v \ge a$ and $b \ge v^{k-1}$, and thus $\frac{b}{a} \ge \frac{v^{k-1}}{u-v}$ (recall that $q > 1$, that is, $u > v \ge 1$). You thus obtain

$$n = 1 + \frac{u^{k-1}-v^{k-1}}{v^{k-1}} \cdot \frac{b}{a} \ge 1 + \frac{u^{k-1}-v^{k-1}}{u-v}$$
$$= 1 + \left(u^{k-2} + u^{k-3}v + \cdots + uv^{k-3} + v^{k-2}\right)$$
$$\ge 1 + \left(2^{k-2} + 2^{k-3} + \cdots + 2 + 1\right) = 2^{k-1}.$$

In conclusion, note that the bound $n \ge 2^{k-1}$ cannot be improved: The arithmetic sequence $1, 2, 3, \ldots, 2^{k-1}$ of length 2^{k-1} contains the geometric sequence $1, 2, 2^2, \ldots, 2^{k-1}$ of length k.

(x) For every n. Set $b_k = (1 + 2\varepsilon)^k$ and $a_k = 1 + (2k - 1)\varepsilon$, $1 \le k \le n + 1$, where $\varepsilon > 0$ is chosen small enough so that $(1 + 2\varepsilon)^n < 1 + \frac{1}{2n}$ (for any fixed n such an ε clearly exists). To verify the required chain of inequalities, use on the one hand Bernoulli's inequality $(1 + 2\varepsilon)^k > 1 + 2\varepsilon k$ (see, e.g., [5], Chapter 2, 3.3.(iii)), and on the other hand the inequalities $(1 + 2\varepsilon)^k = 1 + 2\varepsilon(1 + (1 + 2\varepsilon) + \cdots + (1 + 2\varepsilon)^{k-1}) \le 1 + 2\varepsilon k(1 + 2\varepsilon)^{k-1} \le 1 + 2\varepsilon k(1 + 2\varepsilon)^n < 1 + 2\varepsilon k \left(1 + \frac{1}{2n}\right) \le 1 + \varepsilon(2k + 1)$ for all $k \in \{1, 2, \ldots, n\}$.

(xi) The sequence $x_1, x_2, \ldots, x_n$ is antiarithmetic if none of the numbers $\frac{x_i + x_j}{2}$, $1 \le i < j \le n$, is among its terms. The desired subsequence X_k of length $4 \cdot 2^k$ chosen from $1, 2, \ldots, 5 \cdot 3^k$ can be constructed recursively: $X_0 = (1, 2, 4, 5)$, $X_{k+1} = X_k \cup Y_k$, where $Y_k = (x + 10 \cdot 3^k : x \in X_k)$ for all $k \ge 0$. Assume that X_k is an antiarithmetic sequence for some k. To show that X_{k+1} is also antiarithmetic, choose any two terms $x, y \in X_{k+1}$, $x < y$. Then $\frac{x+y}{2} \notin X_{k+1}$ clearly holds if $x, y \in X_k$ or $x, y \in Y_k$. There remains the case where $x \in X_k$ and $y \in Y_k$; then $1 \le x \le 5 \cdot 3^k$ and $1 + 10 \cdot 3^k \le y \le 15 \cdot 3^k$; that is, $1 + 5 \cdot 3^k \le \frac{x+y}{2} \le 10 \cdot 3^k$. The number $\frac{x+y}{2}$ lies neither in X_k nor in Y_k, and therefore also not in X_{k+1}.

(xii) The desired property is displayed by an infinite sequence that can be obtained as the union of all sequences X_k from the solution of (xi), namely, the sequence $X_0 \cup Y_0 \cup Y_1 \cup Y_2 \cup \cdots \cup Y_n \cup \cdots$. Show by induction on k that every number $x \in \{1, 2, \ldots, 5 \cdot 3^k\}$ that does not lie in X_k is of the form $x = \frac{u+v}{2}$, where $u, v \in X_k$. For $k = 0$ this is true: $3 = \frac{1+5}{2}$; for the step from k to $k+1$ it clearly suffices to consider those x that lie in the interval $[a+1, 2a]$, where $a = 5 \cdot 3^k$. Now show that for each $j = 1, 2, \ldots, a$ the number $a+j$ has the representation $a + j = \frac{x+(y+2a)}{2}$ for appropriate $x, y \in X_k$ (recall that from $y \in X_k$ it follows that $y + 2a \in Y_k$, so both numbers $x, y + 2a$ lie in $X_{k+1} = X_k \cup Y_k$), that is, $j = \frac{x+y}{2}$. Indeed, if $j \in X_k$, then choose $x = y = j$, and if $j \notin X_k$, then numbers $x, y \in X_k$ with the property $j = \frac{x+y}{2}$ exist by the induction hypothesis.

(xiii) If the difference d of the given sequence is 0, then the assertion is trivial; when $d \geq 1$, by the binomial theorem for all $m \geq 1$ the number $b_m = a_1(1 + d)^m$ is a term of the original sequence $a_1, a_1 + d, \ldots$, and furthermore, $\mathcal{P}(b_m) = \mathcal{P}(a_1) \cup \mathcal{P}(1+d)$; that is, the set $\mathcal{P}(b_m)$ is the same for all $m \geq 1$.

(xiv) For a fixed $d \in \mathbb{N}$ divide the set of all prime numbers into d classes according to their remainders upon division by d. Since there are infinitely many primes, one of the classes must contain infinitely many elements $p_1 < p_2 < p_3 < \cdots$. All these primes p_i lie in the arithmetic sequence $a_k = p_1 + (k-1)d$, since $d \mid p_i - p_1$ by construction of the classes.

(xv) Use induction: For $n = 3$ an appropriate sequence is $2^2, 10^2, 14^2$. If for some $n \geq 3$ the numbers $a_i = m_i^{k_i}$ ($1 \leq i \leq n$) form an arithmetic sequence of length n of perfect powers with difference $d = a_{i+1} - a_i > 0$ ($1 \leq i < n$), then construct a sequence $b_1, b_2, \ldots, b_{n+1}$ as follows: $b_i = (a_n + d)^k a_i$ for $i = 1, 2, \ldots, n$, and $b_{n+1} = (a_n + d)^{k+1}$, where k is an arbitrary (for instance the smallest) common multiple of the numbers $k_1, k_2, \ldots, k_n$. This new sequence is clearly arithmetic with difference $d' = (a_n + d)^k d$, and each of its terms is a perfect power.

(xvi) Assume to the contrary that such an arithmetic sequence with terms $a_k = a_1 + (k-1)d$ exists, where $a_1, d \in \mathbb{N}$. By Dirichlet's theorem (see the note following the solution of 2.9.(v)) there exists an infinite subsequence of the form $mp_1, mp_2, \ldots$, where m is the greatest common divisor of the numbers a_1, d, and where $p_1, p_2, \ldots$ is a sequence of distinct prime numbers. By hypothesis every number mp_i is a perfect power, and thus $p_i \mid m$. This last relation, however, can hold only for finitely many primes p_i.

A different proof (also by contradiction, but without using Dirichlet's theorem), follows from the result of 2.10.(xviii), where you set $n = a_1 + d$; indeed, in any set of $a_1 + d$ consecutive positive integers there is at least one term $a_k = a_1 + (k-1)d$.

(xvii) For every $a_1 > 1$ you can set $d = a_1^2$. From the unique factorization theorem in $\mathbb{N}$ it follows that if the relatively prime integers a, b are greater than 1 and if the product ab is a perfect power, then the numbers a, b are also perfect powers (see, e.g., [5], Chapter 3, 2.6.(iii)). Therefore, the number $a_{k+1} = a_1 + ka_1^2 = a_1(1 + ka_1)$ can be a perfect power only if the two relatively prime numbers $a_1, 1 + ka_1$ are perfect powers; by hypothesis, however, a_1 does not have this property. For $a_1 = 1$ the assertion is not true: The sequence $a_{k+1} = 1 + kd$ contains the perfect power $(1 + d)^2$, no matter how the number $d \in \mathbb{N}$ is chosen.

(xviii) Choose n distinct prime numbers $p_1, p_2, \ldots, p_n$. By the Chinese remainder theorem (see, e.g., [5], Chapter 3, 4.8.(x)) the system of congruences $x \equiv p_i - i \pmod{p_i^2}$, $1 \leq i \leq n$, has a solution $x \in \mathbb{N}$. Then for $i = 1, 2, \ldots, n$ the number $x + i$ is divisible by p_i, but not by p_i^2, and is therefore not a perfect power.

(xix) Show by induction that every triple standing in the sequence in an odd, respectively an even, position, is of the form $\{2^{k+1}3^k5^k, 2^k3^{k+1}5^k, 2^k3^k5^{k+1}\}$, respectively $\{2^k3^{k+1}5^{k+1}, 2^{k+1}3^k5^{k+1}, 2^{k+1}3^{k+1}5^k\}$ (ignore the order of the numbers within a triple; it is not important for the given construction). The number $2^m3^n5^k$ is a perfect power if and only if the exponents m, n, k have a common divisor $d > 1$; however, the numbers $k, k+1$ are relatively prime for all $k \geq 1$.

2.12 (i) Note that $x_i \leq 2^{i-1}$ and $x_{i+1} - x_i \leq x_i$ for all $i = 1, 2, \ldots, 99$. Hence

$$S = x_1 + (x_3 - x_2) + (x_5 - x_4) + \cdots + (x_{99} - x_{98}) - x_{100}$$

$$\leq x_1 + x_2 + x_4 + \cdots + x_{98} - x_{100} \leq 1 + 2 + 2^3 \cdots + 2^{97} = \frac{2^{99} + 1}{3},$$

where equality occurs if and only if $x_i = 2^{i-1}$ $(1 \leq i \leq 99)$ and $x_{100} = 0$.

(ii) No. Since $b_2 < b_1 < b_8 < b_3 < b_6 < b_9 = b_{12} < b_4 < b_7 < b_{10} = b_{11} < b_5 < b_{13} < b_{14} = b_{15}$, you would also have $a_2 < a_1 < a_8 < a_3 < a_6 < a_9 = a_{12} < a_4 < a_7 < a_{10} = a_{11} < a_5 < a_{13} < a_{14} = a_{15}$, because $b_i < b_j$ implies $a_i < a_j$ and $b_i = b_j$ implies $a_i = a_j$ for all indices i, j. From the chain of inequalities for a_i it follows that the sequence of the numbers b_i has the form $1, 0, 3, 7, 11, 4, 8, 2, 5, 9, 9, 5, 12, 13, 13$, which is a contradiction.

(iii) If you construct the sequence $b_1, b_2, \ldots, b_{15}$ from the given one, you obtain the same sequence again. Exactly those sequences $a_1, a_2, \ldots, a_{15}$ have this property for which the following holds: If you write the terms a_i in nondecreasing order $a_1' \leq a_2' \leq \cdots \leq a_{15}'$, then $a_1' = 0$, and for all $i = 1, 2, \ldots, 14$ the inequality $a_i' < a_{i+1}'$ implies $a_{i+1}' = i$. To prove this, note that to a change in the order of the numbers a_i there corresponds the same change in the order of the numbers b_i. Verify that sequences with the given property can also be described as follows: They are exactly those that arise from some sequence $a_1, a_2, \ldots, a_{15}$ in the construction described in (ii).

(iv) Set $s_k = a_1 + a_2 + \cdots + a_k$ and prove by induction on $k \geq 1$ that if $1 \leq n \leq s_k$, then the number n can be written as a sum of some terms from $a_1, a_2, \ldots, a_k$. In the induction step from $k - 1$ to k distinguish the cases $1 \leq n \leq s_{k-1}$ and $s_{k-1} < n \leq s_k$; then in the second case use the induction hypothesis for the number $n' = n - a_k$.

(v) For $K = \{0\}$ the answer is yes, and for $K = \{0, 1\}$ it is no. In the case $K = \{0\}$ an appropriate example is $a_i = (-2)^{i-1}$; use induction on n to show that the expression (12), where $i_k \leq n$, has every integer (except 0) from the interval $[\alpha_n, \beta_n]$, where $\alpha_1 = \beta_1 = 1$ and $\alpha_n = -\left(2 + 2^3 + 2^5 + \cdots\right)$, $\beta_n = 1 + 2^2 + 2^4 + \ldots$ (in both sums the terms 2^i are added for $i \leq n - 1$). The uniqueness of (12) is then derived from the fact that $2^{i_1 - 1}$ is the largest power of 2 dividing the number x.

In the case $K = \{0, 1\}$ assume the existence of an appropriate sequence $a_1, a_2, \ldots$ and set $I(x) = \{i_1, i_2, \ldots, i_k\}$ if (12) holds; this is a correct definition for all $x \in \mathbb{Z} \setminus \{0, 1\}$. Clearly, $a_i \notin \{0, 1\}$ for each i. Find an index p such that $a_p \neq -1$, and then an index q such that $a_q \neq -1$ and $q \notin I(1 + a_p) \cup I(1 - a_p)$. Clearly, $q \in I(1 + a_p + a_q) \cap I(1 - a_p + a_q)$, since you obtain expressions for both numbers $1 + a_p + a_q$, $1 - a_p + a_q$ in the form (12) if you adjoin to the expressions of the numbers $1 + a_p$, $1 - a_p$ the (still missing) summand a_q. Distinguish now whether or not $p \in I(1 + a_q)$; in the affirmative case you have $I(1 - a_p + a_q) = I(1 + a_q) \setminus \{p\}$, while in the negative case $I(1 + a_p + a_q) = I(1 + a_q) \cup \{p\}$. However, in both cases you have $q \in I(1 + a_q)$, since q lies in both the sets $I(1 - a_p + a_q)$ and $I(1 + a_p + a_q)$. But from $q \in I(1 + a_q)$ it follows that the number 1 also has an expression (12) (simply delete the summand a_q in the expression of the number $1 + a_q$), and this is a contradiction.

Note that the example given for the set $K = \{0\}$ describes p-adic expansions of nonzero integers with the unusual base $p = -2$.

(vi) An exceptional sequence exists exactly when $n - k$ is an odd integer, it is unique for such fixed n, k, and it has the form $a_i = \frac{1}{2}$ ($1 \leq i \leq n$). Proof: By choosing $n_1 = n - k + 1$ and $n_i = 1$ ($2 \leq i \leq k$) you obtain $a_1(n - k + 1) \in \mathbb{N}$, which implies $a_1(n - k) = a_1(n - k + 1) - a_1 \notin \mathbb{Z}$. Hence with the choice $n_1 = n - k$, $n_2 = 2$, and $n_i = 1$ ($3 \leq i \leq k$) you assert that $2a_2 \in \mathbb{N}$, that is, $a_2 = \frac{1}{2}$. Similarly, by changing the order of the numbers $n_1, n_2, \ldots, n_k$ you get $a_i = \frac{1}{2}$ for all $i = 1, 2, \ldots, k$. But then $a_1(n - k + 1) = \frac{n - k + 1}{2}$ is an integer only when $n - k$ is odd. In this case the sequence obtained is indeed exceptional, since at least one of the numbers n_i in any expansion $n = n_1 + n_2 + \cdots + n_k$ is then even (note that $n_i \equiv 1 \pmod 2$ for all $i = 1, 2, \ldots, k$ implies $n \equiv k \pmod 2$).

(vii) Directly from the definition of the sequence you obtain the chain of inequalities $x_3 \geq x_4 \geq x_5 \geq \cdots$. Therefore, the bounds $x_5 \leq |x_3 - x_4| = x_3 - x_4$, $x_6 \leq |x_4 - x_5| = x_4 - x_5$ hold, and in general $x_n \leq x_{n-2} - x_{n-1}$ for all $n \geq 5$. Use this for a lower bound for the numbers x_{n-2} with the help of x_{n-1} and x_n: $x_{n-2} \geq x_{n-1} + x_n$. Suppose that $x_{21} \neq 0$. Then $x_{20} \geq x_{21} \geq 1$, and consecutively $x_{19} \geq x_{20} + x_{21} \geq 2$, $x_{18} \geq x_{19} + x_{20} \geq 3$, $x_{17} \geq x_{18} + x_{19} \geq 5$, etc. This is a lower bound obtained with the help of the Fibonacci numbers F_i: $x_{21} \geq F_1$, $x_{20} \geq F_2$, $x_{19} \geq F_3$, $\ldots$, $x_3 \geq F_{19}$. Recall that these numbers are the terms of the sequence $F_1 = F_2 = 1$, $F_3, F_4, \ldots$ defined by the recurrence relation $F_{i+2} = F_{i+1} + F_i$ ($i \geq 1$). A routine calculation gives $F_{19} = 4171$, $F_{20} = 6745$, and $F_{21} = 10\,916$. Since a change in the order of the numbers x_1, x_2, x_3 has no influence on the calculation of the further numbers $x_4, x_5, \ldots$, you may assume from the outset that $10\,000 \geq x_1 \geq x_2 \geq x_3$ and complete the argument as follows: $x_2 \geq x_3 + x_4 \geq F_{20}$, $x_1 = x_2 + x_3 \geq F_{21}$, which is a contradiction, since $F_{21} > 10\,000$.

(viii) On the left-hand side of (13) the same number of summands is equal to $+1$ or -1, and therefore $n = 2k$. Since $x_i x_{i+1} = -1$ if and only if $x_{i+1} = -x_i$, the integer k is equal to the number of neighboring pairs in the sequence $x_1, x_2, \ldots, x_n, x_1$ such that the two terms have opposite signs. If there are exactly p such pairs in some sequence $y_1, y_2, \ldots, y_m$ of the numbers ± 1, then clearly $y_m = (-1)^p y_1$. Hence in our case $x_1 = (-1)^k x_1$, and therefore the number k is even, which implies that $4 \mid n$.

3.2 (i) For no $n \geq 2$. The numbers $r_1, r_2, \ldots, r_n, u, v$ of an arbitrary such array have the same parity as n and lie in the interval $[-n, n]$; therefore, they take on at most $n+1$ different values, and by the pigeonhole principle two of them have to be the same.

(ii) If you enter the numbers a_{ij} $(1 \leq i \leq 2n, 1 \leq j \leq 2n)$ according to the rule

$$a_{ij} = \begin{cases} -1 & \text{if } i + j > 2n + 1, \\ 0 & \text{if } i + j = 2n + 1 \text{ and } i > n, \\ 1 & \text{otherwise,} \end{cases}$$

then the row and column sums r_i and c_j satisfy

$$r_i = \begin{cases} 2n + 2 - 2i & (1 \leq i \leq n), \\ 2n + 1 - 2i & (n < i \leq 2n), \end{cases} \qquad c_j = \begin{cases} 2n + 1 - 2j & (1 \leq j \leq n), \\ 2n + 2 - 2j & (n < j \leq 2n). \end{cases}$$

(iii) It is possible to set, for instance, $a_{ij} = (i-1)n + p_{i+j}$ $(1 \leq i \leq n, 1 \leq j \leq n)$, where $p_1, p_2, \ldots, p_n$ is an arbitrary arrangement of the numbers $1, 2, \ldots, n$, with $p_{n+i} = p_i$, $1 \leq i \leq n$. Any column sum is then equal to $(0 + 1 + \cdots + (n-1))n + p_1 + p_2 + \cdots + p_n = \frac{n(n^2+1)}{2}$.

(iv) $\frac{1}{4} mnS$. If $a_1, a_2, \ldots, a_n$ and $b_1, b_2, \ldots, b_n$ are the arithmetic sequences of the elements of the first, respectively the last, row, then

$$a_1 + a_2 + \cdots + a_n = \frac{(a_1 + a_n)n}{2} \quad \text{and} \quad b_1 + b_2 + \cdots + b_n = \frac{(b_1 + b_n)n}{2}.$$

Similarly, the ith column sum satisfies $c_i = \frac{(a_i + b_i)m}{2}$; the desired sum of all numbers in the array is therefore

$$c_1 + c_2 + \cdots + c_n = \frac{(a_1 + b_1 + a_2 + b_2 + \cdots + a_n + b_n)m}{2}$$
$$= \frac{(u_1 + u_n + b_1 + b_n)mn}{4} = \frac{mnS}{4}.$$

(v) Only for $n = 3$. In the case $n = 3$ you have $S = 3k$, where k is the common value of the eight sums u, v, r_i, and c_i $(i = 1, 2, 3)$. From

the identity $4k = r_2 + c_2 + u + v = r_1 + r_2 + r_3 + 3a_{22} = 3k + 3a_{22}$
it follows that $k = 3a_{22}$, that is, $3 \mid k$ and $9 \mid S$. For each $n > 3$ you
clearly have $n \mid S$, but in general not $n^2 \mid S$; indeed, consider the following
example of an $n \times n$ array with the required properties and $S = n$: $a_{11} =$
$a_{n-1,2} = a_{n,3} = a_{24} = a_{35} = a_{46} = \cdots = a_{n-2,n} = 1$ (if $n \geq 4$ is even),
$a_{11} = a_{n-1,2} = a_{n,3} = a_{n-2,4} = a_{25} = a_{36} = \cdots = a_{n-3,n} = 1$ (if $n \geq 5$ is
odd), and finally $a_{ij} = 0$ for all the other pairs of indices $i, j \in \{1, 2, \ldots, n\}$.

(vi) Assume that such an array exists. Then the sum $S_n = 1 + 2 + \cdots +$
$n^2 = \frac{n^2(n^2+1)}{2}$ is divisible by 2^k, the *smallest* of all the row and column
sums of the array. Clearly, $2^k \geq 1 + 2 + \cdots + n > \frac{n^2}{2}$. Verify that $2 \mid S_n$
only for even n, but then $n^2 + 1$ is odd, and thus from $2^k \mid S_n$ it follows
that $2^k \mid \frac{n^2}{2}$, which is a contradiction to $2^k > \frac{n^2}{2}$.

(vii) $c - r = n^2 - n$ (independent of the placement of the numbers). To
prove this, let i_k, j_k denote the pair of indices for the row and column in
which the number k lies, $1 \leq k \leq n^2$. By hypothesis you have $j_1 = i_2$,
$j_2 = i_3, \ldots, j_{n^2-1} = i_{n^2}$. Since both sequences of length n^2, namely, i_1,
$i_2, \ldots, i_{n^2}$, and $j_1, j_2, \ldots, j_{n^2}$, differ only in the order of their terms (in
both of them each index $\alpha \in \{1, 2, \ldots, n\}$ occurs exactly n times), this
means that $i_1 = j_{n^2}$. Therefore, if $1, x_2, x_3, \ldots, x_n$ are all the numbers in
the same row, then $n^2, x_2 - 1, x_3 - 1, \ldots, x_n - 1$ are all the numbers in
the same column. It suffices to add that $r = 1 + x_2 + x_3 + \cdots + x_n$ and
$c = n^2 + (x_2 - 1) + (x_3 - 1) + \cdots + (x_n - 1)$.

(viii) For all even $n \geq 4$. The number $2r_1 = r_1 + r_2 = 1 + 2 + \cdots + 2n =$
$n(2n + 1)$ is even, and thus $2 \mid n$. The number $n = 2$ is easily excluded; for
all even $n = 2k \geq 4$ you obtain an appropriate distribution from the initial
array

$$\begin{pmatrix} 1 & 2 & 3 & \cdots & 2k-1 & 2k \\ 4k & 4k-1 & 4k-2 & \cdots & 2k+2 & 2k+1 \end{pmatrix} \qquad (48)$$

by mutual changes of the numbers $j \leftrightarrows 4k + 1 - j$ in some appropriately
chosen columns. In the case of even k choose exactly the $\frac{k}{2}$ first and the $\frac{k}{2}$
last columns, and in the case of odd $k \geq 3$ choose exactly the columns with
indices $j \in \{2, 3, \ldots, \frac{k+1}{2}, k + 1, k + \frac{k+1}{2}, k + \frac{k+1}{2} + 1, \ldots, 2k\}$. Determine
the row sums r_1, r_2 of the initial array (48) and then consider how their
values change under each switch $j \leftrightarrows 4k + 1 - j$, and finally add the changed
entries in both rows for the suggested values of j.

3.4 (i) Yes. Choose, for instance, $a_{ij} = 2^i 3^j$, $i, j \in \{1, 2, \ldots, n\}$.

(ii) If $n = pk + r$ $(p, r \in \mathbb{N}, 0 < r < k)$, then an appropriate array consists
of the numbers

$$x_{ij} = \begin{cases} -b & \text{if } k \mid i \text{ and } k \mid j, \\ a & \text{otherwise,} \end{cases}$$

where the constants $a, b \in \mathbb{R}^+$ satisfy the inequalities $a\left(k^2 - 1 + \frac{r(2n-r)}{p^2}\right) >$
$b > a\left(k^2 - 1\right)$. Such constants exist, since $k^2 - 1 > 0$ and $r(2n - r) > 0$.

The sum of the numbers in each $k \times k$ subarray is equal to $-b+a\left(k^2-1\right)$, and the sum of all n^2 numbers of the array is $-p^2 b + \left(n^2-p^2\right)a$; show that both numbers have the required signs.

(iii) 3^{2n-1} ways. Explain how each such array is determined by the $2n-1$ numbers in the first row and the first column. Furthermore, these $2n-1$ numbers can be arbitrarily chosen.

(iv) $4^n + 2 \cdot 3^n - 2^{n+2} + 1$ arrays. First show that an array T with numbers a_{ij} has the required property if and only if the auxiliary array T' with numbers $b_{ij} = (-1)^{i+j} a_{ij} \in \{-1,0,1\}$ satisfies the condition $b_{ij} = b_{1j} + b_{i1} - b_{11}$, $i,j \in \{1,2,\ldots,n\}$. That occurs if and only if the ith row of T' is obtained from the first row of T' by adding some constant c_i to all of its elements, $c_i = b_{i1} - b_{11}$ for $i = 2,3,\ldots,n$. The array T' is then determined by the first row and the numbers $c_2, c_3, \ldots, c_n$. Let M denote the set of all the numbers from $\{-1,0,1\}$ that occur in the first row of T'. Assume that you have already chosen the first row of T' and decide how many possibilities there are for choosing the numbers $c_2, c_3, \ldots, c_n$. If M is one of the sets $\{-1\}, \{0\}, \{1\}$ (there are three such first rows), then each of the $n-1$ numbers c_i can be chosen in three ways; you therefore obtain $3 \cdot 3^{n-1} = 3^n$ arrays T'. If M is one of the sets $\{-1,0\}$ or $\{0,1\}$ (there are $2(2^n - 2)$) such first rows), then each number c_i can be chosen in two ways; you therefore obtain $2(2^n - 2) \cdot 2^{n-1} = 4^n - 2^{n+1}$ arrays T'. The remaining case is $\{-1,1\} \subseteq M$; then necessarily $c_i = 0$ $(2 \le i \le n)$, and there are exactly as many arrays T' as there are rows of length n with the numbers $-1,0,1$, in which the two numbers -1 and 1 occur simultaneously, so $3^n - 2 \cdot 2^n + 1$ of them. The desired total number is therefore $3^n + \left(4^n - 2^{n+1}\right) + 3^n - 2 \cdot 2^n + 1 = 4^n + 2 \cdot 3^n - 2^{n+2} + 1$.

(v) If $k \mid n$, then $S = 0$. If $n = pk + q$, where $p, q \in \mathbb{N}$ and $q < k$, then the largest possible sum S is equal to $(p+1)\mu(q) + p\mu(k-q)$, where μ is the function defined by $\mu(t) = \min\left\{t^2, k^2 - t^2\right\}$ for a fixed k. First derive a generalization of (15): The sum S of all n^2 numbers in the array can be expressed in the form $S = S_A - S_B$, where S_A, respectively S_B, is the sum of the numbers in all squares A, respectively B (see Figure 2), which alternate in "hanging" along the diagonal of the array. Here the $p+1$ squares A have size $q \times q$, and the p squares B have size $(k-q) \times (k-q)$. (To prove this, use induction on $p \ge 1$; it is convenient to begin with the value $p = 0$.) Then explain why the absolute value of the sum of all t^2 numbers of any $t \times t$ subarray of our array does not exceed $\mu(t)$ for any $t \in \{1,2,\ldots,k\}$.

The bounds $|S_A| \le (p+1)\mu(q)$ and $|S_B| \le p\mu(k-q)$ imply the required bound for the sum S. It remains to give an example for an $n \times n$ array with the given property and with sum of elements

$$S = (p+1)\mu(q) + p\mu(k-q). \tag{49}$$

Subdivide the whole array into subarrays of type A, B, C, as indicated in

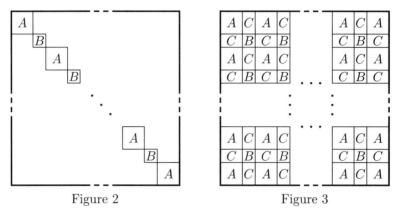

Figure 2   Figure 3

Figure 3. Each part A has size $q \times q$, part B has size $(k - q) \times (k - q)$, and C has size $q \times (k - q)$ or $(k - q) \times q$. Into every field of the whole array write one of the three numbers a, b, c according to whether the field lies in $A, B,$ or C (the values of a, b, c will be determined later). It is easy to see that the sum of the k^2 numbers in any $k \times k$ subarray is equal to $\sigma = aq^2 + 2cq(k - q) + b(k - q)^2$, while the sum of all n^2 numbers of the array is equal to $S = p^2\sigma + (2p + 1)aq^2 + 2cpq(k - q)$. If you set the values

$$a = \frac{\mu(q)}{q^2}, \quad b = \frac{-\mu(k - q)}{(k - q)^2}, \quad c = \frac{\mu(k - q) - \mu(q)}{2q(k - q)},$$

you find that $\sigma = 0$ and that (49) holds. Finally, show that the chosen values of the numbers a, b, c lie indeed in $[-1, 1]$.

3.6 (i) Explain why the following holds: The sum of the tens (units) of the numbers in each row (column) is equal to zero.

(ii) Independent of the way the numbers were chosen the sum is equal to $10(0 + 1 + 2 + \cdots + 9) + (0 + 1 + 2 + \cdots + 9) = 495$. Add the tens and the units of the chosen numbers separately.

(iii) If $a_{ij} = b_i + c_j$, $i, j \in \{1, 2, \ldots, n\}$, then the sum of the numbers in any generalized diagonal is equal to $b_1 + b_2 + \cdots + b_n + c_1 + c_2 + \cdots + c_n$. On the other hand, let all these sums of an $n \times n$ array with numbers a_{ij} be equal. Define the numbers $b_i = a_{i1}$, $c_j = a_{1j} - a_{11}$ for all $i, j \in \{1, 2, \ldots, n\}$. The identity $a_{pq} = b_p + c_q$ clearly holds if one of the indices p, q $(p, q \in \{1, 2, \ldots, n\})$ is equal to 1; let therefore $1 < p \leq n$ and $1 < q \leq n$. Since the two pairs of fields with numbers a_{11}, a_{pq}, respectively a_{p1}, a_{1q}, can be completed to generalized diagonals with the same $n - 2$ fields, the equality $a_{11} + a_{pq} = a_{p1} + a_{1q}$ has to hold, which implies $a_{pq} = a_{p1} + a_{1q} - a_{11} = (b_p + c_1) + (b_1 + c_q) - (b_1 + c_1) = b_p + c_q$.

(iv) By choosing $i = j = k$ you obtain $3a_{ii} = 0$, that is, $a_{ii} = 0$, and by choosing $i \neq j = k$ you get $0 = a_{ij} + a_{jj} + a_{ji} = a_{ij} + a_{ji}$, that is, $a_{ji} = -a_{ij}$. Finally, for a fixed pair (i, j) take consecutively $k = 1, 2, \ldots, n$, and add the n equalities obtained from the condition of the problem: $na_{ij} + (a_{j1} + a_{j2} + \cdots + a_{jn}) - (a_{i1} + a_{i2} + \cdots + a_{in}) = 0$. Therefore, one can set $t_i = \frac{1}{n}(a_{i1} + a_{i2} + \cdots + a_{in})$ for all i.

(v) This is impossible for all $n \geq 3$. Show that the numbers a, b, c, d from any part of the array that has a shape as shown in Figure 4 would have to satisfy the congruences $a \equiv b \pmod 9$ and $c \equiv d \pmod 9$. (To do this, it suffices to consider the remainder upon division by 9 of the sum of the numbers in the five fields formed by each one of these parts of the array.) But this would mean that every number a_{ij} of such an $n \times n$ array would give the same remainder upon division by 9 as one of the numbers $a_{11}, a_{12}, a_{13}, a_{21}, a_{22}, a_{23}, a_{32}$ (if $n = 3$) or $a_{11}, a_{12}, a_{13}, a_{14}, a_{21}, a_{22}, a_{23}, a_{24}$ (if $n \geq 4$), and therefore all the numbers of the array would give at most eight different remainders.

3.8 (i) By the approach from 3.7.(i), any numbers x, y in the array under consideration satisfy the bound $|x - y| \leq (100 + 100 - 2) \cdot 20 = 3960$. Therefore the array has at most 3961 different numbers. Since $2 \cdot 3961 = 7922 < 10^4$, this means by the pigeonhole principle that at least three numbers of the array have to be identical.

(ii) Such a field does not exist in general. You can give an example of an array in which the number in each of its fields is less than exactly four of the eight numbers in the surrounding fields. This is satisfied by every array whose elements in each row, column, and diagonal form a strictly monotonic (that is, increasing or decreasing) doubly infinite sequence; for instance, the array with the numbers $a_{ij} = 3^i 2^j$ $(i, j \in \mathbb{Z})$ has this property.

(iii) There exists a field whose number is less than at least two of the three numbers in neighboring fields, and less than at least six of the twelve numbers in the surrounding fields. Proof: Choose an arbitrary part of the array in the form of a regular hexagon consisting of 24 fields (see Figure 5) and choose from it the field with the smallest number. In this array of 24 fields each one is surrounded by at least 6 fields, of which at least two are neighbors of the given field.

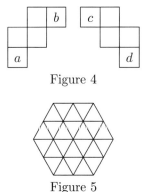

Figure 4

Figure 5

		... 10	9	...		
	... 10	8	7	9	...	
	... 10	8	6	5	7	9 ...
... 10	8	6	4	3	5	7 9 ...
10 8	6	4	2	1	3	5 7 9
11 9	7	5	3	2	4	6 8 10
... 11	9	7	5	4	6	8 10...
	... 11	9	7	6	8	10...
	... 11	9	8	10	...	
		... 11	10	...		

Figure 6

(iv) For a given $n \geq 1$ choose an arbitrary $2n \times 2n$ subarray. Suppose that $|x - y| \leq n$ holds for the numbers x, y of any two of its neighboring fields. By the trick used in 3.7.(i) this then means that any numbers x, y of this array satisfy the bound $|x - y| \leq (4n - 2)n$, and so there are at most $4n^2 - 2n + 1$ different numbers among them. However, there are more numbers in the array, namely $4n^2$, and this is a contradiction.

(v) $N = 2$. The number $N = 1$ does not have the required property (consider the four numbers in the fields that are neighbors of the field with the number 1), and for $N = 2$ an appropriate example is shown in Figure 6. The entire array consists of four "quadrants," in each of which the numbers in every row and every column form an arithmetic sequence with difference $d = 2$.

3.10 (i) With the notation of 3.9.(i) choose, for instance, $a = -1, b = c = \cdots = i = 1$.

(ii) A cyclic shift can be defined as the following change in the order of the rows of an array: The first row is shifted to the place of the second row, the second row to the place of the third row, etc., up to the nth row, which takes the place of the first row. If u_k denotes the diagonal sum of the array obtained after k cyclic shifts $(1 \leq k \leq n)$, then in the sum $S = u_1 + u_2 + \cdots + u_n$ each element of the array occurs exactly once. By hypothesis, $S > 0$, and thus $u_i > 0$ for some $i \in \{1, 2, \ldots, n\}$. In general, you cannot achieve that both sums u, v are positive: The $n \times n$ array with the numbers $a_{11} = n^2$ and $a_{ij} = -1$ for the remaining pairs $i, j \in \{1, 2, \ldots, n\}$ has the sum of elements $n^2 + (n^2 - 1)(-1) = 1$, and for any change in the order of the rows one of the diagonal sums has value $-n$ (for all $n > 1$).

(iii) For every $n \geq 1$ the following assertion holds: After $4n$ transformations every number of the array is divisible by 2^n. Explain why it suffices to analyze only the case $n = 1$, which was in essence already done in the solution to Problem 3.9.(i), where the number 1, respectively -1, can be considered as a symbol for even, respectively odd, numbers.

(iv) Note that the following array A becomes the array B_k after $2k + 1$ transformations $(k = 0, 1, 2, \ldots)$:

$$A = \begin{pmatrix} 0 & 0 & 0 \\ 0 & 1 & 0 \\ 0 & 0 & 0 \end{pmatrix}, \quad B_k = \begin{pmatrix} 0 & 2^{3k} & 0 \\ 2^{3k} & 0 & 2^{3k} \\ 0 & 2^{3k} & 0 \end{pmatrix}.$$

(v) Show that for any $k > 1$ an initial $n \times n$ array can be chosen such that it is not possible to achieve the desired goal. The number of all $n \times n$ residue arrays (modulo k) is exactly k^{n^2}; show that a fixed residue array can be reached through the transformations by at most k^{n^2-1} other arrays. This can be done in a similar way to that in the solution to 3.9.(iii) if you first verify that all transformations can be obtained by repeating the transformation in fixed $n^2 - 1$ pairs of neighboring fields (the number of all such pairs is larger, namely $2\left(n^2 - n\right)$). Let therefore R_{ij} denote the transformation of the pair of fields (i, j) and $(i, j + 1)$ of the residue array, and C_{ij} the transformation of the pair of fields (i, j) and $(i + 1, j)$. Show that each of the $2\left(n^2 - n\right)$ allowable transformations R_{ij} ($1 \le i \le n$, $1 \le j \le n - 1$) and C_{ij} ($1 \le i \le n - 1$, $1 \le j \le n$) can be obtained by appropriate repetitions of some of the $n^2 - 1$ transformations R_{ij} ($1 \le i \le n$, $1 \le j \le n - 1$) and C_{i1} ($1 \le i \le n - 1$). This is easy, since for each $j > 1$ you obtain the transformation C_{ij} if you carry out consecutively $R_{i,j-1}, R_{i+1,j-1}$ and $(k - 1)$-times $C_{i,j-1}$; therefore, you can use induction on the index j.

Note that this exercise can also be solved in a different way: Color the fields of the $n \times n$ array like a classical chessboard and consider the difference between two sums, namely the sum of the numbers in the white fields and that of the numbers in the black fields.

(vi) Begin with the sequence of transformations $P^1_{n,1}$, $P^2_{n,2}$, ..., $P^{n-1}_{n,n-1}$, which annul the diagonal elements a_{ii} ($1 \le i \le n - 1$). Then take any pair $i, j \in \{1, 2, \ldots, n - 1\}$, $i \ne j$, and carry out the sequence $P^j_{j,i}$, $P^i_{i,j}$, $P^j_{i,j}$, $P^i_{n,i}$, $P^i_{j,n}$, which annuls the element a_{ij} and does not change the remaining elements (with the exception of the last row and column). Since after each transformation the row and column sums remain equal to zero, you will finally obtain an array with only zeros.

3.12 (i) Set $a_{ii} = 2^{i-1}$; then $a_{ij} = 2^{i-1} 3^{j-i}$ for all i, j.

(ii) The sum of the numbers in the kth row is $2^{k-2}(n + 2 - k)n$, $1 \le k \le n + 1$. Hint: Verify by induction that the ith number (from the left) in the kth row is equal to $2^{k-2}(2i + k - 3)$, $1 \le i \le n + 2 - k$, $1 \le k \le n + 1$.

(iii) The last row consists of the number $\sum_{k=0}^{n} \binom{n}{k} a_k$. Hint: The desired expression has to be of the form $\sum_{k=0}^{n} \alpha_k a_k$; to determine the coefficient α_k for a fixed k, consider the array with first row $a_i = 0$ ($0 \le i \le n$, $i \ne k$), $a_k = 1$, in which part of Pascal's triangle appears.

(iv) In each of the "slanted rows" $X_i = (a_{2i-1,i}, a_{2i,i}, a_{2i+1,i}, \ldots)$, $Y_i = (a_{2i-1,i}, a_{2i,i+1}, a_{2i+1,i+2}, \ldots)$, where $i \ge 2$, the numbers form an increasing sequence. One can show by induction that $a_{2i-1,i} \ge 2^{i-1}$ for all $i \ge 2$. Thus, if $x^2 \le 2^{k+1}$ and $x \in X_i \cup Y_i$ holds for some i, then also $a^2_{2i-1,i} \le 2^{k+1}$, which is possible only when $2^{2i-2} \le 2^{k+1}$, that is, when $i \le \frac{k+3}{2}$. Since in each of the rows X_i, Y_i a number x occurs at most once and the rows $X_2, Y_2, X_3, Y_3, \ldots$ contain all the numbers of the array that are greater than 1, the number $v(x)$ of all occurrences of each number $x > 1$ in the given array satisfies the following implication: If $x^2 \le 2^{k+1}$, then $v(x) \le 2\left(\frac{k+3}{2} - 1\right) = k + 1$, where equality $v(x) = k + 1$ is not possible, since for $i = \frac{k+3}{2}$ the number x occurs in $X_i \cup Y_i$ at most once. Therefore, $v(x) \le k$.

3.14 (i) If $m_i = \min_j(a_{ij})$ and $M_j = \max_i(a_{ij})$, then for any pair of indices i, j you have $m_i \le a_{ij} \le M_j$, which implies $m_i \le M_j$. Hence

$\max_i m_i \leq \min_j M_j$, where equality occurs if and only if $m_i = M_j$ $(= a_{ij})$ for some pair of indices i, j. In words: Some field of the array contains a number that is the smallest in its row and at the same time the largest in its column (a "saddle field").

(ii) For each pair of columns of the array use the result of Problem 1.3.(ii).

(iii) $S_{\min} = 342$, $S_{\max} = 351$. Explain why the sequence $b_1, b_2, \ldots, b_{99}$ of the relevant numbers on the die that are to be added has the following property: Between two neighboring occurrences of any number x there is exactly one occurrence of the number $7 - x$. Therefore, if you arrange the summands in the sum $S = b_1 + b_2 + \cdots + b_{99}$ in pairs $\{x, 7-x\}$, $1 \leq x \leq 3$, then you obtain the bounds (note that at most three summands can remain unpaired) $4 + 5 + 6 + \frac{99-3}{2} \cdot 7 \geq S \geq 1 + 2 + 3 + \frac{99-3}{2} \cdot 7$, that is, $351 \geq S \geq 342$. Both extreme values can in fact be achieved:

$$S = 351, \quad \text{if } (b_1, \ldots, b_{99}) = (6, 5, 4, 1, 2, 3, 6, 5, 4, 1, 2, 3, \ldots, 6, 5, 4),$$
$$S = 342, \quad \text{if } (b_1, \ldots, b_{99}) = (1, 2, 3, 6, 5, 4, 1, 2, 3, 6, 5, 4, \ldots, 1, 2, 3).$$

(Verify that the "orientation" of a usual die makes such a path from the lower left to the upper right corner possible.)

(iv) First show that in each such array

$$\begin{pmatrix} a_1 & a_2 & a_3 & \ldots & a_n \\ b_1 & b_2 & b_3 & \ldots & b_n \\ c_1 & c_2 & c_3 & \ldots & c_n \\ d_1 & d_2 & d_3 & \ldots & d_n \end{pmatrix} \tag{50}$$

you have $b_n = d_n$ (by shortening the array you then obtain the identity $b_k = d_k$ also for all $k < n$). For each $i \in \mathbb{N}$ let α_i denote the number of occurrences of the integer i in the sequence $b_1, b_2, \ldots, b_n$, that is, in the second row of (50). Next set $b_n = p$, $c_n = q$ and show that

$$\alpha_1 \geq \alpha_2 \geq \cdots \geq \alpha_p = q > \alpha_{p+1} \geq \alpha_{p+2} \geq \alpha_{p+3} \geq \cdots . \tag{51}$$

Indeed, the identity $\alpha_p = q$ follows from the definition of the c_n; for every $i \geq 1$ the integer α_i is equal to the number of elements of the set M_i of all those integers that occur in the first row of (50) at least i times. Clearly, $M_1 \supseteq M_2 \supseteq M_3 \supseteq \cdots$, where $M_p \not\supseteq M_{p+1}$, since $a_n \in M_p \setminus M_{p+1}$. This proves (51). The task is now to verify the identity $d_n = p$, that is, to show that the integer q appears in the third row of (50) exactly p times. This occurs if and only if there are exactly p different integers that have the property that they occur in the second row of (50) at least q times; however, by (51) these are exactly the integers $1, 2, \ldots, p$. This proves the assertion.

4.2 (i) $k_{\min} = n + 2$. The example of the set $\{1, 3, 5, \ldots, 2n + 1\}$ shows that $k = n + 1$ does not have the required property. Order the elements of any $(n+2)$-element set X according to size: $1 \leq x_1 < x_2 < \cdots < x_{n+2} \leq 2n+1$. Then each one of the $2n+2$ numbers $x_2 - x_1, x_3 - x_1, \ldots, x_{n+2} - x_1, x_2, x_3, \ldots, x_{n+2}$ lies in the $(2n+1)$-element set $\{1, 2, \ldots, 2n+1\}$; by the pigeonhole principle two of these numbers must be equal. Hence $x_i - x_1 = x_j$ for appropriate indices $i > j \geq 2$; set $a = x_1$, $b = x_j$, $c = x_i$.

(ii) Enumerate the elements of X according to size, $1 \leq x_1 < x_2 < \cdots < x_{10} \leq 25$, and use the pigeonhole principle for the 25 numbers $x_i - x_j$ $(6 \leq i \leq 10, 1 \leq j \leq 5)$ that lie in the set $\{1, 2, \ldots, 24\}$.

(iii) The example $X = \{1, 2, 3\}$ shows that the assertion does not hold for $n = 3$. For $n \geq 4$ prove the assertion by contradiction: Assume the existence of an n-element set $X \subseteq \mathbb{R}^+$ that does not have the required property. If you denote its elements by $x_1 < x_2 < \cdots < x_n$, then in view of $x_n + x_i > x_n$ you have $x_n + x_i \notin X$, and thus $x_n - x_i \in X$ $(1 \leq i < n)$, which together with the inequalities $x_n > x_n - x_1 > x_n - x_2 > \cdots > x_n - x_{n-1}$ gives $x_n - x_i = x_{n-i}$ $(1 \leq i < n)$. If n is even, $n = 2k$, then $x_{2k} - x_k = x_k$, and the choice $a = x_k$, $b = x_{2k}$ leads to a contradiction. If n is odd, $n = 2k+1$ and $k \geq 2$, then repeat the argument: From the inequalities $x_{n-1} + x_i > x_{n-1} + x_1 = x_n$ $(2 \leq i < n-1)$ and $x_{n-2} = x_n - x_2 > x_{n-1} - x_2 > x_{n-1} - x_3 > \cdots > x_{n-1} - x_{n-2}$ it follows that $x_{n-1} - x_i = x_{n-1-i}$, that is, $x_{2k} - x_i = x_{2k-i}$ $(2 \leq i < 2k)$. The choice $a = x_k$, $b = x_{2k}$ leads again to a contradiction.

(iv) $k_{\max} = \left\lceil \frac{n+2}{3} \right\rceil$. Among the chosen numbers there can be neither of the pairs $x, x+1$ and $x, x+2$. If therefore $x_1 < x_2 < \cdots < x_k$ are the chosen numbers, then $x_{i+1} - x_i \geq 3$ for every $i = 1, 2, \ldots, k-1$, which in view of $x_1 \geq 1$ implies $x_k \geq 3k - 2$. At the same time, however, $x_k \leq n$, and thus $k \leq \frac{n+2}{3}$. On the other hand, the chosen numbers $1, 4, 7, \ldots, 3\left\lceil \frac{n+2}{3} \right\rceil - 2$ have the desired property (the difference of any two of them is a multiple of three, while the sum is not).

(v) You can recursively construct examples for all $n \geq 2$. For $n = 2$ the set $\{1, 2\}$ is appropriate; if $\{a_1, a_2, \ldots, a_n\}$ is an n-element set with the given property, then the set $\{s, a_1 + s, a_2 + s, \ldots, a_n + s\}$ has the same property, where s is any common multiple of the $n + \binom{n}{2}$ numbers a_i $(1 \leq i \leq n)$ and $|a_i - a_j|$ $(1 \leq i < j \leq n)$.

(vi) No such infinite set X exists. If $(b - a) \mid (a + b)$ holds for some positive integers $a < b$, then $a + b \geq 2(b - a)$, that is, $b \leq 3a$. If therefore the set $X \subseteq \mathbb{N}$ has the given property and if $m = \min X$, then $x \leq 3m$ holds for all $x \in X$; that is, X has at most $3m - m + 1 = 2m + 1$ elements.

(vii) Proof that $H_k \wedge H_n \implies H_{kn}$: From the given $(2kn - 1)$-element subset of $\mathbb{Z}$ choose according to H_n a group of n numbers with sum nS_1; from the unchosen $2kn - n - 1$ numbers choose according to H_n a second group of n numbers with sum nS_2, etc., and finally choose a $(2k - 1)$th group of n numbers with sum nS_{2k-1} $(S_j \in \mathbb{Z}, 1 \leq j \leq 2k - 1)$. Now from the numbers $S_1, S_2, \ldots, S_{2k-1}$ you can, according to H_k, choose k numbers $S_{j_1}, S_{j_2}, \ldots, S_{j_k}$ $(1 \leq j_1 < j_2 < \cdots < j_k \leq 2k - 1)$ such that $S_{j_1} + S_{j_2} + \cdots + S_{j_k}$ is an integer multiple of k. (Formally, the hypothesis H_k cannot be used in the case where some of the numbers S_j are equal;

however, every number S_j can be changed to $S'_j = S_j + \alpha_j k$, where the $\alpha_j \in \mathbb{N}$ are chosen such that the new numbers S'_j are distinct.) Then the number $nS_{j_1} + nS_{j_2} + \cdots + nS_{j_k}$ is the sum of nk different elements of the initial set, and is at the same time divisible by kn.

The proofs of H_2 and H_3 are easy: For H_2 choose two numbers with the same parity from any triple of numbers, and for H_3 choose from any five numbers three which upon division by 3 give either the same remainder or distinct remainders. Proof of H_5: Let r_j $(1 \le j \le 9)$ denote the remainders upon division by 5 of the given nine integers. The assertion about choosing five integers clearly holds in the two cases when among the numbers r_j at least five are the same, respectively when each of the remainders $0, 1, 2, 3, 4$ among the numbers r_j occurs at most twice, and thus at least once (using the fact that $5 \mid 0+1+2+3+4$). Therefore, it remains to analyze the case where some remainder r occurs among the numbers r_j exactly three times, or exactly four times. You may furthermore assume that $r = 0$ (since the problem does not change if the same number is added to all nine original numbers). Let therefore $r_1 = r_2 = r_3 = 0$ and $r_j \ne 0$ $(5 \le j \le 9)$. From among the five remainders r_j $(5 \le j \le 9)$ you can, for some $k \in \{2, 3, 4, 5\}$, choose k numbers whose sum is divisible by 5. (Hint: Consider the remainders of the numbers $r_5, r_5 + r_6, \ldots, r_5 + r_6 + \cdots + r_9$; if none of them is zero, then by the pigeonhole principle two are the same.) If to these k numbers you adjoin $(5 - k)$ zeros from r_1, r_2, r_3, you obtain the desired quintuple.

(viii) Choose, for instance, the numbers $a_k = \frac{p_1 p_2 \cdots p_n}{p_k}$ $(k = 1, 2, \ldots, n)$, where $p_1, p_2, \ldots, p_n$ are n distinct prime numbers.

(ix) Choose the numbers $a_k = k \cdot n! + 1$ $(k = 1, 2, \ldots, n)$. Note that for each prime divisor p of the difference $a_i - a_j$ $(i \ne j)$ you have $p \le n$, but such a prime number does not divide any of the a_k. Furthermore, note that the sum of m chosen integers is a multiple of m, for any $m \le n$.

(x) For $n = 3$ choose the integers $6, 12, 18$, and for any $n > 3$ choose the n-tuple $1, 2, 3, 3 \cdot 2, 3 \cdot 2^2, \ldots, 3 \cdot 2^{n-3}$. The sum of these n numbers is equal to $3 \cdot 2^{n-2}$, which is a multiple of each one of them. The product of any $n - 1$ chosen integers is of the form $3^p \cdot 2^q$, where $p \ge 1$ and $q \ge n - 3$.

(xi) Choose the numbers $A, 2A, 3A, \ldots, nA$, where $A = (kn)!$. Then any sum of k of the chosen numbers is of the form mA, where $m \le kn$, and thus divides $(kn)!A = A^2$. At the same time the product of any k of the chosen numbers is divisible by A^k. It remains to note that $A^2 \mid A^k$, since $k \ge 2$.

(xii) The infinite set $X = \{2, 2^2 \cdot 3, 2^2 \cdot 3^2 \cdot 5, \ldots, 2^2 \cdot 3^2 \cdots p_{n-1}^2 \cdot p_n, \ldots\}$ has the required property, where $p_1, p_2, \ldots$ is the (infinite) sequence of all prime numbers. Explain why in the prime power expansion of any sum of different numbers from X there is a prime with exponent 1.

4.4 (i) If $X = \{x_1, x_2, \ldots, x_n\}$ is of type DS, then $x_1 + x_2 + \cdots + x_n$ is the largest of the $2^n - 1$ natural numbers that are equal to the sums of the elements of all nonempty subsets of X.

(ii) It is easy to see that such a set does not exist for any $n \leq 3$. For $n = 4$ an appropriate set is $\{3, 5, 6, 7\}$, and with its help you can recursively construct an appropriate set for each $n > 4$. This can be done by using the following trick: If X is an n-element set of type DS, then $Y = \{1\} \cup \{2x; \ x \in X\}$ is an $(n+1)$-element set of type DS such that $\max Y = 2 \cdot \max X$.

(iii) For each $n > 1$ there is the unique such set $X = \{1, 2, 4, \ldots, 2^{n-1}\}$. Indeed, if $x_1 < x_2 < \cdots < x_n$ are the elements of X, then the two smallest sums of nonempty elements of subsets of X are x_1 and x_2, which implies $x_2 = x_1 + 1$. The third-smallest sum is the smaller of the two integers x_3 and $x_1 + x_2$, and by hypothesis it is equal to $x_1 + 2$. Distinguish therefore the two cases

$$\text{(a) } x_1 + x_2 = x_1 + 2, \quad \text{(b) } x_3 = x_1 + 2.$$

The case (a) (which necessarily occurs when $n = 2$) is easy. Since $x_2 = x_1 + 1$, you immediately obtain $x_1 = 1$ and $x_2 = 2$. Further, it follows by induction that $x_i = 2^{i-1}$ for all $i \leq n$; it suffices to use the identity $1 + 2^1 + \cdots + 2^i = 2^{i+1} - 1$ for $i = 1, 2, \ldots, n-1$. In case (b) the sets $\{x_1, x_1 + 3\}$ and $\{x_2, x_3\}$ have the same sums of elements, which means that $x_1 + 3 \notin X$, and this leads to the identity $x_1 + 3 = x_1 + x_2 = 2x_1 + 1$. Hence necessarily $x_1 = 2$, $x_2 = 3$, and $x_3 = 4$. From this it easily follows that $n > 3$ and $x_4 = 8$, which similarly implies $n > 4$ and $x_5 = 16$, etc. You can then verify by induction that the sums of the elements of the subsets $\{2, 3, 4, 8, \ldots, 2^k\}$ are exactly the numbers $2, 3, 4, 5, \ldots, 2^{k+1} - 1, 2^{k+1} + 1$. This implies that X is an infinite set, which is a contradiction. Case (b) is therefore not possible.

(iv) Denote the elements of the given set X by $0 < x_1 < x_2 < \cdots < x_n$. Then in the sequence $x_1, x_2, \ldots, x_n, x_1 + x_n, x_2 + x_n, \ldots, x_{n-1} + x_n,$ $x_1 + x_{n-1} + x_n, x_2 + x_{n-1} + x_n, \ldots, x_{n-2} + x_{n-1} + x_n, \ldots, x_1 + x_3 + x_4 + \cdots + x_n, x_2 + x_3 + \cdots + x_n, x_1 + x_2 + \cdots + x_n$, which has length $\frac{n(n+1)}{2}$, each term is larger than the preceding one. The example of the set $\{1, 2, \ldots, n\}$ shows that $\frac{n(n+1)}{2}$ is the largest possible number.

(v) Assume that such a 7-element set $X \subseteq \{1, 2, \ldots, 26\}$ exists. Consider all $\binom{7}{1} + \binom{7}{2} + \binom{7}{3} + \binom{7}{4} = 98$ nonempty subsets of X with at most four elements. The sum of elements in each such set Y is less than $23 + 24 + 25 + 26 = 98$ (note that X cannot at the same time contain all four numbers $23, 24, 25, 26$, since $23 + 26 = 24 + 25$). Hence, by the pigeonhole principle, at least two of the sets under consideration have the same sum of elements, but this is a contradiction to X being of type DS.

4.6 (i) Exactly those $k < \frac{n(n+1)}{2}$ that have the same parity as $\frac{n(n+1)}{2}$. The number $1 + 2 + \cdots + n - k$ is equal to twice the smaller of the two sums; it is therefore an even positive integer. This condition is also sufficient: If $s \in \mathbb{N}$ is such that $2s + k = \frac{n(n+1)}{2}$, then the number s is the sum of the elements of a nonempty subset $X \subseteq \{1, 2, \ldots, n\}$. The proof of this assertion can be carried out by induction on $s = 1, 2, \ldots, \frac{n(n+1)}{2}$.

(ii) (a) 50. The numbers $2, 4, 6, \ldots, 100$ (50 of them) must lie in different classes. A partition into 50 classes can be done, for instance, by $T_k = \{2k - 1, 2k\}$, $1 \le k \le 50$.

(b) $26 = 1 + |P|$, where $P = \{2, 3, 5, 7, \ldots, 97\}$ is the set of all primes less than 100. One of the classes is necessarily of the form $\{1\}$, and different numbers from P have to lie in different classes. If $p(n)$ denotes the smallest prime number dividing the integer $n > 1$, then a partition into $1 + |P|$ classes is, for instance, of the form $\{1\} \cup T_2 \cup T_3 \cup T_5 \cup \cdots \cup T_{97}$, where $T_p = \{n : p \le n \le 100 \wedge p(n) = p\}$ for every $p \in P$.

(iii) Assume that there is a partition $\{1, 2, \ldots, 5\} = A \cup B$ such that the equation $x - y = z$ has no solution in A, nor in B. Since $2 - 1 = 1$ and $4 - 2 = 2$, the numbers $1, 2$, as well as the numbers $2, 4$, do not lie in the same class. Hence the numbers $1, 4$ lie in the same class, say $\{1, 4\} \subseteq A$, which implies $3 \in B$ and $5 \in B$, and at the same time also $2 = 5 - 3 \in B$; this is a contradiction.

(iv) In the case $\alpha = 2$ such a partition exists if and only if $n = 3k$ or $n = 3k - 1$. Note that the sum of the numbers in each class is a multiple of 3, so the number $1 + 2 + \cdots + n = \frac{n(n+1)}{2}$ and therefore one of $n, n+1$ have the same property. An appropriate example for $n = 3k$ is $T_1 = \{1, 2, 3\}, T_2 = \{4, 5, 6\}, \ldots, T_k = \{3k - 2, 3k - 1, 3k\}$, and for $n = 3k - 1$, $T_1 = \{1, 2\}, T_2 = \{3, 4, 5\}, \ldots, T_k = \{3k - 3, 3k - 2, 3k - 1\}$.

In the case $\alpha = 3$ such a partition exists if and only if $n = 8k$ or $n = 8k - 1$. The necessity of these conditions follows from the fact that the sum $1 + 2 + \cdots + n$ is a multiple of 4. Any set of eight consecutive integers can be partitioned into two appropriate classes $\{a + 1, a + 3, a + 4, a + 8\}$ and $\{a + 2, a + 5, a + 6, a + 7\}$, since $3(a + 4) = (a + 1) + (a + 3) + (a + 8)$ and $3(a + 5) = (a + 2) + (a + 6) + (a + 7)$. Hence it is clear how to construct an appropriate partition of $\{1, 2, \ldots, n\}$ for $n = 8k$ into $2k$ four-element classes. In the case $n = 8k - 1$ it suffices to note that the set of the first seven integers $\{1, 2, \ldots, 7\}$ has the desired property, $1 + 2 + \cdots + 6 = 3 \cdot 7$, and the $8(k - 1)$ integers $8, 9, \ldots, 8k - 1$ can be partitioned into $2(k - 1)$ four-element classes, as already shown above.

(v) Assume to the contrary that such a partition $\mathbb{Z} = T_1 \cup T_2 \cup T_3$ exists, and let $x \sim y$ indicate that x, y lie in the same class T_i. For every $x \in \mathbb{Z}$ use the choice $t = x$, $t = x + a$, and $t = x + b$ in the condition of the problem to establish that neither $x + a + b \sim x + a$ nor $x + a + b \sim x + b$ holds, and thus $x + a + b \sim x$ holds. From this it follows by induction that $0 \sim p(a + b)$ for all $p \in \mathbb{Z}$, and especially $0 \sim a(a + b)$. On the other hand, it is easy to show that $0 \sim qa$ for some $q \in \mathbb{Z}$ if and only if $3 \mid q$; indeed, considering the triple $x + a, x + 2a, x + a + b$, the relation $x + a + b \sim x$, which was obtained above, implies that $x, x + a, x + 2a$ is a triple of representatives of the three classes T_1, T_2, T_3 for all $x \in \mathbb{Z}$. Then the relation $0 \sim a(a+b)$ is possible only if $3 \mid a+b$.

For the sake of interest we add that the problem can also be studied in the case where $3 \mid a+b$. For any $a, b \in \mathbb{Z}$ a relevant partition exists if and only if $a = 3^\alpha \cdot a_1$ and $b = 3^\alpha b_1$, where $\alpha \in \mathbb{N}_0$ and the integers a_1, b_1 satisfy $a_1 \equiv -b_1 \not\equiv 0 \pmod 3$.

(vi) For $k = 1$ one can put $A(1, n, d) = (n-1)d$; otherwise, such a number A does not exist for any $k > 1$ and any $n \geq 2$. Indeed, for a chosen $d \geq 1$ you can use induction on A ($A \geq k$) to construct a partition $\{1, 2, \ldots, A\}$ into k classes $T_1, T_2, \ldots, T_k$ such that for each $x \in T_i$ you have $(x+d) \notin T_i$, $1 \leq i \leq k$.

(vii) Any partition of the set $\{1, 2, 3, \ldots, 2^A\}$ into classes $T_1, T_2, \ldots, T_k$ generates a partition of $\{1, 2, 3, \ldots, A\}$ into k' ($k' \leq k$) classes formed by those sets $T_j' = \{x \in \mathbb{N} : 2^x \in T_j\}$, $1 \leq j \leq k$, which are nonempty. If a class T_j' contains according to van der Waerden's theorem an arithmetic sequence $x_1, x_2, \ldots, x_n$ with difference $d \neq 0$, then $2^{x_1}, 2^{x_2}, \ldots, 2^{x_n}$ is a geometric sequence with quotient $2^d \neq 1$, which lies in the class T_j of the original partition. Hence it is possible to set $G(k, n) = 2^{A(k,n)}$.

(viii) Such an assertion is in general not true. The class $T_1 = \{x_1, x_2, \ldots\}$ does not contain a three-term arithmetic sequence if $x_{n+1} \geq 2x_n$ for all $n \geq 1$. This holds, for instance, for $x_n = n! + n$; and furthermore, for any $a_1, d \in \mathbb{N}$ the arithmetic sequence with the terms $a_k = a_1 + (k-1)d$ contains infinitely many elements x_n (it suffices to consider the indices $n = a_1 + jd$ for $j \in \mathbb{N}$). Hence the class $T_2 = \mathbb{N} \setminus T_1$ does not contain any infinite arithmetic sequence.

(ix) Suppose that all the cards are placed in n boxes, and let $\overline{ab_1}$, $\overline{ab_2}$, $\ldots$, $\overline{ab_d}$ be all the numbers of the boxes used that start with a chosen digit a. Choose this digit such that the number $d \geq 1$ is the *smallest* possible (if there is more than one such a, then choose an arbitrary one). A card with number $\overline{apq}$, where both digits p, q are different from $b_1, b_2, \ldots, b_d$, must be placed in the box numbered $\overline{pq}$, and there are exactly $(10 - d)^2$ such boxes. Now, by the choice of the number d at least another d^2 boxes are used whose number begins with one of the digits $b_1, b_2, \ldots, b_d$. Therefore, $n \geq (10 - d)^2 + d^2 = 2(d - 5)^2 + 50 \geq 50$, which together with 4.5.(iv) shows that the smallest number of boxes used is 50.

(x) If $d_1 + d_2 + \cdots + d_{k-1} = s$, then partition the s-element set of digits into $(k - 1)$ classes with $d_1, d_2, \ldots, d_{k-1}$ elements each. The distribution of all s^k cards over $d_1^2 + \cdots + d_{k-1}^2$ boxes is then done in a similar way as in the solution of 4.5.(iv); note that by the pigeonhole principle at least two of the digits of the number on each card belong to the same class. For the smallest number $M(k, s)$ of boxes required then prove the bound $M(k, s) \geq \min \{M(k - 1, s - d) + d^2, d = 1, 2, \ldots, s\}$, with $M(k, 0) = 0$. To do this, choose an approach similar to that in the previous solution (ix): If the digits $a, b_1, b_2, \ldots, b_d$ have the same meaning as in (ix), then $M(k - 1, s - d)$ is the number of boxes required for the placement of the numbers $\overline{a p_1 p_2 \cdots p_{k-1}}$, where $p_1, p_2, \ldots, p_{k-1}$ are any digits different from $b_1, b_2, \ldots, b_d$.

We add that the minimum value of the sum $d_1^2 + d_2^2 + \cdots + d_{k-1}^2$ is equal to $(k-1)q^2 + r(2q+1)$, where $q = \left[\frac{s}{k-1}\right]$ and $r = s - (k-1)q$.

(xi) You can prove the following assertion: If the numbers $\{1, 2, \ldots, n\}$ can be partitioned into k classes such that in none of them does the equation $x + y = z$ have a solution, then the numbers from the set $\{1, 2, \ldots, 3n+1\}$ can be partitioned into $k+1$ classes with the same property. From this you easily obtain the solution of the given problem, since the numbers n_k from its statement satisfy $n_{k+1} = 3n_k + 1$ for all $k \geq 1$; furthermore, $n_1 = 1$, and in the one-element set $\{1\}$ there are no solutions of the equation $x + y = z$.

Proof of the assertion: If $\{1, 2, \ldots, n\} = X_1 \cup X_2 \cup \cdots \cup X_k$ is a partition with the given property, then the desired partition $\{1, 2, \ldots, 3n+1\} = Y_1 \cup Y_2 \cup \cdots \cup Y_{k+1}$ can be constructed as follows: $Y_i = X_i \cup \{2n+1+x : x \in X_i\}$ $(1 \leq i \leq k)$, and $Y_{k+1} = \{n+1, n+2, \ldots, 2n+1\}$. Since for $x, y \in Y_{k+1}$ you have $x + y \geq 2(n+1) > 2n+1 = \max Y_{k+1}$, the equation $x + y = z$ does not have a solution in Y_{k+1}. Next, assume that this equation has a solution $x, y, z \in Y_i$ for some $i \leq k$, and that $x \leq y$. Distinguish three cases:

(a) $y < 2n+1$,   (b) $x < 2n+1 < y$,   (c) $2n+1 < x$.

In case (a) you have $z = x + y \leq 2n$, so $z \in Y_i$ means $z \in X_i$; the class X_i therefore contains all three numbers x, y, z, which is a contradiction. In case (b) you have $x \in X_i$, $y = 2n+1+y'$, and $z = 2n+1+z'$, where $y', z' \in X_i$; the equation $x + y' = z'$ is again in contradiction to the property of the class X_i. In case (c), finally, you have $z = x + y > 4n + 2 > 3n + 1 \geq z$, and this is again a contradiction.

(xii) If a_i is the first term and d_i is the difference in the sequence of numbers from T_i $(1 \leq i \leq k)$, then the number d_j divides a_j for the index j for which $d_1 d_2 \cdots d_k \in T_j$.

(xiii) If N is a common multiple of the numbers $d_1, d_2, \ldots, d_k$ defined in the solution of (xii), and if m is sufficiently large ($m > \min T_i$, $1 \leq i \leq k$), then the set $\{m+1, m+2, \ldots, m+N\}$ contains exactly $\frac{N}{d_i}$ terms from T_i. Thus

$$N = \frac{N}{d_1} + \frac{N}{d_2} + \cdots + \frac{N}{d_k},$$

which upon division by N gives the desired identity.

(xiv) To construct a partition, use the decimal expansion of a positive integer $n = c_k 10^k + c_{k-1} 10^{k-1} + \cdots + c_0$. Use the digits c_i to define the sets $A = \{n : 0 = c_1 = c_3 = c_5 = \cdots\}$ and $B = \{n : 0 = c_0 = c_2 = \cdots\}$. An appropriate partition then has the form $T_1 = \{1\} \cup \{1+n : n \in A\}$, $T_{k+1} = \{b_k + n : n \in T_1\}$, where $b_1, b_2, \ldots$ is the sequence of all elements of B. Each $x \in \mathbb{N}$ can therefore be written (in a unique way) in one of the forms $x = 1$, $x = 1 + a$ ($a \in A$), $x = 1 + b$ ($b \in B$), $x = 1 + a + b$ ($a \in A$, $b \in B$).

(xv) The numbers $d_i = 2^{m+i}$ satisfy the inequality in question for all $n \geq 1$ if the constant $m = m(\varepsilon)$ is chosen such that $2^{-m} \leq \varepsilon$. The classes

T_i can be constructed recursively: Set $T_1 = \{1 + jd_1 : j \in \mathbb{N}_0\}$, and if for some $i > 1$ the classes $T_1, T_2, \ldots, T_{i-1}$ are already constructed, then choose $T_i = \{a_i + jd_i : j \in \mathbb{N}_0\}$, where a_i is the smallest positive integer that does not lie in any of the sets $T_1, T_2, \ldots, T_{i-1}$ (such an a_i exists, since the identity $T_1 \cup T_2 \cup \cdots \cup T_{i-1} = \mathbb{N}$ is, according to (xiii), possible only if $\frac{1}{d_1} + \frac{1}{d_2} + \cdots + \frac{1}{d_{i-1}} = 1$). Since $d_i \mid d_j$ and $a_i < a_j$ for any $i < j$, it easily follows that $T_i \cap (T_1 \cup T_2 \cup \cdots \cup T_{i-1}) = \emptyset$ for all $i > 1$, and furthermore you have clearly $n \in T_1 \cup T_2 \cup \cdots \cup T_n$ for all n. Hence $T_1, T_2, \ldots$ does indeed form a partition of $\mathbb{N}$.

(xvi) For a fixed n, partition the elements of K_n into two classes according to whether their digit representation has an even or odd number of twos. Then any two different numbers from the same class have different digits in at least two places, so their sum has the digit 3 in these places.

(xvii) No such partition exists for any $n \geq 3$: The four numbers $11 \ldots 1$, $11 \ldots 12$, $11 \ldots 121$, and $11 \ldots 122$ from K_n would have to lie in different classes.

(xviii) For each $a \in K_n$ let $M(a)$ denote the set consisting of a and of all $b \in K_n$ that are obtained from the number a by means of the change of one of its digits; clearly, $|M(a)| = n + 1$ for all $a \in K_n$. If $X = \{a_1, a_2, \ldots, a_k\}$ is any k-element subset of K_n such that the sum of any two distinct elements of X has among its digits at least three threes, then the sets $M(a_1), M(a_2), \ldots, M(a_k)$ are disjoint, and therefore $2^n = |K_n| \geq |M(a_1) \cup M(a_2) \cup \cdots \cup M(a_k)| = k(n + 1)$, which implies $k \leq \frac{2^n}{n+1}$.

(xix) $p_n = 2^{n+1}$. Proof: Clearly, $p_1 = |K_2| = 4$. Prove the bound $p_n \geq 2^{n+1}$ by induction, using the following notation: If $a \in K_\alpha$ and $b \in K_\beta$, then ab denotes the number in $K_{\alpha+\beta}$ that is obtained by adjoining to the expansion of a the expansion of b from the right; let a^* denote the number in K_α that is obtained from a by replacing in its expansion every digit 1 by the digit 2, and vice versa. The induction step then is done as follows: If $a_1, a_2, \ldots, a_s$, where $s = 2^{n+1}$, are the chosen numbers from K_{2^n} with the required property (that is, any two of them have different digits in at least 2^{n-1} places), then the $2s = 2^{n+2}$ distinct numbers $a_1 a_1, a_1 a_1^*, a_2 a_2, a_2 a_2^*, \ldots, a_n a_n, a_n a_n^*$ lie in $K_{2^{n+1}}$, and any two of them differ in at least 2^n places. Finally, assume that for some $n \geq 1$ you can choose $s > 2^{n+1}$ numbers $a_1, a_2, \ldots, a_s$ from K_{2^n} with the desired property. By the pigeonhole principle there are at least $2^n + 1$ numbers among them that have the same first digit from the left; let those be the numbers in $M = \{a_1, a_2, \ldots, a_{2^n+1}\}$. Since two different elements of M have the same digits in at most $2^n - 2^{n-1} = 2^{n-1}$ places, the total number S of such agreements for all pairs in M is bounded from above by $\binom{2^n+1}{2} \cdot 2^{n-1} = 2^{3n-2} + 2^{2n-2}$. Now the number S includes the $\binom{2^n+1}{2}$ agreements in the highest place; you can show that in each of the remaining $2^n - 1$ places the number of agreements is at least 2^{2n-2} (this leads to the bound $S \geq \binom{2^n+1}{2} + (2^n - 1)2^{2n-2} = 2^{3n-2} + 2^{2n-2} + 2^{n-1}$, which is a contradiction to the inequality $S \leq 2^{3n-2} + 2^{2n-2}$ obtained earlier). Indeed, if in some fixed place exactly k numbers from M have the digit 1, then exactly $2^n + 1 - k$ numbers from M have the digit 2 there, and the number of agreements of the digits in this place is equal to

$$\binom{k}{2} + \binom{2^n + 1 - k}{2} = k^2 - k\left(2^n + 1\right) + 2^{2n-1} + 2^{n-1}$$

$$= \left(k - \frac{2^n + 1}{2}\right)^2 + 2^{2n-2} - \frac{1}{4} \geq 2^{2n-2},$$

since $\left|k - \frac{2^n+1}{2}\right| \geq \frac{1}{2}$.

(xx) Denote the given numbers by $0 < x_1 < x_2 < \cdots < x_n$, and set $m_k = \frac{1}{2}(x_1 + x_2 + \cdots + x_k)$, $M_k = x_1 + x_2 + \cdots + x_k$ $(1 \leq k \leq n)$. For each $k = 1, 2, \ldots, n$ form the class T_k of exactly those sums s of numbers x_i for which $m_k \leq s \leq M_k$ holds (if for a sum s there is more than one such index k, then choose an arbitrary one of them). Since $M_k = 2m_k$ $(1 \leq k \leq n)$, every class T_k has the desired property. It remains to show that every sum s falls into some class T_k. Suppose that there is a sum s for which $s \notin T_k$ $(1 \leq k \leq n)$. Since $M_1 < s < M_n$, there is an index k such that $M_k < s < M_{k+1}$. From $s > M_k$ it follows that $s \geq x_i$ for some $i > k$, which further implies $2s = s + s > x_i + M_k \geq x_{k+1} + M_k = 2m_{k+1}$, that is, $s > m_{k+1}$. Therefore, $m_{k+1} < s < M_{k+1}$, and thus $s \in T_{k+1}$, which is a contradiction.

4.8 (i) For sets with elements of any arithmetic sequence of length nk with nonzero difference.

(ii) For $n = 4k$ or $n = 4k - 1$. The necessary condition $2 \mid (1+2+\cdots+n)$ is satisfied for exactly these n. For $n = 4k$ an appropriate partition is, for instance, $\{1, 4, 5, 8, \ldots, 4k - 3, 4k\} \cup \{2, 3, 6, 7, \ldots, 4k - 2, 4k - 1\}$, and for $n = 4k - 1$, $\{1, 2, 4, 7, 8, 11, \ldots, 4k - 4, 4k - 1\} \cup \{3, 5, 6, 9, 10, \ldots, 4k - 3, 4k - 2\}$. Both constructions are based on the same sum-balanced partition $\{x, x + 1, x + 2, x + 3\} = \{x, x + 3\} \cup \{x + 1, x + 2\}$.

(iii) $n = 3k$ $(k \geq 2)$ or $n = 3k + 2$ $(k \geq 1)$. The necessary condition $3 \mid (1 + 2 + \cdots + n)$ is satisfied only when $3 \mid n$ or $3 \mid (n - 2)$, and at the same time it is clear why you need $n \geq 5$. Based on the sum-balanced partition $\{x + 1, x + 2, x + 3, x + 4, x + 5, x + 6\} = \{x + 1, x + 6\} \cup \{x + 2, x + 5\} \cup \{x + 3, x + 4\}$ you can do an induction step from n to $(n + 6)$. It remains to give appropriate partitions for $n \in \{5, 6, 8, 9\}$: $\{1, 4\} \cup \{2, 3\} \cup \{5\}$, $\{1, 6\} \cup \{2, 5\} \cup \{3, 4\}$, $\{1, 2, 3, 6\} \cup \{4, 8\} \cup \{5, 7\}$, and $\{1, 2, 3, 4, 5\} \cup \{6, 9\} \cup \{7, 8\}$.

(iv) If there is such a partition into two classes with one and four elements, then the assertion is clear (the largest element must then form the one-element class of each sum-balanced partition). Hence suppose that the elements x_i of the given set X satisfy $x_1 + x_2 + x_3 = x_4 + x_5$. If you denote the value of both sides of this identity by S, then clearly $\max\{x_1 + x_2, x_1 + x_3, x_2 + x_3\} < S$, $x_i + x_4 \neq S$, and $x_i + x_5 \neq S$ for all $i \in \{1, 2, 3\}$. This means that $X = \{x_1, x_2, x_3\} \cup \{x_4, x_5\}$ is the unique sum-balanced partition of X into two classes with two and three elements.

(v) Apply the identity $x^2 + (x + 3)^2 + (x + 5)^2 + (x + 6)^2 = (x + 1)^2 + (x + 2)^2 + (x + 4)^2 + (x + 7)^2$ for all $x = 1 + 8n$, where $n = 0, 1, 2, \ldots, 124$.

(vi) The first two of the sets $M_1(x) = \{x^2, (x + 5)^2, (x + 7)^2\}$, $M_2(x) = \{(x + 1)^2, (x + 3)^2, (x + 8)^2\}$, $M_3(x) = \{(x + 2)^2, (x + 4)^2, (x + 6)^2\}$ have the same sum of elements, which is larger by 18 than the sum of the elements of the third. Therefore, the three nine-element sets $M_1(x) \cup M_2(x + 9) \cup M_3(x + 18)$, $M_2(x) \cup M_3(x + 9) \cup M_1(x + 18)$, $M_3(x) \cup M_1(x + 9) \cup M_2(x + 18)$ form a sum- and cardinality-balanced partition of the set $\{x^2, (x + 1)^2, \ldots, (x + 26)^2\}$. Use this for $x = 1, 28, 55, \ldots, 27n - 26$.

(vii) The condition $m \geq n$ is necessary, since the number m gives the sum of the elements in each of the k classes of a sum-balanced partition of the set $X_n = \{1, 2, \ldots, n\}$, and some class contains the number n. The sufficiency of the condition can be verified by constructing an appropriate partition of X_n with the help of induction on n. The cases $n = 1$, $m = n$, and $m = n + 1$ are easy, so let $n > 1$ and $m > n + 1$. To construct a partition of X_n, distinguish three cases.

(a) $n + 1 < m < 2n$, where m is odd. Then $1 + 2 + \cdots + (m - n - 1) = m \cdot k'$, where $k' = k - \frac{2n - m + 1}{2}$. Because of $n > m - n - 1$ and in view of $m \geq m - n - 1$ there exists, by induction, a sum-balanced partition of X_{m-n-1} into k' classes (the sum of the elements of the classes is equal to m); the remaining $2n - m + 1 = 2(k - k')$ numbers from $X_n \setminus X_{m-n-1}$ can then be partitioned into $k - k'$ two-element classes such that the sum of the two numbers in each class is again equal to m: $\{m - n, n\}, \{m - n + 1, n - 1\}, \ldots, \{\frac{m-1}{2}, \frac{m+1}{2}\}$.

(b) $n + 1 < m < 2n$, where m is even. Then $1 + 2 + \cdots + (m - n - 1) = \frac{m}{2} \cdot k'$, where $k' = 2k - 2n + m - 1$. Because of $n > m - n - 1$ and in view of $\frac{m}{2} \geq m - n - 1$ there exists, by induction, a sum-balanced partition of X_{m-n-1} into k' classes (the sum of the elements of the classes is equal to $\frac{m}{2}$); since k' is odd, adjoin the number $\frac{m}{2}$ to one of the classes, and take the unions of pairs of the remaining classes. You thus obtain a partition of $X_{m-n-1} \cup \{\frac{m}{2}\}$ into $\frac{k'+1}{2} = k - n + \frac{m}{2}$ classes with the same sum of elements (equal to m), and the remaining numbers from X_n can be partitioned into the $n - \frac{m}{2}$ pairs $\{m - n, n\}, \{m - n + 1, n - 1\}, \ldots, \{\frac{m}{2} - 1, \frac{m}{2} + 1\}$.

(c) $m \geq 2n$. Since $1 + 2 + \cdots + (n - 2k) = m' \cdot k$ holds, where $m' = m - 2n + 2k - 1 \geq n - 2k$ since $m' - (n - 2k) = \frac{(m - 2n)(m - n - 1)}{m} \geq 0$, there exists, by induction ($n > n - 2k$), a sum-balanced partition of X_{n-2k} into k classes. If you now adjoin to each of the classes one of the pairs $\{n - 2k + 1, n\}, \{n - 2k + 2, n - 1\}, \ldots, \{n - k, n - k + 1\}$, you obtain a sum-balanced partition of X_n into k classes.

4.10 (i)–(iii) Equality holds.

(iv) $A + A \supseteq 2 \cdot A$; counterexample $A = \{1, 2\}$.

(v) $(A + B) \cdot C \subseteq (A \cdot C) + (B \cdot C)$; counterexample $A = B = \{1\}$, $C = \{1, 2\}$.

(vi) $(A \cap B) \cdot C \subseteq (A \cdot C) \cap (B \cdot C)$; counterexample $A = \{1, 3\}$, $B = \{2, 3\}$, and $C = \{1, 2\}$.

(vii) In general, the sets are not comparable: For $A = \{1\}$, $B = \{1, 2\}$, and $C = \{1, 3\}$ you have $(A + B) \cap C = \{3\}$ and $(A \cap C) + (B \cap C) = \{2\}$.

(viii)–(ix) Equality holds.

(x) $A \subseteq (A + B) - B$; counterexample $A = B = \{1, 2\}$.

(xi) This holds exactly for all one-element sets. Suppose that A contains two different numbers x and y; choose them such that $|x - y| > 0$ has the *smallest* possible value. Since $x + y \in A + A = 2 \cdot A$, there is a $z \in A$ such that $x + y = 2 \cdot z$, that is, $z = \frac{x+y}{2}$. But then $0 < |z - x| = \frac{1}{2}|x - y| < |x - y|$, which is a contradiction to the choice of the pair x, y.

(xii) This holds, for instance, for $B = \{1\}$.

(xiii) The counterexample $A = \{-2, -1, 1, 2\}$ disproves the converse.

4.12 (i) Let $k = k(n)$ denote the largest integer $k \geq 0$ for which $c^k \cdot n \in \mathbb{N}$ (note that from $0 < c < 1$ it follows that $c^k \cdot n < 1$ for sufficiently large k). Then $n \in A$ if and only if $k(n)$ is even.

(ii) Enumerate the elements of the given k-element set C according to size, $0 < c_1 < c_2 < \cdots < c_s < 1 < c_{s+1} < c_{s+2} < \cdots < c_k$, and recursively construct the sets $A_1, A_2, \ldots, A_{k+1}$. Let, for instance, $1 \in A_1$, and suppose that the numbers $1, 2, \ldots, n - 1$ have already been placed. Then place the number n into that one of the $k + 1$ sets $A_1, A_2, \ldots, A_{k+1}$ that does not contain any of the k numbers $c_1 n, c_2 n, \ldots, c_s n, n/c_{s+1}, n/c_{s+2}, \ldots, n/c_k$ (if there is more than one such set, choose an arbitrary one).

(iii) Suppose that for some $n \in \mathbb{N}$ such sets A, B exist. Let, for instance, $1 \in A$. Then $\{n + 1, n + 2\} \subseteq B$; this implies $\{2n + 1, 2n + 2, 2n + 3\} \subseteq A$, which further implies $\{3n + 1, 3n + 2, 3n + 3, 3n + 4\} \subseteq B$, etc. One of the sets A, B then contains all numbers $n^2 + 1, n^2 + 2, \ldots, n^2 + (n + 1)$; however, the difference between the first and the last of these is n, which is a contradiction. Hence for any n such sets A, B do not exist.

(iv) If you set $X_n = \{1, 2, \ldots, n\}$, then the sets $A = X_n + 3n \cdot \mathbb{N}_0$, $B = n + X_n + 3n \cdot \mathbb{N}_0$, and $C = 2n + X_n + 3n \cdot \mathbb{N}_0$ give a solution.

(v) With the conditions $\mathbb{N} = A \cup B$, $2 \notin A - A$, $2 \notin B - B$ you can easily show that $\{A, B\}$ is either the pair of sets $\{\{1, 2\} + 4 \cdot \mathbb{N}_0, \{3, 4\} + 4 \cdot \mathbb{N}_0\}$ or the pair $\{\{1, 4\} + 4 \cdot \mathbb{N}_0, \{2, 3\} + 4 \cdot \mathbb{N}_0\}$. (Hint: Distinguish whether or not the numbers 1, 2 belong to one of the sets A, B.) In both cases the set C has the same form $C = 2 + 4 \cdot \mathbb{N}_0$.

(vi) The number $12\,131 = 2^2 \cdot 3^2 \cdot 337 - 1$ is not an element of the set $K = \{2^\alpha \cdot 3^\beta - 1 : \alpha, \beta \in \mathbb{N}_0\}$, which has the desired property: $\{1, 2\} \subseteq K$, and the relation $K * K \subseteq K$ is easily verified with the identity $a * b + 1 = (a + 1) \cdot (b + 1)$.

(vii) For infinitely many values $k \in \mathbb{N}$ the number $p = 2k + 1$ is a prime; if you assume that some such k lies in $\mathbb{N} * \mathbb{N}$, then from the identity

$k = a + b + 2ab$ you obtain for appropriate $a, b \in \mathbb{N}$ the expression $p = 2k + 1 = 2(a + b + 2ab) + 1 = (2a + 1)(2b + 1)$, which contradicts the fact that p is a prime number.

(viii) From $1 \in A + A$ it follows that $\{0, 1\} \subseteq A$, that is, $a_0 = 0$ and $a_1 = 1$. For every fixed $n > 1$ among the numbers

$$a_1 + a_1, a_1 + a_2, \ldots, a_1 + a_{n-1},$$
$$a_2 + a_2, a_2 + a_3, \ldots, a_2 + a_{n-1}, \ldots, a_{n-1} + a_{n-1}, \tag{52}$$

there have to be all elements from the set $B_n = \{1, 2, \ldots, a_n\} \setminus \{a_1, a_2, \ldots, a_n\}$. Since B_n has exactly $a_n - n$ elements and in (52) exactly $\binom{n}{2}$ (not necessarily distinct) numbers are listed, you have $a_n - n \leq \binom{n}{2}$, and the assertion follows.

(ix) If $x = c_k 10^k + c_{k-1} 10^{k-1} + \cdots + c_1 10 + c_0$ is the decimal expansion of a number $x \in \mathbb{N}_0$, then

$$7x = \sum_{m=1}^{9} \left(\sum_{j=0}^{k} d_{jm} 10^j \right),$$

where $d_{jm} = 7$ for $1 \leq m \leq c_j$, and $d_{jm} = 0$ for $c_j < m \leq 9$. Thus, for instance, for the number $1029 = 7 \cdot 147$ you obtain the expansion $1029 = 777 + 77 + 77 + 77 + 7 + 7 + 7 + 0 + 0$.

(x) Write the number n^2 as the sum of n copies of the number n. Multiplying the expansions $x = a_1 + a_2 + \cdots + a_n$ and $y = b_1 + b_2 + \cdots + b_m$ corresponding to the numbers $x, y \in K$ you obtain an appropriate expansion of xy as sum of mn summands $a_i b_j$, which implies $K \cdot K \subseteq K$. Next show that $p \cdot K + q \subseteq K$ holds for some $p, q \in \mathbb{N}$ if there are numbers $c_1, c_2, \ldots, c_n \in \mathbb{N}$ such that $c_1 + c_2 + \cdots + c_n = q$ and $\frac{p-1}{p} = \frac{1}{c_1} + \frac{1}{c_2} + \cdots + \frac{1}{c_n}$. For these n-tuples $(c_1, c_2, \ldots, c_n)$ choose consecutively (2), $(4, 4)$, $(3, 6)$, $(4, 8, 8)$, $(3, 3)$, $(2, 6)$, $(2, 4)$, $(3, 4, 6)$, and $(2, 3)$.

We add for the sake of interest that from what you proved above you can furthermore find that $\{1, 4, 9, 10, 11, 16, 17, 18, 20, 22, 24, 25, 26, \ldots\} \subseteq K$.

(xi) The elements of an appropriate set $K = \{a_1, a_2, a_3, \ldots\}$ can be determined recursively by pairs. Set $(a_1, a_2) = (1, 2)$ and suppose that for some $k \geq 1$ the pairs $(a_1, a_2), (a_3, a_4), \ldots, (a_{2k-1}, a_{2k})$ have already been chosen such that $a_1 < a_2 < \cdots < a_{2k}$ and among the numbers $a_i - a_j$, $1 \leq j < i \leq 2k$, no two are equal. Then set $a_{2k+1} = 2a_{2k} + 1$ and $a_{2k+2} = 2a_{2k} + d + 1$, where d is the smallest number of the set $\mathbb{N} \setminus \{a_i - a_j : 1 \leq j < i \leq 2k\}$. The desired properties of the new pair (a_{2k+1}, a_{2k+2}) follow from the relations $a_{2k+1} - a_{2k} = a_{2k} + 1 > 0$, $a_{2k+2} - a_{2k+1} = d > 0$, $a_{2k+2} - a_i > a_{2k+1} - a_i = a_{2k} + 1 + (a_{2k} - a_i) > a_{2k}$ $(1 \leq i \leq 2k)$, $a_{2k+2} - a_i \neq a_{2k+1} - a_j$ $(i, j \in \{1, 2, \ldots, 2k\})$. Finally, the property $\mathbb{N} \subseteq K - K$ follows from the way the number d was chosen: For every n you have $n \in \{a_i - a_j : 1 \leq j < i \leq 2n\}$.

(xii) If some of the sets A_1, A_2, A_3 contain 0, say $0 \in A_1$, then $A_2 = A_3$ (for then the relations $A_1 + A_3 \subseteq A_2$ and $A_1 + A_2 \subseteq A_3$ lead to $A_3 \subseteq A_2$, respectively $A_2 \subseteq A_3$). The case $0 \in A_1$ occurs especially when $A_2 \cap A_3 \neq \emptyset$, since $A_2 - A_3 \subseteq A_1$ and $0 = x - x$ for any $x \in A_2 \cap A_3$. Also note that $x \in A_1$ always holds if and only if $-x \in A_1$ (it suffices to use the identity $-x = y - (x + y)$ for any $y \in A_2$: Indeed, from $x \in A_1$ it follows that $x + y \in A_3$, and thus $-x \in A_2 - A_3$). To prove the assertion $A_i = A_j$ for some $i \neq j$ it suffices to consider the case where $0 \notin A_1 \cup A_2 \cup A_3$, that is, where $A_i = B_i \cup (-B_i)$, where B_1, B_2, B_3 are nonempty disjoint subsets of $\mathbb{N}$. Set $m_i = \min B_i$ and choose indices i, j, k such that $m_i > m_j > m_k$. But then from $A_j - A_k \subseteq A_i$ it follows that $m_j - m_k \in A_i$, but since $m_i > m_j - m_k > 0$, this means that $m_j - m_k$ is an element of B_i that is less than m_i, and this is a contradiction. This proves the assertion.

You can now describe all such triples of sets. By the above one can choose indices such that $0 \in A_1$ and $A_2 = A_3$; if $0 \in A_2$, then $A_1 = A_2 = A_3$, and from $A_1 - A_1 \subseteq A_1$ it follows according to 4.11.(iii) that $A_1 = A_2 = A_3 = m \cdot \mathbb{Z}$ for an appropriate $m \in \mathbb{N}_0$. Let therefore $0 \notin A_2$; then $A_1 = \{0\} \cup B \cup (-B)$ and $A_2 = A_3 = C \cup (-C)$, where B, C are nonempty disjoint subsets of $\mathbb{N}$. Set $n = \min B$ and $m = \min C$. From $m + m \in B$ it follows that $2m \geq n$. On the other hand, $n - m \in A_2$, and therefore $|n - m| \in C$, which implies $|n - m| \geq m$, but this is possible only if $n \geq 2m$. Therefore, $n = 2m$. From $2m \in A_1$ and $m \in A_2 = A_3$ one can easily derive that $2m \cdot \mathbb{Z} \subseteq A_1$ and $m \cdot (2 \cdot \mathbb{Z} + 1) \subseteq A_2 = A_3$. Thus, the identities $2m \cdot \mathbb{Z} = A_1$ and $m \cdot (2 \cdot \mathbb{Z} + 1) = A_2 = A_3$ can be derived by contradiction. Indeed, assume that, for instance, A_1 contains a number of the form $2km + c$, where $0 < |c - m| < m$. Then $c - m = (2km + c) - (2k + 1)m \in A_1 - A_2$, that is, $|c - m| \in C$. The number $|c - m|$ is then an element of C, and it is less than m, which is a contradiction. In a similar way you obtain a contradiction in the case where A_2 contains a number of the form $(2k + 1)m + c$, where $0 < |c - m| < m$.

To summarize the result: All triples of sets A_1, A_2, A_3 are of the form $A_1 = A_2 = A_3 = m \cdot \mathbb{Z}$, where $m \in \mathbb{N}_0$, respectively $A_i = 2m \cdot \mathbb{Z}$, $A_j = A_k = m \cdot (2 \cdot \mathbb{Z} + 1)$, where $m \in \mathbb{N}$ and $\{i, j, k\} = \{1, 2, 3\}$. Note that the sets $2m \cdot \mathbb{Z}$ and $m \cdot (2 \cdot \mathbb{Z} + 1)$ are disjoint for all $m \in \mathbb{N}$.

(xiii) Let m, respectively m', denote the *least* positive, respectively the *largest* negative, number of the given set A. Since $m + m' \in A$ and $m' < m + m' < m$, you have $m + m' = 0$, that is, $m' = -m$. Now it is easy to verify by induction that $km \in A$ for all $k \in \mathbb{Z}$, that is, $m \cdot \mathbb{Z} \subseteq A$. Every number $x \in A$ can be written in the form $x = km + q$, where $k \in \mathbb{Z}$ and $q \in \{0, 1, \ldots, m - 1\}$. From $-km \in A$ it follows that $q = x + (-km) \in A$, and thus $q = 0$. Therefore, $A = m \cdot \mathbb{Z}$.

(xiv) The set A is clearly infinite, since from $A + A \subseteq A$ it follows that A does not have a largest element. Let m denote the *smallest* positive element of the set $\{x - y : x \in A \wedge y \in A\}$ and choose fixed numbers $x, y \in A$ for which $x - y = m$. If r denotes the greatest common divisor of

x and y, then $x = pr$ and $y = qr$, where $p, q \in \mathbb{N}$ are relatively prime. By Bézout's theorem (see, e.g., [5], Chapter 3, Section 1.8) there are numbers $a, b, c, d \in \mathbb{N}$ such that $ap - bq = 1$ and $cq - dp = 1$. Now show that for each $k \geq pd + qb$ there are numbers $u, v \in \mathbb{N}_0$ such that

$$kr = ux + vy. \tag{53}$$

To do this, use induction on k. Since $x = pr$ and $y = qr$, for $k = pd + qb$ it is possible to set $u = d$ and $v = b$ in (53). If (53) holds for some $k \geq pd + qb$, then $(k+1)r = (u+a)x + (v-b)y$ and $(k+1)r = (u-d)x + (v+c)y$. One of these identities is the desired expression (53) for the number $(k+1)r$, since not both $v - b$ and $u - d$ can be negative; indeed, if $v < b$ and at the same time $u < d$, then you would have $kr = ux + vy < dx + by = (pd + qb)r$, which contradicts the hypothesis $k \geq pd + qb$. This proves the assertion about the existence of (53).

Now note that from $A + A \subseteq A$ and $x, y \in A$ it follows that $ux + vy \in A$ for any $u, v \in \mathbb{N}_0$ with the exception of the pair $u = v = 0$. This, together with the above, means that $kr \in A$ for all $k \geq pd + qb$. From this and the definition of m it follows that $m \leq r$, but at the same time $r \mid m$, since $m = x - y$ and r is a common divisor of x and y. Hence $r = m$. You have thus shown that the set A contains all the numbers $M, M+m, M+2m, \ldots$, where $M = (pd+qb)m$; from the definition of m it is also clear that A cannot contain any other number $x > M$. It remains to verify that $A \subseteq m \cdot \mathbb{N}$. Therefore, for any $z \in A$ first find an $\alpha \in \mathbb{N}$ such that $z(m + 1)^\alpha > M$. Since $z \in A$ implies $z(m + 1)^\alpha \in A$, you have $m \mid z(m + 1)^\alpha$ by what was already shown; therefore, $m \mid z$ (and thus $z \in m \cdot \mathbb{N}$), since the numbers m and $(m + 1)^\alpha$ are relatively prime.

(xv) Let $A_1, A_2, \ldots$ be any Fibonacci sequence of sets. Choose elements $a_1 \in A_1$ and $a_2 \in A_2$, and use the relation $a_{n+2} = a_{n+1} + a_n$ $(n \geq 1)$ to define a Fibonacci sequence of numbers. Clearly, $a_n \in A_n$ for all n. Define a new sequence of sets $B_n = A_n - a_n$; this is also a Fibonacci sequence (verify this), and furthermore, $0 \in B_n$ for all n. Since $0 \in B_2$, the identities $B_1 + B_2 = B_3$ and $B_3 - B_2 = B_1$ imply that $B_1 \subseteq B_3$ and $B_3 \subseteq B_1$, that is, $B_1 = B_3$. Similarly, one can show that $B_2 = B_4$. Thus $B_4 = B_2 + B_3 = B_2 + B_1 = B_3 = B_1$. The identity $B_1 = B_3 - B_2$ can therefore be rewritten as $B_1 = B_1 - B_1$, which by 4.11.(iii) means that $B_1 = k \cdot \mathbb{Z}$ for an appropriate $k \in \mathbb{N}_0$. Since it is clear that $B_n = B_1$ for all $n \geq 1$, this proves the assertion concerning the existence of the representation (26). The converse is trivial.

4.14 (i) The examples of the sets $\{0, 1, 3\}$ and $\{0, 1, 3, 4\}$ show that the assertion does not hold for $n = 3$, respectively $n = 4$. However, it does hold for all $n \geq 5$; indeed, the chain of the $2n - 7$ numbers $x_2 + x_3 < x_2 + x_4 < \cdots < x_2 + x_{n-1} < x_3 + x_{n-1} < \cdots < x_{n-2} + x_{n-1}$ has to be a subchain of the chain (27) of $2n - 3$ numbers. Since $x_1 + x_3 < x_2 + x_3$ and $x_{n-2} + x_{n-1} < x_{n-2} + x_n$, the identities $x_1 + x_4 = x_2 + x_3$, $x_1 + x_5 = x_2 + x_4$,

$\ldots, x_1+x_n = x_2+x_{n-1}, x_2+x_n = x_3+x_{n-1}, \ldots, x_{n-3}+x_n = x_{n-2}+x_{n-1}$
must hold, from which in the case $n \geq 5$ it easily follows that $x_1, x_2, \ldots, x_n$
is an arithmetic sequence.

(ii) Such integers do not exist. Indeed, in the affirmative case the sum of
the 36 integers $n + (n+1) + \cdots + (n+35) = 18(2n+35)$ would have to be
8 times the sum $a_1 + a_2 + \cdots + a_9$; however, the number $2n+35$ is odd.

(iii) $\min p_k(X) = k(n-k)+1$, $\max p_k(X) = \binom{n}{k}$. The bound $p_k(X) \leq \binom{n}{k}$
is trivial, and equality occurs, for instance, for $X = \{1, q, q^2, \ldots, q^{n-1}\}$,
where $q \geq 2$ is an integer (the proof is comparable with that of the case
$k = 2$; see the solution of 4.13.(v)). The bound $p_k(X) \geq k(n-k)+1$
can be demonstrated as follows: If $x_1 < x_2 < \cdots < x_n$ are the elements
of X, then among the sums of k elements from X there are $k(n-k)+1$
distinct values $s_0 < s_1 < s_2 < \cdots < s_{k(n-k)}$, with $s_0 = x_1 + x_2 + \cdots + x_k$
and $s_{ik+j} = (x_{i+1} + x_{i+2} + \cdots + x_{i+k+1}) - x_{i+k+1-j}$, where $1 \leq j \leq k$,
$0 \leq i \leq n - k - 1$. On the other hand, in the case $x_i = i$ $(1 \leq i \leq n)$ any
sum s of k elements from X satisfies the bound $s_0 \leq s \leq s_{k(n-k)}$, where
now $s_0 = 1 + 2 + \cdots + k = \frac{k(k+1)}{2}$ and $s_{k(n-k)} = (n-k+1) + (n-k+2) + \cdots + n = k\left(n - \frac{k-1}{2}\right)$; hence the number of different values s is at most
$s_{k(n-k)} - s_0 + 1 = k(n-k) + 1$.

(iv) The set A contains at most one number greater than 24, since for
$y > x \geq 25$ you have $|y - x| = y - x < y \leq \frac{xy}{25}$. Let $A = \{x_1, x_2, \ldots, x_n\}$,
where $1 \leq x_1 < x_2 < \cdots < x_n$ and $x_{n-1} < 25$. If you set $d_j = x_{j+1} - x_j$
$(1 \leq j \leq n-1)$, then

$$d_j \geq \frac{x_j x_{j+1}}{25} = \frac{x_j(x_j + d_j)}{25} \quad \text{or} \quad d_j \geq \frac{x_j^2}{25 - x_j} \quad (1 \leq j \leq n-1).$$

Since the function $h(x) = \frac{x^2}{25-x}$ is increasing in the interval $[1, 24]$, you
obtain consecutively $x_5 \geq 5$, $d_5 \geq h(5) > 1$ or $x_6 \geq 7$, $d_6 \geq h(7) > 2$
or $x_7 \geq 10$, $d_7 \geq h(10) > 6$ or $x_8 \geq 17$, $d_8 \geq h(17) > 36$ or $x_9 \geq 54$. Therefore, you must have $n \leq 9$. At the same time you can see that
$\{1, 2, 3, 4, 5, 7, 10, 17, 54\}$ is an appropriate nine-element set.

(v) The set $X = \{0, \frac{1}{2}, 1, 2\}$, for instance, is a solution, since $0 \cdot \frac{1}{2} + 1 \cdot 2 = 2$,
$0 \cdot 1 + \frac{1}{2} \cdot 2 = 1$, and $0 \cdot 2 + \frac{1}{2} \cdot 1 = \frac{1}{2}$ are all different values of the sum
$ab + cd$ in question. The inclusion $A \subseteq \mathbb{R}^+$ cannot hold. Indeed, assume to
the contrary that such a set $A \subseteq \mathbb{R}^+$ does exist. Then $x > 1$ for at most one
$x \in X$ (consider the sum $ab + cd$ with the two largest elements $a, b \in X$).
Now show that X cannot be of the form $X = \{a, b, c, d\}$, where $a > 1 \geq b >
c > d$; indeed, from the inequalities $ab + cd > ac + bd > ad + bc > ad > d$ it
would follow that $ab + cd = a$, $ac + bd = b$, $ad + bc = c$, which would imply
$a - c = (ab + cd) - (ad + bc) = (a - c)(b - d) < a - c$, a contradiction.

The above shows that the inequality $x \leq 1$ holds for at least four different
values $x \in X$. If you therefore find a pair $a, b \in X$, $a > b$, with *minimal*
difference $a - b$, then there are numbers $c, d \in X$, $c \leq 1$, $d \leq 1$, such that
a, b, c, d are distinct. Then $(ac + bd) - (ad + bc) = (a - b)(c - d) \neq 0$, and

$|(ac + bd) - (ad + bc)| = (a - b)|c - d| < a - b$, but at the same time $ac + bd, ad + bc \in X$. This is a contradiction to the choice of the numbers a, b.

(vi) Such sets exist. First show that for any $\alpha \in \mathbb{Z}$ the following holds: If $\{\alpha, \alpha + 2\} \in X$, then also $\{\alpha - 6, \alpha - 4, \alpha + 6, \alpha + 8\} \in X$. From this derive that $X + 6 \subseteq X$, $X - 6 \subseteq X$ (where sum and difference are to be understood as in 4.9). Then the conditions of the problem can be satisfied, apart from the empty set, only by the sets $M_i = \{x \in \mathbb{Z} : x - i \equiv \pm 1$ (mod 6)\} $(0 \le i \le 5)$ and $M_i \cup M_j$ $(j \in \{1, 3, 5\}, i \in \{0, 2, 4\})$. It is easy to verify that these sets do satisfy the conditions. Indeed, for each $n \in \mathbb{N}$ you have $2^n \equiv 2$ (mod 6) or $2^n \equiv 4$ (mod 6); hence for any $x \in \mathbb{Z}$ the pair of integers $x + 2^n$, $x - 2^n$ is congruent modulo 6 to the pair $x + 2$, $x - 2$ in some order, which means that for each n exactly one of the integers $x + 2^n$, $x - 2^n$ lies together with x in the set in question, that is, exactly one y with the property $|x - y| = 2^n$.

(vii) For the given set $K = \{k_1, k_2, \dots\}$ put

$$X = \left\{ \textstyle\sum_{i \in I} k_i : I \subset \mathbb{N} \text{ is finite and nonempty} \right\}.$$

Choose fixed $k_p \in X$ and $x = k_{i_1} + k_{i_2} + \cdots + k_{i_s} \in X$, $1 \le i_1 < i_2 < \cdots < i_s$. If $p \notin \{i_1, i_2, \dots, i_s\}$, then $x + k_p \in X$ and (yet to be shown) $x - k_p \notin X$. To prove this last relation, assume to the contrary that $k_{i_1} + k_{i_2} + \cdots + k_{i_s} - k_p = k_{j_1} + k_{j_2} + \cdots + k_{j_t}$ $(1 \le j_1 < j_2 < \cdots < j_t)$; you may clearly assume that the index sets $\{i_1, i_2, \dots, i_s\}$ and $\{j_1, j_2, \dots, j_t\}$ are disjoint. The above identity then contradicts the condition (28) for $n = \max\{i_s, j_t, p\}$. If $p \in \{i_1, i_2, \dots, i_s\}$, then $x - k_p \in X$ and $x + k_p \notin X$ (this last one can be shown by contradiction, similarly to the first case).

(viii) Each number $x \in \{1, 2, \dots, 2n\}$ can be written in the form $x = 2^\alpha \cdot y$, where $\alpha \in \mathbb{N}_0$ and y is an odd number lying in the n-element set $\{1, 3, \dots, 2n - 1\}$. Thus by the pigeonhole principle the $(n + 1)$-element set X contains two different numbers $2^\alpha \cdot y$ and $2^\beta \cdot y$, whose quotient is an integer power of 2. Next consider the n-element set $\{n + 1, n + 2, \dots, 2n\}$.

(ix) Proof by contradiction: Assume that 100 numbers can be chosen as described, but such that they do not have the desired property. By the argument in the solution of (viii) these numbers are of the form $2^{\alpha_1} \cdot 1, 2^{\alpha_3} \cdot 3, 2^{\alpha_5} \cdot 5, \dots, 2^{\alpha_{99}} \cdot 99, 101, 103, \dots, 199$, where $\alpha_i \in \mathbb{N}_0$ for all i. Clearly, $\alpha_1 > \alpha_3 > \alpha_9 > \alpha_{27} > \alpha_{81}$, $\alpha_5 > \alpha_{15} > \alpha_{45}$, $\alpha_7 > \alpha_{21} > \alpha_{63}$, $\alpha_{11} > \alpha_{33} > \alpha_{99}$, $\alpha_{13} > \alpha_{39}$. This implies $2^{\alpha_1} \ge 16$, $2^{\alpha_3} \cdot 3 \ge 24$, $2^{\alpha_i} \cdot i \ge 4i$ $(5 \le i \le 11)$, $2^{\alpha_{13}} \cdot 13 \ge 26$, and $2^{\alpha_{15}} \cdot 15 \ge 30$. But this means that none of the 100 numbers chosen is less than 16, which is a contradiction.

(x) The assertion is false with 17, as the following counterexample shows: $\{2^4, 2^3 \cdot 3, 2^3 \cdot 5, 2^3 \cdot 7, 2^2 \cdot 9, \dots, 2^2 \cdot 21, 2 \cdot 23, \dots, 2 \cdot 65, 67, \dots, 199\}$.

(xi) Let $1 \le a_1 < a_2 < \cdots < a_n \le 2n - 1$ be the chosen numbers. In a similar way as in (viii) you have $a_i = 2^{\alpha_i} \cdot b_i$ $(1 \le i \le n)$, where $b_1, b_2, \dots, b_n$ is some arrangement of the numbers of the set $L = \{1, 3, \dots, 2n - 1\}$. First of all, consider the $k + 1$ numbers a_i for which $b_i \in \{1, 3, 3^2, \dots, 3^k\}$ and show that they satisfy the bounds $a_i \geq 2^k$. Next assume that $a_1 = 2^{\alpha_1} \cdot b_1 < 2^k$. Then $b_1 \neq 3^\beta$, $b_1 \geq 5$, $k - \alpha_1 \geq 3$, $3^{\alpha_1 + 1} \cdot b_1 < 3^{\alpha_1 + 3} \cdot 2^{k - \alpha_1 - 3} < 3^k < 2n$, and therefore $L^* = \{3b_1, 3^2 b_1, \ldots, 3^{\alpha_1 + 1} \cdot b_1\}$ is a subset of L. Define $I = \{i; \ b_i \in L^*\}$. If $\alpha_i \geq \alpha_1$ for all $i \in I$, then $a_1 \mid a_i$, which is a contradiction. Hence necessarily $0 \leq \alpha_i < \alpha_1$ $(i \in I)$, and by the pigeonhole principle among the $\alpha_1 + 1$ numbers α_i $(i \in I)$ some two are identical, $\alpha_i = \alpha_j$ $(i, j \in I, \ i \neq j)$, which means that either $a_i \mid a_j$ (if $b_i < b_j$), or $a_j \mid a_i$ (if $b_i > b_j$), but this is a contradiction.

(xii) For the numbers $x_1, \ldots, x_n$ and $y_1, \ldots, y_N$ under consideration, set

$$\sigma_k = \sum_{\substack{\{(i_1, i_2, \ldots, i_k): \\ 1 \leq i_1 < i_2 < \cdots < i_k \leq n\}}} x_{i_1} x_{i_2} \cdots x_{i_k}, \quad s_k = \sum_{i=1}^{n} x_i^k, \quad \text{and} \quad t_k = \sum_{i=1}^{N} y_i^k,$$

where $1 \leq k \leq n$. From the theory of polynomials it is known that the collection of numbers $x_1, x_2, \ldots, x_n$ is uniquely determined by the values $\sigma_1, \sigma_2, \ldots, \sigma_n$ of the elementary symmetric polynomials. By Newton's theorem (see, e.g., [8], page 323), $s_k - s_{k-1}\sigma_1 + s_{k-2}\sigma_2 - \cdots + (-1)^{k-1} s_1 \sigma_{k-1} = (-1)^{k-1} k \sigma_k$, where $1 \leq k \leq n$, and furthermore it follows that the numbers $\sigma_1, \sigma_2, \ldots, \sigma_n$ are uniquely determined by the power sums $s_1, s_2, \ldots, s_n$. Therefore, it suffices to show (by induction on k) that the number s_k is uniquely determined by the values of the sums $t_1, t_2, \ldots, t_k$, $1 \leq k \leq n$. Indeed, you have $s_1 = \frac{t_1}{n-1}$, and in the case $2^k \neq 2n$,

$$s_k = \frac{1}{2n - 2^k} \left[2t_k - \sum_{j=1}^{k-1} \binom{k}{j} s_j s_{k-j} \right].$$

This last formula follows from the identity

$$2t_k + 2^k s_k = \sum_{i=1}^{n} \sum_{j=1}^{n} (x_i + x_j)^k$$

by using the binomial theorem for the powers on the right. Finally, for the values $n = 2^k$ you can recursively construct a counterexample of two disjoint n-element sets $A_n = \{x_1, \ldots, x_n\}$ and $B_n = \{y_1, \ldots, y_n\}$ for which the sums $x_i + x_j$, respectively $y_i + y_j$, form two identical $\binom{n}{2}$-element collections: Set $A_2 = \{0, 3\}$, $B_2 = \{1, 2\}$, $A_{2n} = A_n \cup (c + B_n)$, and $B_{2n} = B_n \cup (c + A_n)$, where the number c (depending on n) is chosen large enough so that the sets with index $2n$ have $2n$ elements and are disjoint.

5.2 (i) Assume to the contrary that after n steps all $2k + 1$ numbers were even, and take the *smallest* such n. Then in the preceding step all numbers would be odd, and in the step before that any two neighboring numbers would have different parities. This is not possible, since $2k + 1$ is an odd number.

(ii) Since after each step the sum of all 25 numbers doubles, after 100 steps it will be equal to -2^{100}. It remains to verify that $2^{100} > 25 \cdot 10^{28}$, or $2^{72} > 5^{30}$. You even have $2^{70} > 5^{30}$, since $2^7 = 128 > 125 = 5^3$.

(iii) If $k = 0$, then n is odd; if then the n-tuple $(1, 1, \ldots, 1)$ is not the initial one, then its first occurrence comes after the n-tuple $(-1, -1, \ldots, -1)$; this already is the initial n-tuple, since $(x_1 x_2)(x_2 x_3) \cdots (x_n x_1) = x_1^2 x_2^2 \cdots x_n^2 = 1 \neq (-1)^n$. Then base the induction step from $n = 2^k \ell$ to $n = 2^{k+1} \ell$ on considering the two sequences (29), as in the solution of 5.1.(iii).

5.4 (i) If you change the signs of the numbers ± 1, then the product of the numbers in any 2×2 subarray is an invariant. If you compare the 2×2 subarrays in the lower right corner, you find that the arrays in Figure 13 are mutually nonattainable.

(ii) The assertion is true. Use the invariant $I = (a_1 a_4 a_7 a_{10})(a_3 a_6 a_9 a_{12})$.

(iii) The assertion is true. Use the invariant $I = a_2 a_3 a_5 a_6 a_8 a_9 a_{11} a_{12}$.

(iv) Show that the sum $I(M) = \sum 2^{-x-y+2}$ is an invariant, where the summation is extended over all pairs $(x, y) \in M$. For the initial set M you have $I(M) = 1$. Explain why in any attainable set M' there is at most one pair $(1, y)$ and at most one pair $(x, 1)$. If you therefore assume that M' has the property required in the statement of the problem, then you obtain

$$I(M') \leq \frac{1}{8} + \frac{1}{8} + \sum_{(x,y) \in D \cap M'} 2^{-x-y+2},$$

where $D = \{(x, y) \in \mathbb{N} \times \mathbb{N} : x \geq 2 \wedge y \geq 2 \wedge x + y \geq 5\}$. Then you easily obtain the infinite series

$$\sum_{(x,y) \in D} 2^{-x-y+2} = \frac{3}{4}$$

which implies $I(M') < \frac{2}{8} + \frac{3}{4} = 1$, but this is a contradiction.

(v) Let $p = qd$. After one transformation the sign of exactly q factors in the product defining each of the numbers s_k changes; that is, the value of s_k changes its sign exactly when q is odd. Hence after any number of steps you have either $s'_k = s_k$ $(1 \leq k \leq d)$ or (only in the case of odd q and an odd number of steps) $s'_k = -s_k$ $(1 \leq k \leq d)$. Next show that by an appropriate iteration (with an even number of steps, so that the numbers s_k do not change) one can achieve changes in the signs of exactly two of the numbers a_i, a_j $(i \neq j)$ in the cases $i - j = p$, $p \mid (i - j)$ and even $d \mid (i - j)$. Therefore, if you have $s'_k = s_k$ $(1 \leq k \leq d)$, then you can transform the n-tuple $(a_1, a_2, \ldots, a_n)$ into the n-tuple $(a''_1, a''_2, \ldots, a''_n)$ such that $s''_k = s_k$ $(1 < k < d)$ and $a''_k = a'_k$ $(d < k \leq n)$. But then from the identity $a''_k a''_{k+d} \cdots a''_{k+n-d} = s''_k = s_k = s'_k = a'_k a'_{k+d} \cdots a'_{k+n-d}$ it also follows that $a''_k = a'_k$ $(1 \leq k \leq d)$. The case where $s'_k = -s_k$ $(1 \leq k \leq d)$ holds for odd q can be reduced to the preceding case by one transformation (in an arbitrary way).

5.6 (i) The last one to grow is a lemon; their number is odd every morning.

(ii) The number of pieces of paper always satisfies $n \equiv 1 \pmod 9$, while $1991 \equiv 2 \pmod 9$.

(iii) The identity $\left(c \cdot 10^k + x\right) - (c + x) = c\left(10^k - 1\right)$ shows that the remainder upon division by 9 is an invariant of the transformation. Since 9 does not divide 2^{1991}, the digit sum of the number A cannot be equal to $0 + 1 + 2 + \cdots + 9 = 45$.

(iv) The equation $p_1' + p_2' + \cdots + p_n' = N$ is a preservation law for the number of larks; the necessity of the congruence condition was shown in 5.5.(iv). To prove sufficiency, begin by explaining how from any initial position of N larks you can, through a sequence of allowable transformations, achieve that at least $N-1$ larks sit in the tree with number n. The remaining Nth lark in this situation sits in the jth tree (the possibility $j = n$ is not excluded), where the index j is uniquely determined by the initial situation $(p_1, p_2, \ldots, p_n)$ by way of the congruence $p_1 + 2p_2 + \cdots + np_n \equiv j + (N-1)n \pmod n$. Finally, use the fact that if the state p' is attainable from the state p, then also p is attainable from p'.

(v) For the position $(p, q, r) = (2, 0, 0)$, which was not tabulated in the solution of 5.5.(iii), all three sums s_i are even; therefore, from the given initial situation at most one of the four positions mentioned can be achieved. Now carry out the operations according to the following strategy: At each step adjoin the digit with the least number of occurrences (if this digit is not unique, then choose the smallest of them). Since the sum $p + q + r$ decreases by 1 after each step, you finally reach a position (p', q', r') where no further operation is possible. Clearly, p', q', r' are the numbers $n, 0, 0$ in some order ($n \geq 1$). If $n = 1$, then everything is in order; in the case $n \geq 2$ the position (p', q', r') follows one consisting of the numbers $1, 1, n - 1$, which in view of your strategy means that $n - 1 \leq 1$, that is, $n = 2$. This is then the position $(1, 1, 1)$, which is followed by the final position $(2, 0, 0)$.

(vi) Exactly those k, $0 \leq k \leq n$, for which $k \equiv 1 + 2 + \cdots + n = \frac{n(n+1)}{2}$ $\pmod 2$ holds can remain on the board. Indeed, on the one hand, the sum of all numbers on the board does not change parity after an operation, since $a + b \equiv |a - b| \pmod 2$ for any $a, b \in \mathbb{Z}$. On the other hand, for any such k you can carry out the operations in the following order (distinguish the cases (a) n even, k even; (b) n even, k odd; (c) n odd, k odd; (d) n odd, k even; write down all pairs (a, b) of consecutive erased numbers):

(a) $(2, 3), (4, 5), \ldots, (k-2, k-1), (k+1, k+2), (k+3, k+4), \ldots, (n-1, n)$;

(b) $(2, 3), (4, 5), \ldots, (k-1, k), (k+2, k+3), (k+4, k+5), \ldots, (n-1, n)$, $(k+1, 1)$;

(c) $(1, 2), (3, 4), \ldots, (k-2, k-1), (k+1, k+2), (k+3, k+4), \ldots, (n-1, n)$;

(d) $(1, 2), (3, 4), \ldots, (k-1, k), (k+2, k+3), (k+4, k+5), \ldots, (n-1, n)$, $(k+1, 1)$.

In all cases you then obtain on the board the number k and an even number of the integer 1, and it is clear how one has to proceed further so that only the number k remains. (The assertion about the parity of the number of ones follows from considering the invariant mentioned above.)

(vii) All congruences are modulo m. If you substitute $x_{k+1} \equiv x_1 + x_2 + \cdots + x_k$ into the congruence $\alpha_1 x_1 + \alpha_2 x_2 + \cdots + \alpha_k x_k \equiv \alpha_1 x_2 + \alpha_2 x_3 + \cdots + \alpha_k x_{k+1}$, you obtain the condition $(\alpha_k - \alpha_1)x_1 + (\alpha_k + \alpha_1 - \alpha_2)x_2 + \cdots + (\alpha_k + \alpha_{k-1} - \alpha_k)x_k \equiv 0$, which holds for any $x_1, x_2, \ldots, x_k$ if and only if $0 \equiv \alpha_k - \alpha_1 \equiv \alpha_k + \alpha_1 - \alpha_2 \equiv \cdots \equiv \alpha_k + \alpha_{k-1} - \alpha_k$, which is equivalent to the system $\alpha_i \equiv i\alpha_1$, $(1 \le i \le k-1)$, $\alpha_{k-1} \equiv 0$, and $\alpha_k \equiv \alpha_1$. Therefore, $\alpha_1 \not\equiv 0$ and $(k-1)\alpha_1 \equiv 0$ have to hold, which is possible if and only if the integers $k-1$ and m are not relatively prime.

5.8 (i) The sum of all n numbers is a nonincreasing valuation.

(ii) The desired number is 2^{-10}. Note that the quotient $\frac{a}{2^n}$, where a is the smallest positive number and n the number of zeros on the board, is a nondecreasing valuation.

(iii) The valuation $J = x_1^2 + x_2^2 + \cdots + x_n^2$ is nondecreasing; after the first transformation the value of J increases.

(iv) Denote the nth iteration by (a_n, b_n, c_n, d_n) and assume cyclicity. The numbers $A_n = a_n + b_n + c_n + d_n$ satisfy $A_n = 2^n A_0$; since there are two identical numbers in the sequence $A_0, A_1, \ldots$, we have $A_0 = a+b+c+d = 0$. The numbers $B_n = a_n^2 + b_n^2 + c_n^2 + d_n^2$ satisfy $B_{n+1} = 2B_n + 2(a_n + c_n)(b_n + d_n) = 2B_n + 2A_{n-1}^2 = 2B_n$ for all $n \ge 1$, so $B_n = 2^{n-1}B_1$ $(n \ge 2)$, which implies $B_1 = 0$. But this means that $a_1 = b_1 = c_1 = d_1 = 0$, which leads to $a = -b = c = -d$.

(v) The assertion does not hold for any $n = 3k$: Show that the initial n-tuple $(1, -1, 0, 1, -1, 0, \ldots, 1, -1, 0)$ is transformed to itself after 6 iterations.

We add without proof that for all $n \ne 3k$ the following holds: If in some sequence of iterations two elements (that is, two ordered n-tuples) are identical, then the initial n-tuple is of the form $(a, -a, a, -a, \ldots, a, -a)$ or $(0, 0, \ldots, 0)$ according to whether n is even or odd.

(vi) For each $n \ge 6$ the answer is negative. A suitable counterexample is easier to construct in the field of complex numbers. Consider the n-tuple $(1, \varepsilon, \ldots, \varepsilon^{n-1})$, where $\varepsilon = \cos \frac{2\pi}{n} + i \sin \frac{2\pi}{n}$. Since $\varepsilon^n = 1$, it is not difficult to verify that the kth iteration is of the form $((1 - \varepsilon)^k, (1 - \varepsilon)^k \varepsilon, \ldots, (1 - \varepsilon)^k \varepsilon^{n-1})$ for all $k \ge 1$. Each number in this n-tuple has absolute value equal to

$$\left|(1 - \varepsilon)^k \varepsilon^j\right| = |1 - \varepsilon|^k = \left(\left(1 - \cos \tfrac{2\pi}{n}\right)^2 + \sin^2 \tfrac{2\pi}{n}\right)^{\frac{k}{2}} = \left(2 \sin \tfrac{\pi}{n}\right)^k,$$

which does not exceed 1 for any $k \ge 1$ if $n \ge 6$. A suitable example in real numbers is obtained by taking the real parts of the above complex numbers: $\left(1, \cos \frac{2\pi}{n}, \cos \frac{4\pi}{n}, \ldots, \cos \frac{2(n-1)\pi}{n}\right)$.

(vii) Exactly the triples $[a, a, 0]$, where $a \ge 0$. Let $[a_k, b_k, c_k]$ be the kth iteration of the triple $[a, b, c]$, and suppose that $[a_n, b_n, c_n] = [a, b, c]$ for some $n \geq 1$. Then clearly $a, b, c \in \mathbb{R}_0^+$. Next show, as in 5.7.(iv), that the numbers $A_k = \max\{a_k, b_k, c_k\}$ satisfy $A_0 \geq A_1 \geq \cdots \geq A_n$. Hence from $A_0 = A_n$ it follows that $A_0 = A_1$, which is possible only when $0 \in \{a, b, c\}$. Similarly, from $A_1 = A_2$ it follows that $0 \in \{a_1, b_1, c_1\}$ (which holds also when $n = 1$). Then two of the numbers a, b, c are identical. Therefore, $[a, b, c]$ is of the form $[a, 0, 0]$ or $[a, a, 0]$ for appropriate $a \geq 0$. Consider the iterations of both forms.

(viii) An appropriate m-tuple $(1, \alpha, \alpha^2, \ldots, \alpha^{m-1})$, where $\alpha > 1$, has a kth iteration of the form $(\beta^k, \beta^k \alpha, \beta^k \alpha^2, \ldots, \beta^k \alpha^{m-1})$ as long as

$$\frac{\alpha - 1}{1} = \frac{\alpha^2 - \alpha}{\alpha} = \cdots = \frac{\alpha^{m-1} - \alpha^{m-2}}{\alpha^{m-2}} = \frac{\alpha^{m-1} - 1}{\alpha^{m-1}} = \beta > 0.$$

This occurs if and only if $\beta = \alpha - 1$ and $\alpha > 1$ is a root of the equation $F_m(x) = 0$, where $F_m(x) = x^{m-1} - x^{m-2} - \cdots - x - 1$. The equation $F_m(x) = 0$ has a root in the interval $(1, 2)$ for every $m \geq 3$, since $F_m(1) < 0$ and $F_m(2) > 0$.

(ix) Independently of the initial m-tuple, you obtain after a certain number of steps an m-tuple of even numbers $(2b_1, 2b_2, \ldots, 2b_m)$; this follows from the result of 5.1.(iii) if you consider the numbers 1 and -1 as symbols for an even, respectively odd, number. Repeat the same argument for the m-tuple $(b_1, b_2, \ldots, b_m)$, etc.; hence for every $k \in \mathbb{N}$ you obtain after a certain number of steps the m-tuple $(2^k c_1, 2^k c_2, \ldots, 2^k c_m)$, where $c_i \in \mathbb{N}_0$, $1 \leq i \leq m$. On the other hand, the numbers in all m-tuples are bounded from above by twice the largest of the numbers $|a_1|, |a_2|, \ldots, |a_m|$. For sufficiently large k this means that $c_1 = c_2 = \cdots = c_m = 0$.

(x) Call an arrow pointing to the left (respectively to the right) a left (respectively right) arrow. The numbers L and R of left, respectively right, arrows on the line are invariants; since a transformation can be carried out if and only if some left arrow lies to the right of some right arrow, the final state is uniquely determined: Going from left to right, there are first L left and then R right arrows. Next, consider the valuation J defined by the number of (unordered) pairs of arrows with opposite directions and pointing to each other (they don't have to be neighbors). An allowable change of some pair of neighboring arrows is possible if and only if $J > 0$; after each such transformation the value of J decreases by 1. This means that the transformation can (independently of the particular sequence) be repeated exactly J_0 times, where J_0 is the initial value of J.

(xi) The desired condition has the form

$$a_1 + a_2 + \cdots + a_n = 2^m \cdot d, \quad m \in \mathbb{N}_0. \tag{54}$$

Indeed, the value of d either remains unchanged after a transformation or is doubled; this implies the necessity of (54). Its sufficiency can be proven by induction on the exponent m. If $m = 0$, then $n = 1$ and everything is trivial. If (54) holds for some $m \geq 1$, then under the assumption that

$d = 1$ (this is no loss of generality, since the pebbles are shifted in integer multiples of d) the number of odd integers among $a_1, a_2, \ldots, a_n$ is even, say $2k$, where $k \in \mathbb{N}$. Therefore, the piles with odd numbers of pebbles can be "paired," and in each one of these k pairs you can carry out a transformation. After these k transformations you obtain $n'(\leq n)$ piles with numbers of pebbles $b_1, b_2, \ldots, b_{n'}$, all of which are even. Therefore, $2^m = a_1 + a_2 + \cdots + a_n = b_1 + b_2 + \cdots + b_{n'} = 2^{m'} \cdot d'$ ($m' \in \mathbb{N}_0$), where d' is the greatest common divisor of the numbers $b_1, b_2, \ldots, b_{n'}$. Since $d' = 2^i$ for some $i \geq 1$, you have $m' \leq m - 1$, and therefore you can use the induction hypothesis.

(xii) Consider the two valuations

$$J_1(a_1, a_2, \ldots, a_n) = \sum_{i;\ a_i \leq i} 2^{a_i}, \quad J_2(a_1, a_2, \ldots, a_n) = \sum_{i;\ a_i \geq i} 2^{-a_i}.$$

If the indices c, k are as in the statement and if $|c - k| \geq 2$, then the inequalities $2^{k+1} + 2^{k+2} + \cdots + 2^{c-1} < 2^c$ (if $k < c$) or $2^{-c-1} + 2^{-c-2} + \cdots + 2^{-k+1} < 2^{-c}$ (if $k > c$) imply that after the transformation (32) the value of J_1 does not decrease and the value of J_2 increases, while after the transformation (33) the value of J_1 increases and that of J_2 does not decrease. Moreover, these properties of J_1 and J_2 are evident if $|c - k| = 1$. Therefore, $J = J_1 + J_2$ is an increasing valuation. Since there is a finite number of arrangements of the integers $1, 2, \ldots, n$, the value of J can increase only a finite number of times.

5.10 (i) The conclusion is not true for any $n > 1$. Consider a sequence of length $2n + 1$ formed with $n + 1$ copies of 1 and n copies of -1.

(ii) Show that $J = p(x_1^2 + x_2^2 + x_3^2 + x_4^2) + q(x_1 x_2 + x_2 x_3 + x_3 x_4 + x_4 x_1) + r(x_1 x_3 + x_2 x_4)$ is a suitable valuation if $2p - 2q + r = 2q - 2r = c > 0$, that is, if $q = r + \frac{c}{2}$ and $p = c + \frac{r}{2}$. By choosing $r = -2$ and $c = 4$ you obtain $J = 2\left(x_1^2 + x_2^2 + x_3^2 + x_4^2\right) + (x_1 - x_3)^2 + (x_2 - x_4)^2$.

(iii) Suppose that the index $k = 1$ is disallowed. Note below how the numbers $S_i = x_1 + x_2 + \cdots + x_i$ ($1 \leq i \leq n$) will change. When a transformation with the triple of numbers x_{k-1}, x_k, x_{k+1} is carried out, then the following holds: If $1 < k < n$, then the transformation is allowable only when $S_k < S_{k-1}$, and as a result, S_{k-1} and S_k are interchanged; if $k = n$, then the transformation is allowable only when $S_n < S_{n-1}$, in which case the n-tuple $(S_1, S_2, \ldots, S_n)$ is changed to $(S_1 + x_n, S_2 + x_n, \ldots, S_{n-2} + x_n, S_n + x_n, S_{n-1} + x_n)$. Each transformation of a triple x_{k-1}, x_k, x_{k+1}, where $1 < k < n$, then decreases the number of inversions in the n-tuple $(S_1, S_2, \ldots, S_n)$, that is, the number of pairs of indices (i, j) such that $1 \leq i < j \leq n$ and $S_i > S_j$. This proves the assertion on the finite number of possible transformations in the case where the index $k = 1$ is disallowed.

(iv) Show that for the valuation

$$J(x_1, x_2, \ldots, x_n) = \sum_{k=0}^{n-1} p_k(x_1 x_{1+k} + x_2 x_{2+k} + \cdots + x_n x_{n+k}) \qquad (55)$$

(where $x_{n+j} = x_j$ for $j = 1, 2, \ldots, n-1$) with constants $p_k = p_{n-k}$ you have $J(x_1 + x_2, -x_2, x_3 + x_2, x_4, \ldots, x_n) - J(x_1, x_2, \ldots, x_n) = 2x_2[(p_0 - 2p_1 + p_2)(x_1 + x_2 + x_3) + (p_3 - 2p_2 + p_1)x_4 + (p_4 - 2p_3 + p_2)x_5 + \cdots + (p_{n-1} - 2p_{n-2} + p_{n-3})x_n] = 2x_2 c \cdot S$, if $c = p_2 - 2p_1 + p_0 = p_3 - 2p_2 + p_1 = \cdots = p_{n-1} - 2p_{n-2} + p_{n-3}$. (In view of symmetry it is not necessary to consider other transformations.) The last identities hold if and only if the sequence $p_1 - p_0, p_2 - p_1, \ldots, p_{n-1} - p_{n-2}$ is arithmetic with difference c, that is, $p_k - p_{k-1} = p_1 - p_0 + (k-1)c$ $(1 \le k \le n-1)$, from which by summing you obtain $p_k = p_0 - k(p_0 - p_1) + \frac{k(k-1)}{2}c$ $(0 \le k \le n-1)$. By substituting into $p_1 = p_{n-1}$ you will obtain $p_0 - p_1 = \frac{n-1}{2} \cdot c$, and thus

$$p_k = p_0 - \frac{k(n-k)}{2}c \quad (0 \le k \le n-1). \qquad (56)$$

Verify that p_k defined by (56) possesses all the desired properties. If you choose any $p_0 \in \mathbb{R}$ and $c \in \mathbb{R}^+$, then by substituting p_k from (56) into (55) you obtain a decreasing valuation J. Next you must consider whether J is a nonnegative function for some pair p_0, c. If you set, for instance, $p_0 = \frac{n^2+n}{2}$ and $c = 2$, then $p_k = \binom{n-k+1}{2} + \binom{k+1}{2}$ for all $k \in \{1, 2, \ldots, n-1\}$, and $J = S_1 + S_2 + \cdots + S_n$, with $S_j = \sum_{i=1}^{n}(x_i + x_{i+1} + \cdots + x_{i+j-1})^2 \ge 0$ for $1 \le j \le n$.

(v) Assume that there is an infinite iterative sequence. From the properties described in (iv) it follows that the corresponding infinite sequence of valuations of the n-tuples of our iterative sequence is decreasing and bounded, so convergent to a number L. Therefore, you can choose an iterative term with a valuation J_0 satisfying $L < J_0 < L + \frac{4S^2}{n}$ and leave out all the previous terms. So, without loss of generality, you can assume that these inequalities are satisfied already by the initial n-tuple $(x_1, \ldots, x_n)$. Then the infinite sequence consisting of the negative central terms $x_k, x_k', x_k'', \ldots$ of all transformed triples satisfies

$$(-x_k) + (-x_k') + (-x_k'') + \cdots = \frac{1}{4S}(J_0 - L) < \frac{S}{n}. \qquad (57)$$

The largest of the numbers $x_1, \ldots, x_n$ in the initial n-tuple is at least $\frac{S}{n}$; let, for instance, $x_1 \ge \frac{S}{n}$. From (57) it then follows that in the place of x_1 after an arbitrary number of steps there will be a nonnegative number. Hence the index $k = 1$ is disallowed in the sense of (iii), which means a finite number of transformations, which is a contradiction.

(vi) A suitable constant K is, for instance, of the form

$$K = \max_{1 \le k \le n} \left\{ \max_{0 \le i \le n-1} |x_k + x_{k+1} + \cdots + x_{k+i}| \right\}.$$

Show that the expression on the right is a nonincreasing valuation; to do this, use inequalities of the type $x_2 < x_1 + 2x_2 + x_3 + \cdots + x_n < x_1 + x_2 + \cdots + x_n$ (if $x_2 < 0$).

5.12 (i) A final distribution exists only when $n = 1 + 2 + \cdots + d$ for some $d \in \mathbb{N}$, and it is of the form $[d, d-1, \ldots, 1]$. Proof: Comparing the distribution $[a_1, a_2, \ldots, a_m]$ with $[m, a_1 - 1, a_2 - 1, \ldots, a_m - 1]$ under the assumption $a_1 \geq a_2 \geq \cdots \geq a_m$ you obtain $a_m = 1$. If $m > 1$, then $a_{m-1} - 1 = a_m$, that is, $a_{m-1} = 2$, etc.

(ii) It suffices to modify the conclusion of Section 5.11 after the inequalities (44), from which in the case $\frac{(d-1)d}{2} < n < \frac{d(d+1)}{2}$ it follows by adding that $m_0 = d$ and $a_j(k) = d - k + \varepsilon_j(k)$, with $\varepsilon_j(k) \in \{0, 1\}$, $1 \leq k \leq d$, $j \in \mathbb{N}_0$, where $a_j(d) = 0$ when $m_j < d$, that is, when $m_j = d - 1$. Show that $(\varepsilon_0(1), \varepsilon_0(2), \ldots, \varepsilon_0(d))$ can be any d-tuple $(\varepsilon_1, \varepsilon_2, \ldots, \varepsilon_d)$ of the numbers 0 and 1 satisfying $\varepsilon_1 + \varepsilon_2 + \cdots + \varepsilon_d = n - \frac{(d-1)d}{2}$, and that for each $j \geq 1$ you have $(\varepsilon_{j+1}(1), \varepsilon_{j+1}(2), \ldots, \varepsilon_{j+1}(d)) = (\varepsilon_j(d), \varepsilon_j(1), \varepsilon_j(2), \ldots, \varepsilon_j(d-1))$, that is, $\varepsilon_j(k) = \varepsilon_{j+k}$ ($1 \leq k \leq d$, $j \in \mathbb{N}_0$), if you set $\varepsilon_{d+j} = \varepsilon_j$ for all $j \geq 1$. This implies a description of all periodic p-tuples of the iterations $a_j(k) = d - k + \varepsilon_{j+k}$ ($1 \leq k \leq d$, $j \in \mathbb{N}_0$) and the assertion about its length p being equal to the (smallest) period of the infinite sequence $\varepsilon_1, \varepsilon_2, \ldots, \varepsilon_d, \varepsilon_1, \varepsilon_2, \ldots, \varepsilon_d, \ldots$, since $n - \frac{(d-1)d}{2}$ copies of 1 in the d-tuple $(\varepsilon_1, \varepsilon_2, \ldots, \varepsilon_d)$ are distributed into $\frac{d}{p}$ copies of the p-tuple $(\varepsilon_1, \varepsilon_2, \ldots, \varepsilon_p)$.

(iii) Set $m = n - \frac{(d-1)d}{2}$. From the analysis done in the solution of (ii) it follows that the repeating distributions are exactly the d-tuples $[d - 1 + \varepsilon_1, d - 2 + \varepsilon_2, \ldots, 1 + \varepsilon_{d-1}, \varepsilon_d]$, where $\varepsilon_1, \varepsilon_2, \ldots, \varepsilon_d \in \{0, 1\}$ and $\varepsilon_1 + \varepsilon_2 + \cdots + \varepsilon_d = m$. The period of the repetition is then equal to the smallest $k \in \mathbb{N}$ such that $\varepsilon_i = \varepsilon_{i+k}$ (here, obviously, put $\varepsilon_{d+j} = \varepsilon_j$ for all $j \in \mathbb{N}$). For the given m, d ($1 \leq m < d$), and any positive integer p, let $f(p)$ denote the number of distributions with the period p, and $g(p)$ the number of distributions with an arbitrary period q, $q \mid p$. By (ii) you have $f(p) = 0$ whenever $p \nmid d$ or $d \nmid pm$. If $p \mid d$ and $d \nmid pm$, then for any divisor $q \mid p$ we also have $d \nmid qm$, so $g(p) = 0$. On the other hand, if $p \mid d$ and $d \mid pm$, then $g(p)$ is equal to the number of all ordered p-tuples of $\frac{pm}{d}$ ones and $\frac{p(d-m)}{d}$ zeros; that is, $g(p) = \binom{p}{pm/d}$. It follows directly from the definition that

$$g(p) = \sum_{q \mid p} f(q).$$

Then by the Möbius inversion formula (see Chapter 1, 6.13.(iv)), for $p \mid d$ and $d \mid pm$ we obtain (where $\mu(q)$ is the Möbius function)

$$f(p) = \sum_{q \mid p} \mu(q) g\left(\tfrac{p}{q}\right) = \sum_{q \mid p} \mu\left(\tfrac{p}{q}\right) g(q) = \sum_{\substack{q \mid p \\ d \mid qm}} \mu\left(\tfrac{p}{q}\right) \binom{q}{qm/d}.$$

3   Hints and Answers to Chapter 3

1.3 (i) Let k denote the largest number of acute angles. The sum $S_n = (n-2) \cdot 180°$ of interior angles of the polygon is then less than $k \cdot 90° + (n-k) \cdot 180°$. From the inequality $k \cdot 90° + (n - k) \cdot 180° > (n - 2) \cdot 180°$ it follows that $k < 4$. Polygons with three acute angles clearly exist for each $n \geq 3$ (you should come up with your own construction).

(ii) Show that two such *nonneighboring* vertices exist; do it by induction on $n \geq 4$. Use one of the constructed diagonals to divide the whole polygon into two polygons with a smaller number of sides. Any further diagonal is a diagonal of one of these two polygons.

(iii) If k is the desired smallest number, then the inequality $k \geq n - 2$ follows from the fact that the diagonals emanating from one of the vertices of the polygon divide this polygon into $(n-2)$ triangles. The value $k = n-2$ is possible: Given a convex n-gon $A_1 A_2 \ldots A_n$, for every $j = 2, 3, \ldots, n-1$ choose an interior point X_j of the triangle that arises as the intersection of the two triangles $A_1 A_j A_n$ and $A_{j-1} A_j A_{j+1}$. Then every $\triangle A_p A_q A_r$ ($p < q < r$) contains the chosen point X_q.

(iv) Distinguish two cases: If the point P lies on some diagonal $A_i A_j$, then the lines PA_i, PA_j are identical and do not cross any side of the polygon; the remaining $(2n - 2)$ lines cross at most $(2n - 2) < 2n$ sides in an interior point. If the point P does not lie on any diagonal, then it is an interior point of one of the $(n + 1)$-gons $\mathcal{M}_1 = A_1 A_2 \ldots A_{n+1}$, $\mathcal{M}_2 = A_1 A_{n+1} \ldots A_{2n}$. Assume that $P \in \mathcal{M}_1$. Then none of the $(n + 1)$ lines $PA_{n+1}, PA_{n+2}, \ldots, PA_{2n}, PA_1$ crosses any of the n sides $A_{n+1} A_{n+2}, A_{n+2} A_{n+3}, \ldots, A_{2n} A_1$; therefore each one of these sides can be crossed by only one of the remaining $(n-1)$ lines $PA_2, PA_3, \ldots, PA_n$, and by the pigeonhole principle at least one of the sides $A_{n+1} A_{n+2}, \ldots, A_{2n} A_1$ is not crossed by any of the lines.

1.5 (i) pq. [Each of the p lines of the first bundle has exactly q intersections with the lines of the second bundle.]

(ii) $3m^2$. [On each of the $3m$ lines there are $2m$ intersections.]

(iii) The line determined by a pair of given points A, B has at most $2 \cdot \binom{n-2}{3} + 2 \cdot \binom{n-2}{2}$ intersections, different from A, B, with the given circles. There are $\binom{n}{2}$ such lines, and therefore there are at most $n + \binom{n}{2}\left(2 \cdot \binom{n-2}{3} + 2 \cdot \binom{n-2}{2}\right) = n + \binom{n}{3}(n - 1)(n - 3)$ points in question.

(iv) Choose any one of the $\binom{n}{3}$ circles, and let A, B, C be the corresponding given points. There are a total of $\binom{n-3}{3}$ of the circles that pass through none of the points A, B, C, each of which has at most two intersections with the chosen circle. Observe that $3 \cdot \binom{n-3}{2}$ circles go through exactly one of the points A, B, C, and each of them intersects the chosen circle in at most one point different from A, B, C. Thus on the given circle there are $2 \cdot \binom{n-3}{3} + 3 \cdot \binom{n-3}{2} = \frac{(n-3)(n-4)(2n-1)}{6}$ intersections different from A, B, C.

Altogether, you then obtain at most

$$\frac{1}{2}\binom{n}{3} \cdot \frac{(n-3)(n-4)(2n-1)}{6} = \frac{5(2n-1)}{3}\binom{n}{5}$$

intersections that are different from the given n points. If you add them in, then the number N of intersections of all the circles satisfies the bound

$$N \leq n + \frac{5(2n-1)}{3} \cdot \binom{n}{5}.$$

(v) Rotate the given broken line until its segments are horizontal and vertical; then 7 segments are horizontal and 7 are vertical. Denote the vertical segments from left to right by $r_1, r_2, \ldots, r_7$; if you determine the number of intersections p_i on the segment r_i, then for the desired number N of intersections you have $N = p_1 + p_2 + \cdots + p_7$. On the segments r_1 and r_7 there can be no intersections, and on each of

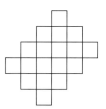

Figure 7

the segments r_2 and r_6 there are at most 2 intersections (to the left of r_2, respectively to the right of r_6, there are 2 vertices of the broken line); similarly, $p_3 \leq 4$, $p_5 \leq 4$, and $p_4 \leq 6$. In the last inequality you cannot have equality: Each of the vertices of the segments r_i $(i = 1, 2, 3)$ would be connected by a horizontal segment with a vertex of a segment r_j $(j = 5, 6, 7)$ and vice versa, and therefore the vertices of the segment r_4 would not be connected with the others by any horizontal segment. Hence $p_4 \leq 5$, and $N \leq 2(2+4) + 5 = 17$. Figure 7 shows an example for $N = 17$.

(vi) First determine the number of all intersections of the diagonals that are different from the vertices of $\mathcal{M}$ (here a diagonal will be seen as a straight line that contains the diagonal as a segment). According to (1) there are $u_n = \frac{n(n-3)}{2}$ diagonals in $\mathcal{M}$, and each diagonal AB intersects with all diagonals of $\mathcal{M}$ that do not go through the points A, B; therefore, it has at most $\frac{n(n-3)}{2} - 2(n-3) + 1 = \frac{n^2-7n+14}{2}$ intersections different from A, B. The number of intersections on all diagonals is no greater than $\frac{1}{2} \cdot \frac{n(n-3)}{2} \cdot \frac{n^2-7n+14}{2} = \frac{n(n-3)(n^2-7n+14)}{8}$. By (3) you know that $\binom{n}{4}$ of these are "interior," so the number p of "exterior" intersections satisfies the bound

$$p \leq \frac{n(n-3)(n^2 - 7n + 14)}{8} - \binom{n}{4} = \frac{n(n-3)(n-4)(n-5)}{12}.$$

As in 1.2.(i) show that equality can occur in this inequality.

(vii) Proceed as in 1.4.(iii). To each of the given points construct a total of $\binom{n-1}{2}$ perpendiculars. Choose any two points A and B (this can be done in $\binom{n}{2}$ ways) and determine the number of intersections on the perpendiculars leading through these points. Divide the perpendiculars that go through the point A into groups S_1, respectively S_2, according to whether they are perpendicular to lines that go, respectively do not go, through the point B. Then $|S_1| = (n-2)$, and on each of these lines there are $\binom{n-1}{2}$ intersections with perpendiculars going through the point B; similarly, $|S_2| = \binom{n-2}{2}$, and on each of these lines there are $\binom{n-1}{2} - 1$ intersections with perpendiculars that go through B. In the system of points there are $\binom{n}{3}$ triangles, and therefore the total number N of all intersections is no greater than

$$\binom{n}{2}\left[(n-2) \cdot \binom{n-1}{2} + \binom{n-2}{2}\frac{n^2 - 3n}{2}\right] - 2 \cdot \binom{n}{3}$$

$$= \binom{n}{3} \cdot \frac{3n^3 - 12n^2 + 9n + 4}{4}.$$

1.7 (i) $\binom{n}{3} - \binom{p}{3}$.   (ii) $\binom{p}{2}q + p \cdot \binom{q}{2} = \frac{pq}{2}(p + q - 2)$.
(iii) $22n^2(13n - 3)$; $12n^2(5n - 3)$. [See 1.6.(i).]
(iv) None for odd n, and $n(\frac{n}{2} - 1)$ for even n. [The hypotenuse of each triangle is a diameter of a Thales circle that circumscribes the regular n-gon.]
(v) $n \cdot [\frac{n-1}{2}] - \frac{2n}{3}([\frac{n}{3}] - [\frac{n-1}{3}])$.   (vi) $n \cdot [\frac{n-1}{4}]$.
(vii) For $n < 3s + 3$ such triangles do not exist. Let now $n \geq 3s + 3$, and set $\mathcal{M} = A_1 A_2 \ldots A_n$. To determine the number of triangles with vertex A_1, choose the other two vertices from among $A_{s+2}, A_{s+3}, \ldots, A_{n-s}$, of which there are $(n - 2s - 1)$. The number of allowable choices is $\binom{(n-2s-1)-s}{2} = \binom{n-3s-1}{2}$ (combine the first chosen vertex with the following s vertices to one new "point"). Hence the number of desired triangles is $\frac{n}{3}\binom{n-3s-1}{2}$.

1.9 (i) $n^4 + 4 \cdot 3 \cdot n^2 \binom{n}{2} + \binom{4}{2}\binom{n}{2}^2 = \frac{n^2}{2}(17n^2 - 18n + 3)$.
(ii) For n divisible by 4 there are $\frac{n}{4}$; otherwise, there are none.
(iii) $\binom{n+1}{2}^2$. [Each rectangle is formed with two pairs of parallel line segments.]
(iv) For each $k = 1, 2, \ldots, n$ there are $(n - k + 1)^2$ squares with side length k, so the total number is $n^2 + (n - 1)^2 + \cdots + 1^2 = \frac{n(n+1)(2n+1)}{6}$.
(v) $\sum_{k=3}^{n} \binom{n}{k} = 2^n - \frac{n^2+n+2}{2}$ for all $n \geq 3$.
(vi) If you consider any polygon that does not contain the vertex A, then by adjoining this vertex you can construct a polygon that does contain A. The inverse operation can be done with a k-gon only if $k \geq 4$; therefore, $m_1 > m_2$. The integer $m_1 - m_2$ gives the number of triangles with A as a vertex: $m_1 - m_2 = \binom{n-1}{2}$.
(vii) Construct the half-lines $A_9 A_i$ ($i = 5, 6, 7, 8$) and distinguish two cases: Inside one of the angles formed by a pair of neighboring half-lines there are two vertices of the square (for instance the vertices A_1, A_2 inside the angle $A_5 A_9 A_6$; see Figure 8a). Then the pentagon with vertices A_1, A_2, A_5, A_6, A_9 is convex. In the opposite case, inside the convex an-

gle determined by the half-lines A_9A_5 and A_9A_7 there are two vertices of the square, say A_2 and A_3 (Figure 8b). Then the pentagon with vertices A_2, A_3, A_5, A_7, A_9 is convex.

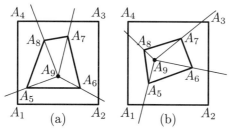

Figure 8

(viii) For each of the $\binom{n}{5}$ different quintuples of points chosen from the given n there exists, according to 1.8.(ii), at least one convex quadrilateral with vertices from that quintuple. Each such quadrilateral is counted at most $(n-4)$ times. Hence the number N of these quadrilaterals satisfies the inequality $(n-4) \cdot N \geq \binom{n}{5}$.

(ix) For $n < k(s+1)$ such a k-gon does not exist, and for $n \geq k(s+1)$ there are $\frac{n}{k} \cdot \binom{n-ks-1}{k-1}$ of them. [Generalize the argument from 1.7.(viii).]

1.11 (i) For $n = 3$ the assertion is clear. For $n \geq 4$ let $\mathcal{S}$ denote the set of intersections of the lines from $\mathcal{L}$, and from all the distances between the points in $\mathcal{S}$ and the lines in $\mathcal{L}$ choose the *smallest positive* distance, say from the point $X \in \mathcal{S}$ to the line $l \in \mathcal{L}$. If you assume that some three lines $l_A, l_B, l_C \in \mathcal{L}$ pass through the point X, and if you denote their intersections with the line l by A, B, and C, respectively, then on one of the two opposing half-lines into which the line l is divided by the orthogonal projection Y of the point X there are two of the three points A, B, C. For this pair derive, as in 1.10.(i), a contradiction to the choice of X and l.

(ii) It does not hold. A suitable counterexample is the countable set of all lattice points in the plane, that is, all points with integer coordinates in a given Cartesian coordinate system.

(iii) Suppose that not all n points lie on one line. Then there is a quadruple of noncollinear points A, B, C, D, some three of which (say A, B, and C) lie on a line l. Now show by contradiction that l also contains any other given point $E \neq D$. Assume that $E \notin l$. Then from the quadruple A, B, D, E you can choose two triples of noncollinear points A, B, D and A, B, E, so either the triple A, D, E, or the triple B, D, E is collinear. In the first case one cannot choose a collinear triple from the points B, C, D, E, and in the second case this is true for the points A, C, D, E.

(iv) Proceed as in 1.10.(iii). If A_4 lies inside $\triangle A_1A_2A_3$, then consider the angles at A_4; if A_1, A_2, A_3, A_4 form a convex quadrilateral, consider its interior angles. Every rectangle satisfies the conditions of the second part of the problem.

(v) Choose any point X in the plane and draw through it n lines that are parallel to the given ones. These lines divide the full angle at X into $2n$ angles, at least one of which is therefore no greater than $\frac{360°}{2n} = \frac{180°}{n}$.

(vi) If the planes from airports A and B land at O, then AB is the longest side of $\triangle ABO$, and therefore $|\sphericalangle AOB| > 60°$. Assume that planes from the airports $A_1, A_2, \ldots, A_n$ land at O. Then the half-lines $OA_1, OA_2, \ldots, OA_n$ are distinct, and each of the angles $A_i O A_j$ is greater than $60°$, so $n \cdot 60° < 360°$, which means that $n < 6$.

(vii) Clearly, $|\mathcal{S}| > 4$. An example of a suitable five-element set is the set of vertices of a regular pentagon.

1.13 (i) Proceed as in 1.12.(i); assume that assertion (a) does not hold and derive the truth of assertion (b). To do this, use the notation from there, and for each $k = 1, 2, \ldots, n$ find among the segments $I_{(k-1)n+1}, I_{(k-1)n+2}, \ldots, I_{kn}$ the segment U_k that is disjoint from each of the I_j, $j \geq kn + 1$. You then obtain $(n + 1)$ disjoint segments $U_1, U_2, \ldots, U_n, I_{n^2+1}$.

(ii) A point with the desired property is, for instance, the right-hand endpoint of the segment with the leftmost right-hand endpoint.

(iii) Change each arc $\widehat{A_i B_i}$ to the circular segment on the base $A_i B_i$, and apply Helly's theorem 1.1.(iii) to this system of circular segments. The assertion does not hold for any $\alpha > 180°$: As the four arcs consider two pairs of semicircles constructed on a pair of different diameters of the circle k.

(iv) If a point A cuts some of the given arcs in half, then none of the arcs contains the point B that lies on the circle "opposite" A. In the point B you can "cut" the circle, straighten it out into a line segment, and apply (ii).

(v) You may clearly assume that in the given system $\mathcal{S}$ of arcs there is none of size $180°$ or more. Choose any point A inside some arc $l_0 \in \mathcal{S}$ and assume that there is an arc $l \in \mathcal{S}$ that contains neither A nor its opposite point A'; it is clear that all such arcs l (forming a nonempty subsystem $\mathcal{S}' \subsetneq \mathcal{S}$) lie on the same semicircle p with endpoints A, A'. Subdivide the arcs $l \in \mathcal{S} \setminus \mathcal{S}'$ into groups $\mathcal{S}(A)$ and $\mathcal{S}(A')$ according to whether they contain the point A or A'. By the choice of A you have $\mathcal{S}(A) \neq \emptyset$; from all the arcs $l \cap p$, where $l \in \mathcal{S}(A)$, now choose the *smallest* one, and let it be the arc $\widehat{AB}$. Explain why $B \in l$ for all $l \in \mathcal{S}(A) \cup \mathcal{S}'$ and why $B \notin l$ implies $B' \in l$ for every arc $l \in \mathcal{S}(A')$, where B' is the point opposite B. Hence B, B' is a suitable pair of points.

(vi) (a) Construct a circle k with center A and radius 1, and map any point $X \neq A$ to the point X' that is the intersection of the half-line AX with the circle k. Note that the image of the given system of polygons is a system of arcs of the circle k that satisfies the conditions of Exercise (v).

(b) Construct an arbitrary half-line with endpoint A, and map any point $X \neq A$ to the point X' that is the intersection of this half-line with the circle k with center A and radius $|AX|$. Then use Exercise (ii).

(vii) Generalize the argument of the second solution of 1.12.(ii) and divide the set M of all m segments into at most $(n-1)$ classes $M_1, M_2, \ldots, M_{n-1}$ such that all segments in the same class M_i have a common point A_i. The construction of the classes M_k and the points A_k with coordinates γ_k can be done as follows: Set $\gamma_1 = \min\{\beta : [\alpha, \beta] \in M\}$ and $M_1 = \{[\alpha, \beta] \in M : \alpha \leq \gamma_1\}$; if for some $k \geq 1$ the set $M_k' = M \setminus (M_1 \cup M_2 \cup \cdots \cup M_k)$ is nonempty, then set $\gamma_{k+1} = \min\{\beta : [\alpha, \beta] \in M_k'\}$ and $M_{k+1} = \{[\alpha, \beta] \in M_k' : \alpha \leq \gamma_{k+1}\}$. Explain why $M_k' \neq \emptyset$ can occur only for $k < n - 1$.

(viii) The given situation cannot occur: Consider a straight line l that is not perpendicular to any of the segments under consideration, and construct orthogonal projections of the endpoints of all segments onto the line l. Then neither of the two outermost projections can be a projection of some interior point on one of the given segments.

(ix) Consider all systems of segments A_iB_j, where $1 \leq i, j \leq n$ and each one of the given points is the endpoint of exactly one of the segments. There are finitely many of them (namely $n!$), and therefore among them there is one for which the sum of the lengths of all the segments is *minimal* (if there is more than one such system, choose an arbitrary one from among them). Then no two segments of this system have a point in common. Indeed, if this were not the case, then two segments A_iB_p and A_jB_q would form the diagonals of a convex quadrilateral; these could then be replaced by the pair of opposite sides A_iB_q, A_jB_p, where you would have $|A_iB_q| + |A_jB_p| < |A_iB_p| + |A_jB_q|$, which contradicts our choice of the system of segments.

(x) $N = 100$. To prove that $N \leq 100$, consider a diameter AB of c and build a set $\mathcal{S}$ from 11 points close to A and 10 points close to B; the number of desired arcs is then $\binom{11}{2} + \binom{10}{2} = 100$. To prove that $N \geq 100$, apply repeatedly the following lemma (namely, with $k = 21, 20, 19, \ldots, 4, 3$): Given any k-element set $\mathcal{P}$ of points on a circle c, there exists a point $X \in \mathcal{P}$ such that at least $[(k-1)/2]$ arcs $\widehat{XY}$ ($Y \in \mathcal{P}$, $Y \neq X$) are of sizes that do not exceed $120°$. Proof of the lemma: Choose any point $X \in \mathcal{P}$ and divide the circle c into three arcs c_1, c_2, c_3 of sizes $120°$ such that X is the common endpoint of c_1 and c_2. If the point X does not possess the required property, then at most $[(k-1)/2]$ points of $\mathcal{P}$ (including the point X) lie on $c_1 \cup c_2$, hence at least $1 + [(k-1)/2]$ points of $\mathcal{P}$ lie on c_3, and each of them may serve as the desired point X.

1.15 (i) and (ii) Use induction on the number n of the sides of the polygon $\mathcal{M}$. In the induction step use "interior" diagonals, whose existence is guaranteed by 1.14.(iii).

(iii) Let k denote the number of acute angles of $\mathcal{M}$. The sum s of interior angles satisfies, on the one hand, $s = (n-2) \cdot 180°$ by (ii), and on the other hand, the bound $s < k \cdot 90° + (n-k) \cdot 360°$, which implies the inequality $3k < 2n + 4$. Hence for $n = 3m$ you have $k \leq 2m + 1$, for $n = 3m + 1$ it is $k \leq 2m + 1$, and for $n = 3m + 2$, finally, $k \leq 2m + 2$. From Figures 9 (a)–(c) for $m = 3$ it follows that equality may occur in the above inequalities, and furthermore, it is clear how to construct such examples for each $m \geq 3$.

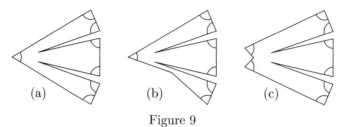

Figure 9

(iv) If the line l cuts every side of the polygon $\mathcal{M}$, this means that the sequence of vertices on its boundary alternates between the two half-planes formed by l. This is not possible if n is odd. From Figure 10 it is obvious how to construct a suitable example for each even $n \geq 4$.

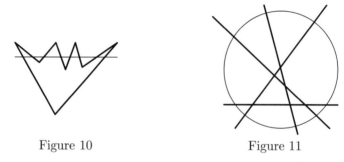

Figure 10   Figure 11

2.2 (i) $p_1(1) = 2$, $p_1(n + 1) = p_1(n) + 1$; therefore, $p_1(n) = n + 1$.

(ii) $p_2(1) = 2$, $p_2(n + 1) = p_2(n) + 2$; therefore, $p_2(n) = 2n$.

(iii) Any two neighboring unbounded regions are separated by half-lines; on the other hand, each one of two half-lines that are parts of a "dividing" line divides exactly two unbounded regions. Hence $N_2 = 2n$, which means that $N_1 = p(n) - 2n = \frac{1}{2}(n^2 - 3n + 2)$, where the expression for $p(n)$ from the solution of 2.1.(i) has been used.

(iv) All intersections of the given lines can be placed on some circular disk with sufficiently large radius (Figure 11), where the boundary circle is divided into $4k$ arcs by these lines. Denote the endpoints of the arcs, in order, by $X_1, X_2, \ldots, X_{4k}$ (the given lines are then $X_1 X_{2k+1}, X_2 X_{2k+2}, \ldots, X_{2k} X_{4k}$). It is clear that two neighboring arcs cannot belong to angular regions; hence the number of such regions is at most $2k$. If you assume that the number of angular regions is equal to $2k$, then you can choose a numbering of points such that the arcs $\widehat{X_1 X_2}, \widehat{X_3 X_4}, \widehat{X_5 X_6}, \ldots, \widehat{X_{4k-1} X_{4k}}$ belong to the angular regions. But then the lines $X_1 X_{2k+1}$ and $X_2 X_{2k+2}$ bound angular regions containing the arcs $\widehat{X_1 X_2}$ and $\widehat{X_{2k+1} X_{2k+2}}$, which contradicts the general position of the given $n = 2k$ lines.

(v) (a) For the number $P_1(m, n)$ of regions into which the plane is divided by the lines, you have by (6) the identity $P_1(m, n) = mn + 2(m + n) - 1 = (m + 2)(n + 2) - 5$. From $P_1(m, n) = 1987$ it follows that $(m + 2)(n + 2) = 1992 = 2^3 \cdot 3 \cdot 83$. The possible products are $1992 = 2 \cdot 996 = 3 \cdot 664 = 4 \cdot 498 = 6 \cdot 332 = 8 \cdot 249 = 12 \cdot 166 = 24 \cdot 83$. From this you get the following 12 possibilities for (m, n): $(1, 662)$, $(2, 496)$, $(4, 330)$, $(6, 247)$, $(10, 164)$, $(22, 81)$, and six more obtained by switching the orders.

(b) For the number $P_2(m, n)$ of regions into which the plane is divided, the relations $P_2(0, n) = 2n$ and $P_2(m, n) = P_2(m - 1, n) + n + 1$ for $m \geq 1$ imply that $P_2(m, n) = 2n + m(n + 1) = (m + 2)(n + 1) - 2$. From $P_2(m, n) = 1987$ it then follows that $(m + 2)(n + 1) = 1989 = 3^2 \cdot 13 \cdot 17$. The possible products are $1989 = 3 \cdot 663 = 9 \cdot 221 = 13 \cdot 153 = 17 \cdot 117 = 39 \cdot 51$, and another five obtained by switching the order of the factors. Hence $(m, n) \in \{(1, 662), (7, 220), (11, 152), (15, 116), (37, 50), (49, 38), (115, 16), (151, 12), (219, 8), (661, 2)\}$.

(vi) Denote the numbers of lines of the bundle with center A (respectively B, respectively C) by a, b, c, respectively. Let $P(a, b, c)$ be the maximal number of regions into which the plane is divided by these $a + b + c = 18$ lines. From the relations $P(a, 0, 0) = 2a$, $P(a, b, 0) = ab + 2(a + b) - 1$, $P(a, b, 1) = P(a, b, 0) + a + b + 1$, $P(a, b, c) = P(a, b, c - 1) + a + b + 2$ for $c > 1$ you can derive for $a, b, c \in \mathbb{N}$ the expression $P(a, b, c) = ab + bc + ca + 2(a + b + c) - 2$. From the inequality $ab + bc + ca \leq \frac{1}{3}(a + b + c)^2$ (which you should verify), in which equality occurs for $a = b = c$, you obtain the estimate $P(a, b, c) \leq \frac{1}{3}(a + b + c)^2 + 2(a + b + c) - 2$. If $a + b + c = 18$ for positive a, b, and c, then $P(a, b, c) \leq \frac{1}{3} \cdot 18^2 + 2 \cdot 18 - 2 = 142$. Furthermore, $P(18, 0, 0) = 36$, and for positive a and b, $P(a, b, 0) \leq 116$ if $a + b = 18$ (this follows from $ab \leq \frac{1}{4}(a + b)^2$). The number of regions will be equal to 142 if and only if $a = b = c = 6$, that is, when each of the bundles consists of exactly 6 lines and no three of them have a common point different from A, B, C.

2.4 (i) If you denote the desired number by N, then $3 \leq N \leq 10$, where equality $N = k$ can hold for any $k = 3, 4, \ldots, 10$. [As in 2.3.(ii), derive the estimate $N \leq 1 + 1 + (4 + 4)$, and for all possible values of N give an example of a suitable configuration.]

(ii) Use an approach similar to that in 2.3.(ii) to derive the estimate $M(n) \leq 2n + 2$. The existence of the value $M(n) = 2n + 2$ can be shown as follows: To a regular n-gon with center S inscribe (respectively circumscribe) the circle k_1, respectively k_2, with center S and radius r_1, respectively r_2. The circle k with center S and radius $r = \frac{1}{2}(r_1 + r_2)$ and the boundary of the n-gon divide the plane into $2n + 2$ regions.

(iii) Let $N(p, k)$ denote the desired largest possible number of regions divided by p lines and k circles. By 2.3.(i) you have $N(0, k) = k^2 - k + 2$ for $k \geq 1$. If you add the first line, then the k circles divide it into at most $2k + 1$ parts, which implies $N(1, k) \leq N(0, k) + 2k = k^2 + k + 2$

(recall that both unbounded parts of the line divide the same region). By considering the addition of new lines you obtain for all $p \geq 2$ the estimate $N(p, k) - N(p - 1, k) \leq 2k + p$, and thus for any $p, k \in \mathbb{N}$, $N(p, k) \leq k^2 + k + 2 + \sum_{i=2}^{p}(2k + i) = k^2 - k + 1 + 2kp + \frac{1}{2}p(p + 1)$, where equality occurs if there is a distribution of p lines and k circles such that no three of these $(p + k)$ curves go through the same point, no two of the straight lines are parallel, and every circle has exactly two points in common with each one of the remaining $(p + k - 1)$ curves. That can be constructed as follows: First construct p straight lines in general position, and choose a circular disk $\mathcal{D}$ such that all $\binom{p}{2}$ intersections of these lines lie in the interior of $\mathcal{D}$. Then choose the centers for the circles on an arbitrary diameter AB of $\mathcal{D}$, and as their common radius choose $|AB|$.

2.6 (i) The segments emanating from B divide the triangle into $(n + 1)$ smaller triangles. Each new segment emanating from C increases the number of regions by $(n + 1)$. Hence the desired number is $(n + 1)^2$.

(ii) Let us continue with the solution of (i). Having drawn all the segments emanating from B and C, each new segment emanating from A increases the number of regions by $(2n + 1)$. The answer is $(n + 1)^2 + n(2n + 1) = 3n^2 + 3n + 1$.

(iii) If you consider the given points as vertices of a convex n-gon (whose sides are the chords) inscribed into a circular disk $\mathcal{D}$, then the number $d(n)$ of regions into which $\mathcal{D}$ is divided satisfies the estimate

$$d(n) \leq f(n) + n = \frac{(n - 1)(n - 2)(n^2 - 3n + 12)}{24} + n = 1 + \binom{n}{2} + \binom{n}{4},$$

where by (8), $f(n)$ is the greatest number of regions into which a convex n-gon is divided by its diagonals (see 2.5.(ii)). The value $d(n) = 1 + \binom{n}{2} + \binom{n}{4}$ is possible; consider a polygon $\mathcal{M}$ inscribed in $\mathcal{D}$, such that no three of its diagonals meet at the same interior point of $\mathcal{M}$ (its existence follows from the construction in 1.2.(i)).

(iv) After each break the number of pieces increases by 1. Since initially you are given one piece of chocolate, and in the end you obtain mn pieces, $mn - 1$ breaks are necessary, independently of the strategy of breaking.

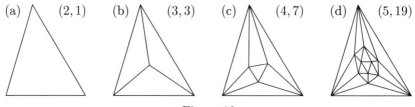

(a) $(2, 1)$ (b) $(3, 3)$ (c) $(4, 7)$ (d) $(5, 19)$

Figure 12

(v) Let n denote the desired number of triangles, p the number of endpoints of the segments under consideration, each of which is the endpoint of

k "dividing" segments, and denote the total number of these segments by h. You clearly have $p \cdot k = 2h$, and by considering the interior angles of the triangles you obtain $180° \cdot n = 180° + 360° \cdot (p-3)$, that is, $2p = n+5$. All h segments are sides of the $(n+1)$ triangles, so $3(n+1) = 2h$. Altogether, you get $n = \frac{5k-6}{6-k}$, which in view of $n > 0$ means that $k \in \{2,3,4,5\}$; then n takes on the values 1,3,7,19, respectively. The largest n is therefore 19, and Figure 12 (a)–(d) shows the solutions for all possible pairs (k,n).

(vi) Proceed by induction on k; in the induction step do not forget to distinguish between two possible positions of a point X to be added: It can lie either inside a previously obtained triangle PQR, or on one of its sides. In both cases the number of triangles increases by 2, as is clear from Figure 13.

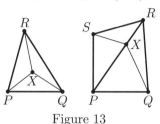

Figure 13

(vii) Assume that such a division of a square $ABCD$ with side a is possible; then each region has area $\frac{a^2}{9}$. Let P be one the two given points. Both triangles ABP, CDP are divided into several regions, and the sum of their areas is $k \cdot \frac{a^2}{9}$, where $k \in \mathbb{N}$. But the sum of these areas must be $\frac{a^2}{2}$, that is, $k = \frac{9}{2}$, which is a contradiction.

(viii) First note that each one of the lines cuts two opposite sides of the square. From the given ratio derive that each line goes through one of the four points that divide the connecting line segments between the centers of opposite sides in the ratio $2 : 3$. By the pigeonhole principle there is at least one of these points through which three of the given lines will go.

(ix) Use any of the methods described in 2.5.(iv).

(x) This is not possible. Connect the centers of neighboring fields on the outer rows and columns by 28 line segments. If you assume that a division with the required property is possible, then each of the line segments must be cut by one of the 13 given lines. However, each line can intersect at most two of the 28 segments, so by the pigeonhole principle some segment is not cut by any of the lines. The centers of the two fields connected by this segment therefore lie in the same part.

2.8 (i) $q(1) = 2$, $q(n) = q(n-1) + 2(n-1)$; therefore, $q(n) = n^2 - n + 2$.

(ii) $q(1) = 2$, $q(n) = q(n-1) + 2$; therefore, $q(n) = 2n$.

(iii) The planes are in general position; see 2.7.(i). Therefore, the desired number is $r(4) = 15$, where (15) has been used.

(iv) Let $P(p,q,r)$ denote the number of regions into which space is divided by three systems of parallel planes with p (respectively q, respectively r) planes, and such that no two planes of different systems are parallel. Derive the relation $P(p,q,r) = (p+1)(q+1)(r+1)$. In particular, $P(2,2,2) = 27$.

(v) Generalize the arguments from 2.1.(ii): First show that there are at most three planes such that the intersections of any three of the planes from the problem lie in one half-space bounded by the given plane. For such a plane consider the triple of planes whose intersection has the least distance from this plane. These four planes determine a tetrahedron disjoint from all other given planes. Similarly, the remaining planes give at least two tetrahedra (in opposite half-spaces). Each tetrahedron constructed in this way is associated with at most 4 planes. From these considerations you obtain for the number N of tetrahedra the estimate $N \geq \frac{2(n-3)+3}{4} = \frac{2n-3}{4}$, which implies that $N \geq \frac{n-1}{2}$.

3.2 (i) Divide the object $\mathcal{M}_0 = \mathcal{M}_1 \cup \mathcal{M}_2 \cup \cdots \cup \mathcal{M}_n$ into 2^n disjoint parts, just as in 3.1 in the proof of (17). Then use the fact that for a given $i \in \{1, 2, \ldots, n-1\}$ you have

$$\sum_{r=1}^{i}(-1)^{r+1}\binom{n}{r} = \sum_{r=1}^{i}(-1)^{r+1}\left(\binom{n-1}{r-1} + \binom{n-1}{r}\right)$$

$$= \sum_{r=1}^{i}(-1)^{r+1}\binom{n-1}{r-1} + \sum_{r=1}^{i}(-1)^{r+1}\binom{n-1}{r}$$

$$= \binom{n-1}{0} + (-1)^{i+1}\binom{n-1}{i} = 1 + (-1)^{i+1}\binom{n-1}{i}.$$

(ii) Use the inequality (19) with $n = 3$ and $i = 2$.

(iii) Use the inequality (19) with $n = 4$ and $i = 2$.

(iv) Use the inequality (19) with $n = 7$ and $i = 2$.

(v) Use the inequality (21) with $n = 3$; from this it follows that the common part of the three "inner" rectangles has an area of at least 1 cm^2.

(vi) The inequality (19) with $n = 5$ and $i = 2$ is not sufficient for the required proof. The assertion follows from the inequality

$$\sum_{\{i,j\}} |\mathcal{M}_i \cap \mathcal{M}_j| \geq 2 \cdot \sum_{i=1}^{5} |\mathcal{M}_i| - 3 \cdot \left|\bigcup_{i=1}^{5} \mathcal{M}_i\right|,$$

where on the left there are $\binom{5}{2} = 10$ summands, which correspond to all choices of two-element subsets of the set $\{1, 2, \ldots, 5\}$. The inequality can be shown in a similar way as (21) in 3.1.(ii), by using the estimate $\binom{r}{2} \geq 2r - 3$, $r \in \{2, 3, 4, 5\}$, and the inequality $0 > 2 - 3$ for $r = 1$.

(vii) It is even true that if $\mathcal{S}_n$ is an arbitrary n-element subsystem of the system $\mathcal{S}_{2n}$, then one of the systems $\mathcal{S}_n$ or $\mathcal{S}'_n = \mathcal{S}_{2n} \setminus \mathcal{S}_n$ has the required property.

(viii) If $\mathcal{S}_n = \{\mathcal{U}_1, \mathcal{U}_2, \ldots, \mathcal{U}_n\}$ is the system in question, then there are indices $j_1, j_2, \ldots, j_k$ such that the objects $\mathcal{U}_{j_1}, \mathcal{U}_{j_2}, \ldots, \mathcal{U}_{j_k}$ satisfy

$$|\mathcal{U}_{j_1} \cup \mathcal{U}_{j_2} \cup \cdots \cup \mathcal{U}_{j_k}| \geq \frac{k}{n}|\mathcal{U}_1 \cup \mathcal{U}_2 \cup \cdots \cup \mathcal{U}_n| \geq \frac{k}{n}|\mathcal{M}|$$

(and therefore you may set $\mathcal{S}_k = \{\mathcal{U}_{j_1}, \mathcal{U}_{j_2}, \dots, \mathcal{U}_{j_k}\}$). This assertion follows from the inequality

$$\sum |\mathcal{U}_{j_1} \cup \mathcal{U}_{j_2} \cup \cdots \cup \mathcal{U}_{j_k}| \geq \binom{n-1}{k-1} \cdot |\mathcal{U}_1 \cup \mathcal{U}_2 \cup \cdots \cup \mathcal{U}_n|, \qquad (25)$$

where the sum on the left consists of $\binom{n}{k}$ summands, which correspond to the choice of all k-element subsets $\{j_1, \dots, j_k\}$ of $\{1, 2, \dots, n\}$. The inequality (25) already implies the existence of a k-tuple of indices $j_1, j_2, \dots, j_k$ satisfying the estimate

$$|\mathcal{U}_{j_1} \cup \mathcal{U}_{j_2} \cup \cdots \cup \mathcal{U}_{j_k}| \geq \frac{\binom{n-1}{k-1}}{\binom{n}{k}} |\mathcal{U}_1 \cup \mathcal{U}_2 \cup \cdots \cup \mathcal{U}_n| = \frac{k}{n} |\mathcal{U}_1 \cup \mathcal{U}_2 \cup \cdots \cup \mathcal{U}_n|.$$

It remains to prove the inequality (25): Divide the set $\mathcal{M}_0 = \mathcal{U}_1 \cup \mathcal{U}_2 \cup \cdots \cup \mathcal{U}_n$ into 2^n disjoint parts of the type $\mathcal{X}_1 \cap \mathcal{X}_2 \cap \cdots \cap \mathcal{X}_n$, where $\mathcal{X}_i = \mathcal{U}_i$ or $\mathcal{X}_i = \mathcal{M}_0 \setminus \mathcal{U}_i$. Since on the right of (25) the area of each such part is counted $\binom{n-1}{k-1}$ times, it remains to show that it is counted at least as often on the left. For each nonempty part $\mathcal{X}_1 \cap \mathcal{X}_2 \cap \cdots \cap \mathcal{X}_n$ there is an index t such that $\mathcal{X}_t = \mathcal{U}_t$, and therefore the area of each such part is counted in every summand of the type $|\mathcal{U}_{j_1} \cup \mathcal{U}_{j_2} \cup \cdots \cup \mathcal{U}_{j_k}|$, where t is one of the indices $j_1, j_2, \dots, j_k$. There are exactly $\binom{n-1}{k-1}$ sets of such indices, since for each fixed index t one has to choose another $(k-1)$ from the remaining $(n-1)$.

3.4 (i) In the case of an obtuse or a right triangle the longest side is a diameter of the desired disk. In the case of an acute triangle ABC it is the disk $\mathcal{D}(S, R)$ determined by the corresponding circle circumscribing the triangle ABC. [If there were a disk $\mathcal{D}(S', r)$ covering $\triangle ABC$ such that $r < R$, you would have $S' \in \mathcal{D}(A, r) \cap \mathcal{D}(B, r) \cap \mathcal{D}(C, r)$. However, this intersection is empty: Note that from $S' \in \mathcal{D}(A, r) \cap \mathcal{D}(B, r)$ and $r < R$ it follows that $|\sphericalangle AS'B| > |\sphericalangle ASB|$.]

(ii) By 3.3.(i) it suffices to prove the assertion for $\mathcal{M} = \{X_1, X_2, X_3\}$. If the points X_1, X_2, X_3 form the vertices of a right or an obtuse triangle, then as diameter of the covering disk choose its largest side and use the bound $1 < \frac{2}{\sqrt{3}}$. If $\triangle X_1 X_2 X_3$ is acute, cover it with the disk bounded by the circle circumscribing $\triangle X_1 X_2 X_3$. By the law of sines for the diameter $2R$ of such a disk you get $2R = \frac{a}{\sin \alpha}$, where α denotes the largest angle of $\triangle X_1 X_2 X_3$, and a the length of the opposite side. You have $90° > \alpha \geq 60°$ (that is, $\sin \alpha \geq \frac{\sqrt{3}}{2}$) and $a \leq 1$, which implies $2R = \frac{a}{\sin \alpha} \leq \frac{2}{\sqrt{3}}$.

(iii) Let T denote the centroid of $\triangle ABC$. Then one of the disks $\mathcal{D}_1, \mathcal{D}_2, \mathcal{D}_3$ contains some two of the points A, B, C, T (by the pigeonhole principle), which means that $2R \geq |AT| = \frac{\sqrt{3}}{3} a$. The value $R = \frac{\sqrt{3}}{6} a$ is possible, and three disks with radius R constructed on AT, BT, CT as diameters cover the whole $\triangle ABC$.

(iv) Assume that $|BC| \leq |AC|$, that is, $\alpha \leq 45°$, and distinguish the following two cases:

(a) $30° \leq \alpha \leq 45°$. If D is the midpoint of the hypotenuse AB, then $|AD| = |BD| = |CD| = \frac{c}{2} \leq |BC| \leq |AC|$ (Figure 14a), and therefore any two of the four points A, B, C, D have distance between them at least $\frac{c}{2}$, which means that $R \geq \frac{c}{4}$. You obtain a covering for $R = \frac{c}{4}$ if you construct disks with AD, BD, and CD as diameters; they cover $\triangle AKD, \triangle BDL$, and the rectangle $CKDL$, respectively, where K, L are the midpoints of the sides AC and BC.

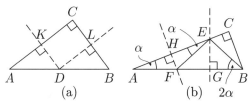

Figure 14

(b) $0 < \alpha < 30°$. Now $2\alpha < |{\sphericalangle}ABC|$, and therefore on the line segment AC there is an interior point E such that $|{\sphericalangle}ABE| = 2\alpha$ (Figure 14b). Set $d = |BE|$. Since $|{\sphericalangle}AEB| = 180° - 3\alpha > \alpha$, there is a point F on the segment AB such that $|{\sphericalangle}AEF| = \alpha$. Then $d = |BE| = |EF| = |FA|$, $d < |AE|$, $d < |BF|$, and therefore any two distinct points from the quadruple A, B, E, F have a distance between them of at least d, and thus $R \geq \frac{d}{2}$. On the other hand, it is easy to show that the disks that have BE, EF, and AF as diameters will cover the whole $\triangle ABC$. Hence $R = \frac{d}{2} = \frac{c}{2(1+2\cos 2\alpha)}$.

(v) $R = \frac{a \cdot \sqrt{5}}{4}$. [If two neighboring vertices of the square lie in different disks, use the pigeonhole principle for these two vertices and the midpoint of the opposite side (otherwise, the assertion is clear).]

(vi) $R = \frac{a \cdot \sqrt{2}}{4}$. [Use the pigeonhole principle for five points, namely, the center of the square and its vertices.]

(vii) It is possible. [An example of a suitable covering is shown in Figure 15, where the square $ABCD$ is divided into five nonoverlapping rectangles with the same diagonal $u = \frac{\sqrt{2257}}{72}a < \frac{a\sqrt{2}}{2}$. If you circumscribe a circle around each of these rectangles, then you obtain a covering by the corresponding disks with radius $\frac{u}{2}$.]

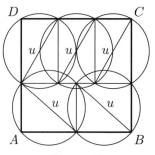

Figure 15

(viii) If the disk $\mathcal{D}$ with radius R is covered by three congruent disks with radius r, then $R \leq \frac{2r}{\sqrt{3}}$; the value $R = \frac{2r}{\sqrt{3}}$ is possible, for instance, when the disk $\mathcal{D}$ circumscribes an equilateral triangle with side length $2r$, and three disks with radius r have their centers in the midpoints of the sides of this triangle. [Each of the covering disks $\mathcal{D}_1, \mathcal{D}_2, \mathcal{D}_3$ cuts an arc from the boundary circle of $\mathcal{D}$, and the union of these is the whole circle; therefore, one of the arcs (with endpoints A, B, say) has size $\alpha \geq 120°$. Then in the case $\alpha < 180°$ you have by the law of sines $2r \geq |AB| = 2R \sin \frac{\alpha}{2} \geq 2R \sin 60° = R \cdot \sqrt{3}$, while in the case $\alpha \geq 180°$ the arc AB contains the endpoints of some diameter of the disk $\mathcal{D}$; that is, you even have $r \geq R$.]

(ix) If the given $\triangle ABC$ is acute or right, then it is easy to show that $R = r$, where r is the diameter of the circle circumscribing $\triangle ABC$ (its center S must lie in one of the covering disks). If $\triangle ABC$ is obtuse, then you can show that $R \geq \frac{a}{2\cos\beta}$, where a is the second-longest side, and β the smallest angle of $\triangle ABC$ (adjacent to the side a; see Figure 16).

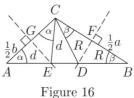

Figure 16

If D is the intersection of the perpendicular bisector of the side BC with the side AB, then $|AD| \geq |BD| = |CD| = \frac{a}{2\cos\beta}$. The point D must lie in one of the disks $\mathcal{D}_i$, and therefore $R \geq \min(|AD|, |BD|, |CD|) = \frac{a}{2\cos\beta}$. Next you show that three disks with radius $R = \frac{a}{2\cos\beta}$ can indeed cover the whole of $\triangle ABC$. You can do this by dividing the triangle into smaller triangles (some further points are introduced in Figure 16). It is clear that $\triangle BDF$ is covered by the disk $\mathcal{D}_2(B, R)$, and similarly $\triangle CDF$ is covered by $\mathcal{D}_3(C, R)$. The longest side of $\triangle AEG$ is $d = \frac{b}{2\cos\alpha}$, and with an easy calculation (using the inequality $\alpha \geq \beta$) you can see that $d \leq R$, and thus $\triangle AEG$ and $\triangle CEG$ are covered by $\mathcal{D}_1(A, R)$ and $\mathcal{D}_3(C, R)$, respectively. There remains $\triangle CED$, for which $|CD| = R$ and $|CE| = d \leq R$ hold. The disk $\mathcal{D}_3(C, R)$ therefore contains the points C, D, E, and thus the whole of $\triangle CDE$.

(x) Consider the n nonoverlapping disks $\mathcal{D}_i$ with diameter 1 whose centers lie in $\mathcal{M}$. Replace every disk $\mathcal{D}_i$ with a concentric disk $\mathcal{D}_i^*$ with radius 1. If there is a point $X \in \mathcal{M}$ not covered by the system of disks $\mathcal{D}_i^*$, then its distance to the center of every disk is at least 1. Therefore, the disk $\mathcal{D}(X, \frac{1}{2})$ does not overlap with any of the $\mathcal{D}_i$; however, this is a contradiction to the definition of n. Therefore, the n disks $\mathcal{D}_i^*$ cover the polygon $\mathcal{M}$, and thus $m \leq n$.

3.6 (i) Let the given disks be $\mathcal{D}_i(S_i, r)$, assume that $A \notin \{S_1, S_2, \ldots, S_6\}$, and consider the line segments AS_i as well as the angles between them. Then one of the angles $S_i A S_j$ is at most equal to $60°$. Hence $|S_i S_j| \leq \max\{|AS_i|, |AS_j|\} \leq r$; that is, $S_j \in \mathcal{D}_i$.

(ii) Solve the following more general problem: Show that the area of any parallelogram $ABCD$ enclosed in the triangle KLM is not greater than half of the area of this triangle. Extend two opposite sides AD and BC of the parallelogram such that they cut the sides of $\triangle KLM$; at least two of the four intersection points with the boundary of $\triangle KLM$ lie on one of its sides, say $A_1, B_1 \in KL$. Construct the points D_1, C_1 on the half-lines $A_1 D, B_1 C$, respectively, such that the parallelograms $ABCD$ and $A_1 B_1 C_1 D_1$ have the same area. Denote by X and Y the intersections of the line C_1D_1 with the sides KM and LM, respectively. On the side KL consider the point Z such that $KZYX$ is a parallelogram; its area is clearly not less than that of the given parallelogram. It remains to show that the area of the parallelogram $KZYX$ is not greater than half of the area of $\triangle KLM$. If $|LY| < \frac{1}{2}|LM|$, denote by L' the point symmetric to L with respect to Y, and $K' \in KM$ such that $KLL'K'$ is a trapezoid with an area twice that of the parallelogram $KZYX$ (use the properties of the median of the trapezoid). If $|LY| > \frac{1}{2}|LM|$, consider an analogous trapezoid $KMM'K'$; if finally Y is the midpoint of LM, then the area of $KZYX$ equals half of that of $\triangle KLM$.

(iii) Solve the following equivalent problem: "Inscribe a square of maximal area into a given triangle." Use the result of (ii), where equality in $2S_1 \leq S_2$ is possible for a right isosceles triangle with lateral sides of length $2a$. The corresponding covering is then easy to describe.

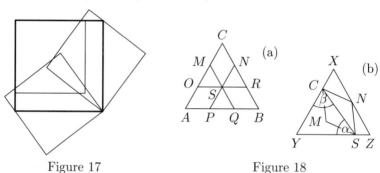

Figure 17   Figure 18

(iv) This is possible; Figure 17 shows an example of a suitable covering.

(v) The answer is $b = \frac{3}{2}a$. [An example of a suitable covering is given by $\triangle AQM$, $\triangle BNP$, and $\triangle COR$ in Figure 18a. Prove the inequality $b \leq \frac{3}{2}a$ by contradiction. Assume that for $b > \frac{3}{2}a$ the equilateral triangles $\mathcal{T}_1, \mathcal{T}_2, \mathcal{T}_3$ with side length a cover the equilateral triangle ABC with side length b. Note that each of the points A, B, C is then covered by a different triangle. Without loss of generality you may then assume that the centroid S of the triangle ABC lies, along with C, in $\mathcal{T}_1$. Then neither of the points M and N (see Figure 18a) can lie in $\mathcal{T}_2$ or in $\mathcal{T}_3$, and thus $\mathcal{T}_1$ covers the rhombus $CMSN$ with side length $\frac{b}{3}$. Let XYZ be the triangle $\mathcal{T}_1$ or its homothetic "reduction" such that the points C, S, and N lie on the sides XY, YZ, and ZX, respectively, while the point M belongs to $\triangle CSY$, as shown in Figure 18b (interchange the points M and N if necessary). Since the sizes $\alpha = |\sphericalangle CSY|$ and $\beta = |\sphericalangle SCY|$ satisfy $\alpha \geq |\sphericalangle CSM| = 30°$, $\beta \geq |\sphericalangle SCM| = 30°$, and $\alpha + \beta = 120°$, we have $\alpha \leq 90°$. Using the law of sines you obtain

$$a \geq |YZ| = |YS| + |SZ| = \sin(120° - \alpha)\frac{|CS|}{\sin 60°} + \sin(\alpha - 30°)\frac{|SN|}{\sin 60°},$$

and thus

$$a \geq \tfrac{4b}{3\sqrt{3}}\left(\sin 60° \cos(\alpha - 30°) + \cos 60° \sin(\alpha - 30°)\right)$$
$$= \tfrac{4b}{3\sqrt{3}} \sin(\alpha + 30°) \geq \tfrac{2b}{3},$$

since $\alpha \in [30°, 90°]$, which is a contradiction.]

(vi) Not always. [Proceed as in 3.5.(v) and find an infinite sequence $\mathcal{R}_1, \mathcal{R}_2, \ldots$ of rectangles in which no one rectangle can be covered by the remaining ones. One possible example can be described as follows: For any $n \geq 1$ let $\mathcal{R}_n$ be the rectangle with sides $a_n = 4^{n-1}$, $b_n = \tfrac{1}{4^{2n-1}}$, and with area $S_n = \tfrac{1}{4^n}$. As in 3.5.(v) show that the area of $\mathcal{R}_n$ that can be covered by the rectangles $\mathcal{R}_1, \mathcal{R}_2, \ldots, \mathcal{R}_{n-1}$ is less than $\tfrac{1}{3}S_n$, and also that the sum $S_{n+1} + S_{n+2} + \cdots$ of the areas of the remaining rectangles satisfies

$$S_{n+1} + S_{n+2} + \cdots = \frac{1}{4^{n+1}} + \frac{1}{4^{n+2}} + \cdots = \frac{1}{3} \cdot \frac{1}{4^n} = \frac{1}{3}S_n.]$$

(vii) From all the triangles with vertices in $\mathcal{M}$ (there is a finite number of them), choose $\triangle ABC$ of *largest* area (if there is more than one, choose one at random) and show that the set $\mathcal{M}$ is covered by a triangle that has the points A, B, C as midpoints of its sides.

3.8 (i) Proceed as in 3.7.(i); the result follows from the inequality $20 \cdot (1 + 4 + \pi) < 13^2$.

(ii) Consider the orthogonal projections of all the circles onto one of the sides of the square; the sum of the lengths in centimeters of all projections is $\tfrac{10}{\pi} > 3 \cdot 1$, which means that on that side there is a point X that lies on the projection of at least four circles. The perpendicular to this side going through the point X then has the desired property.

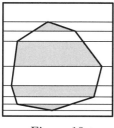

Figure 19

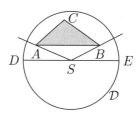

Figure 20

(iii) Draw lines through all the vertices of $\mathcal{M}$ parallel to a chosen side of the square (see Figure 19). This divides $\mathcal{M}$ into a system of trapezoids and triangles. It suffices to show that the base of some trapezoid or the side of some triangle parallel to the chosen side of the square is longer than 0.5 cm. If you assume that such a situation does not occur, then the sum S of the areas (in cm²) of all these parts satisfies $S < 0.5 \cdot 1 = 0.5 < S_0$, where S_0 is the area of the polygon $\mathcal{M}$, but this is impossible.

(iv) Assume that the two nonoverlapping triangles $\mathcal{T}_1, \mathcal{T}_2$ with areas $S_i >$ 1 cm^2 can be placed inside the disk $\mathcal{D}(S, 1$ cm$)$. You may furthermore assume that the center S is not an interior point of the triangle $\mathcal{T}_1$; then $\mathcal{T}_1$ lies in an angular region of size $\alpha \leq 180°$ with vertex in S (see Figure 20). Let the vertices A, B of this triangle lie on the legs of the angle, and let DE be the diameter of the disk $\mathcal{D}$ parallel to AB. Since $|DE| \geq |AB|$, the area of $\triangle DEC$ (where C is the third vertex of $\mathcal{T}_1$) is not less than the area of $\mathcal{T}_1$, so it is greater than 1 cm^2. However, the height h in $\triangle DEC$ on the side DE of length 2 cm satisfies $h \leq 1$ cm; the area of the triangle is therefore not greater than 1 cm^2, and this is a contradiction.

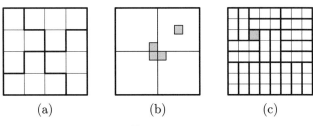

(a)   (b)   (c)

Figure 21

3.10 (i) In Figure 21a one-quarter of the chessboard is tiled in the required fashion.

(ii) Proceed by induction on n; as induction step divide the chessboard into four "quarters" of size $2^n \times 2^n$. Then tile, as in Figure 21b, those three "quarters" that do not contain the cut-out square.

(iii) A solution is given in Figure 21c.

(iv) This is not possible. Consider a system $\mathcal{P}$ of five horizontal and five vertical lines going through the segments that divide the chessboard into its individual fields. Then for any tiling of the chessboard there is some $p_0 \in \mathcal{P}$ that does not cut through any domino. [To obtain a contradiction, assume that every line $p \in \mathcal{P}$ cuts through some domino. Explain why each line cuts through an even number of dominoes, and thus through at least two of them. But this is a contradiction, since $|\mathcal{P}| = 10$, while the total number of dominoes is only $\frac{36}{2} = 18$.]

2	3			3		2	2	3			3		2	3			3
3		2	2			3			2	2			3		2	2	
		2 --- 2						2 --- 2						2 --- 2			
3		2	2			3			2	2			3		2	2	
(a)		$6k+2$				(b)		$6k+4$				(c)		$6k+5$			

Figure 22

(v) For $n \in \{2, 4, 5\}$ you can easily construct a required tiling. For $n > 6$ distinguish the four cases $n = 6k + i$, where $i \in \{1, 2, 4, 5\}$, and consider these cases separately. For $n = 6k+1$ the assertion follows from the solution of 3.9.(ii). For the induction step always assume that the square that can be tiled according to the induction hypothesis covers the part of the chessboard from which the corner piece is removed. In all three cases it is necessary to find a tiling of a square A of size 6×6 and of two congruent rectangles B, C of size $6 \times (6k + i)$, where $i \in \{2, 4, 5\}$; however, this is not difficult to do (see Figures 22 a–c).

3.12 (i) Tile the given rectangle with 49 squares of size 5×5 and five 7×1 rectangles. By the pigeonhole principle one of these rectangles, which have congruent diagonals of length $\sqrt{50}$, contains two of the given 55 points.

(ii) Using lines connecting the midpoints of the sides, divide the given triangle $\mathcal{T}$ into four congruent equilateral triangles that tile $\mathcal{T}$.

(iii) Tile the square with 25 squares of side length 0.2; each one of them can be covered with a disk of radius $0.1 \cdot \sqrt{2}$, and in one of these disks there are at least three of the given points. In view of the inequality $0.1 \cdot \sqrt{2} < \frac{1}{7}$, this completes the proof.

(iv) By the result of Exercise 2.6.(vi), with $k = 7$, the square can be tiled with $2 \cdot 7 + 2 = 16$ triangles with the desired property. Then one of them has an area of at most $\frac{1}{16}$.

(v) Tile the chessboard with 16 squares of size 2×2; in each of these squares at least two fields must be covered. Hence at least $2 \cdot 16 = 32$ fields must be covered, for which at least 11 pieces are required. This number is also sufficient for the desired placement (see Figure 23).

Figure 23

4.2 (i) In the cases (a) and (b) such points may not exist, while in case (c) the answer is affirmative.

(a) Consider a coloring where any two "opposite" points on the circle have different colors.

(b) Since each point of the circle c is a vertex of exactly one inscribed equilateral triangle, the whole circle can be divided into disjoint triples of vertices of such triangles, and in each triple both colors can be used.

(c) Assume that there is a coloring of the circle such that no inscribed isosceles triangle has all its 3 vertices of the same color. Consider an inscribed regular 12-gon $A_1 A_2 \ldots A_{12}$; assume that in the triple A_1, A_5, A_9 the points A_1, A_5 have the same color, and successively establish the colors of the points $A_9, A_3, A_6, A_{12}, A_7, A_{11}$. You will then find that $\triangle A_3 A_7 A_{11}$ has vertices of the same color, which is a contradiction.

(ii) Choose two different points A, B of the same color, and three different straight lines p, q, r parallel to AB. Assume that on two of them there are points of the same color as that of A, B, for instance $C \in p$ and $D \in q$ (otherwise, the assertion is obvious). Choose points $E, F \in p$ and $G, H \in q$ such that $\overrightarrow{AB} = \overrightarrow{CE} = \overrightarrow{DG} = \overrightarrow{FC} = \overrightarrow{HD}$ (see Figure 24) and consider consecutively the parallelograms $ABEC$, $ABCF$, $ABGD$, $ABDH$, and $EGHF$.

(iii) Consider the following example of a coloring of the plane with 7 colors where the distance between any two points of the same color is not equal to 1. Divide the plane into congruent hexagons with side a and color them according to Figure 25 (points that lie on the boundary of two or three hexagons can have the colors of any of these hexagons). The greatest distance between points of the same color and from the same hexagon is not greater than $2a$, while the smallest distance between points of the same color and belonging to different hexagons is at least $|AB|$ (see Figure 25). Note that $|AB|^2 = |AC|^2 + |BC|^2 = 4a^2 + 3a^2 = 7a^2 > (2a)^2$. If therefore $2a < 1 < \sqrt{7}a$, that is, if $\frac{1}{\sqrt{7}} < a < \frac{1}{2}$, then the distance between any two points of the same color cannot be equal to 1.

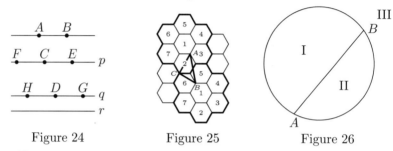

Figure 24   Figure 25   Figure 26

4.4 (i) Proceed recursively, in a similar way as in 4.3.(i); upon adding a new circle c, change the coloring of exactly those regions that lie inside c.

(ii) Proceed recursively. Upon adding a new circle with chord AB change the colors a, b, c of the regions in the parts I, II, III (see Figure 26) according to the following scheme: In I: $a \to b$, $b \to c$, $c \to a$; in II: $a \to c$, $b \to a$, $c \to b$; and in III leave the coloring unchanged. Explain why the condition on the chords formulated in the statement is essential for this proof.

(iii) Prove the assertion by induction on the number n of triangles. For $n \le 3$ this is trivial. Assume that $n > 3$ and that the assertion holds for any division of any polygon into $k \le n - 1$ triangles. From the given polygon $\mathcal{M}$, which is divided into n triangles, now remove an arbitrary triangle $\mathcal{T}$, one of whose sides lies on the circumference of $\mathcal{M}$; then apply the induction hypothesis. Do not forget that upon removing the triangle $\mathcal{T}$ the remainder of the polygon $\mathcal{M}$ may disintegrate into two polygons.

4.6 (i) If in some square there are k black and $(4 - k)$ white fields, then upon recoloring you obtain $(4 - k)$ black and k white fields. The number of black fields changes by $(4 - k) - k = 2(2 - k)$, and therefore the situation described cannot occur.

(ii) Number the rows and columns $1, 2, \ldots, n$ and color those fields that lie at the intersections of rows and columns with odd indices. Such a coloring satisfies the conditions of the problem, and $\left[\frac{n+1}{2}\right]^2$ fields are colored. More fields cannot be colored: Tile the chessboard with 2×2 squares and, when n is odd, with $(n-1)$ rectangles of size 2×1 and one 1×1 square in addition. In each of these parts one can color at most one field, and therefore the number of colored fields does not exceed $\left(\frac{n}{2}\right)^2$ when n is even, or $\left(\frac{n-1}{2}\right)^2 + (n-1) + 1 = \left(\frac{n+1}{2}\right)^2$ when n is odd.

(iii) Let p arbitrary fields be colored on the chessboard, and for $i = 0, 1, \ldots, 4$ let n_i denote the number of colored fields that abut exactly i colored fields (notice that $p = \sum_{i=0}^4 n_i$). Denote by N the number of sides that are shared by two colored fields. Each such side is shared by exactly two colored fields, and therefore you have $N = \frac{1}{2}(1 \cdot n_1 + 2 \cdot n_2 + 3 \cdot n_3 + 4 \cdot n_4) = \frac{1}{2}(n_1 + n_3) + n_2 + n_3 + 2n_4$. Since N is an integer, the number $(n_1 + n_3)$ is even; that is, the number of colored fields that have an odd number of colored neighbors is always even. Hence the desired coloring ($n_0 = n_2 = n_4 = 0$) does not exist for any odd p.

(iv) Proceed as in 4.5.(ii), where $k = 498$.

(v) The desired coloring is possible: Divide the chessboard into $4 \cdot 4 = 16$ squares with side length k, and color them according to Figure 27a.

(vi) The three fields in one column can be colored in $2^3 = 8$ ways, and therefore by the pigeonhole principle some two columns are colored in the same way. The fields with the color that occurs at least twice in these columns then have the desired property.

(vii) The chessboard has 25 fields, and by the pigeonhole principle at least 13 of them have the same color, say white. If all five fields of some column are white, then one of the remaining four columns has at least two white fields. If four of the fields of some column are white, then one of the remaining four columns contains at least three of the remaining nine white fields. If all columns contain at most three white fields, then there are two possible distributions of the numbers of the 13 white fields in the different columns: $3, 3, 3, 2, 2$, or $3, 3, 3, 3, 1$. Then from among those one can form $3 \cdot \binom{3}{2} + 2 \cdot \binom{2}{2} = 11$ (respectively $4 \cdot \binom{3}{2} = 12$) pairs of white fields in the same column. Since pairs of fields in the same column can be chosen in $\binom{5}{2} = 10$ ways, by the pigeonhole principle at least two pairs of white fields from different columns must lie in the same rows. This completes the proof. A counterexample of a coloring of a 4×4 chessboard is shown in Figure 27b.

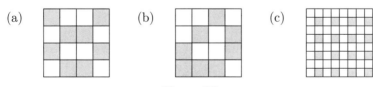

(a)   (b)   (c)

Figure 27

(viii) 2^{48} is the number of allowable colorings, where none of the fields indicated in Figure 27c is colored.

(ix) Proceed as in 4.5.(iv) and derive the recurrence relation $Q(n) = 7 \cdot Q(n-1) + 4 \cdot Q(n-2)$, valid for any $n \geq 3$, where $Q(1) = 15$, $Q(2) = 113$. Then use induction to prove the assertion.

4.8 (i) If the chessboard is colored in the classical way, then 13 fields are colored with one color (say, black), and 12 fields are then white. After the spiders have moved, one black field will remain empty, since only spiders from white fields will have moved to black fields.

(ii) Color the chessboard with two colors, as indicated in Figure 28. Since both m and n are even, the numbers of fields of both colors are also even. It suffices to show that the number of "vertical" pieces is even. Assume the contrary; then all the "vertical" pieces cover an odd number of black fields. It therefore remains to cover an odd number of black fields with "horizontal" pieces; this, however, is impossible, since all "horizontal" pieces cover an even number of black fields.

Figure 28

(iii) Assume that the desired tiling is possible, and consider a classical coloring of the chessboard. Then every piece covers either one or three black fields, that is, always an odd number of black fields. The given chessboard would be tiled with 25 pieces, so the total number of black fields covered would be odd. But this is a contradiction, since the chessboard has 50 black fields.

(iv) This is not possible. [Color the chessboard with four colors (with the colors numbered as in Figure 29a). It is easy to figure out that 26 fields have color no. 2, while only 24 fields have color no. 4. Since each tetromino covers one field of each color, the desired tiling does not exist.]

(v) Color the chessboard according to Figure 29b. Each 2×2 piece has a common field with exactly one of the colored 2×2 squares, and there are 100 such squares in total on the chessboard.

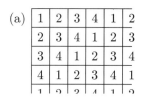
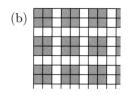
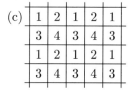

Figure 29

(vi) Color the squares of the checkered paper with four colors, according to Figure 29c. By the pigeonhole principle, among the marked squares at least n have the same color, and no two of them have a point in common.

(vii) Color the triangle according to Figure 30; then for the number n_1 (respectively n_2) of black (respectively white) triangles you obtain $n_1 = 1 + 2 + \cdots + n = \frac{n(n+1)}{2}$; $n_2 = 1 + 2 + \cdots + (n-1) = \frac{n(n-1)}{2}$.

It is clear that two triangles with consecutive numbers have different colors. Hence the number of black numbered triangles can be at most 1 more than the number of white numbered triangles; then the total number m of numbered triangles is at most $2 \cdot \frac{n(n-1)}{2} + 1 = n^2 - n + 1$.

Figure 30

It is now easy to find a numbering with $m = n^2 - n + 1$ (in each row, leave one black triangle unnumbered, alternating between the "left" and the "right" sides of the triangle, and starting with the second row from "above").

4.10 (i) Divide the given line segment into five segments of length 2, and then use the result of 4.9.(i) for each of the five parts.

(ii) Proceed as in 4.9.(ii), and begin by rotating the circles from such a position that the colored arcs of the two circles coincide. Consider the points marked by the painter when some two arcs that coincided at the beginning of the rotation overlap. These points lie on an arc whose midpoint is the position of the painter at the beginning of the rotation and whose length does not exceed $\frac{2\pi}{k^2-k+1}$. In a similar way as in 4.9.(ii) show that the sum of the lengths of the remaining arcs colored by the painter is less than $2k(k-1) \cdot \frac{\pi}{k^2-k+1}$; hence there exists a point not colored by the painter.

(iii) Let the segment AC_1 (respectively C_1C_2) have the color a (respectively b). It certainly suffices to consider the case where $p = |AC_1| < \frac{1}{3}$, and where in the set of all points with color a for some (from now on fixed) number x ($p < x < \frac{1}{3}$) there are no two points of distance x. Now choose an arbitrary d ($0 < d < \frac{1}{3}$) and show that then $|EF| = d$ for some two points E, F having the color b. To do this, consider the points G, H, I given by the equations

$$\overrightarrow{AG} = (p+x)\overrightarrow{AB}, \quad \overrightarrow{AH} = (p+d)\overrightarrow{AB}, \quad \text{and} \quad \overrightarrow{AI} = (p+x+d)\overrightarrow{AB}.$$

Since $|C_1G| = x$, G has the color b. Therefore, if the point I also has the color b, you can set $\{E, F\} = \{G, I\}$; therefore, let I have the color a. Since $|HI| = x$, H has the color b, and one can set $\{E, F\} = \{C_1, H\}$. This completes the proof.

A stronger assertion does not hold in general (that is, independently of the choice of the points $C_1, \ldots, C_n$) for any $\alpha > \frac{1}{3}$. Counterexample: To each $\alpha > \frac{1}{3}$ find an ε ($0 < \varepsilon < \frac{1}{3}$) and construct two points C_1, C_2 such that

$$\tfrac{1}{3} + 3\varepsilon < \alpha, \quad |AC_1| = \tfrac{1}{3} - \varepsilon, \quad \text{and} \quad |AC_2| = \tfrac{2}{3} + \varepsilon.$$

Then the set $AC_1 \cup C_2B$ does not contain any two points of distance $\frac{1}{3}$, while the segment C_1C_2 does not contain any two points of distance $\frac{1}{3} + 3\varepsilon$.

(iv) First show that the hypothesis can hold only for $\alpha = 120°$. To do that, divide the circle c with the points $A_1, A_2, \ldots, A_{3m}$ into $3m$ congruent arcs of size $\beta_m = \frac{120°}{m}$, and give the same color to the arcs in the sets $\left\{ \widehat{A_1 A_2}, \widehat{A_4 A_5}, \ldots, \widehat{A_{3m-2}A_{3m-1}} \right\}$, $\left\{ \widehat{A_2 A_3}, \widehat{A_5 A_6}, \ldots, \widehat{A_{3m-1}A_{3m}} \right\}$, $\left\{ \widehat{A_3 A_4}, \widehat{A_6 A_7}, \ldots, \widehat{A_{3m}A_1} \right\}$. Then there does not exist an arc of size α with endpoints of the same color if α lies in one of the intervals

$$(\beta_m, 2\beta_m), (4\beta_m, 5\beta_m), \ldots, ((3m-2)\beta_m, (3m-1)\beta_m).$$

Convince yourself that these "forbidden" intervals for $m = 1, 2, \ldots$ cover the whole interval $(0, 360°)$ with the exception of the values $120°$ and $240°$. Proof of the hypothesis for $\alpha = 120°$: For any coloring of the arcs $\overarc{A_i A_{i+1}}$ find a point A_i having two colors (if it doesn't exist, then the whole circle c is colored with one color) and consider the coloring of the vertices of the equilateral triangle $A_i P Q$ inscribed in the circle c.

Bibliography

A. Books Cited in the Text

[1] S.W. Golomb, *Polyominoes*, Scribner, New York, 1965.

[2] R.P. Grimaldi, *Discrete and Combinatorial Mathematics*, fourth edition, Addison-Wesley, Reading, MA, 1999.

[3] J. Gross and J. Yellen, *Graph Theory and Its Applications*, CRC Press, Boca Raton, 1999.

[4] H. Hadwiger and H. Debrunner, *Combinatorial Geometry in the Plane*, Holt, Rinehart and Winston, New York, 1964.

[5] J. Herman, R. Kučera, and J. Šimša, *Equations and Inequalities: Elementary Problems and Theorems in Algebra and Number Theory*, Springer-Verlag, New York, 2000.

[6] A.Y. Khinchin, *Three Pearls of Number Theory*, Dover Publications, Mineola, NY, 1998.

[7] S.V. Konyagin et al., *Zarubezhnyye matematicheskiye olimpiady*, Nauka, Moskva, 1987.

[8] A.G. Kurosh, *Higher Algebra*, Mir Publishers, Moscow, 1972.

[9] N.Ya. Vilenkin, *Combinatorics*, Academic Press, New York-London, 1971.

B. Additional English-Language Books

[10] E.J. Barbeau, M.S. Klamkin, and W.O.J. Moser, *Five Hundred Mathematical Challenges*, Mathematical Association of America, Washington, DC, 1995.

[11] C. Berge, *Principles of Combinatorics*, Academic Press, New York–London, 1971.

[12] M. Doob (ed.), *The Canadian Mathematical Olympiad, 1969–1993*, Canadian Mathematical Society, Ottawa, 1993.

[13] H. Dörrie, *100 Great Problems of Elementary Mathematics*, Dover Publications, 1965. Reprinted 1989.

[14] A. Engel, *Problem-Solving Strategies*, Springer-Verlag, New York, 1998.

[15] M.J. Erickson and J. Flowers, *Principles of Mathematical Problem Solving*, Prentice-Hall, 1999.

[16] A. Gardiner, *The Mathematical Olympiad Handbook. An Introduction to Problem Solving Based on the First 32 British Mathematical Olympiads 1965–1996*, Oxford University Press, 1997.

[17] G.T. Gilbert, M.I. Krusemeyer, and L.C. Larson, *The Wohascum County Problem Book*, Mathematical Association of America, Washington, DC, 1993.

[18] S.L. Greitzer, *International Mathematical Olympiads 1959–1977*, Mathematical Association of America, Washington, D.C., 1978.

[19] V. Gusev, V. Litvinenko, and A. Mordkovich, *Solving Problems in Geometry*, Mir Publishers, Moscow, 1988.

[20] K. Hardy and K.S. Williams, *The Red Book of Mathematical Problems*, Dover Publications, Mineola, NY, 1996.

[21] K. Hardy and K.S. Williams, *The Green Book of Mathematical Problems*, Dover Publications, Mineola, NY, 1997.

[22] R. Honsberger, *From Erdős to Kiev. Problems of Olympiad Caliber*, Mathematical Association of America, Washington, DC, 1996.

[23] M.S. Klamkin, *USA Mathematical Olympiads 1972–1986*, Mathematical Association of America, Washington, DC, 1988.

[24] J. Kürschák, *Hungarian Problem Book I, Based on the Eötvös Competitions, 1894–1905*, Mathematical Association of America, Washington, DC, 1963.

[25] J. Kürschák, *Hungarian Problem Book II, Based on the Eötvös Competitions, 1906–1928*, Mathematical Association of America, Washington, DC, 1963.

[26] L.C. Larson, *Problem-Solving Through Problems*, Springer-Verlag, New York, 1983.

[27] D.J. Newman, *A Problem Seminar*, Springer-Verlag, New York, 1982.

[28] G. Pólya, *Mathematical Discovery. On Understanding, Learning, and Teaching Problem Solving*, reprint in one volume, John Wiley & Sons, Inc., New York, 1981.

[29] G. Pólya, *How to Solve It. A New Aspect of Mathematical Method*, Princeton University Press, Princeton, NJ, 1988.

[30] J. Riordan, *Combinatorial Identities*, Wiley, New York, 1968.

[31] D.O. Shklarsky, N.N. Chentzov, and I.M. Yaglom, *The USSR Olympiad Problem Book*, W.H. Freeman, San Francisco, 1962. Reprinted by Dover Publications, New York, 1993.

[32] H. Steinhaus, *One Hundred Problems in Elementary Mathematics*, Basic Books, Inc., New York, 1964. Reprinted by Dover Publications, New York, 1979.

[33] Ch.W. Trigg, *Mathematical Quickies*, Dover Publications, New York, 1985.

[34] W.A. Wickelgren, *How to Solve Mathematical Problems*, Dover Publications, Mineola, NY, 1995.

[35] I.M. Yaglom and V.G. Boltyanskii, *Convex Figures*, Holt, Rinehart and Winston, New York, 1961.

[36] A.M. Yaglom and I.M. Yaglom, *Challenging Mathematical Problems with El-ementary Solutions. Vol. I. Combinatorial Analysis and Probability Theory*, Dover Publications, New York, 1987.

[37] A.M. Yaglom and I.M. Yaglom, *Challenging Mathematical Problems with Elementary Solutions. Vol. II. Problems from Various Branches of Mathematics*, Dover Publications, New York, 1987.

[38] P. Zeitz, *The Art and Craft of Problem Solving*, Wiley, New York, 1999.

C. East European Sources

[39] I.L. Babinskaya, *Zadachi matematicheskikh olimpiad*, Nauka, Moskva, 1975.

[40] V.I. Bernik, I.K. Zhuk, and O.V. Melnikov, *Sbornik olimpiadnykh zadach po matematike*, Narodnaya osveta, Minsk, 1980.

[41] L. Boček and A. Vrba, *Vybrané úlohy matematické olympiády (kategorie B)*, SPN, Praha, 1984.

[42] O. Dunkel, *Izbranniye zadachi iz zhurnala "American Math. Monthly,"* Mir, Moskva, 1977.

[43] E. Fuchs, *Kombinatorika a teorie grafů*, učební text, PřF UJEP, Brno, 1986.

[44] G.A. Galperin and A.K. Tolpygo, *Moskovskiye matematicheskiye olimpiady*, Prosveshcheniye, Moskva, 1986.

[45] E.Ya. Gik, *Shakhmaty i matematika*, Nauka, Moskva, 1983.

[46] L.I. Golovina and I.M. Yaglom, *Induktsiya v geometrii*, GITTL, Moskva, 1956.

[47] K. Horák et al., *Úlohy mezinárodních matematických olympiád*, SPN, Praha, 1986.

[48] J. Kautský, *Kombinatorické identity*, Veda, Bratislava, 1973.

[49] Kvant, Popular-scientific physical-mathematical journal, Nauka, Moscow. Founded 1970, appears monthly. (English edition: *Quantum*, National Science Teachers Association, 1990–2001; see http://www.nsta.org/quantum/.)

[50] J. Molnár and J. Kobza, *Extremální a kombinatorické úlohy z geometrie*, SPN, Praha, 1990.

[51] J. Nešetřil, *Teorie grafů*, SNTL, Praha, 1979.

[52] *N-tý ročník matematické olympiády*, SPN, Praha, (from 1953 to 1993).

[53] O. Odvárko et al., *Metody řešení matematických úloh I*, učební text MFF UK, Praha, 1977.

[54] O. Pokluda, *Fibonacciho posloupnosti množin*, práce SOČ, Gymnázium tř. Kpt. Jaroše, Brno, 1987.

[55] V.V. Prasolov, *Zadachi po planimetrii II*, second edition, Nauka, Moskva, 1991.

[56] V.V. Prasolov and I.F. Sharygin, *Zadachi po stereometrii*, Nauka, Moskva, 1989.

[57] V.A. Sadovnichij et al., *Zadachi studencheskikh matematicheskikh olimpiad*, Izd. MGU, Moskva, 1987.

[58] J. Šedivý et al., *Metody řešení matematických úloh II*, učební text MFF UK, Praha, 1978.

[59] J. Sedláček, *Faktoriály a kombinační čísla*, Mladá fronta, edice ŠMM, Praha, 1964.

[60] Z. Šimková, *Rovinné úlohy o pokrytí*, Dipl. práce, PřF MU, Brno, 1989.

[61] D.O. Shklarsky, N.N. Chentzov, and I.M. Yaglom, *Geometricheskie neraven-stva i zadachi na maksimum i minimum*, Nauka, Moskva, 1970.

[62] D.O. Shklarsky, N.N. Chentzov, and I.M. Yaglom, *Geometricheskie otsenki i zadachi iz kombinatornoy geometrii*, Nauka, Moskva, 1974.

[63] S. Straszewicz and J. Browkin, *Polskiye matematicheskiye olimpiady*, Mir, Moskva, 1978.

[64] N.V. Vasilyev and A.A. Yegorov, *Zadachi vsesoyuznykh matematicheskikh olimpiad*, Nauka, Moskva, 1988.

[65] N.V. Vasilyev et al., *Zaochnyye matematicheskiye olimpiady*, Nauka, Moskva, 1986.

[66] A. Vrba, *Princip matematické indukce*, Mladá fronta, edice ŠMM, Praha, 1977.

[67] A. Vrba, *Kombinatorika*, Mladá fronta, edice ŠMM, Praha, 1980.

[68] A. Vrba and K. Horák, *Vybrané úlohy MO kategorie A*, SPN, Praha, 1988.

[69] V.A. Vyshenskiy et al., *Sbornik zadach kiyevskikh matematicheskikh olimpiad*, Vishcha shkola, Kiyev, 1984.

[70] J. Vyšín, *Konvexní útvary*, Mladá fronta, edice ŠMM, Praha, 1964.

[71] B. Zelinka, *Rovinné grafy*, Mladá fronta, edice ŠMM, Praha, 1977.

Index